CFD Applications in Renewable Energy Systems

CFD Applications in Renewable Energy Systems

Guest Editors

Omar Darío López Mejía
Santiago Laín

MDPI

Basel • Beijing • Wuhan • Barcelona • Belgrade • Novi Sad • Cluj • Manchester

Guest Editors
Omar Darío López Mejía
Department of Mechanical Engineering
Universidad de los Andes
Bogotá
Colombia

Santiago Laín
Mechanical Engineering Department
Universidad Autónoma de Occidente
Cali
Colombia

Editorial Office
MDPI AG
Grosspeteranlage 5
4052 Basel, Switzerland

This is a reprint of the Special Issue, published open access by the journal *Processes* (ISSN 2227-9717), freely accessible at: https://www.mdpi.com/journal/processes/special_issues/0B269BR3KC.

For citation purposes, cite each article independently as indicated on the article page online and as indicated below:

Lastname, A.A.; Lastname, B.B. Article Title. *Journal Name* **Year**, *Volume Number*, Page Range.

ISBN 978-3-7258-5719-7 (Hbk)
ISBN 978-3-7258-5720-3 (PDF)
https://doi.org/10.3390/books978-3-7258-5720-3

Contents

About the Editors

Omar Dario Lopez Mejia

Omar Dario Lopez Mejia (PhD) is an associate professor in the Department of Mechanical Engineering at Universidad de los Andes in Bogotá, Colombia. He obtained his PhD degree in mechanical engineering at the University of Texas in Austin in 2009. His research interests are focused on the application of Computational Fluid Dynamics (CFD) in thermal-fluid systems. Specifically, in the problems involving turbulence modelling applied to aerodynamics and hydrodynamics.

Santiago Lain

Santiago Lain (PhD. Dr.-Ing. Habil.) is a Fluid Mechanics professor in the Department of Mechanical Engineering at Universidad Autónoma de Occidente Cali, Colombia. He is a Physicist and Mathematician, with a PhD in Physical Sciences from Universidad of Zaragoza, Spain, and Dr.-Ing. Habil. of Martin Luther University Halle-Wittenberg, Germany. His research interests are focused on numerical methods and Computational Fluid Dynamics applied to turbulence modelling and simulation, multiphase flow, and renewable energy systems.

Preface

The ongoing global energy transition represents one of the most significant technological and societal transformations of the 21st century. Driven by the urgent need to mitigate climate change and reduce greenhouse gas emissions, this transition aims to replace fossil fuel-based systems with cleaner, renewable alternatives. Technologies such as wind, solar, hydro, tidal, geothermal, and bioenergy are emerging as central components of a sustainable and diversified energy mix. Their integration is fundamental to achieving a decarbonized and resilient energy infrastructure capable of meeting the growing global demand while minimizing environmental impact. However, the variable and nonlinear nature of these systems introduces complex physical challenges that require advanced tools for analysis, design, and optimization.

In this context, numerical simulation—and, in particular, Computational Fluid Dynamics (CFD)—has become an indispensable tool for understanding and improving the performance of renewable energy devices. The steady growth in computational power, together with advances in numerical algorithms, has enabled high-fidelity simulations that capture detailed flow, heat transfer, and multiphysics interactions within renewable systems. These methods allow researchers and engineers to analyze turbine aerodynamics and wake dynamics, optimize the thermal behavior of solar collectors, and model multiphase flows in hydrogen or biofuel production processes. As a result, CFD has significantly reduced the reliance on costly experimental testing and accelerated innovation cycles in clean energy technologies.

Looking forward, the synergistic integration of Artificial Intelligence (AI) with classical numerical simulation is expected to redefine the landscape of renewable energy research. AI-enhanced models—such as physics-informed neural networks and machine learning-based optimization frameworks—will enable real-time prediction, adaptive control, and design automation. This convergence between data-driven intelligence and physics-based modeling will pave the way toward more efficient, reliable, and sustainable energy systems, ultimately supporting the global pursuit of a carbon-neutral future.

This reprint of the Special Issue presents recent advances and practical applications of Computational Fluid Dynamics (CFD) simulations aimed at enhancing the understanding, design, and optimization of renewable energy systems. The contributions encompass a wide range of technologies, including wind, hydraulic, and solar energy systems, as well as hydrogen combustion processes. Collectively, they demonstrate how high-fidelity numerical modeling can provide deeper physical insight, support performance improvement, and guide innovation in clean energy technologies. The editors believe that this reprint serves as a valuable reference for researchers and graduate students seeking reliable methodologies for the prediction and analysis of renewable energy system performance through advanced numerical simulations.

Omar Darío López Mejía and Santiago Laín
Guest Editors

processes

MDPI

Editorial

Special Issue on "CFD Applications in Renewable Energy Systems"

Omar D. Lopez Mejia [1,*] and Santiago Laín [2,*]

1 Department of Mechanical Engineering, Universidad de los Andes, Cra 1 Este N 19A-40, Bogotá 111711, Colombia
2 PAI+ Group, Mechanical Engineering Department, Faculty of Engineering, Universidad Autónoma de Occidente, Cali 760030, Colombia
* Correspondence: od.lopez20@uniandes.edu.co (O.D.L.M.); slain@uao.edu.co (S.L.)

The global energy landscape is undergoing a critical transformation driven by the urgent need to mitigate climate change, reduce greenhouse gas (GHG) emissions, and ensure long-term energy security. Fossil fuel-based energy systems remain the dominant source of global electricity and heat, contributing more than 70% of total carbon dioxide emissions worldwide [1]. To counteract their impact, renewable energy—primarily solar, wind, hydro, and marine—must rapidly replace traditional energy sources across all sectors, including electricity generation, transportation, and hard-to-abate sectors. Yet, integrating and optimising renewable energy systems poses significant engineering and physical challenges due to their inherent variability, nonlinear behaviour, and complex fluid-dynamic interactions. In this context, Computational Fluid Dynamics (CFD) has emerged as a powerful modelling approach enabling detailed simulations of flow, turbulence, heat transfer, chemical reactions, cavitation, and multiphase dynamics. These capabilities are essential for improving the design, efficiency, and reliability of renewable energy systems while also reducing experimental costs and development time. CFD enables researchers and engineers to analyse complex flow phenomena in renewable energy systems, including wind turbine wakes, vortex shedding in tidal and hydro turbines, flame dynamics in hydrogen combustors, or convective and radiative heat transfer in solar collectors, among others. Furthermore, advances in high-performance computing, turbulence modelling, and integration of optimization and machine learning algorithms have greatly expanded the predictive capabilities of CFD [2,3].

The aim of this Special Issue is to showcase recent developments and novel applications of CFD tools in the simulation of renewable energy systems. The accepted contributions are organised into five topics: (1) tidal energy, (2) hydraulic energy, (3) wind energy, (4) combustion processes, including hydrogen, and (5) solar energy.

Regarding tidal energy, da Silviera et al. investigated the performance of overtopping wave energy converters (WECs) using CFD simulations under both regular and irregular wave conditions along the southern coast of Brazil [4]. They demonstrated that device geometry—particularly, the ramp height-to-length ratio—strongly influences water capture volume and energy conversion efficiency. Optimising this geometric parameter significantly improved performance under realistic sea states, consistent with earlier CFD-based studies that emphasised the importance of wave focusing and overtopping thresholds in WEC efficiency [5,6]. A second contribution from the same research group introduces and validates the WaveMIMO methodology for generating realistic irregular waves in CFD simulations, using spectral data from Brazilian coastal sites. The study systematically examines key parameters—mesh refinement, time step, velocity boundary discretization—and provides theoretical guidance to improve accuracy. Validation results show that the optimised setup reproduces irregular waves with up to 8% better fidelity compared to previous approaches, enhancing the reliability of coastal and wave energy systems modelling [7]. An additional noteworthy contribution by Thangaraj et al. evaluated seven hybrid Wells turbine designs (based on a NACA0015 airfoil) across a range of inlet fluid velocities representative of coastal regions. CFD was used to assess torque, drag, and hydrodynamic pressure, ultimately selecting the best-performing design for further study. This optimal configuration was then refined through profile modifications (such as stepped-back or zigzag airfoils), which delivered about 15.19% higher hydropower output compared to the baseline design under identical conditions. The study also coupled hydrodynamic modelling with structural, modal, and vibration analyses (including piezoelectric vibration harvesting patches) to ensure that the design maintains structural integrity and offers additional energy harvesting potential [8].

In hydro energy, Lu et al. conducted a comprehensive CFD and experimental study on cavitation in pump–turbines operating under part-load conditions, with a focus on entropy generation and vortex dynamics [9]. Using advanced flow diagnostics, such as Proper Orthogonal Decomposition (POD), Dynamic Mode Decomposition (DMD), and the Q-criterion, the study reveals the formation and evolution of small-scale vortices that intensify cavitation and energy dissipation. The results show that the entropy production rate correlates directly with hydraulic loss mechanisms, providing valuable insight into system inefficiencies. In a related study, Yang et al. studied the effect of sediment concentration and particle size on cavitation in a Venturi flow channel, combining high-speed photography with a gas–liquid–solid numerical mixture model. Their findings demonstrate that increasing the sediment concentration has a stronger influence than increasing the particle size on reducing the cavitation number, extending the cavitation cloud length, accelerating the cavitation evolution cycle, and intensifying shearing and re-entrant jet effects near walls. Numerical predictions align reasonably well with experimental observations, although simulations tend to overestimate the duration of the cavitation cloud evolution cycle [10].

Three contributions in this Special Issue focus on advancing the understanding of wind energy devices. Malael and Stratila conducted Large Eddy Simulations to analyse the dynamic starting behaviour of a vertical-axis wind turbine (VAWT) enhanced with passive vortex control [11]. Their results show that incorporating a vortex-trap design improves lift generation, accelerates self-starting, and enhances torque output. This study builds on earlier CFD research investigating passive flow control mechanisms for improving both vertical- and horizontal-axis wind turbines' starting process [12]. Xu et al. evaluated the aerodynamic effects of dielectric barrier discharge (DBD) plasma actuators installed on wind turbine blade rudders. By varying the input frequency, they demonstrate that plasma excitation increases lift at low angles of attack and modifies flow separation at higher angles, especially under crosswind conditions. [13]. Contreras et al. used CFD with FENSAP-ICE 2022 R1in ANSYS to evaluate the impact of surface roughness and turbulence models on the aerodynamic performance of the S809 airfoil under both rime and glaze ice scenarios [14]. Their findings revealed that ice accretion significantly increases drag and reduces lift, primarily due to roughness-induced surface modifications, while the choice of turbulence model resulted in relatively minor differences (under ~10%). The beading roughness model was shown to more accurately capture aerodynamic changes during ice growth, especially in representing the spatial variation of roughness across the iced airfoil.

Hydrogen combustion in micro-scale systems presents unique challenges due to high diffusivity, flame instability, and temperature gradients. In this Special Issue, CFD was applied to simulate hydrogen-fuelled micro-planar combustors with dual cylindrical geometry and recirculating channels [15]. The study reports significant improvements in combustion efficiency, temperature uniformity, and wall cooling, including a 580 K reduction in exhaust temperature and a 14% increase in radiation efficiency. These results highlight the potential of CFD to support the development of safer and more efficient hydrogen combustion systems for portable and stationary applications [16,17]. Two additional contributions focus on diesel-based fuels. Zhou et al. developed a detailed chemical-kinetics mechanism (187 species, 735 reactions) to model the oxidation behaviour of ternary fuel blends of hydrogenated biodiesel, ethanol, and diesel under varied temperatures (900–1400 K); an equivalence ratio of 1.0; and a pressure of 1.0 MPa [18]. Their results show that, at lower temperatures, increasing the biodiesel and ethanol content reduces OH radical production and slows reactivity, while, at higher temperatures, the same increase enhances OH formation and promotes oxidation. This study also identified oleic and stearic methyl esters as better surrogates for hydrogenated biodiesel kinetics than methyl decanoate and highlights the central role of the $CH_2O \rightarrow HCO \rightarrow CO$ pathway in both low- and high-temperature oxidation. Valiño et al. implemented a 3D CFD model using an Eulerian–Eulerian two-phase flow formulation combined with a probability density function (PDF) to represent particle size distribution and deposition in a BOSCH automotive diesel filter. The model employs the Brinkman–Darcy approximation for flow through the filter media, showing that under typical operating conditions, this approach simplifies porous medium modelling without significant loss of accuracy. The resulting simulations qualitatively (and, in early stages, quantitatively) reproduce the deposition patterns seen in used filters, particularly the preferential deposition of particles in fold pleats and near inlet slots [19].

Regarding solar energy, a CFD-assisted experimental study investigated the thermal performance of two flat-plate solar air heaters, emphasising the influence of geometry on air temperature distribution and collector efficiency. Numerical results were validated against experimental data, showing good agreement and confirming that geometry-induced flow patterns significantly influence heat transfer [20].

In summary, this Special Issue highlights the essential role of CFD in advancing renewable energy technologies. Across different energy sources—tidal, hydraulic, wind, combustion, and solar—the selected studies demonstrate how CFD delivers critical insights into fluid–thermal interactions, energy conversion processes, and system-level optimization. Looking ahead, the integration of CFD with AI-driven optimization, experimental validation, and real-time data from digital twins offers significant potential. Future research should also address key challenges, such as uncertainties in turbulence models, computational cost reduction, and scale-bridging simulations. We hope this collection inspires further innovation in CFD applications for clean energy.

We thank all the contributions sent to this Special Issue and the support of all the editors at MDPI that helped us in this process.

Author Contributions: O.D.L.M. and S.L. contributed equally to this work. All authors have read and agreed to the published version of the manuscript.

Funding: This research received no external funding.

Conflicts of Interest: The authors declare no conflicts of interest.

References

1. International Energy Agency (IEA). *World Energy Outlook 2023*; IEA: Paris, France, 2023. Available online: https://www.iea.org/reports/world-energy-outlook-2023 (accessed on 24 September 2025).
2. Szubel, M.; Filipowicz, M.; Papis-Fraczek, K.; Krys, M. *Computational Fluid Dynamics in Renewable Energy Technology*, 1st ed.; CRC Press: Boca Raton, FL, USA, 2023. [CrossRef]

3. Abu Sofian, A.D.A.B.; Lim, H.R.; Halimatul, H.S.; Ma, Z.; Chew, K.W.; Show, P.L. Machine learning and the renewable energy revolution: Exploring solar and wind energy solutions for a sustainable future including innovations in energy storage. *Sustain. Dev.* **2023**, *32*, 3973–3978. [CrossRef]
4. Paiva, M.d.S.; Mocellin, A.P.G.; Oleinik, P.H.; dos Santos, E.D.; Rocha, L.A.O.; Isoldi, L.A.; Machado, B.N. Geometrical Evaluation of an Overtopping Wave Energy Converter Device Subject to Realistic Irregular Waves and Representative Regular Waves of the Sea State That Occurred in Rio Grande—RS. *Processes* **2025**, *13*, 335. [CrossRef]
5. Musa, M.A.; Maliki, A.Y.; Ahmad, M.F.; Sani, W.N.; Yaakob, O.; Samo, K.B. Numerical Simulation of Wave Flow Over the Overtopping Breakwater for Energy Conversion (OBREC) Device. *Procedia Eng.* **2017**, *194*, 166–173. [CrossRef]
6. Cao, D.; Zeng, H.; He, J.; Liang, H.; Yang, Z.; Ancha, V.R.; Chen, H. A numerical investigation of irregular wave overtopping on a 2D fixed overtopping wave energy converter. *Eng. Appl. CFD* **2025**, *19*, 2446658. [CrossRef]
7. Paiva, M.d.S.; Mocellin, A.P.G.; Oleinik, P.H.; dos Santos, E.D.; Rocha, L.A.O.; Isoldi, L.A.; Machado, B.N. Theoretical Recommendations and Validation for the Generation of Realistic Irregular Waves Through the WaveMIMO Methodology. *Processes* **2025**, *13*, 1395. [CrossRef]
8. Thangaraj, J.; Madasamy, S.K.; Rajendran, P.; Zulkifli, S.; Rajapandi, R.; AL-bonsrulah, H.A.Z.; Stanislaus Arputharaj, B.; Jeyaraj, H.P.; Raja, V. Design, Multi-Perspective Computational Investigations, and Experimental Correlational Studies on Conventional and Advanced Design Profile Modified Hybrid Wells Turbines Patched with Piezoelectric Vibrational Energy Harvester Devices for Coastal Regions. *Processes* **2023**, *11*, 2625. [CrossRef]
9. Lu, J.; Li, J.; Zhang, C.; Zhou, Y.; He, Y. Research on the Cavitation Characteristics of Pump Turbines Based on Mode Decomposition. *Processes* **2025**, *13*, 732. [CrossRef]
10. Yang, Z.; Yu, G.; Zhao, Q. Experimental Study and Numerical Simulation of Sediment's Promoting Effect on Cavitation Based on Flow Field Analysis. *Processes* **2025**, *13*, 1299. [CrossRef]
11. Mălăel, I.; Strătilă, S. High-Precision Numerical Investigation of a VAWT Starting Process. *Processes* **2024**, *12*, 2263. [CrossRef]
12. Fatahian, E.; Mishra, R.; Jackson, F.F.; Fatahian, H. Optimization and analysis of self-starting capabilities of vertical axis wind turbine pairs: A CFD-Taguchi approach. *Ocean. Eng.* **2024**, *302*, 117614. [CrossRef]
13. Xu, J.; Zhao, J.; Chang, J. Effect of Different Voltage Frequencies of Plasma Actuators on Wind Turbine Blade Lift and Rudder Efficiency. *Processes* **2025**, *13*, 1032. [CrossRef]
14. Contreras Montoya, L.T.; Ilinca, A.; Lain, S. Influence of Surface Roughness Modeling on the Aerodynamics of an Iced Wind Turbine S809 Airfoil. *Processes* **2023**, *11*, 3371. [CrossRef]
15. Almutairi, F. A CFD Study of Thermodynamics and Efficiency Metrics in a Hydrogen-Fueled Micro Planar Combustor Housing Dual Heat-Recirculating Cylindrical Combustors for MTPV Applications. *Processes* **2025**, *13*, 1142. [CrossRef]
16. Hua, J.; Wu, M.; Kumar, K. Numerical simulation of the combustion of hydrogen–air mixture in micro-scaled chambers Part II: CFD analysis for a micro-combustor. *Chem. Eng. Sci.* **2005**, *60*, 3507–3515. [CrossRef]
17. Bagheri, G.; Hosseini, S.E.; Wahid, M.A. Effects of bluff body shape on the flame stability in premixed micro-combustion of hydrogen-air mixture. *Appl. Therm. Eng.* **2014**, *67*, 266–272. [CrossRef]
18. Zhou, J.; Chen, L.; Zhang, R.; Zhao, W. Study on Oxidation Activity of Hydrogenated Biodiesel–Ethanol–Diesel Blends. *Processes* **2024**, *12*, 462. [CrossRef]
19. Valiño, L.; Mustata, R.; Hierro, J.; Hernández, J.L.; García, M.J.; Blasco, C.; Chen, Y.-T.; Chen, L.-W.; Roy, P. Particle Deposition Pattern on an Automotive Diesel Filter Using an Eulerian Probability Density Function Method. *Processes* **2023**, *11*, 1100. [CrossRef]
20. El-Sebaey, M.S.; Ellman, A.; El-Din, S.S.; Essa, F.A. Thermal Performance Evaluation for Two Designs of Flat-Plate Solar Air Heater: An Experimental and CFD Investigations. *Processes* **2023**, *11*, 1227. [CrossRef]

processes

MDPI

Article

Particle Deposition Pattern on an Automotive Diesel Filter Using an Eulerian Probability Density Function Method

Luis Valiño [1], Radu Mustata [1,*], Juan Hierro [2], Juan Luis Hernández [3], María José García [3], Carlos Blasco [3], Yi-Tung Chen [4], Lung-Wen Chen [5] and Prosun Roy [4,5]

1 ICB/CSIC, María de Luna 10, E-50018 Zaragoza, Spain; l.valino@csic.es
2 Centro Universitario de la Defensa, Carretera Huesca s/n, E-50090 Zaragoza, Spain; hierro@unizar.es
3 Robert Bosch España Gasoline Systems, Raso de la Estrella, s/n, E-28300 Aranjuez, Spain; juanluis.hernandezcarabias@es.bosch.com (J.L.H.); mariajose.garciavargas@es.bosch.com (M.J.G.); carlos.blasco@es.bosch.com (C.B.)
4 Department of Mechanical Engineering, University of Nevada Las Vegas, 4505 S Maryland Pkwy, Las Vegas, NV 89154, USA; yitung.chen@unlv.edu (Y.-T.C.); royp3@unlv.nevada.edu (P.R.)
5 Department of Environmental and Occupational Health, University of Nevada Las Vegas, Harry Reid Center (HRC) for Environmental Studies, 4505 S Maryland Pkwy, Las Vegas, NV 89154, USA; antony.chen@unlv.edu
* Correspondence: radu.mustata@csic.es

Abstract: A full 3D numerical simulation of the two-phase flow made up of (bio)diesel and particles, has been carried out to reproduce the deposition pattern of particles in a BOSCH automotive filter. From a probability density function (PDF), a simple Eulerian-Eulerian two-phase model is proposed for diesel and particles. The proposed formulation allows for a detailed description of the relationship between the velocity and size of the particles. A Brinkman-Darcy approximation has been considered for the flow through the filtering paper and is proved to be sufficient for the typical filter working conditions. The new tool is able to reproduce the deposition pattern shown by the used filters.

Keywords: filter; two-phase flows; Eulerian probability density function; particle size distribution

1. Introduction

The "trial and error" method is still the usual way to design new filters. This method is both time- and cost-consuming, as it implies the design and construction of various experimental test grids and prototypes that render the process difficult or impossible to systematize [1]. At the most basic level, filtration is the process of separation from a fluid suspension of undesired (or desired) particles while passing through a porous medium (the filtering medium). The filter design can vary widely depending on the application, in terms of both the internal pore structure (microscale design) and how the filtering medium is deployed (macroscale design). However, not just any simple design can be deemed fit for this purpose. The challenge for the automotive industry is not only to design filters with higher efficiency but also, with the ever-increasing limits in space, to design filters with smaller physical sizes. The most common design used to achieve this tradeoff is the single-pass pleated filtering media approach (see Figure 1a); such a design is subject to study in this present paper.

In the automotive industry, computational fluid dynamics (CFD) simulation programs are already an established tool in the R&D departments, aimed at reducing the development and time costs of designing different vehicle parts, such as injectors, inlet and outlet manifold design, exhaust systems, etc. Although these tools have been used for more than 20 years in the automotive world, CFD programs are still not standard tools for fuel filter development. The first numerical tools have been created only recently, with the lack of both appropriate models and algorithms being the factors behind this delay [2,3]. In order to reduce prototyping cycles, it is convenient to have a specific numerical simulation tool on hand that can model particle filtration during the design stage.

(a)

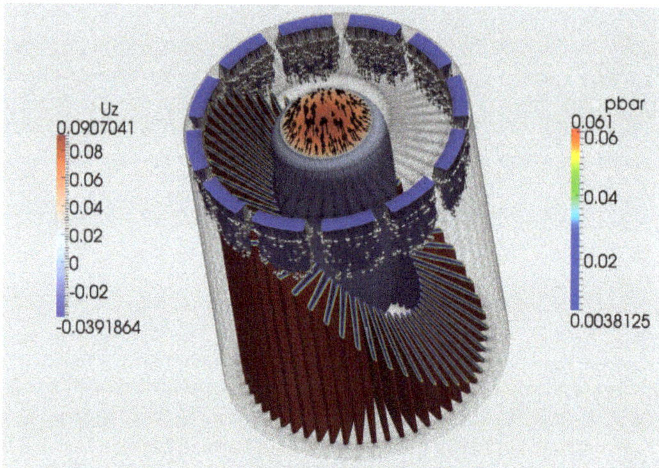

(b)

Figure 1. (a) Bosch filter device with the external holder removed to see the filtering medium. The flow enters vertically through the upper-side external slots and exits through the upper-side central duct; (b) 3.3 M cell mesh with pressure drop and velocity pattern.

Modern injection systems are very sensitive to the possible presence of solid particles in the fuel (diesel or gasoline) since these particles can damage the injectors due to corrosion, abrasion, or other effects [4]. Therefore, it is necessary to guarantee a certain level of particle cleanliness to ensure that injection systems work properly throughout the entire operating lifetime.

Diesel fuels contain a higher amount of water than gasoline (up to 500 ppm). In the case of biodiesel, this is even worse. Consequently, it is necessary to filtrate an extra amount of water [5] since water can cause corrosion problems in the injectors. Fuel filters for diesel engines have to be able to filtrate the particles contained in the fuel and separate the water droplets, which normally are collected at the bottom of the filter and purged periodically.

In diesel filtering, the considered flow is made up of a liquid for the continuous phase and a solid for the discrete phase. Other relevant situations in particle filtering in the automotive industry involve gas-solid combinations, such as in the exhaust gas manifold [6].

Filtering media must provide a certain level of efficiency depending on customer requirements and/or particular industrial standards. In addition, the filter must have a specific particle retention capacity. This capacity is defined by means of a change interval or lifetime, which represents the guaranteed amount of kilometers before the filter gets clogged due to the retained particles and has to be changed.

The use of either a filter with a wrong design or a filter medium that leads to lower particle efficiency or a shorter lifetime, may have negative consequences, such as injector damage or filter clogging. Therefore, the engine system must include a fuel filter, especially for diesel engines, whose fuel contains a higher amount of particles.

From the numerical point of view, the simulation of multi-phase flows is not only an interesting scientific challenge but, due to the number of industrial applications, such as chemical reactors, combustion devices, medical inhalers, etc, it is also motivated by economical and safety considerations. In the study of the single-pass filters for (bio)diesel-from now on, diesel-vehicles, the difficulty is enhanced due to the presence of the fluid flow through a porous medium coupled with the interaction between the filtering medium and the particles. On the other hand, the flow rates in normal operating conditions render small Reynolds numbers, i.e., laminar flow conditions. In any case, once the models have been established, a code is needed to adequately discretize such models in the flow domain and to numerically solve the resulting linear equations.

This paper presents one new CFD approach to solve the flow in a diesel filter device, with the initial application to numerically reproduce the deposition pattern of particles shown on a filter under real conditions. The paper is structured in various sections. In Section 1, a brief description of an automotive diesel filter in a general situation is given. In Section 2, a model is presented for the two-phase flow composed of diesel and particles, based on a probability density function (PDF) approach [7] for the diluted phase position and size. In Section 3, the Brinkman-Darcy approximation for the flow through the filtering paper [8] is shown to be adequate for normal operating conditions of diesel filters. In Section 4, the numerical results are depicted, which show a very good qualitative agreement in terms of the deposition of particles on a used filter. Conclusions are drawn, and future work is suggested.

2. Eulerian-Eulerian Statistical Approach

In this section, the modeling approach of (solid) particles in a diesel flow is explained. Water filtering has a different behavior and will be the subject of another study.

There are several alternatives for predicting the evolution of two-phase flows, depending on the approximation used to solve each phase. In Eulerian-Lagrangian approaches, the continuous phase (diesel, in our case) is solved in the usual Eulerian frame. The dispersed phase elements (solid particles, in our case) are tracked throughout the simulation, and their interactions with the continuous phase are taken into account locally. These methods, such as the Discrete Phase Model [9], are known to be able to reproduce accurately the behavior of the system, but they can be numerically expensive, as a large number of discrete elements must be used. Moreover, a mixed Lagrangian-Eulerian code is needed.

In Eulerian-Eulerian approaches, both phases are treated as being continuous after introducing some kind of averaging procedure (time, space, ensemble, etc.) in the discrete phase. The exact knowledge of the position of the particles is lost, which is, in principle, a disadvantage for the modeling of the interaction between both phases, although, depending on the averaging procedure, the necessary information for local interaction may be kept (as in the present paper). As the formulation is purely Eulerian, the usual discretization and numerical algorithms can be used in opposition to Eulerian-Lagrangian formulations.

In the present paper, an Eulerian-Eulerian approach is used. To choose the most convenient average of the discrete phase, an analysis of the relevant interaction phenomena in the present problem is pertinent. The void fraction of particles (the fraction of the total volume occupied by particles) is extremely small. That means that both the particle-particle interaction and the influence of the particles on the diesel can be neglected. Therefore, the flow of diesel can be solved independently. In the present case, the flow is laminar (the Reynolds number is well below 200 in all cases), so there is no need for any turbulence modeling, although other difficulties arise, as shown in Section 3.

On the other hand, the effect of the fluid on the particles should be considered, along with the gravity field. The key parameter is the particle size since the forces acting on the particles are directly related to their size. At this point, it should be stated that only spherical particles are to be considered in this study. Of the various forces that can act on the particles, such as drag, history (Basset), and buoyancy, the drag is considered the most relevant under the present highly diluted conditions. The effect of gravity/buoyancy may also be important for the bigger particles, given the density difference between the fluid and the particles, so it should be taken into account as well. As the density of the particles used to verify the filter quality is constant, the buoyancy is proportional to the cube of the radius. In the end, the particle radii and velocities are the only relevant magnitudes needed to characterize their evolution.

With these simplifying assumptions, the transport equations of the magnitudes characterizing the discrete phase as a function of the particle radius may be obtained. In strictly obtaining this, a probabilistic description should be considered. The reason is that boundary conditions for the particles are only known in a probabilistic sense. One knows that the particles entering the filter follow a specified size distribution, but no information about the specific position, velocity, or size of any single particle in the flow is available. Even this statistical information about the discrete phase at the boundaries is rather incomplete, although some plausible assumptions make this limited information sufficient. The details are found in Appendix A.

The resulting transport equations are, however, quite intuitive; firstly, we have a continuity-like equation to express the conservation of the number density for the particles of each radius (there is no agglomeration or wear of the particles):

$$\frac{\partial n(r,\mathbf{x};t)}{\partial t} = -\frac{\partial}{\partial x_j}\left[u_j^p(r,\mathbf{x};t)n(r,\mathbf{x};t)\right] \tag{1}$$

where $n(r,\mathbf{x};t)$ is the expected number density of particles of the size r at position $\mathbf{x}$, and $\mathbf{u}^p(r,\mathbf{x};t)$is their corresponding velocity. Secondly, the momentum conservation adopts the form of the evolution of the velocity of small spherical particles in a laminar flow, considering only the dominant terms, drag, and buoyancy in our case [10]:

$$\begin{aligned}(\rho^p + 0.5\rho)\frac{4\pi r^3}{3}\frac{du_i{}^p}{dt} &= (\rho - \rho^p)\frac{4\pi r^3}{3}g_i \\ &\quad -3\rho\pi r\nu(u_i{}^p - u_i) \\ &\quad -\frac{9\rho\pi r^2}{4}\|u_i{}^p - u_i\|(u_i{}^p - u_i)\end{aligned} \tag{2}$$

where ρ^p and ρ are the densities of the particles and the fluid, respectively; $4\pi r^3/3$ is the volume of a spherical particle with diameter r; ν is the kinematic viscosity; $\mathbf{u}$ is the fluid velocity; and d/dt stands for the total time derivative of the velocity of the particle, that is to say, its total acceleration or its Lagrangian time derivative. In Equation (2), the left term represents the linear momentum variation of a spherical particle and its added mass (because of the equivalent volume of fluid, which moves with the particle), whereas, on the right side, there is first a buoyancy contribution and, next, a drag contribution up to the second order to the relative velocities (the velocity of the particle relative to that of the fluid) with an approximation valid for those velocity particles whose Reynolds number, based on the relative velocity, is less than one, or the theoretical observed drag.

Values for the fluid properties and particle sizes are taken from those of Shell Fluid 41 doped with the fine test dust A2 ISO 12103-1. The details are shown in Table 1.

Table 1. Flow properties and geometrical data. The intake area is the area of the annulus of the filter, taking out the small section of the internal separating walls. Filtering paper K11B40A. Fluid values for Shell Fluid 41. Fine test dust A2 ISO 12103-1 approximated by a Weibull distribution.

Geometrical	Inflow area	9.13 cm^2
	Filter height	89.7 mm
	Outer annulus diameter	80 mm
	Inner annulus diameter	70 mm
	Exit duct diameter	25 mm
Filtering Media	Permeability K	2.81×10^{-13}
	Porosity ε	0.517
	Paper thickness L	0.65 mm
Flow	Flow rate	120 L/h
	Fluid density	850 kg/m^3
	Fluid kinematic viscosity	13.6 cSt
	Particle density	2650 kg/m^3
	Dust concentration	50 mg/L
	Prob. density function	$\beta \frac{x^{(\beta-1)}}{xref^{\beta}} e^{-\left(\frac{x}{xref}\right)^{\beta}}$
	Size parameter	$xref = 14.3$ µm
	Shape parameter	$\beta = 1.1$

The boundary conditions at the inlet are given by the expected density number for each particle size following a prescribed Weibull probability density function for the A2 dust, which is equally distributed in space and with particles that match the fluid velocity at the domain entry. At the exit, zero gradient conditions are used, while non-slip conditions are prescribed at walls. Notice that the void fraction information is known from $n(r, \mathbf{x}; t)$.

3. Diesel in the Paper Filter: Darcy-Brinkman Approximation

From the diesel point of view, as the flow inside the filter is not affected by the particles during this short time (the permeability evolves very slowly in comparison to the flow characteristic time), a single-phase flow in a porous medium is to be studied. This is the object of the present section.

We use the approximations of Ochoa-Tapia and Whitaker [11] as the most convenient way to describe the flow in the free and porous medium:

$$\frac{\partial u_i}{\partial t} + \frac{\partial u_i u_j}{\partial x_j} = -\frac{1}{\rho}\frac{\partial P}{\partial x_i} + \frac{\partial}{\partial x_k}\left(\nu \frac{\partial u_i}{\partial x_k}\right) \tag{3}$$

$$\frac{1}{\varepsilon}\frac{\partial u_i}{\partial t} + \frac{1}{\varepsilon^2}\frac{\partial u_i u_j}{\partial x_j} = -\frac{1}{\rho}\frac{\partial P}{\partial x_i} + \frac{\partial}{\partial x_k}\left(\frac{\nu}{\varepsilon}\frac{\partial u_i}{\partial x_k}\right) - \frac{\nu}{K} u_i, \tag{4}$$

where P is the fluid pressure without the hydrostatic contribution, ε is the porosity (the empty volume fraction available for the fluid), and K is Darcy's permeability in units of $[L]^2$. The flow is assumed to be incompressible and to have a constant density; that is to say, the null divergence constraint is applied on the velocity field, $\nabla \cdot \mathbf{u} = u_{i,i} = 0$. The gravitational effects on the fluid may be considered canceled with the hydrostatic pressure gradient.

In Equation (4), the velocity is really a superficial volume average,

$$u_i \equiv \langle u_i \rangle_S = \frac{1}{V}\int_V u_i \, dV = \varepsilon \frac{1}{V_{free}}\int_V u_i \, dV = \varepsilon \langle u_i \rangle_I, \tag{5}$$

whereas the pressure is an intrinsic volume average.

$$p \equiv \langle p \rangle_I = \frac{1}{V_{free}}\int_V p \, dV = \frac{1}{\varepsilon}\frac{1}{V}\int_V p \, dV = \frac{1}{\varepsilon}\langle p \rangle_S. \tag{6}$$

These two types of averages, related by the porosity ε, are chosen because they match the velocity and pressure variables, respectively, from Equation (3) at the interface [11]. For this reason, no average operator is written explicitly in the paper.

It should be noted that in Equation (4), one could add a second-order term, the Forchheimer correction, to its right side, as follows: [12]

$$-\frac{F}{K^{1/2}}(u_j u_j)^{1/2} u_i \tag{7}$$

where F is a non-dimensional coefficient. However, the usual velocity magnitudes present in the filtering problem with which we are dealing are small enough, and this leads to neglect of Forchheimer's correction from the formulation. One should also be aware that the permeability is, in general, a second-order tensor, so $K^{-1}u_i$ should really be $K_{ij}^{-1}u_j$. Again, the filtering paper is assumed to be isotropic at least along its flow-normal direction; that is to say, $K_{ij}^{-1}u_j = K^{-1}u_n$, and the work can be performed with a scalar permeability.

Averages are computed over a mesoscopic (larger than the pore size, smaller than the large-scale variations of the fluid) porous zone. In this sense, the previous equations should be understood as valid on that scale.

The fluid and porous zones may be integrated as two different domains using Equations (3) and (4), respectively, plus the incompressibility constraint, which is linked by a proper boundary condition [13] as follows:

$$\begin{aligned} \mathbf{u}_f &= \mathbf{u}_p \\ P_f &= P_p \\ \nu\frac{\partial \mathbf{u}_f}{\partial x_i} n_i &= \frac{\nu}{\varepsilon}\frac{\partial \mathbf{u}_p}{\partial x_i} n_i \end{aligned} \tag{8}$$

where $\mathbf{n}$ represents the vector normal to the fluid-porous interface, and the sub-indexes f and p stand for the fluid and porous media at the interface, respectively.

There is also the possibility of modeling the coupling between the porous and fluid media by assuming that the whole computational domain represents a heterogeneous fictitious porous medium [14] where $\varepsilon = 1$ and $k = \infty$ in the free zones. This approach is justified [14] at least in the Stokes case (without advection). Equation (4) is applied in the full domain with the incompressibility constraint [15] and variable values of ε and K, with a gradient singularity at the inner interface between the porous and fluid media. That is the approach followed in the present paper.

4. Numerical Results

Due to the purely Eulerian approximation applied, the two-phase formulation proposed here has been incorporated into a standard CFD solver, OpenFOAM. OpenFOAM [16] is a software package that solves partial differential equations discretized on 3D unstructured meshes by finite volume methods. It is open-source, so the code can be modified to the needs of each particular problem. Usually, some of the available pre-configured solvers are used as a template and adapted, as in the present filtering problem. A computational fluid dynamics SIMPLE algorithm was appropriate for the present case.

To represent the porous filter, a specific permeability value has been chosen by fitting the pressure drop of a clean filter. This means that a snapshot of the behavior of the filter at a particular time has been used. In fact, a different value does not change the general pattern of the flow of particles, whose study is the main objective of this paper.

A typical filtering element without clogging is depicted in Figure 1a. The numerical results for the pressure drop through the filter and the vertical component of the flow velocity are presented in Figure 1b. An optimized all-hex 3.3 M cell mesh is used for this purpose. A mesh detail from a cut through the filtering media is presented in Figure 2a whilst the flow through the filtering media is detailed in Figure 2b. The mesh in the normal direction of the interface(s) (free media (blue wireframe)/porous media (green/pink wireframes)) has been kept at a comparable size on either side of the interface(s). A WENO-like approach for pressure/velocity has been used in the treatment of the interface to avoid the under- and overshoots in the velocity profile.

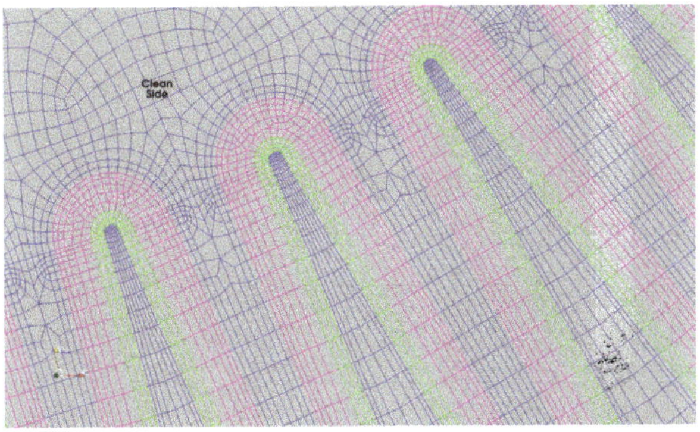

(a)

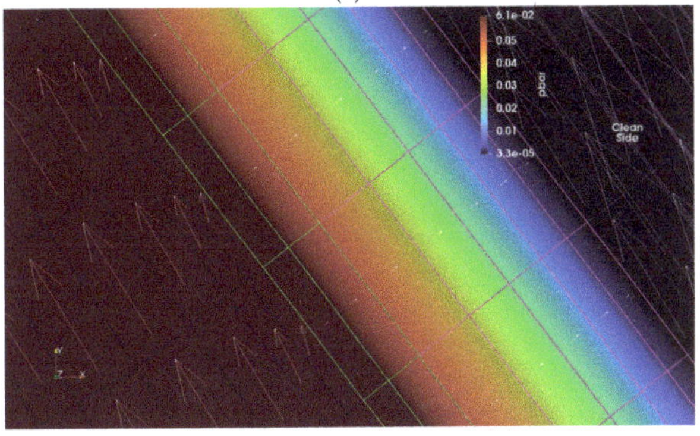

(b)

Figure 2. (**a**) Mesh detail in a plane through the filtering device (normal to z-direction); (**b**) velocity vectors and pressure drop detail through the same plane.

As can be observed in Figures 1b and 2b, most of the pressure drop takes place in the filtering media. The two-layer composition of the filtering media, the hydrophobic (melt-blown), and the cellulose are highlighted in Figure 2b with green and pink wireframes, respectively. However, for the performed 3D numerical simulation, both media were assigned the same properties (permeability and porosity), as given in Table 1. The computational domain was reduced to a sixth of the full filtering device by taking into account its cylindrical symmetry, as depicted in Figure 3, thus enabling reasonable computing times to be achieved (under two hours CPU time on an Intel Core i7-7700K four cores 4.2 GHz).

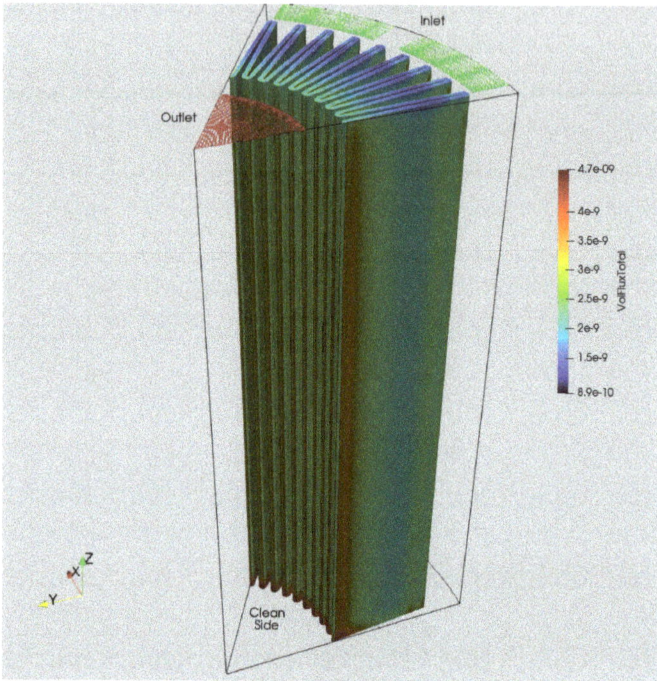

Figure 3. Flux of particle mass arriving at the filtering media.

It is worth pointing out that horizontal velocities tend to become normal across the filter medium, whilst, inside the free medium, they tend to align with the porous-free interface, as can be seen in Figure 2b. This effect is a natural, expected consequence of the presence of a Darcy medium (the filter) with low permeability. High lateral pressure gradients parallel to the porous-free medium interface would imply (see Equation (3)) high fluid accelerations and high velocities inside the free medium, which, in turn, would render the flow incompatible with mass conservation and the provided boundary conditions. As a result, most of the pressure drop inside the porous medium must be normal to those interfaces. With the typical values of permeability in the studied filter, the dominant contributions to the flow inside the porous medium (see Equation (4)) are the pressure gradient and the Darcy one, which cancel each other almost exactly. With regard to the free medium, the presence of a porous-free interface acts as a barrier that leads the flow to try to surround it, until it impinges upon the inner corners of the filter folds. Of course, there is a certain amount of flow passing through the filter before the folds as can be seen in Figure 2b.

In Figure 3, the total mass per unit area and the arrival time at the clean filter are shown. The qualitative agreement with Figure 4 is obvious, and the pattern is well explained using the same considerations as in the previous paragraph: the porous-free interface leads most of the incoming flow towards the folds of the filter through which, subsequently, more particles are going. Eventually, and in longer calculations than those reported in this paper, the flow would be diverted from the folds and become more evenly distributed along the filter: particle deposition starts at the inner side of the folds and propagates upstream from there (see Figure 4).

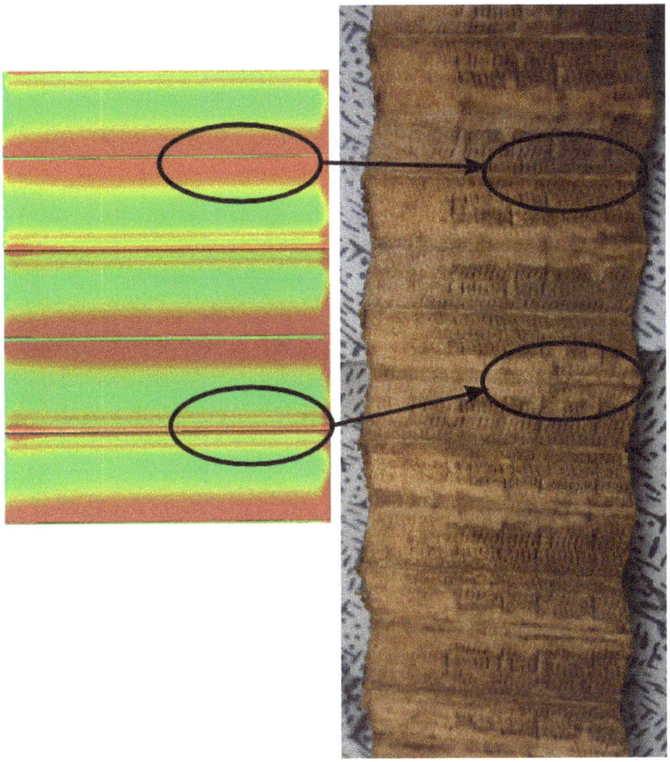

Figure 4. Unfolded view of a typical stained filter and comparison with numerical results.

It should be noticed that the real conditions are not known. This includes, among other things, the type of diesel used, its cleanliness, its particle distribution, its vibrations, etc. Those effects should distort somehow the deposition pattern, which is preserved, nevertheless, in its general characteristics. This is so that the main features of the flow, described above, are not fundamentally affected by those effects.

In Figures 5–7, the same quantity, or the particle number flow through a clean filter, is depicted but for progressively larger sized particles. One can notice the increase in gravitational effects as the particles grow in size. This effect may be noticed in Figure 4, and, in the total mass simulation, Figures 3 and 7, as well; there is not only a tendency for particles to accumulate starting from the folds but also from its bottom area where the larger sized particles aggregate. The most interesting feature of these simulations is that both these cited effects may be explainable without resorting to long calculations with variable permeability. In a clean filter with uniform properties, the flow is diverted so as to increase the flow of filtered particles through the areas that, eventually, would become more clogged.

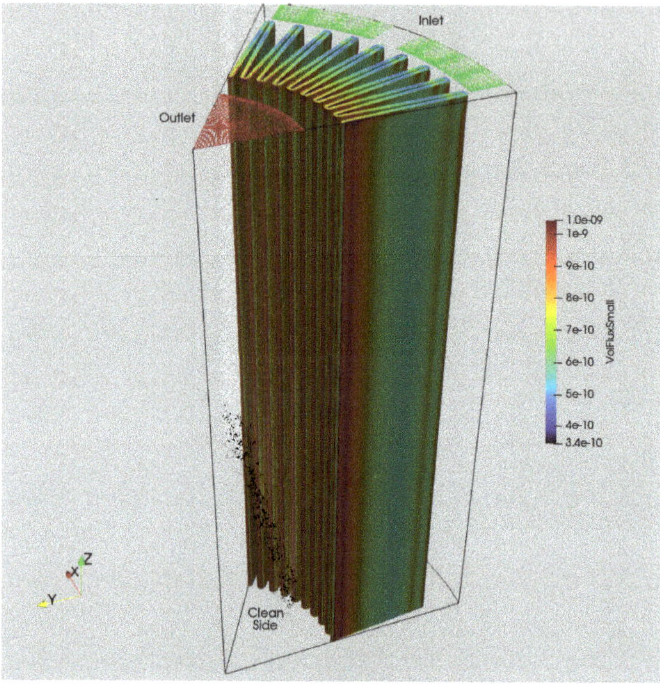

Figure 5. Flux of particle mass for particles with diameters of less than 10 μm.

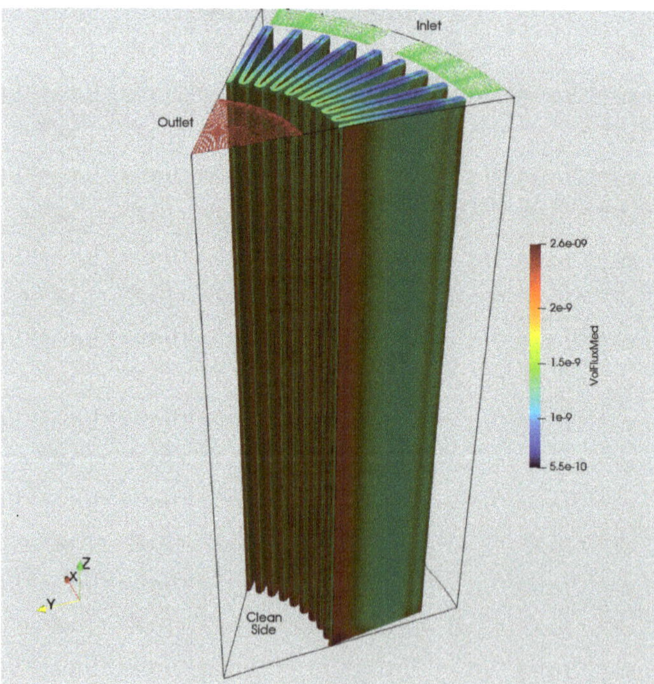

Figure 6. Flux of particle mass for particles with medium-sized diameter ($30 \leq d \leq 40$ μm).

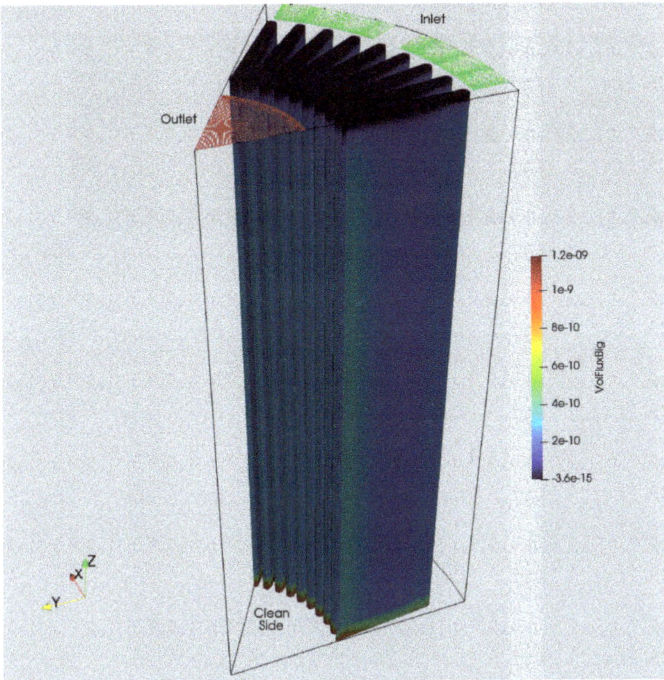

Figure 7. Flux of particle mass for particles with diameters in the upper range of the distribution ($\geq$70 μm).

Quantitative Comparisons

Although some small-scale distribution measurements are available [17], there are no experiments to date that provide measurements of the spatial distribution of the mass of particles absorbed in filters, such as the one studied in this article. In order to quantify the good qualitative agreement observed between the flow pattern computed through a clean filter and its observed amount of particle deposition, one further step has been considered, and the findings are presented in Figure 8. As long as clogging does not significantly affect the filter properties, the spatial distribution of the mass absorption will essentially remain unchanged; only the local values will increase proportionally with time. Therefore, if we normalize by the maximum value, the distribution is independent of time. It is recalled that this is valid for relatively early times in the life of the filter, that is to say, as long as the effect of clogging can be neglected.

For the filter, the picture (Figure 4 or Figure 8a) has been taken with a camera with a linear response, and it is assumed that there is also a linear relationship between the gray intensity level and the absorbed mass. This last assumption will be revisited later in this section. In addition, a linear relationship is established between the numerical results and the gray level of their representation (the grayscale version of the left side of Figure 4). The gray levels are normalized in the $[0, 1]$ range, with 1 representing the clearest value and 0 the most opaque one.

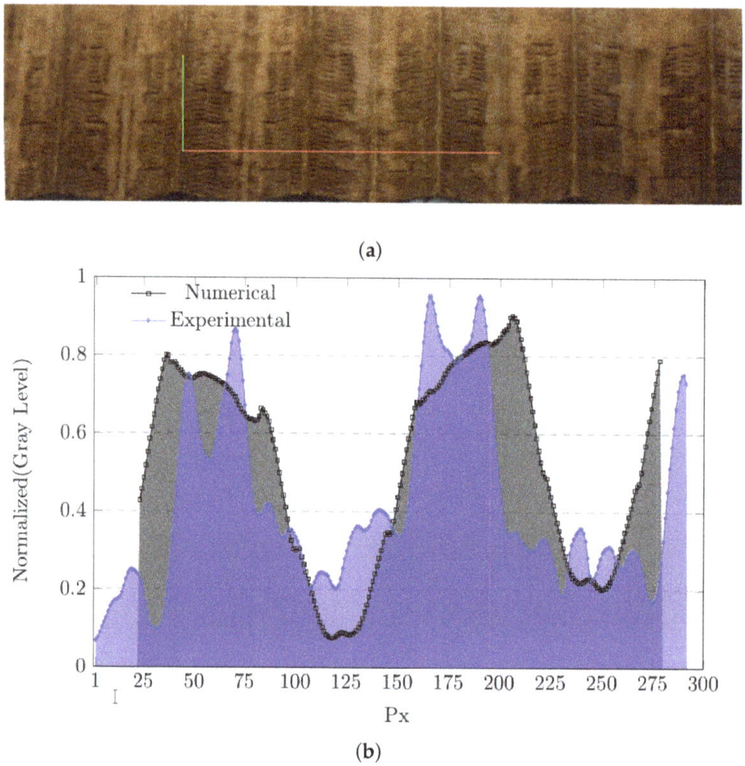

Figure 8. Intensity profile: (**a**) averaging zone; (**b**) comparison. Notice the effect of the pleats that are in contact with the upper side of the real device, which yields a double-peaked profile.

Due to the inherent noisiness of the data, and in order to perform a valid comparative analysis, one has to perform some kind of spatial averaging. A representative area determined by the red and green lines of Figure 8a and its counterpart of the numerical data have been considered for a two-step averaging procedure, resulting in the intensity data profiles represented in Figure 8b. Firstly, the intensity profile was extracted by sweeping across the red line and averaging along the green line. Secondly, the moving averages of a size (15 px), which preserves the main features of the profiles, are calculated over the intensity data obtained in the first step. The experimental irregularities in the pattern for the higher and lower peaks of the profile reflect the impossibility of having perfect pleats without deformations or any contact between their inner sides. The percentage difference between the integrated values of the two curves (shaded areas in Figure 8b), which represent the total mass of particles absorbed by the filtering area considered, amounts to 8.43%. This relatively small difference is consistent with the assumption of the linearity that exists between the gray level and the absorbed mass.

5. Conclusions

The full 3D simulations of particle filtering inside an automotive filter have been presented, in which the continuous phase governing the equations is supplemented with transport equations for the particle size distribution. The deposition pattern of particles on a BOSCH filter used under real conditions has been qualitatively reproduced with good quantitative behavior during the initial phases of clogging. This shows the potential of this tool. In future work, a model relating pressure drop with particle clogging should be developed in order to estimate the evolution of permeability and predict the lifetime of the filtering element.

Author Contributions: Conceptualization, L.V., R.M., J.H., J.L.H., M.J.G., C.B., Y.-T.C., L.-W.C. and P.R.; methodology, L.V., R.M. and J.H.; software, L.V., R.M., J.H. and P.R.; validation, L.V., R.M., J.H., J.L.H., M.J.G. and C.B.; formal analysis, L.V., R.M., J.H., Y.-T.C., L.-W.C. and P.R.; investigation, L.V., R.M., J.H. and P.R.; resources, L.V., R.M., J.H., J.L.H., M.J.G. and C.B.; data curation, L.V., R.M., J.H. and P.R.; writing—original draft preparation, L.V., R.M., J.H. and J.L.H.; writing—review and editing, L.V., R.M., J.H., Y.-T.C., L.-W.C. and P.R.; visualization, L.V., R.M. and J.H.; supervision, L.V.; project administration, L.V. and R.M.; funding acquisition, L.V. and C.B. All authors have read and agreed to the published version of the manuscript.

Funding: This work was partially funded by the Spanish Government through the CENIT PIIBE project and the CSIC Intramural Grant 202260I190.

Data Availability Statement: Data sharing is not applicable to this article.

Conflicts of Interest: The authors declare no conflict of interest. The funders had no role in the design of this study; in the collection, analyses, or interpretation of data; in the writing of the manuscript; or in the decision to publish the results.

Appendix A

For a statistical description of the discrete phase, in which the particles are moving in the diesel, the appropriate quantity is the droplet density function first introduced to describe the droplets' evolution in a spray [18,19]. The droplet density function $n(\mathbf{v}, r, \mathbf{x}; t)$ is defined such that

$$n(\mathbf{v}, r, \mathbf{x}; t)\, d\mathbf{v}\, dr\, d\mathbf{x}$$

is the expected number of particles with velocities in the interval $(\mathbf{v}, \mathbf{v} + d\mathbf{v})$, radii in the interval $(r, r + dr)$, and positions in the interval $(\mathbf{x}, \mathbf{x} + d\mathbf{x})$. This function contains all the statistical information needed to describe our system.

Using conditional statistics, we can express this function as

$$n(\mathbf{v}, r, \mathbf{x}; t) = n(r, \mathbf{x}; t)\, P(\mathbf{v}|r, \mathbf{x}; t), \tag{A1}$$

where $n(r, \mathbf{x}; t)$ is the expected number density of particles with the radius r and position $\mathbf{x}$ at time t, and $P(\mathbf{v}|\mathbf{r}, \mathbf{x}; t)$ is the probability density function (PDF) of the particle velocity, conditional on the radius r and position $\mathbf{x}$ at time t. For the present analysis, we will derive transport equations for the two quantities on the right-hand side of Equation (A1) to obtain the statistical information of the discrete phase.

First, the expected number density $n(r, \mathbf{x}; t)$ is analyzed. This is the probable density of the particles found in the small region $d\mathbf{x}\, dr$. It can be easily expressed by means of the Lagrangian PDF $P_{\mathbf{x}^{\mathbf{P}+} r^+}(\mathbf{x}, r; t)$ of the particle radius r^+ and position $\mathbf{x}^{\mathbf{P}+}$ (where the sign + is used to indicate the Lagrangian frame) [20]:

$$P_{\mathbf{x}^{\mathbf{P}+} r^+}(\mathbf{x}, r; t) = \frac{n(\mathbf{x}, r, t)}{N}, \tag{A2}$$

where N is the total number of particles in the domain, which is supposed to have negligible fluctuations. Integration of the number density n over a finite region in $(\mathbf{x}, r)$ space would then provide the probable number of particles in that region.

Remember that the statistical information of $n(r, \mathbf{x}; t)$ is equivalent to that provided by $P(\mathbf{x}, r; t)$ once the total number of particles is known. As N should be a piece of provided information, next, a transport equation will be derived for $P(\mathbf{x}, r; t)$.

From the theoretical point of view, the description based on this PDF comes from a description of an ensemble average (the average of a set of identical systems). Its transport equation can be derived by standard methods [20]:

$$\begin{aligned}\frac{\partial P(r,\mathbf{x};t)}{\partial t} = & -\frac{\partial}{\partial x_j}\left[\left(\overline{\frac{dx^{p+}_j}{dt}|\mathbf{x},r}\right)P(r,\mathbf{x};t)\right] \\ & -\frac{\partial}{\partial r}\left[\left(\overline{\frac{dr^+}{dt}|\mathbf{x},r}\right)P(r,\mathbf{x};t)\right] \\ & -JP(r,\mathbf{x};t)\end{aligned} \tag{A3}$$

where the overline represents an ensemble averaging, and the terms inside the parentheses are conditional-averaged quantities (in the ensemble average approximation). J is the net flux of the probability of the PDF $P(r,\mathbf{x};t)$ crossing the boundaries of the definition domain of the PDF. Notice that the term in J takes this form to guarantee that the integration of the PDF in the probabilistic space is always equal to one. In the present case, the domain in the physical space would be the zone of the device upstream from the paper filter, and the "radius" space would be all the (positive) sizes except zero. In fact, the PDF is implicitly conditional on being defined in the described region [20], as we are not interested in particles outside that domain or with meaningless size. No increase or decrease in the size of the particles, including coalescence or breaking up, should be considered, given the low void fraction of particles in diesel. An accumulation is to occur for sure in the paper filter, but this is outside our domain.

In steady-state situations, the last term in Equation (A4) is null, as the same number of particles that enter the definition domain should leave it (either by being trapped in the paper filter or crossing to the other side), so the net flux is zero. In the present case, the PDF transport equation could be considered in a quasi-steady state. The reason is that the effect of the particles on the filter changes very slowly compared to the fluid flow characteristic times, which means that temporal variations of the paper filter properties can be neglected in the transport equation. That is, for the duration of a car ride, changes in the filter properties are negligible.

Should this PDF be described in the Lagrangian frame, it would be represented numerically by particles evolving in the flow in a Monte Carlo simulation. This is essentially what an Eulerian-Lagrangian description does. In the present case, our interest lies in the use of an Eulerian-Eulerian formulation, so the PDF is represented as a set of notional Eulerian fields [7]. Notice that the information obtained is identical for both representations. In fact, in this case, the traditional distinction between Eulerian-Lagrangian and Eulerian-Eulerian is really a distinction between two numerical algorithms to solve what essentially is the same quantity, the PDF of the particle radius (and position).

For the particular situation described in this paper, as the flow is laminar, by using the Eulerian-Eulerian approach, we will end up with non-stochastic Eulerian transport equations for the particle-related magnitudes, whose derivation is the purpose of this appendix. That does not mean that we know exactly the position of every particle in the flow, as the boundary conditions are only known in a probabilistic sense.

From Equations (A2) and (A4), a conservation equation for the number density $n(r,\mathbf{x};t)$ is obtained (notice that the last term in Equation (A4) compensates for the one appearing from the time derivative of N [20]):

$$\begin{aligned}\frac{\partial n(r,\mathbf{x};t)}{\partial t} = & -\frac{\partial}{\partial x_j}\left[\left(\overline{\frac{dx^{p+}_j}{dt}|\mathbf{x},r}\right)n(r,\mathbf{x};t)\right] \\ & -\frac{\partial}{\partial r}\left[\left(\overline{\frac{dr^+}{dt}|\mathbf{x},r}\right)n(r,\mathbf{x};t)\right]\end{aligned} \tag{A4}$$

In the previous equation, the terms that are conditional on position $\mathbf{x}$ can be replaced by their Eulerian counterparts:

$$\begin{aligned} \frac{\partial n(r,\mathbf{x};t)}{\partial t} &= -\frac{\partial}{\partial x_j}\left[\left(\overline{u^p{}_j|r}\right)n(r,\mathbf{x};t)\right] \\ &\quad -\frac{\partial}{\partial r}\left[\left(\overline{\dot{r}|r}\right)n(r,\mathbf{x};t)\right] \end{aligned} \tag{A5}$$

In Equation (A6), the variable $\mathbf{x}$ should be taken in the physical space, no longer in the probabilistic space. The velocity of the particle $\mathbf{u}^\mathbf{p}$ and the radius variation $\dot{r}$, as Eulerian quantities, are now functions not only of time but also space.

We have already mentioned that the overline means some kind of ensemble average; for example, we can imagine a set of repeated equivalent systems (filters with diesel and particles). In the case of turbulent flows, as turbulence is a chaotic dynamical system, each individual system will evolve in a different way. In the present case, however, the flow is laminar, which means that the flow is not chaotic, so all the systems would evolve in an identical way. As a consequence, we can safely remove the "overline" from the previous equation. As, in this case, the particles do not change size, Equation (A6), which is particularized for the particles inside the diesel, is rewritten as:

$$\frac{\partial n(r,\mathbf{x};t)}{\partial t} = -\frac{\partial}{\partial x_j}\left[u_j^p(r,\mathbf{x};t)n(r,\mathbf{x};t)\right] \tag{A6}$$

Once the differential equation for $n(r,\mathbf{x};t)$ is set up, the boundary conditions should be prescribed. As the number density is a statistical magnitude, only statistical information is needed. Hence, it is not necessary to prescribe the individual position of each particle, just some kind of PDF information about the boundary.

Now, the second term of the right-hand side of Equation (A1), $P(\mathbf{v}|r,\mathbf{x};t)$, is analyzed. This is the probability density of having a particle velocity value v conditional on its radius r and position $\mathbf{x}$ at time t. In the present case, being a laminar flow with no particle interaction, this is not a stochastic magnitude: the velocity evolution just depends on those magnitudes under the problem conditions. Its value $\mathbf{u}^p(r,\mathbf{x};t))$ is obtained in our problem by solving the transport equation of a small spherical particle in a laminar flow, considering only the dominant, drag, and buoyancy terms [10]:

$$\begin{aligned} (\rho^p+0.5\rho)\frac{4\pi r^3}{3}\frac{du_i{}^p}{dt} &= (\rho-\rho^p)\frac{4\pi r^3}{3}g_i \\ &\quad -3\rho\pi r\nu(u_i{}^p-u_i) \\ &\quad -\frac{9\rho\pi r^2}{4}\|u_i{}^p-u_i\|(u_i{}^p-u_i) \end{aligned} \tag{A7}$$

where ρ^p and ρ are the densities of the particle and the fluid, respectively; $4\pi r^3/3$ is the volume of a spherical particle with a diameter r; ν is the kinematic viscosity; $\mathbf{u}^p$ is the particle velocity; $\mathbf{u}$ is the fluid velocity; and d/dt stands for the total time derivative of the velocity of the particle, that is to say, its total acceleration, or its Lagrangian time derivative. In Equation (A8), the left term represents the linear momentum variation of a spherical particle and its added mass (because of the equivalent volume of fluid, which moves with the particle), whereas, on the right side, there is first a buoyancy contribution and, next, a drag contribution up to the second order to the relative velocities (the velocity of the particle relative to the velocity of the fluid) with an approximation valid for those velocity particles whose Reynolds number, based on the relative velocity, is less than one, or the theoretical observed drag.

Notice that several of these transport equations for the velocity should be solved, once for each radius representative r considered in the particle size distribution.

At this moment, we have a complete and closed system to calculate all of the variables of interest (before entering the paper filter). To calculate the flux of particles entering the filter, we multiply the velocity particle perpendicular to the filter surface times the particle number for each radius. With this knowledge, the flux of the "volume" or mass of the particles (they do have a constant density) is immediately deduced.

References

1. Ulbricht, M. Advanced functional polymer membranes. *Polymer* **2006**, *47*, 2217–2262. [CrossRef]
2. Iliev, O.; Kirsch, R.; Osterroth, S. Combined Depth and Cake Filtration Model Coupled with Flow Simulation for Flat and Pleated Filters. *Chem. Eng. Technol.* **2018**, *41*, 70–78. [CrossRef]
3. Sun, Y.; Sanaei, P.; Kondic, L.; Cummings, L.J. Modeling and design optimization for pleated membrane filters. *Phys. Rev. Fluids* **2020**, *5*, 044306. [CrossRef]
4. Von Stockhausen, A.; Mangold, M.P.; Eppinger, D.; Livingston, T. Procedure for determining the allowable particle contamination for diesel fuel injection equipment (FIE). *SAE Int. J. Fuels Lubr.* **2009**, *2*, 294–304. [CrossRef]
5. Bessee, G.; Hutzler, S. The effects of diesel fuel additives on water separation performance. *SAE Int. J. Fuels Lubr.* **2009**, *2*, 287–293. [CrossRef]
6. Bensaid, S.; Marchisio, D.L.; Fino, D. Numerical Simulation of soot filtration and combustion within diesel particulate filters. *Chem. Eng. Sci.* **2010**, *65*, 357–363. [CrossRef]
7. Valiño, L. A field Monte Carlo formulation for calculating the probability density function of a single scalar in a turbulent flow. *Flow Turbul. Combust.* **1998**, *60*, 157–172. [CrossRef]
8. Valdés-Parada, F.J.; Goyeau, B.; Ochoa-Tapia, J.A. Jump momentum boundary condition at fluid-porous dividing surface: Derivation of the closure problem. *Chem. Eng. Sci.* **2007**, *62*, 4025–4039. [CrossRef]
9. Pathapati, S.; Sansalone, J.J. CFD Modeling of Particulate Matter Fate and Pressure Drop in a Storm-Water Radial Filter. *J. Environ. Eng.-ASCE* **2009**, *135*, 77–85. [CrossRef]
10. Loth, E. Numerical approaches for motion of dispersed particles, droplets and bubbles. *Prog. Energy Combust. Sci.* **2000**, *26*, 161–223. [CrossRef]
11. Ochoa-Tapia, J.; Whitaker, S. Momentum transfer at the boundary between a porous medium and a homogeneous fluid—I. Theoretical development. *Int. J. Heat Mass Transf.* **1995**, *38*, 2635–2646. [CrossRef]
12. Masarotti, N.; Nithiarasu, P.; Zienkiewicz, O. Natural convection in porous medium-fluid interface problems. A finite element analysis by using the CBS procedure. *Int. J. Numer. Methods Heat Fluid Flow* **2001**, *11*, 473–490. [CrossRef]
13. Betchen, L.; Straatman, A.; Thompson, B. A nonequilibrium finite-voume model for conjugate fluid/porous/solid domain. *Numer. Heat Transf. A* **2006**, *49*, 543–565. [CrossRef]
14. Angot, P. Analysis of singular perturbations on the Brinkman problem for fictious domain models of viscous flows. *Math. Meth. Appl. Sci.* **1999**, *22*, 1395–1412. [CrossRef]
15. Iliev, O.; Laptev, V. On numerical simulation of flow through oil filters. *Comput. Vis. Sci.* **2004**, *6*, 139–146. [CrossRef]
16. OpenCFD Ltd. (ESI Group). 2014. Available online: http://www.openfoam.com (accessed on 20 December 2022).
17. Jones, M.P.; Storm, M.; York, A.P.E.; Hyde, T.I.; Hatton, G.D.; Greenaway, A.G.; Haigh, S.J.; Eastwood, D.S. 4D In-Situ Microscopy of Aerosol Filtration in a Wall Flow Filter. *Materials* **2020**, *13*, 5676. [CrossRef] [PubMed]
18. Williams, F.A. Spray combustion and atomization. *Phys. Fluids* **1992**, *1*, 541. [CrossRef]
19. Subramaniam, S. Statistical representation of a spray as a point process. *Phys. Fluids* **2000**, *12*, 2413–2431. [CrossRef]
20. Valiño, L.; Hierro, J. Boundary conditions for probability density function transport equations in fluid mechanics. *Phys. Rev. E* **2003**, *67*, 046310. [CrossRef] [PubMed]

processes

Article

Thermal Performance Evaluation for Two Designs of Flat-Plate Solar Air Heater: An Experimental and CFD Investigations

Mahmoud S. El-Sebaey [1,*], Asko Ellman [2], Sh. Shams El-Din [1] and Fadl A. Essa [3]

1 Mechanical Power Engineering Department, Faculty of Engineering, Menoufia University, Shebin El-Kom 32511, Egypt
2 Faculty of Engineering and Natural Sciences, Tampere University, 33100 Tampere, Finland
3 Mechanical Engineering Department, Faculty of Engineering, Kafrelsheikh University, Kafrelsheikh 33516, Egypt
* Correspondence: mahmoud.el-sebaey@sh-eng.menofia.edu.eg or eng_mahmoudelsebaey@yahoo.com

Abstract: The main objective of this research was to create two different configurations of a flat-plate solar air heater, namely, Conventional-Case A and Modified-Case B, and develop a three-dimensional computational fluid dynamics (CFD) model using ANSYS R15.0. The purpose of the CFD model was to simulate the heat transfer behavior of the proposed solar air heaters under unsteady conditions. The RNG k-ε turbulence model was employed for this CFD study. The experiments were conducted on sunny days, under the same conditions as the Egyptian climate. The results of the experiments show that the simulated CFD model and the measured outlet airflow temperatures, relative humidity, and velocities of the two tested solar air heaters were compared. The developed model made very satisfactory predictions. Moreover, the deviations between the average CFD outlet air temperatures and the experimental results were 7% and 7.8% for Case B and Case A, respectively. The CFD-simulated average relative humidity was reduced by 31.6% when using Case B compared with Case A, and it was reduced by 28.8% when comparing the experimental data to Case B. Additionally, the average CFD thermal efficiencies obtained for Case B and Case A were 28.7% and 21.6%, respectively, while the average experimental thermal efficiencies for the cases were 26.4% and 18.2%, respectively. The proposed model can be used to design and simulate other solar air heater designs.

Keywords: solar air heaters; natural convection; air gap; CFD modeling and thermal efficiency

1. Introduction

Solar energy has a beneficial effect on the development of sustainable and clean energy sources [1]. However, although heating accounts for a significant portion of the global energy consumption, it is given less attention compared to other sectors [2]. Among the solar energy harvesting methods available, solar air heaters (SAHs) are widely recognized as a cost-effective option for various purposes, including drying [3], room heating, and industrial activities [4].

An SAH acts as a solar thermal collector that captures thermal solar energy and converts it into warm air suitable for low-temperature residential and industrial applications [5]. As the energy demand continues to increase, with conditioning systems accounting for a significant portion, SAHs could play a crucial role in reducing the energy consumption of buildings. Conventional SAHs have been widely adopted worldwide for solar thermal processes [6]. Due to its simple design, affordability, and low maintenance costs, the SAH is an ideal choice for various applications, including heating, dehumidification, building heating, energy storage, and drying agricultural products [7].

The thermal efficiency of an SAH is a crucial factor to consider when evaluating its performance [8]. However, the coefficient of heat transfer between the absorber and the airflow inside the SAH is low due to the thermo-physical properties of the airflow, a major weakness that poses a significant limitation in the development of SAHs [9]. In an effort to enhance efficiency and cost-effectiveness, researchers have conducted numerous studies over the years to increase the heat transfer coefficient [10]. A flat-plate solar air heater (FPSAH) is a type of heat exchanger that converts radiated solar energy into heat [11].

Researchers have introduced various approaches to improve the heat transfer coefficient of SAHs, including modifying their shape. For instance, rectangular [12], triangular [13], curved, spiral, corrugated, and circular absorber plates have been investigated [14].

Several studies have investigated the positioning of the absorber plate with respect to the airflow to improve the heat transfer in SAHs [15–17]. Additionally, the addition of baffles or fins has been explored as means of enhancing the heat transfer rate between the absorber plate and the air [18,19].

Researchers have also explored alternative absorber shapes, such as U-shaped corrugation or V-shaped grooves, to enhance the efficiency of SAHs [20]. These studies have demonstrated that the absorber position and air retention time in the collector are crucial factors in improving the efficiency of the system. However, conducting experiments to determine these factors can be time-consuming and expensive. Therefore, modeling and simulation tools have been utilized to address these issues.

Sanjay and colleagues [21] utilized CFD to develop a model of an SAH with rib roughening, examining how different rib arrangements affect the frictional loss and heat transfer properties. Meanwhile, using ANSYS Fluent, Dongxu et al. [22] employed numerical methods to investigate heat transfer in a roughened SAH featuring multiple V-shaped ribs on the absorber plate. Espinosa et al. [23] developed a numerical model of an SAH to estimate its thermal efficiency.

The literature reviewed lacks sufficient articles on simulating a flat-plate solar air heater (FPSAH) using ANSYS Fluent, and there was limited discussion on choosing models that accurately predict the mode and dynamics of heat transfer. This study aims to fill this gap by identifying the required boundary conditions, thermo-physical material and fluid properties, and meteorological conditions to create an accurate ANSYS Fluent model to simulate the FPSAH, which has not been previously explored. Thus, the primary aim of this research is to:

1. Perform experimental investigations on two FPSAH designs under the weather conditions in Shebin El-Kom, Egypt.
2. Evaluate the influence of operating parameters on the thermal performance of the tested FPSAH designs.
3. Create a detailed and dynamic CFD model capable of accurately simulating the real-world operation of FPSAHs under realistic operating conditions.

2. Test Rig Description

A schematic view of a conventional FPSAH (Case A) is presented in Figure 1. The heart of the FPSAH, which is the absorber, was made of a 1 mm galvanized iron sheet with dimensions of 1000 mm × 2000 mm, which was later painted black. The incoming solar radiation penetrates a 3 mm glass plate with the same dimensions as the absorber, with a distance of 100 mm between them. The air enters the solar air heater at a low temperature, and the absorber then transfers the maximum amount of heat to the airflow so that the outlet temperature reaches its maximum value. To minimize the heat lost to the atmosphere, the absorber bottom wall was insulated with 100 mm of sawdust. The device had a tilt angle of 30°, measured from the horizontal in the city of Shebin El-Kom, Egypt, which is located at 30.56° N and 31.00° E. Similar to Case A, Case B refers to a modified FPSAH using an air gap, as indicated in Figure 2. The gap between the two glass layers was 50 mm, with static air to raise the efficiency of the SAH. All the side walls of the upper glass box were glass with a thickness of 3 mm.

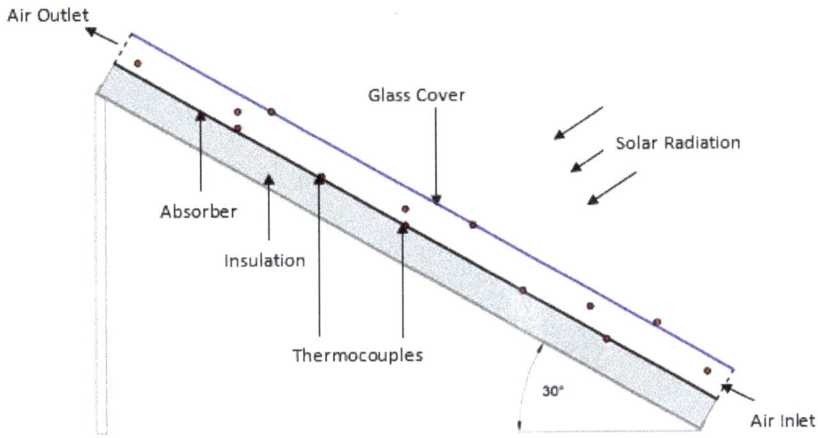

Figure 1. A schematic view of the conventional FPSAH, Case A.

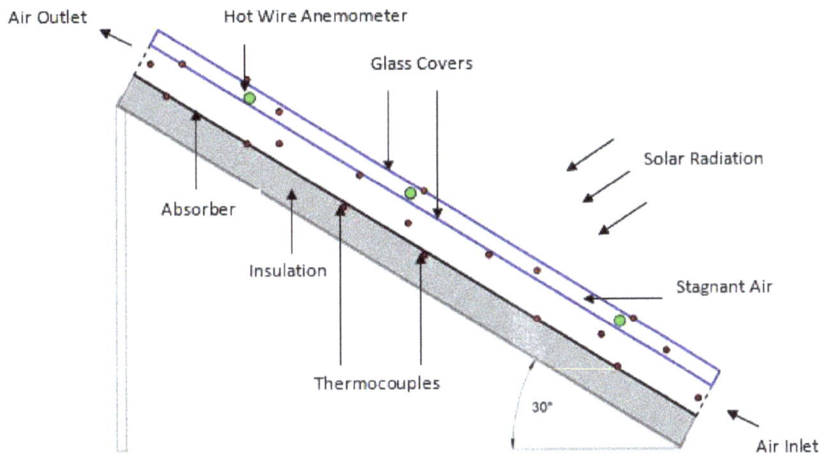

Figure 2. A schematic view of the Modified FPSAH, Case B.

Figure 3 exhibits a photograph of the test rig. The two tested FPSAHs (Case A and Case B) were orientated in a southern direction to receive the most global irradiation through the experiments.

The two tested devices were equipped with a set of calibrated K-type thermocouples (range of −210 to +750 °C; accuracy of ±1 °C; uncertainty of 1.3%) to measure the temperature over time. The thermocouples were installed on each inlet, outlet, absorber, glass, and side walls at different points to record the average values of the temperatures. A calibrated digital pyranometer was used to measure the amount of global irradiation (range of 0 to 2000 W/m^2; accuracy of ±10 W/m^2; uncertainty v1.6%). Additionally, a calibrated hot-wire anemometer was utilized to measure the inside velocity of the airflow at different positions along the two tested solar air heaters (range of 0 to 25 m/s; accuracy of ±0.1 m/s; uncertainty of 1.2%).

Case A

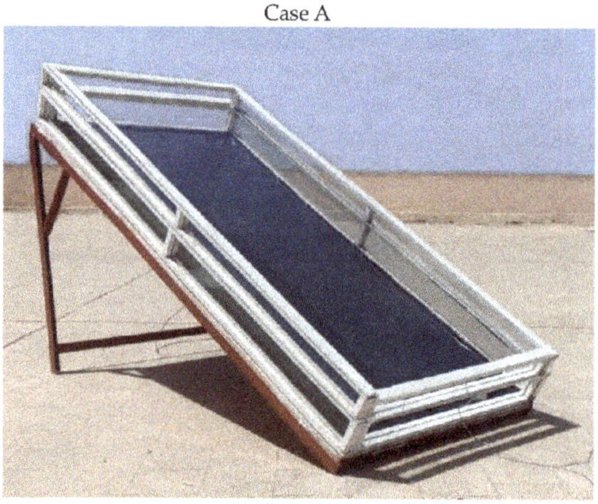

Case B

Figure 3. A photograph of the two FPSAH designs.

3. Thermal Analysis

3.1. Energy Balance

The energy balance of each part of the two tested solar air heaters {Case A and Case B} can be established. The energy balance of each part can be obtained using the classical method. In the chosen part, the total amount of the thermal flux can be written as the sum of all exchanged energy among the parts and the amount of solar energy lost or gained by this part. The thermal flux developed for each case of the solar air heater designed was described as follows:

3.1.1. Conventional Solar Air Heater (Case A)

In case A, the airflow absorbs a portion of global irradiation transmitted through the glass cover and then exchanges energy within both the flat-plate absorber and the glass cover by convection, thus:

$$S_Q = \frac{1}{S.L}\left[A_{gc}\tau_{gc}\alpha_f I(t) + A_{ab}h_{cf-ab}(T_{ab} - T_f) + A_{gc}h_{cf-gc}(T_{gc} - T_f)\right] \quad (1)$$

Glass Cover

A portion of global solar energy is absorbed by the glass cover, which exchanges energy with both the ambient air and airflow by convection and with both the flat-plate absorber and the sky by radiation, thus:

$$\frac{m_{gc}c_{pgc}}{A_{gc}}\frac{\partial T_{gc}}{\partial t} = \alpha_{gc}I(t) + h_{cex}(T_{amb} - T_{gc}) + h_{cf-gc}(T_f - T_{gc}) + h_{rgc-bb}(T_{ab} - T_{gc}) + h_{rgc-sky}\left(T_{sky} - T_{gc}\right) \quad (2)$$

Absorber Plate

The flat-plate absorber absorbs a portion of the transmitted global solar energy via airflow and the glass cover and exchanges energy with the airflow through convection, with the glass cover through radiation, and with the insulation by conduction, thus:

$$\frac{m_{ab}c_{pab}}{A_{ab}}\frac{\partial T_{ab}}{\partial t} = \tau_{gc}\tau_f\alpha_{ab}I(t) + h_{cf-ab}(T_f - T_{ab}) + h_{rab-gc}(T_{gc} - T_{ab}) + U_{cond}(T_{Ins} - T_{ab}) \quad (3)$$

3.1.2. Modified Solar Air Heater (Case B)

In Case B, the airflow partially absorbs the solar radiation, which is transferred by the two glass covers, the stagnant air, and the energy exchanged through convection between the SAH's absorber and the lower glass cover, thus:

$$S_Q = \frac{1}{S.L}\left[A_{gc}\tau_{st}\tau_{gc}{}^2\alpha_f I(t) + A_{ab}h_{cf-ab}(T_{ab} - T_f) + A_{gc}h_{cf-gc2}(T_{gc2} - T_f)\right] \quad (4)$$

Upper Glass Cover

The upper glass cover absorbs a portion of solar irradiation, exchanges the energy with the ambient air through convection, then with the stagnant air, and with the lower cover of the glass and with the sky through radiation, thus:

$$\frac{m_{gc}c_{pgc}}{A_{gc}}\frac{\partial T_{gcu}}{\partial t} = \alpha_{gc}I(t) + h_{cex}(T_{amb} - T_{gcu}) + h_{cgcu-st}(T_{st} - T_{gcu}) + h_{rgcu-gcl}\left(T_{gcl} - T_{gcu}\right) + h_{rgcu-sky}\left(T_{sky} - T_{gcu}\right) \quad (5)$$

Stagnant Air

A portion of the global irradiation transmitted by the upper glass cover is absorbed by the stagnant air, which then exchanges energy with the lower glass cover by convection, thus:

$$\frac{m_{st}c_{pst}}{A_{gc}}\frac{\partial T_{st}}{\partial t} = \alpha_{st}\tau_{gc}I(t) + h_{cgcl-st}\left(T_{gcl} - T_{st}\right) + h_{cgcu-st}(T_{gcu} - T_{st}) \quad (6)$$

Lower Glass Cover

The lower glass cover absorbs a portion of the global solar energy transmitted by the upper glass cover and the stagnant air and exchanges energy with the stagnant air and the airflow by convection. In addition, it exchanges energy with the absorber flat-plate and the upper glass cover by radiation, thus:

$$\frac{m_{gc}c_{pgc}}{A_{gc}}\frac{\partial T_{gcu}}{\partial t} = \alpha_{gc}\tau_{gc}\tau_{st}I(t) + h_{cgcl-st}\left(T_{st} - T_{gcl}\right) + h_{cf-gcl}\left(T_f - T_{gcl}\right) + h_{rgcl-ab}\left(T_{ab} - T_{gcl}\right) + h_{rgcl-gcu}\left(T_{gcu} - T_{gcl}\right) \quad (7)$$

Absorber Plate

The flat-plate absorber absorbs a portion of the transmitted global irradiation through the stagnant air and the two glass covers and exchanges energy with the airflow by convection and with the lower glass cover by radiation and conduction from the insulation, thus:

$$\frac{m_{ab}c_{pab}}{A_{ab}}\frac{\partial T_{ab}}{\partial t} = \tau_{gc}{}^2\tau_{st}\tau_f\alpha_{ab}I(t) + h_{cf-ab}(T_f - T_{ab}) + h_{rab-gcl}\left(T_{gcl} - T_{ab}\right) + U_{cond}(T_{Ins} - T_{ab}) \quad (8)$$

3.2. Calculation of the Relative Humidity

The relative humidity of the airflow though the FPSAH can be calculated as [24]:

$$RH = \frac{e_w - N(1 + 0.00115T_w)(T_d - T_w)}{e_d} \times 100 \quad (9)$$

where T_d and T_w are the dry and wet bulb temperatures, in °C.

N is a constant equal to 0.6687451584.

$$e_w = 6.112e^{\left(\frac{17.502T_w}{240.97+T_w}\right)} \quad (10)$$

$$e_d = 6.112e^{\left(\frac{17.502T_d}{240.97+T_d}\right)} \quad (11)$$

3.3. Thermal Efficiency

The thermal efficiency of the FPSAH can be calculated as [25]:

$$\eta_{th} = \frac{\dot{m}C_{pf}\left(T_{f_{out}} - T_{f_{int}}\right)}{A_{gc}I(t)} \quad (12)$$

The specific heat of the air is provided by:

$$C_{pf} = 1009.26 - 0.0040403T_f - 6.1759 \times 10^{-4}{T_f}^2 - 4.097 \times 10^{-7}{T_f}^3 \quad (13)$$

T_f is in °C.

4. Economic Analysis

An economic analysis presents the performance of the FPSAH in terms of its cost and benefits. The results of this analysis assist in selecting an appropriate device for any specific type of application. The heating levelized cost (HLC) was the parameter utilized to estimate the cost per kWh by considering key factors such as the capital cost, operating cost, maintenance cost, lifetime, inflation rate, and the device efficiency, etc. The HLC is estimated using the following equation [26]:

$$HLC = \frac{AC}{AE} \quad (14)$$

where AE is the average annual energy produced by the collector, while AC represents the total annual cost for the collector and can be calculated as:

$$AC = FC + MC + OC - SF \quad (15)$$

The fixed annual cost (FC) is determined by:

$$FC = CF.CC \quad (16)$$

where CC is the capital cost of the collector, and CF is the capital recovery factor, which can be calculated as:

$$CF = \frac{i.(1+i)^n}{(1+i)^n - 1} \quad (17)$$

where n is the economic collector lifetime.

The annual maintenance cost (MC) per year is determined using:

$$MC = (0.15).FC \quad (18)$$

The annual savage factor (SF) is determined by:

$$SF = FF.S \tag{19}$$

$$FF = \frac{i}{(1+i)^n - 1} \tag{20}$$

$$S = (0.2).CC \tag{21}$$

where i is the annual interest rate, %.

5. Computational Fluid Dynamics (CFD)

CFD is a computer code that can predict and explain the phenomena associated with heat, mass, flow, and energy, such as fluid flow, heat transfer, chemical reaction, etc., involving numerical solutions of the conservation equations for energy, mass, and momentum.

5.1. Geometry Creation and Details of the Meshing

In any problem, the initial step in CFD is the creation of a geometric model. Three-dimensional geometries of Case A and Case B were created using the ANSYS R15.0 Workbench, which provided Design Modeller as a design tool to develop the geometric configurations of the physical problem domain. Figure 4 presents the geometric configurations of both tested SAHs with similar dimensions to the designed experimental configurations, which were imported into the ANSYS meshing module.

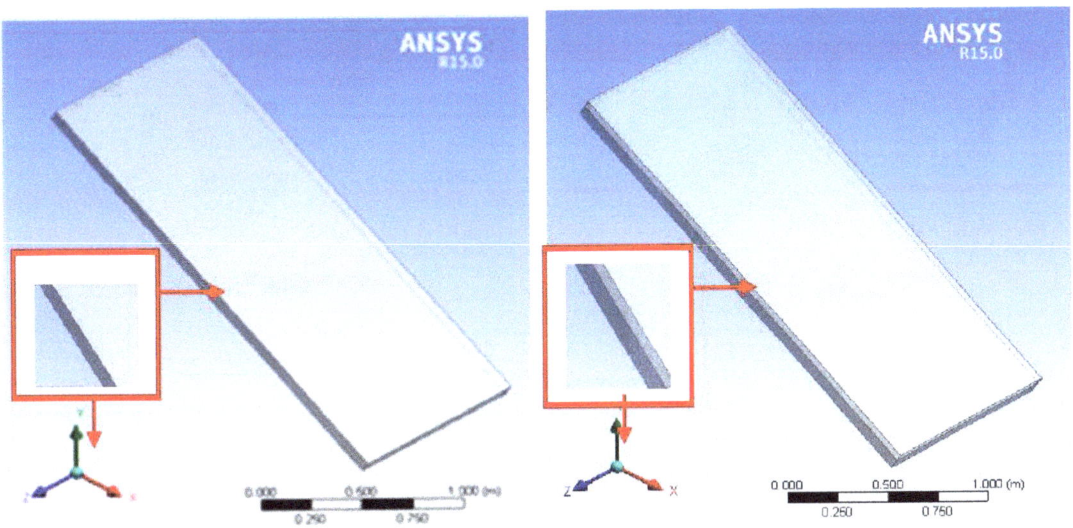

(Case A) (Case B)

Figure 4. Geometric model for the two tested FPSAHs.

Since the geometries of Case A and Case B did not involve any type of curved surfaces, the CutCell meshing method was extremely well-suited [27,28] and could provide accurate solutions with the moderate computational time required to use the ANSYS MESHING workbench, as shown in Figure 5. The total number of nodes for Case A and Case B in the meshed domain was 218,160 and 366,226, respectively, and the total number of nodes for the elements was 206,400 and 343,600, respectively.

After the mesh generation, it was important to assess its quality as it can influence the solution accuracy to a great extent. There were various available parameters in ANSYS for assessing the mesh quality. Several of these important parameters were skewness, aspect ratio, and element quality. In this research, some of these parameters were studied and analyzed. From the survey, the average skewness value should be permanently less than 0.3 for a perfect mesh quality. Therefore, it is important to note that all of the elements had skewness values of less than 0.1 for both the solar air heaters tested. Similarly, the aspect ratio parameter was estimated for the generated mesh quality. The average aspect ratio value for a good quality of mesh should be less than 2. All the elements had an aspect ratio value of less than 1.2. These results show that the mesh, which was generated with mostly hexahedral elements, was a perfect quality mesh for the skewness and aspect ratio.

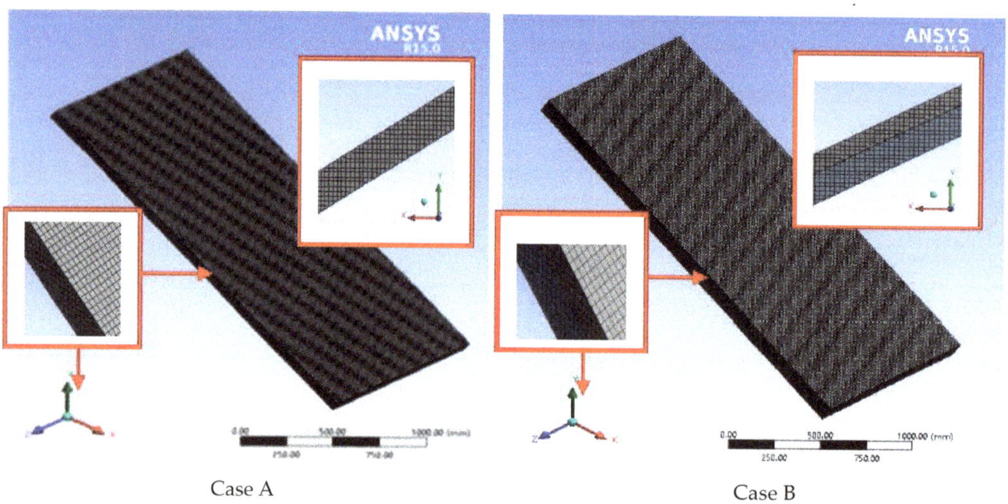

Figure 5. Meshed CFD domain for the two tested FPSAHs.

5.2. Assumptions for Simulation

This type of solution requires some assumptions to simplify the complicated flow geometries, such as solar air heater problems, with a small error value compared to the exact solutions. The following assumptions were considered here for the CFD simulation modeling:

1. As the surrounding air velocities were very low, the influence of the surrounding air velocity was neglected, and free convection was assumed.
2. The bottom of the physical absorber of the system was insulated and was therefore considered adiabatic.
3. As the variation in temperatures was medium, the air properties, such as the thermal conductivity, specific heat, viscosity, and density, were assumed to be piecewise linear, and the temperature and the physical properties of the solid materials were assumed to be constants.
4. The pressure of the outlet air was assumed to be equal to the atmospheric pressure.

5.3. Governing Equations

5.3.1. Mass Conservation Equation

The equation of mass conservation or the continuity equation can be expressed as the following as within a controlled volume, the mass of the flow remains constant with time:

$$\frac{\partial \rho}{\partial t} + \nabla.\left(\rho \vec{V}\right) = 0 \tag{22}$$

where $\vec{V}$ is the overall velocity vector, and ρ is the density.

5.3.2. Momentum Conservation Equation

The change in momentum over time is equal to the sum of the force acting on the system. This Navier–Stokes equation is reflective of the law of conservation of momentum, and it can be expressed as:

$$\frac{\partial}{\partial t}\left(\rho \vec{V}\right) + \nabla.\left(\rho \vec{V}\vec{V}\right) = -\nabla p + \nabla.\bar{\bar{\tau}} + \rho \vec{g} \tag{23}$$

where $\rho \vec{g}$ is the gravitational body force, and $\bar{\bar{\tau}}$ is the viscous stress, which is given by:

$$\bar{\bar{\tau}} = \mu\left[\left(\nabla \vec{V} + \nabla \vec{V}^{T}\right) - \frac{2}{3}\nabla.\vec{V}I\right] \tag{24}$$

5.3.3. Energy Equation

Similarly, the first law of thermodynamics reflects the Navier–Stokes energy equation. The energy conservation law states that the total energy of the system equals the sum of work and heat added to the system.

$$\nabla.(\vec{V}(\rho E + P)) = \nabla.(K_{eff}\nabla T - \textstyle\sum_j h_j.\vec{j}_j + (\bar{\bar{\tau}}_{eff}.\vec{V})) \tag{25}$$

where K_{eff} is the effective conductivity and can be provided by:

$$K_{eff} = k + \frac{c_{pf}\mu_t}{Pr_t} \tag{26}$$

5.4. Boundary Types and Conditions for the Two Tested FPSAH Models

Defining appropriate boundary types and boundary conditions is important to achieving an exact solution for any fluid flow problem. The chosen boundary conditions were a significant stage in the CFD simulation. Any CFD model solves the different equations included in the model based on constraints. The physical or real boundary conditions were simplified and idealized to insert them into the CFD simulation. For example, in this research, the FPSAH bottom, which was insulated physically, was considered to have an adiabatic wall as a boundary condition in the simulation setup. For the inlet and outlet portions, the boundary conditions were taken into account for velocity inlet and pressure outlet. Table 1 indicates the boundary types and conditions for the various parts of the two tested solar air heaters. The description of the upper glass covers and side walls, which were made from transparent window glass, assumed that they were semi-transparent walls, as provided in the CFD setup. Due to the convection heat transfer between the surrounding ambient air and the upper glass cover and side walls, the thermal conditions were assumed to be a convection of 5.7 $W/m^2.k$ from the empirical correlation of zero wind velocity. In addition, the boundary conditions for the inlet of the FPSAHs was assumed to be a velocity inlet with an average value of 1 m/s, and a gauge pressure of zero value was assumed for the outlet boundary condition.

Table 1. Boundary types and conditions for FPSAHs models.

Zone Name	Zone Type	Description	Thermal Conditions	Wall Thickness
Upper Glass Wall	Wall	Semi-Transparent T: 0.89- R: 0.08-A:0.03	Convection (5.7 $W/m^2{\cdot}K$) And Radition	0.003 m
Lower Glass Wall	Wall	Semi-Transparent T: 0.89- R: 0.08-A:0.03	Coupled	0.003 m
Absorber Wall	Wall	Opaque	Convection (9.5 $W/m^2{\cdot}K$) And Radition	0.001 m
Side Walls	Walls	Semi-Transparent T: 0.89- R: 0.08-A:0.03	Convection (5.7 $W/m^2.k$)	0.003 m
Inlet	Velocity Inlet	1 m/s		
Outlet	Pressure Outlet	Gauge Pressure 0 kPa		

where T, R, and A are the transmissivity, reflectivity, and absorption coefficients of glass, respectively.

5.5. Models Selection for Simulation

The operating parameters and models utilized in the simulation of Case A and Case B for the two tested FPSAHs in the FLUENT ANSYS solver are indicated in Table 2. After fixing the required input parameters and models, the solution was initialized. The time step was initiated at 0.001 s and reached 5.0 s, according to the ease of time and convergence needed to complete the solution of the simulation.

Table 2. Input parameters and models of the solver.

Function		Specification	
Solver Setting	Space	3D	
	Time	Unsteady; first-order implicit	
	Viscous Model	Turbulence model; k-epsilon with RNG Enhanced wall treatment with thermal effects	
	Radiation	Rosseland radiation model with solar loading and solar ray tracing Utilizing solar calculator (latitude 30.5° N and longitude 31.01° E) Days: 14:16.06.2022 Start time: 07.00 AM to End time: 07.00 PM North and West directions of the SAH	
Material Properties	Solid	Glass and Galvanized iron	Thermo-physical properties including: density; thermal conductivity; specific heat capacity
	Fluid	Air	
Operating Conditions	Operating Pressure	101.3 kPa	
	Gravity	−9.81 Z-Direction	
	Operating Temperature	288.16 K	

6. Results and Discussion

The performance of the flat-plate solar air heaters was influenced directly by the weather conditions, especially the ambient temperature, global irradiance, and wind speed. Figure 6 shows the average values of the minimum and maximum ambient air temperature for Shebin El-Kom city, Egypt during the year. The high values were obtained during the months of June and August. In the year under review, temperatures were low from November to February. However, the global irradiance intensity increased with an increase in temperature from March to October. Thus, it is possible to employ solar air heating in Shebin El-Kom during this period. Therefore, the experimental tests were conducted on sunny days from 14–16 June 2022. Figure 7 shows the hourly variations in the experimental and CFD data for global irradiance as well as the ambient temperature and the ambient relative humidity for the different sunny days tested. It can be seen from Figure 7 that the results indicate a good agreement and similar trends between the CFD-simulated results and measured data for the intensity of the solar global irradiance all over the tested days (with maximum amounts of discrepancy of 7%). It can also be noted that the CFD-simulated and experimental global irradiance increased gradually through the local daytime and reached peak values at 12:00, after which they decreased according to the weather conditions. In addition, the variations in the ambient air temperature during the day ranged between 18 °C at night and 38 °C during the day. Figure 8 shows the variations in the ambient air velocity throughout the daytime period of the tested sunny days; the ambient velocities fluctuated between increment and decrement. According to Figure 8, the minimum and maximum ambient velocity during the tested days ranged between 0.25 and 3.2 m/s, respectively.

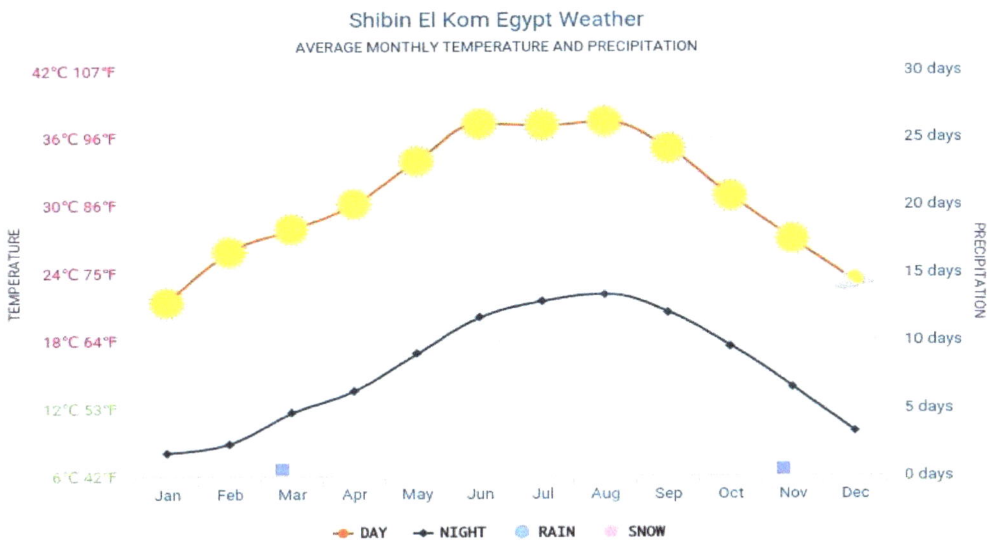

Figure 6. Variations in average values of the minimum and maximum temperatures of the ambient air for Shebin El-Kom city, Egypt during the year.

In the FPSAHs, the temperature of the absorber, the outlet air flow, glass covers, the interior of the air gab, and the stagnant air play a pivotal role in the FPSAH's performance. In general, the effectiveness of the FPSAH depends on the difference in temperature between the absorber and glass cover, as well as the difference in temperature between the inlet and outlet airflows.

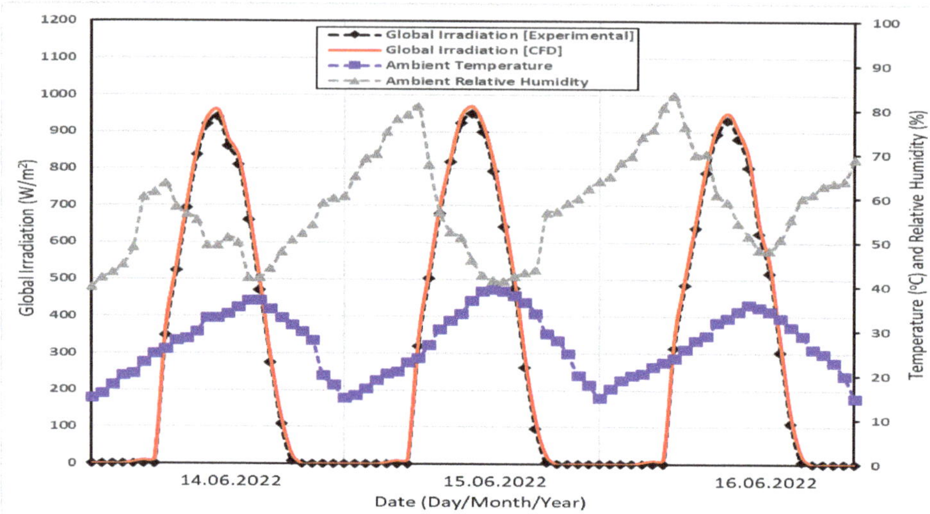

Figure 7. Variations in global irradiance, temperature, and relative humidity of ambient air during experiments.

The variations in the measured temperatures of the absorber, gab, glass covers, stagnation air, inlet, and outlet airflow during the tested sunny days for the two tested solar air heater configurations are indicated in Figure 9. From this Figure, it can clearly be seen that the temperatures of the absorber and outlet airflow for Case B are higher when compared with the temperatures of the absorber and outlet airflow in Case A. Thus, the presence of stagnant air near each absorber plate is important in raising the thermal efficiency of the FPSAH. In addition, it can clearly be seen from Figure 9 that the maximum absorber temperatures obtained at 13:00 ranged between 76 and 73.5 °C for Case B and ranged between 70.3 and 63.7 °C for Case A during the tested days.

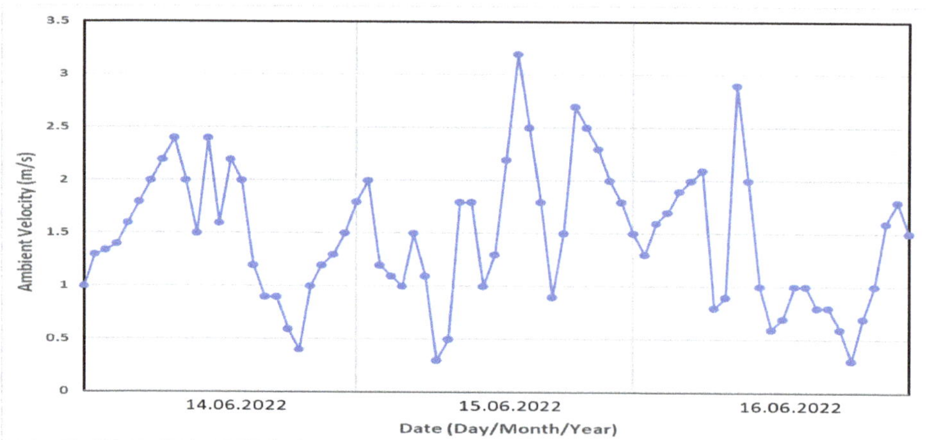

Figure 8. Variations in ambient air velocities during experiments.

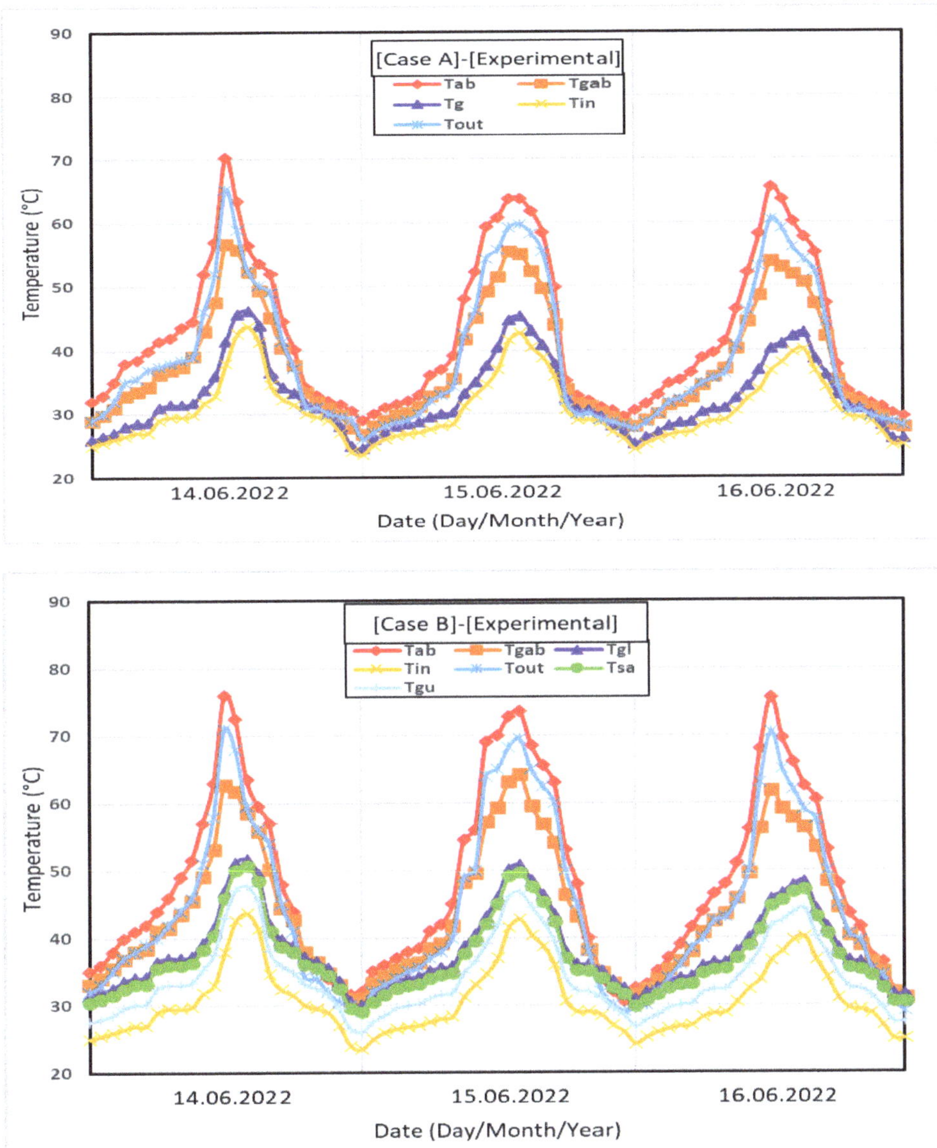

Figure 9. Temperature evolution of the components and the two tested FPSAHs.

The temperature contours for the absorber and upper glass cover for both tested Case A and Case B at different time intervals on 15 June 2022 are shown in Figures 10 and 11, respectively. The coding color for the temperatures is shown in the charts in Kelvin. The color "blue" shows the minimum temperature values, and the color "red" indicates the maximum temperature values. The temperature contour charts of the flat-plate absorber and the upper glass cover suggest that:

- Inside the both tested solar air heaters, the temperatures of the flat-plate absorber and upper glass cover begin to increase as global solar irradiation falls on the SAH. The temperature contours show increments of gradual increase until 14:00; after this point, they decrease steadily.

- Within the tested FPSAH models, the airflow temperature begins to increase as the global irradiation falls on the absorber. After some time, the air flow begins to heat up, and the density of the airflow decreases, leading to an increase in the velocity of the airflow. The interior airflow temperatures inside both tested air heaters gradually increased until 14:00; after this point, they gradually minimized.

Figures 12 and 13 present the CFD and the experimental results of the outlet airflow temperature and the outlet airflow relative humidity, respectively, for the FPSAH models. According to Figures 10 and 11, the results indicate similar trends and a good agreement between the CFD-simulated and experimental results for the outlet air temperature and the outlet airflow relative humidity all over the tested days (the maximum errors were estimated to be 5.6% and 8.2% for the temperature and relative humidity, respectively). The small deviations between the predicted CFD results and the experimental data were explained by the fact of that ANSYS FLUENT takes the ideal characteristics for galvanized iron and glass covers and not their real properties. Additionally, the most likely cause for the deviation was that values of global irradiation used in the CFD simulation were higher than the experimental measured values, as suggested by Figure 7. From these figures, it can also be seen that when temperature increased, the relative humidity decreased. This is because colder air does not require as much moisture to become saturated as warmer air.

The contours of the airflow velocities inside the gap of the two tested FPSAHs, Case A and Case B, at various time intervals throughout the daytime are drawn Figure 14. Figure 14 shows clearly that the high-velocity airflow zones occur in the central section, at the outlet of the tested SAHs. Near the walls (the absorber and glass covers), the velocities are minimized. Additionally, it can be seen from the figure the outlet velocity for Case B was higher than the outlet velocity for Case A for the different time intervals; this is because of the higher temperature values of the SAH's inner the gap, which led to a decrease in the density and therefore an increase in the velocity for a similar mass flow rate. The magnitudes of the velocities are in m/s. Furthermore, the predicted contours of the velocity profile for the tested modified solar air heater (Case B) agreed with the measured velocity inside the stagnant air gab, which measured by calibrated hot-wire anemometers at different positions along the gap section, as can be seen in Figure 2, by making a small hole in the right side of the glass wall and inserting the sensing element (sensor) of the hot-wire anemometer to measure the airflow velocities. We used the silicon rubber to ensure there was no leakage through these holes and the inserted hot-wire anemometers. From the experimental results, the recorded velocities were very small, near to zero.

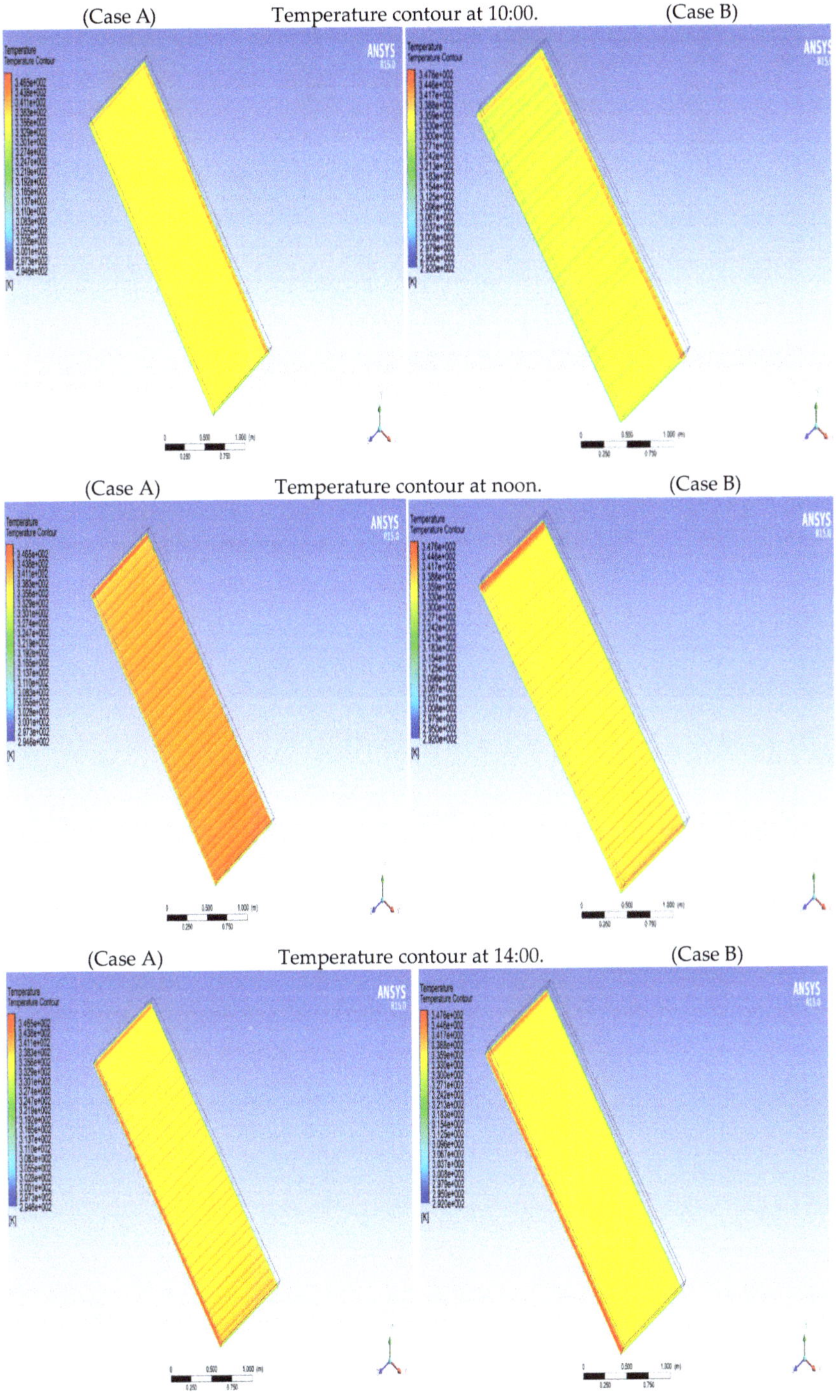

Figure 10. *Cont.*

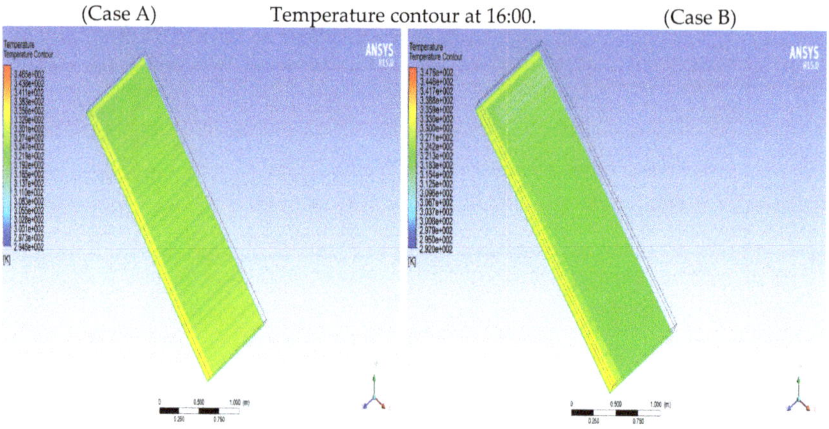

Figure 10. The absorber flat-plate temperature contours at various time intervals.

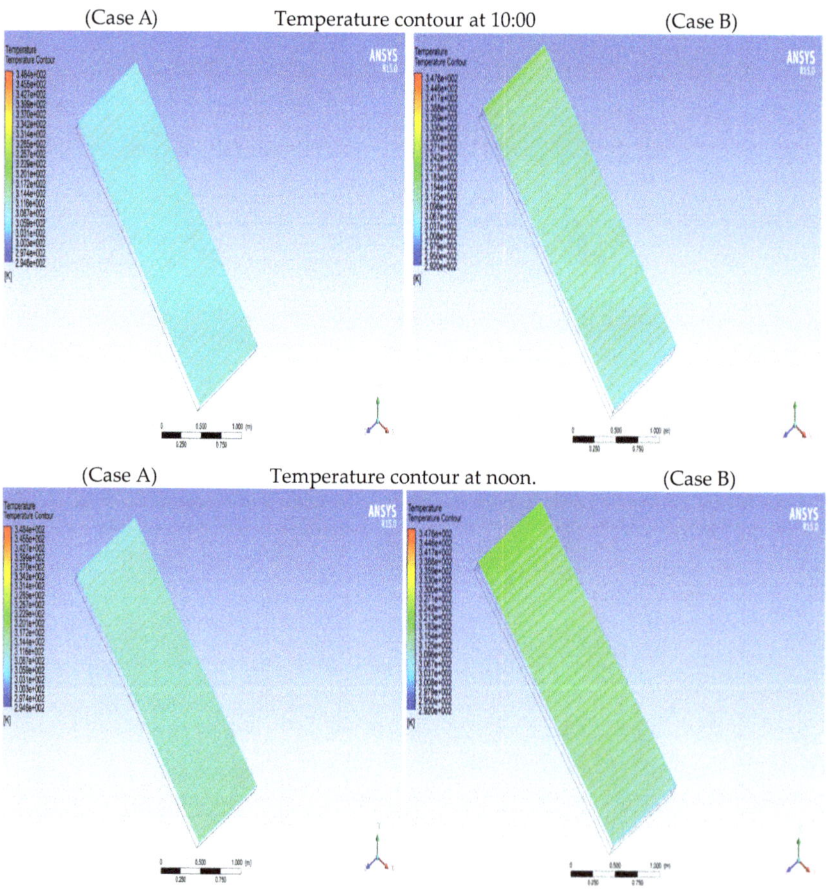

Figure 11. *Cont.*

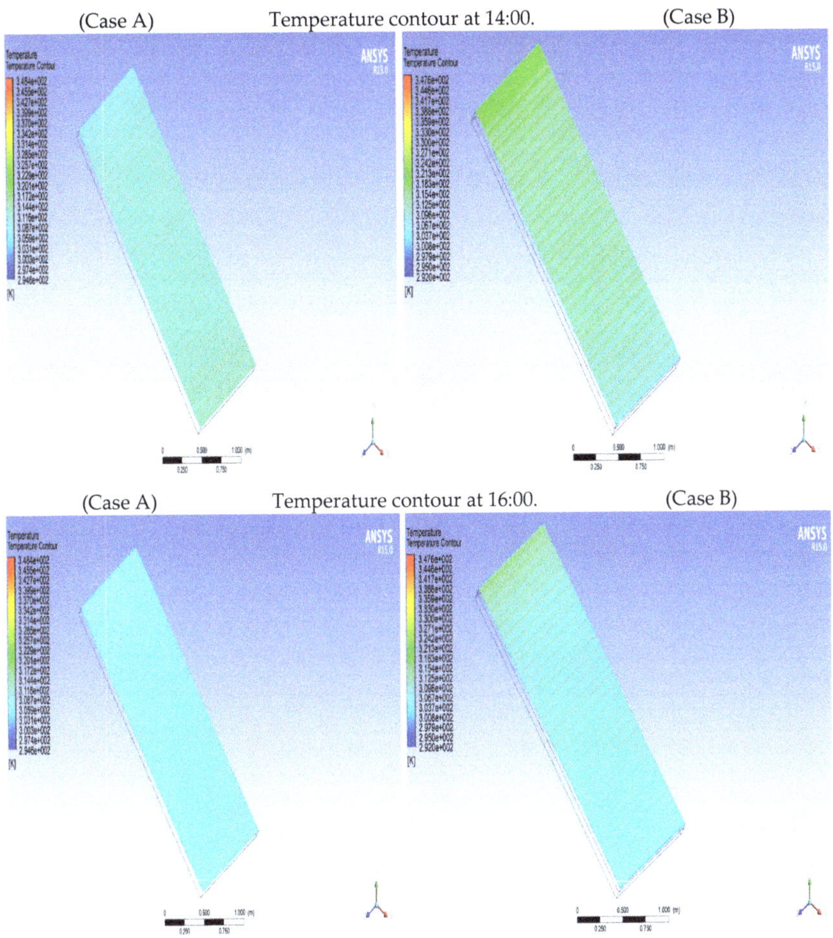

Figure 11. The temperature contours of the upper glass cover at various time intervals.

Figures 15 and 16 indicate the comparison between the CFD simulation and the experimental results of the airflow temperature and the relative humidity inside the two tested SAHs (Case A and Case B) at different time intervals and in different positions on 15 June 2022. In general, there were similar trends and a good agreement between the CFD-simulated and experimental results at the different time intervals and different positions (the maximum errors were estimated to be 4.3% and 7.5% for the temperature and relative humidity, respectively). In addition, the airflow was heated for both tested solar heaters as it traversed along the SAH. The temperature flux increased and decreased with the solar irradiation. Solar air heaters also help to decrease the relative humidity during the day. This is important in the application of SAHs, such as in solar drying, because the reduction in the relative humidity of drying air increases the capacity to absorb humidity from the dried products. From Figures 15 and 16, it can be noted that Case B helped to maintain higher airflow temperatures and a lower airflow relative humidity compared to Case A for the various time intervals and different positions.

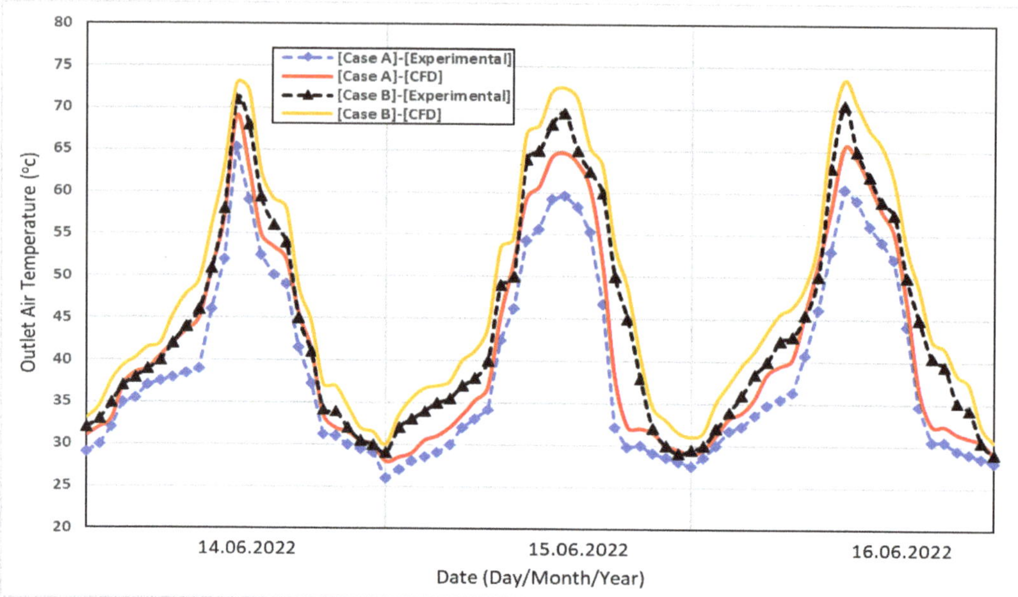

Figure 12. Validation in the CFD outlet airflow temperature variation and the experimental data for the two tested FPSAHs.

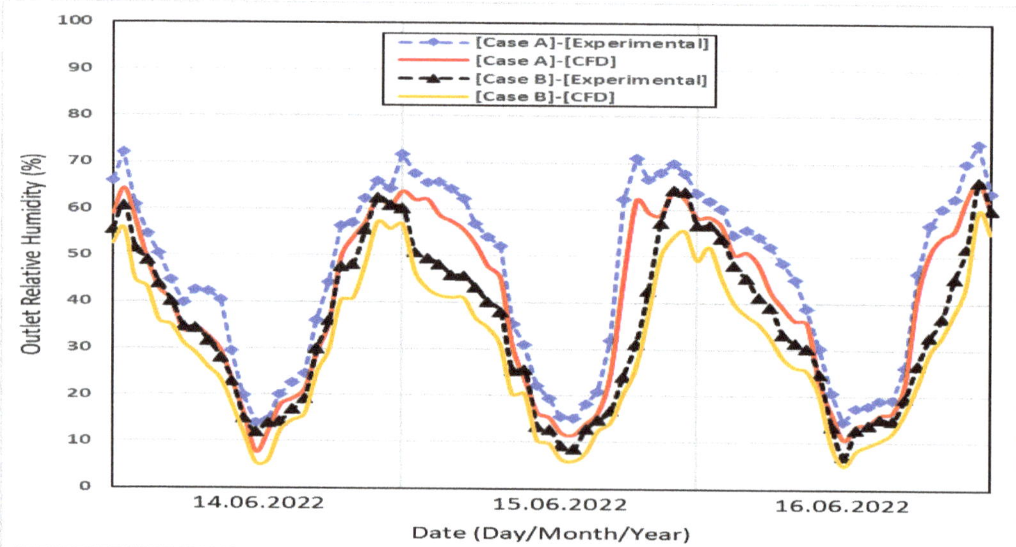

Figure 13. Validation of the CFD outlet airflow relative humidity variation and the experiment data for the two tested FPSAHs.

Figures 17–19 present the CFD and the experimental results of the airflow velocities and the CFD simulation results of the density and pressure ratio along the two tested air heaters (Case A and Case B), respectively, for different time intervals and different positions for 15 June 2022 at 10:00 AM; 1:00 PM, and 4:00 PM. Due to the airflow being heated along with the FPSAH, the density decreased. However, this decrease was due to increased values of solar irradiation. The velocity of the airflow increased along with the FPSAH, but this increase was very low. Depending on the increase, the pressure ratio of the airflow decreased along with the two tested solar air heaters, as indicated in Figure 19.

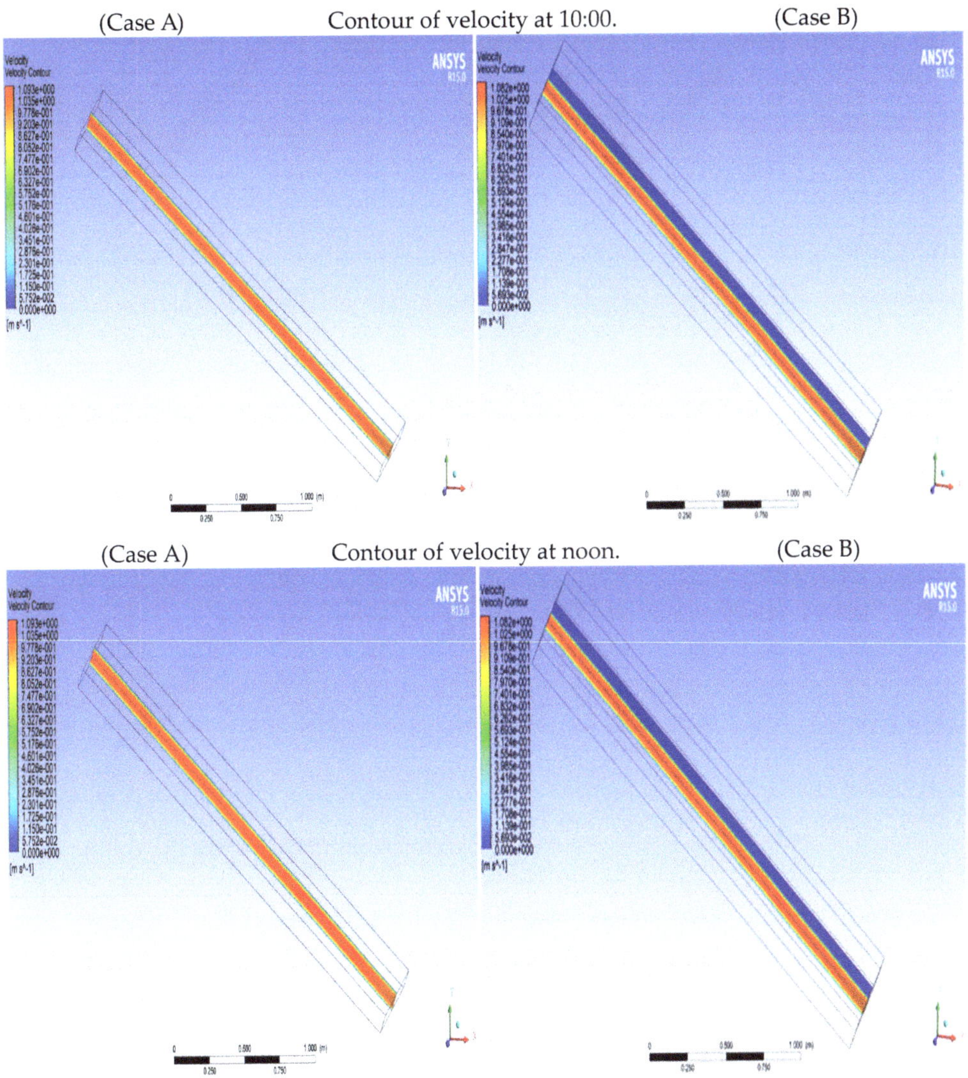

Figure 14. *Cont.*

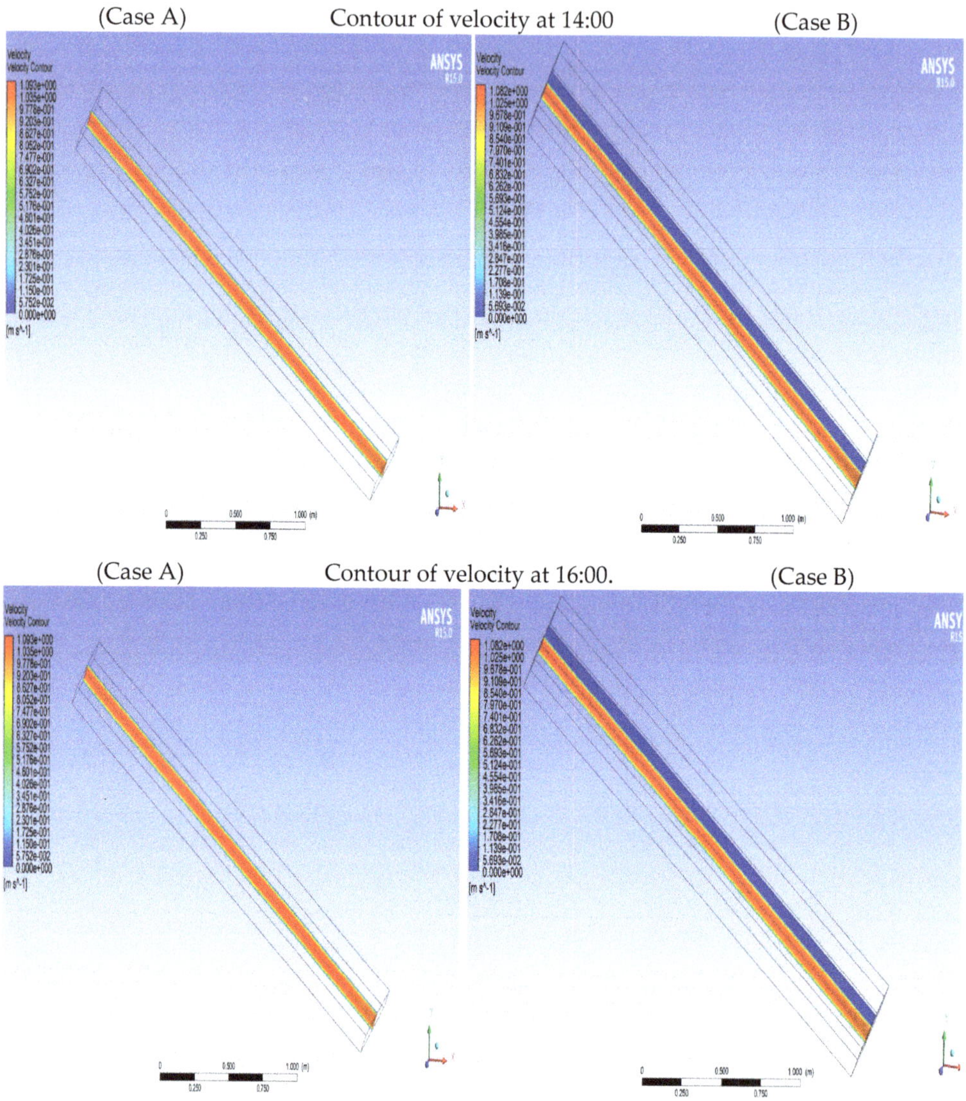

Figure 14. Airflow velocity contours for the tested FPSAHs at various time intervals.

The heating levelized cost (HLC) can be utilized to compare the economic performance of the tested and modified FPSAH (Case B) with the conventional case (Case A) delivering hot air in the desired temperature range. It was computed for all types of cost, such as the capital investment, operating cost, maintenance cost, and inflation rate, etc. The HLC is the parameter that expresses the cost per kWh. It directly relates the cost involved and the thermal performance of the FPSAH. The average lifetime of the two tested FPSAHs was considered to be 20 years. For the modified FPSAH (Case B), the cost per kWh to deliver hot air in the range of 70 to 80 °C is estimated at USD 0.0837, while for the conventional FPSAH (Case A), delivering hot air in the range of 60 to 70 °C, the cost per kWh was estimated at USD 0.0974 for the same tested inlet mass flow rates.

Figures 20–22 present the CFD and experimental values for the average thermal efficiency, average outlet air temperatures and average relative air humidity during the sunny period of the experiment, from 7:00 AM to 7:00 PM, on 16 June 2022. From Figure 20, it can be seen that the average CFD and experimental thermal efficiency values for Case B were higher than that of Case A. The average CFD thermal efficiency values obtained for Case B and Case A were 28.7% and 21.6%, respectively, while the experimental thermal efficiency values for the two cases were 26.4% and 18.2%, respectively. Thus, Case B offers the best thermal efficiency. In addition, from Figure 21, it is clear that the average CFD outlet air temperature values were very close to the experimental results, with deviations of 7% and 7.8% for Case B and the Case A, respectively. Furthermore, from Figure 22, it can be seen that the CFD-simulated average relative humidity was reduced by 31.6% when using case B compared with case A, and reduced by 28.8% for the experimental data compared to Case B.

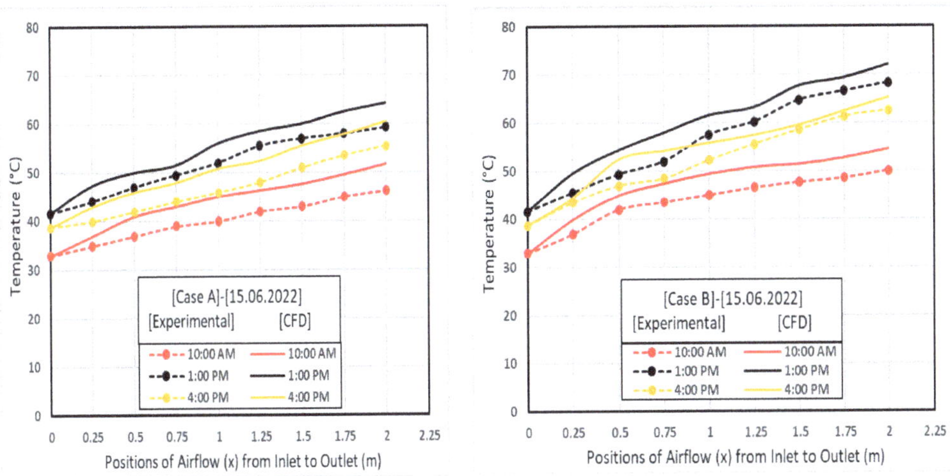

Figure 15. Validation of the CFD airflow temperatures variation and the experimental data for the Case A and Case B FPSAHs tested at various time intervals.

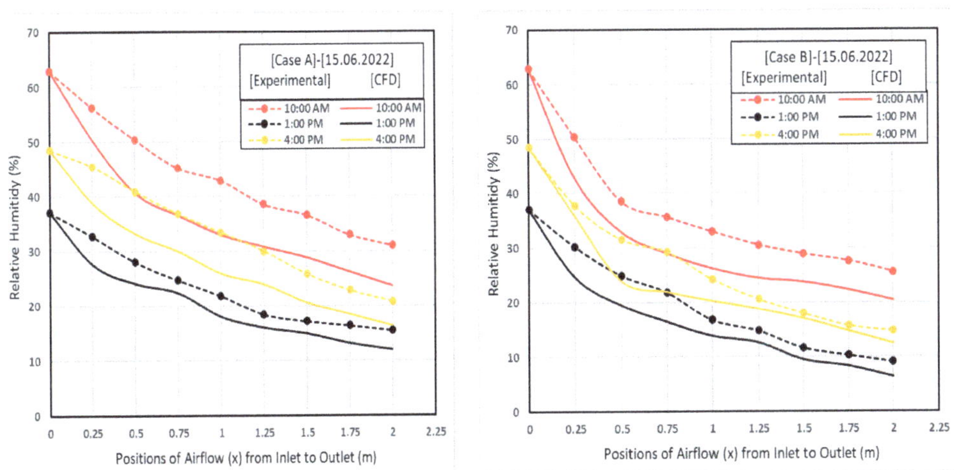

Figure 16. CFD validation of the airflow's relative humidity variation and the experimental data for the Case A and Case B FPSAHs tested at various time intervals.

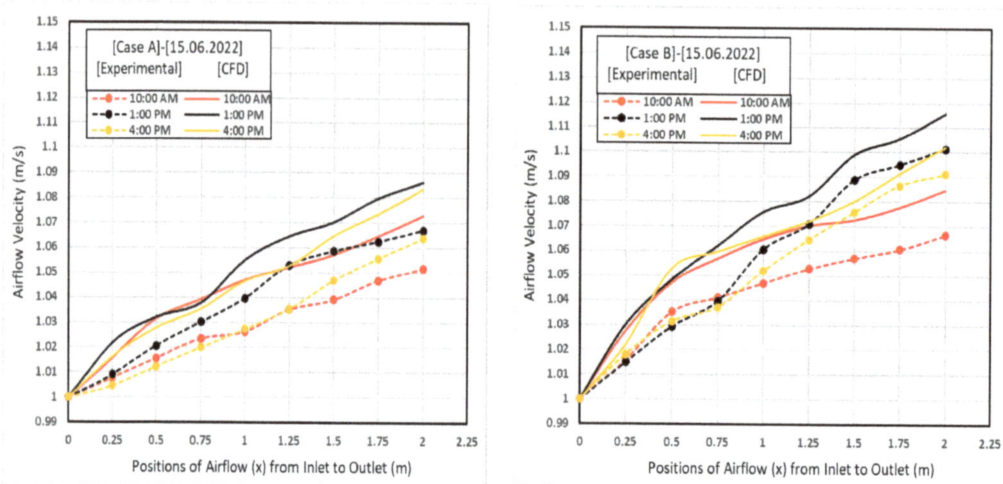

Figure 17. Validation of the variation in the CFD airflow velocity and the experimental data for the Case A and Case B FPSAHs tested at various time intervals.

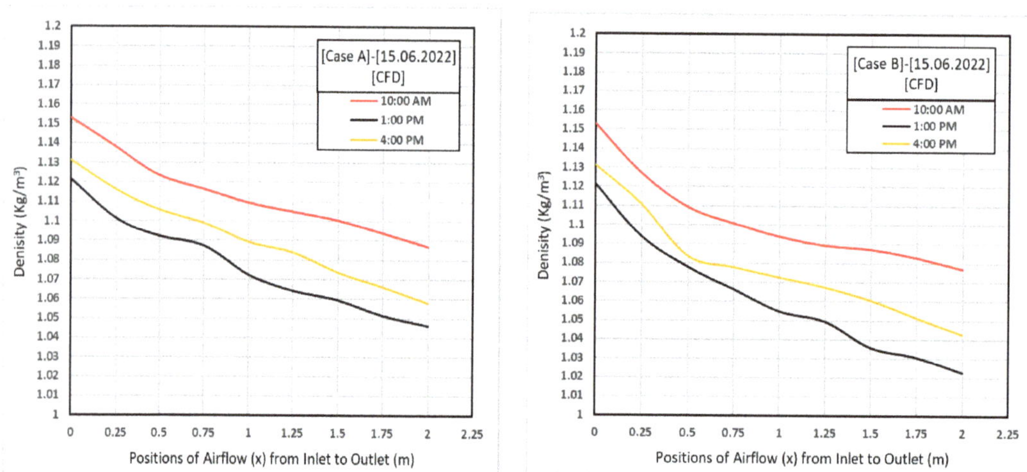

Figure 18. CFD airflow density results for the Case A and Case B FPSAHs tested at various time intervals.

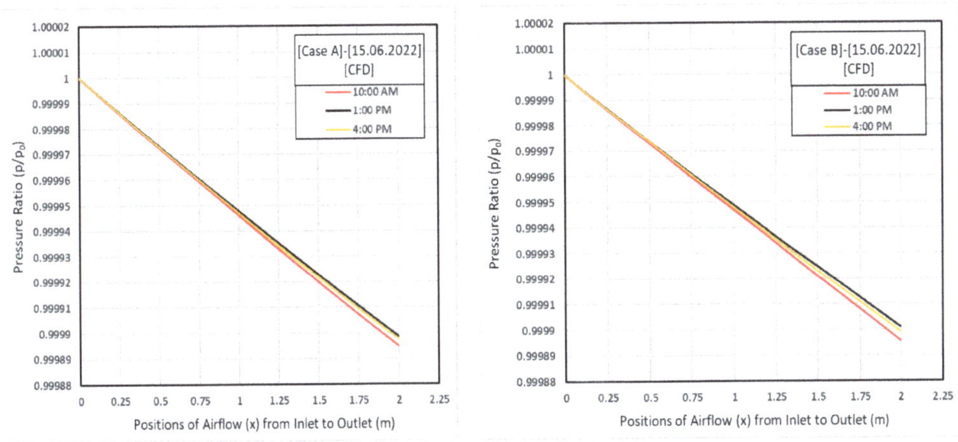

Figure 19. CFD airflow pressure ratio results for the Case A and Case B FPSAHs tested at various time intervals.

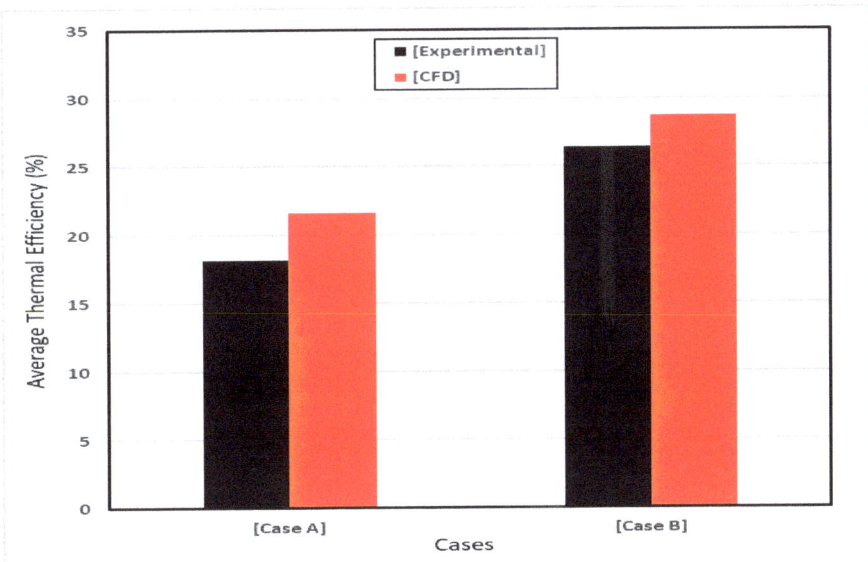

Figure 20. Simulated and experimental average thermal efficiency during the experimental period for the tested Case A and Case B FPSAHs.

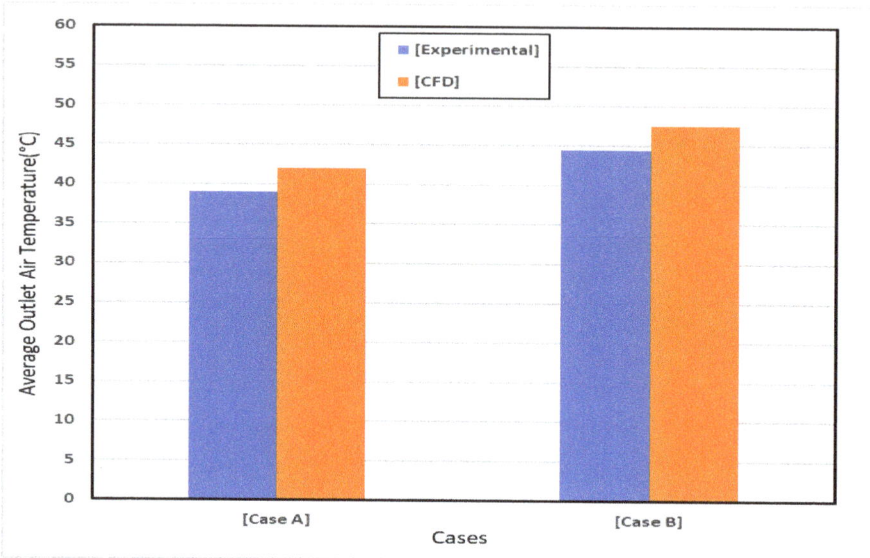

Figure 21. Simulated and experimental average outlet air temperature during the experimental period for the tested Case A and Case B FPSAHs.

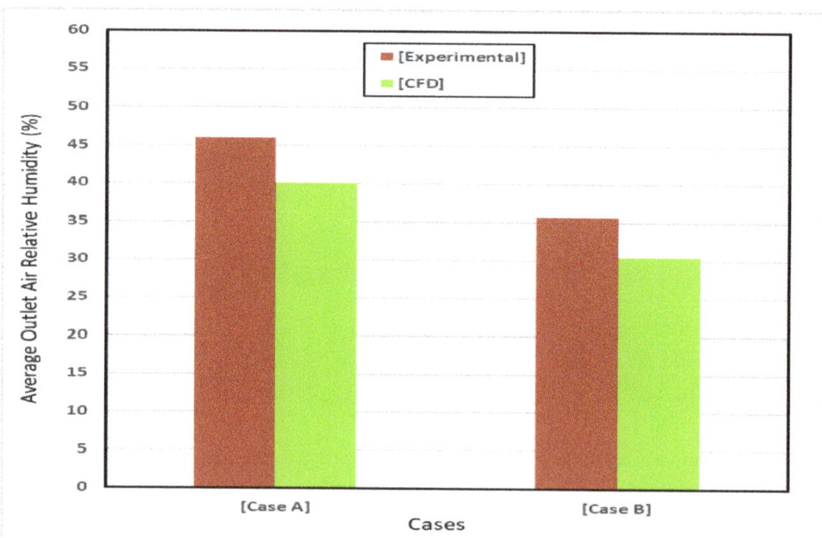

Figure 22. Simulated and experimental average relative air humidity during the experimental period for the tested Case A and Case B FPSAHs.

7. Conclusions

The primary objective of this research was to design and manufacture two distinct models of SAH (conventional—Case A and modified—Case B) and develop a comprehensive three-dimensional CFD model to simulate the performance of the proposed FPSAHs. The conclusions drawn from the CFD simulations and experimental results of this research are as follows:

1. The presence of stagnant air has a beneficial effect on the thermal efficiency of an SAH.
2. CFD simulations provide solutions of satisfactory quality, demonstrating the effectiveness of CFD as a tool for predicting the behavior and performance of FPSAHs.
3. Upon comparison of the CFD-simulated results with the experimental data for both tested SAHs (Case A and Case B), it was observed that the simulated global irradiation, air flow temperature, relative humidity, absorber and glass cover temperatures, and air flow velocities closely matched the corresponding experimental data.
4. The average CFD thermal efficiency values obtained for Case B and Case A were 28.7% and 21.6%, respectively. Meanwhile, the average experimental thermal efficiency values for these cases were 26.4% and 18.2%, respectively. Therefore, it can be concluded that Case B offers the best thermal efficiency.
5. For both Case B and Case A, the average CFD outlet air temperature values demonstrated deviations of 7% and 7.8%, respectively, which were very similar to the corresponding experimental results.
6. Comparing Case B to Case A, the CFD simulation showed a 31.6% reduction in the average relative humidity, while the experimental data exhibited a 28.8% reduction in the relative humidity for Case B compared to the reference case.
7. By conducting a CFD modeling study, designers of FPSAHs can obtain a broad range of efficient information which can help save costs and time before undertaking any expensive and time-consuming experimental investigations.

8. Future Studies

The performance of more complex designs for a solar air heater can be investigated with the help of the presented CFD modeling development approach.

Author Contributions: Designing and manufacturing the experimental setup, performing the CFD simulation, conceptualization, methodology, analyzing the results and writing the original draft, M.S.E.-S.; analysis of the achieved data, cooperating in the scientific discussion and supervision, A.E.; analysis of the achieved data and preparing for the design of the experiments, S.S.E.-D.; validation and contribution to the scientific discussion, F.A.E. All authors have read and agreed to the published version of the manuscript.

Funding: This research received no external funding.

Data Availability Statement: Data may be requested from the corresponding author.

Conflicts of Interest: The authors declare no conflict of interest.

Nomenclature

Variable	Definition	Unit
A_{ab}	Absorber area	m^2
A_{gc}	Glass cover area	m^2
A_{st}	Stagnant air area	m^2
C_{pgc}	The glass cover specific heat	J/kg·K
C_{pst}	The stagnant air specific heat	J/kg·K
C_{pab}	The absorber specific heat	J/kg·K
C_{pf}	The airflow specific heat	J/kg·K
E	The total energy	J
$\vec{g}$	Gravitational acceleration	m/s^2
$h_{cf\text{-}ab}$	The convective coefficient of the heat transfer between the airflow and absorber	$W/m^2{\cdot}K$
$h_{cf\text{-}gcu}$	The convective coefficient of the heat transfer between the airflow and the upper glass cover	$W/m^2{\cdot}K$
$h_{cgcl\text{-}st}$	The convective coefficient of heat transfer between the lower glass cover and the stagnant air	$W/m^2{\cdot}K$

$h_{cgcu\text{-}st}$	The convective coefficient of heat transfer between the upper glass cover and the stagnant air	$W/m^2{\cdot}K$
$h_{rgcl\text{-}gcu}$	The radiative coefficient of heat transfer between the lower glass cover and the upper glass cover	$W/m^2{\cdot}K$
$h_{rgcl\text{-}sky}$	The radiative coefficient of heat transfer between the lower glass cover and the sky	$W/m^2{\cdot}K$
$h_{rgcu\text{-}ab}$	The radiative coefficient of heat transfer between the upper glass cover and the absorber	$W/m^2{\cdot}K$
$h_{rab\text{-}gcu}$	The radiative coefficient of heat transfer between the absorber and the upper glass cover	$W/m^2{\cdot}K$
$h_{rgcu\text{-}gcl}$	The radiative coefficient of heat transfer between the upper glass cover and the lower glass cover	$W/m^2{\cdot}K$
I(t)	Global solar irradiation on the inclined surface	W/m^2
m_{gc}	Mass of glass cover	kg
I	The unit tensor	–
$\vec{j}_j$	The diffusion flux of species j	–
k	The thermal conductivity	W/m.K
m_{st}	Mass of stagnant air	kg
m_{ab}	Mass of absorber	kg
P	The static pressure	Pa
Pr_t	The turbulent Prandtl numbers.	–
S_Q	Volumetric convective thermal power exchanged between airflow and near component	W/m^3
S.L	The volume of airflow located between insulation plate and absorber	m^3
T_{ab}	The absorber temperature	K
T_f	The airflow temperature	K
T_{gcl}	The lower glass cover temperature	K
T_{gcu}	The upper glass cover temperature	K
T_{sky}	The sky temperature	K
T_{st}	The stagnant air temperature	K
T_{amb}	The ambient air temperature	K
T_{fout}	The outlet airflow	K
T_{fint}	The inlet airflow	K
τ_{st}	The stagnant air transmissivity coefficient	–
τ_{gc}	The glass cover transmissivity coefficient	–
τ_f	The airflow transmissivity coefficient	–
α_f	The airflow absorptivity coefficient	–
α_{gc}	The glass cover absorptivity coefficient	–
α_{ab}	The absorber absorptivity coefficient	–
α_{st}	The stagnant air absorptivity coefficient	–
$\vec{V}$	Overall velocity vector	m/s
$\bar{\bar{\tau}}$	The viscous stress	Pa
μ	The molecular viscosity	Pa·s
μ_t	The turbulent viscosity	Pa·s

Abbreviations

Case A	Conventional Solar Air Heater
Case B	Modified Solar Air Heater
CFD	Computational Fluid Dynamics
FPSAH	Flat Plate Solar Air Heater
SAH	Solar Air Heater

References

1. Li, S.; Wang, H.; Meng, X.; Wei, X. Comparative study on the performance of a new solar air collector with different surface shapes. *Appl. Therm. Eng.* **2017**, *114*, 639–644. [CrossRef]
2. Yadav, S.; Saini, R.P. Numerical investigation on the performance of a solar air heater using jet impingement with absorber plate. *Sol. Energy* **2020**, *208*, 236–248. [CrossRef]

3. El-Sebaey, M.S.; Mousavi, S.M.T.; El-Din, S.S.; Essa, F.A. An experimental case study on development the design and the performance of indirect solar dryer type for drying bananas. *Sol. Energy* **2023**, *255*, 50–59. [CrossRef]
4. Akhbari, M.; Rahimi, A.; Hatamipour, M.S. Modeling and Experimental Study of a Triangular Channel Solar Air Heater. *Appl. Therm. Eng.* **2020**, *170*, 114902. [CrossRef]
5. Manjunath, M.S.; Vasudeva Karanth, K.; Yagnesh Sharma, N. Numerical analysis of flat plate solar air heater integrated with an array of pin fins on absorber plate for enhancement in thermal performance. *J. Sol. Energy Eng.* **2019**, *141*, 051008. [CrossRef]
6. Jouybari, N.F.; Lundström, T.S. Performance improvement of a solar air heater by covering the absorber plate with a thin porous material. *Energy* **2020**, *190*, 116437. [CrossRef]
7. Jiaa, B.; Liua, F.; Wang, D. Experimental study on the performance of spiral solar air heater. *Sol. Energy* **2020**, *208*, 236–248. [CrossRef]
8. Wanga, D.; Liua, J.; Liua, Y.; Wanga, Y.; Lib, B.; Liua, J. Evaluation of the performance of an improved solar air heater with "S" shaped ribs with gap. *Sol. Energy* **2020**, *195*, 89–101. [CrossRef]
9. Ojike, O.; Okonkwo, W.I. Study of a passive solar air heater using palm oil and paraffin as storage media. *Case Stud. Therm. Eng.* **2019**, *13*, 100373. [CrossRef]
10. Lakshmia, D.V.N.; Layekb, A.; Kumarc, P.M. Performance analysis of trapezoidal corrugated solar air heater with sensible heat storage material. *Energy Procedia* **2017**, *109*, 463–470. [CrossRef]
11. Kalash, A.R.; Jary, A.M.; Jabe, H.J.; Habeeb, L.J. Implementation of flat plate solar collector design of different design conditions. *J. Mech. Eng. Res. Dev.* **2020**, *44*, 09–26.
12. Chabane, F.; Moummi, N.; Bensahal, D.; Brima, A. Heat Transfer Coefficient and Thermal Losses of Solar Collector and Nusselt Number Correlation for Rectangular Solar Air Heater Duct with Longitudinal Fins Hold under the Absorber Plate. *Appl. Sol. Energy* **2014**, *50*, 19–26. [CrossRef]
13. Moraveja, M. An experimental study of the performance of a solar flat plate collector with triangular geometry. *J. Sol. Energy Res.* **2021**, *6*, 923–936.
14. Shetty, S.P.; Madhwesh, N.; Karanth, K.V. Numerical analysis of a solar air heater with circular perforated absorber plate. *Sol. Energy* **2021**, *215*, 416–433. [CrossRef]
15. Mousavi, S.M.T.; Egelioglu, F. Experimental study of the effect of slit width and slit spacing on the thermal performance of slit-glazed collectors. *Adv. Mech. Eng.* **2017**, *9*, 168781401772847. [CrossRef]
16. Khoukhi, M.; Maruyama, S. Theoretical approach of a flat plate solar collector with clear and low-iron glass covers taking into account the spectral absorption and emission within glass covers layer. *Renew. Energy* **2005**, *30*, 1177–1194. [CrossRef]
17. Abo-Elfadl, S.; Hassan, H.; El-Dosoky, M.F. Study of the performance of double pass solar air heater of a new designed absorber: An experimental work. *Sol. Energy* **2020**, *198*, 479–489. [CrossRef]
18. Nemś, M.; Kasperski, J. Experimental investigation of concentrated solar air-heater with internal multiple-fin array. *Renew. Energy* **2016**, *97*, 722–730. [CrossRef]
19. Khanlari, A.; Güler, H.Ö.; Tuncer, A.D.; Şirin, C.; Bilge, Y.C.; Yılmaz, Y.; Güngör, A. Experimental and numerical study of the effect of integrating plus-shaped perforated baffles to solar air collector in drying application. *Renew. Energy* **2020**, *145*, 1677–1692. [CrossRef]
20. Zhang, H.; Ma, X.; You, S.; Wang, Y.; Zheng, X.; Ye, T.; Zheng, W.; Wei, S. Mathematical modelling and performance analysis of a solar air collector with slit-perforated corrugated plate. *Sol. Energy* **2018**, *167*, 147–157. [CrossRef]
21. Sharma, S.K.; Kalamkar, V.R. Experimental and numerical investigation of forced convective heat transfer in solar air heater with thin ribs. *Sol. Energy* **2017**, *147*, 277–291. [CrossRef]
22. Jin, D.; Quan, S.; Zuo, J.; Xu, S. Numerical investigation of heat transfer enhancement in a solar air heater roughened by multiple V-shaped ribs. *Renew. Energy* **2019**, *134*, 78–88. [CrossRef]
23. Pérez-Espinosa, R.; García-Valladares, O.; Pilatowsky, I. Numerical Modeling and Experimental Validation of Back-Pass Rectangular Ducts Solar Air Heaters. *Appl. Therm. Eng.* **2020**, *171*, 115018. [CrossRef]
24. Relative Humidity Calculator. Available online: https://www.1728.org/relhum.htm (accessed on 12 January 2023).
25. Sureandhar, G.; Srinivasan, G.; Muthukumar, P.; Senthilmurugan, S. Investigation of thermal performance in a solar air heater having variable arc ribbed fin configuration. *Sustain. Energy Technol. Assess.* **2022**, *52*, 102069. [CrossRef]
26. Abuşka, M.; Şevik, S. Energy, exergy, economic and environmental (4E) analyses of flat-plate and V-groove solar air collectors based on aluminium and copper. *Sol. Energy* **2017**, *158*, 259–277. [CrossRef]
27. El-Sebaey, M.S.; Ellman, A.; Hegazy, A.; Ghonim, T. Experimental Analysis and CFD Modeling for Conventional Basin-Type Solar Still. *Energies* **2020**, *13*, 5734. [CrossRef]
28. El-Sebaey, M.S.; Hegazy, A.; Ellman, A.; Ghonim, T. Experimental and CFD Study on Single Slope Double Basin Solar Still. *Eng. Res. J.* **2021**, *44*, 21–32. [CrossRef]

processes

MDPI

Article

Design, Multi-Perspective Computational Investigations, and Experimental Correlational Studies on Conventional and Advanced Design Profile Modified Hybrid Wells Turbines Patched with Piezoelectric Vibrational Energy Harvester Devices for Coastal Regions

Janani Thangaraj [1], Senthil Kumar Madasamy [1], Parvathy Rajendran [2,*], Safiah Zulkifli [2,*], Rajkumar Rajapandi [3], Hussein A. Z. AL-bonsrulah [4,5], Beena Stanislaus Arputharaj [6], Hari Prasath Jeyaraj [1] and Vijayanandh Raja [1]

1 Department of Aeronautical Engineering, Kumaraguru College of Technology, Coimbatore 641049, Tamil Nadu, India; vijayanandh.raja@gmail.com (V.R.)
2 School of Aerospace Engineering, Universiti Sains Malaysia, Engineering Campus, Nibong Tebal 14300, Pulau Pinang, Malaysia
3 Department of Mathematics, Kumaraguru College of Technology, Coimbatore 641049, Tamil Nadu, India
4 Mechanical Power Technical Engineering Department, Al-Amarah University College, Amarah 62001, Iraq
5 Department of Computer Techniques Engineering, Al Safwa University College, Karbala 56001, Iraq
6 Department of Research and Innovation, Saveetha School of Engineering, SIMATS, Chennai 602105, Tamil Nadu, India
* Correspondence: aeparvathy@usm.my (P.R.); safiahzul@usm.my (S.Z.)

Abstract: This work primarily investigates the performance and structural integrity of the Wells turbines for power production in coastal locations and their associated unmanned vehicles. An innovative design procedure is imposed on the design stage of the Wells turbine and thus so seven different models are generated. In the first comprehensive investigation, these seven models underwent computational hydrodynamic analysis using ANSYS Fluent 17.2 for various coastal working environments such as hydro-fluid speeds of 0.34 m/s, 1.54 m/s, 12 m/s, and 23 m/s. After this primary investigation, the best-performing Wells turbine model has been imposed as the second comprehensive computational investigation for three unique design profiles. The imposed unique design profile is capable of enhancing the hydro-power by 15.19%. Two detailed, comprehensive investigations suggest the best Wells turbine for coastal location-based applications. Since the working environments are complicated, additional advanced computational investigations are also implemented on the best Wells turbine. The structural withstanding capability of this best Wells turbine model has been tested through coupled computational hydro-structural analysis for various lightweight materials. This best Wells turbine also enforces the vibrational failure factors such as modal and harmonic vibrational analyses. Finally, advanced and validated coupled engineering approaches are proposed as good methodology for coastal location-based hydropower applications.

Keywords: CFD; FEA; FSI; composite materials; hydro-energy; hybrid energy; forced and free vibrations

1. Introduction

The Wells turbine is a type of turbine that was built specifically for the purpose of harvesting energy from waves. The oscillating water column is the most common application for this material. The Wells turbine, which was designed by Prof. Alan Arthur Wells in Belfast and uses symmetrical blades, can convert the flow of air coming from either direction into the same direction as the turbine's spin. The angle of attack (α) is quite high, despite the fact that its efficiency is somewhat poor. The symmetrical airfoils that are utilized in the construction of the Wells turbine blades are the root cause of the poorer efficiency as well as the high angle of attack. The capability of the Wells turbine to self-rectify the flow of incoming waves is one of the fundamental properties that distinguish it from other types of turbines. Additionally, the Wells turbine is a type of turbine design that has been developed expressly for use in wave energy converters. This design is both highly effective and adaptable, making it an excellent choice for this type of application. Because of its potential to work effectively in bidirectional flow and its ability to self-rectify under a diverse set of wave conditions, it is an excellent option for the generation of electrical power from ocean waves.

The turbine is able to function well despite shifts in either the direction or the magnitude of the waves. This is accomplished by the turbine by angling a set of curved blades in such a way that a pressure difference is created between the top and bottom surfaces of the blade as the wave travels over it. This pressure differential is what causes the blade to rotate, and the rotation is what drives the generator that converts the rotation into usable electrical power. High rotating speed and the ability to be directly connected with electrical generators are the two aspects of the Wells turbine that are considered to be its most important characteristics. It is expected that the Wells turbine will play an increasingly vital role in the development of wave energy technology as the world continues its search for new renewable energy sources.

Relevant Works

Wells turbines with automated multi-objective optimization have been designed. Torque and pressure loss were improved with OPAL++. The highest torque (G1) and the lowest stress drop coefficient (G2) played major roles in the achievement of the highest performance. Compared to its predecessor, the new turbine has a stall point of 0.3. Both the G1 and G2 turbines have better efficiency than the reference turbine [1]. For sub-oscillating flight, Wells turbines employed second-regulation and entropy technology in the form of blades. Computational fluid dynamics (CFD) analyses of NACA airfoils with a sinusoidal waft boundary have been performed. The efficiency of the second law was drastically impacted by the angle of attack. This effect was quantified by the NACA 0015 between 15 and 25 degrees of attack. Airfoil section local entropy viscosity can be predicted via numerical simulations. When considering entropy generation, the Reynolds number is crucial, and NACA0015 is more cost effective and uses less expensive alternative airfoils [2]. The authors' analyzed the coupled dynamics that included the aero-elastic and volatile feedback of the movable blades. The most promising configuration was selected based on the CFD results for reducing strain overall. The major outcome was that the total pressure drop is three times that of a hard, rapid, straight blade [3]. To better depict drift unsteadiness, a simpler, two-equation turbulence version was modified. Using the SST (shear stress transport) transient fluid solver and a two-equation turbulence model, Open FOAM evaluated the efficiency of a Wells turbine [4]. The Wells turbine's suction-slotted blades performed admirably. Four different suction slots were investigated. The pressure decrease caused by the turbine drift coefficient was greatest in slots 4 and 7. Three-slot turbines were efficient. A test was carried out on NACA 0015, and slots in the NACA 0015 airfoil model were also used [5]. Disc actuator blades investigated the evolution of the motor. The "interference factors" between a turbine rotor's series of blades and a single airfoil determined the rotor's carry behaviors also drag. There were greater air velocities inside the blade passages in the CFD studies with the thicker rotors [6].Wells turbine performance modeling and layout optimization using computational fluid dynamics were performed. Wells turbine performance can be predicted with either an experimental, analytic, or CFD-based model. Converting power from W to T waves is simple and inexpensive. A Wells turbine is inferior to an impulse turbine. When compared to other airfoils, NACA0015 is more efficient and produced less global entropy. The proposed work was produced the more using manual vanes and W-Ts with several stages. The cost of improved W-T performance increases [7].

Slots on both sides of an airfoil and glued guiding vanes were used in Wells turbines to increase efficiency and torque. Torque was increased by the imposed turbine slots. The Wells turbine's manual vanes were optimized for speed, rotational autonomy, and resistance to stalling. This action demonstrated that manual vanes function more effectively upstream than downstream. Based on the calculations, the slotted turbine turned over more quickly. Productivity, usability, and initialization are all enhanced with slotted blades and manual vanes [8]. The authors' proposed turbine must operate between 0 and 90 degrees to accommodate the varying airflow needed to harness wave energy. The blades in a Wells turbine's cascade are perpendicular to the axis of rotation, are offset by 90 degrees, and make it difficult to float. Complex stalls with a wide field of view can be simulated [9]. Different blade profiles, including NACA0020, NACA0015, CA9, and HSIM 15-262123-1576, were studied. Circular motion of the Wind CA9 blade profiles with a rotor solidity of 0.64 were used in Wells turbines. Turbogenerators that rotate in a circle exhibited rapid flotation tendencies. The performance curve of constant drift was slowed down by low flow coefficients [10]. The performance of a lightweight Wells turbine in a cutting-edge sea wave energy system was evaluated. Analysis was possible when the three-dimensional Navier–Stokes equations were stabilized. Going with the flow is an incompressible mechanism in SST k-turbulence. The efficiency of a turbine can be predicted by simulating the end clearance [11]. Numerical and practical tests were conducted on a small, single-wing Wells turbine with high solidity and no guide vanes. NACA0015 blades with a constant chord were used in the prototype. After running normal incompressible Reynolds-averaged Navier–Stokes (RANS) equations via finite-element simulations in three dimensions, the numerical evaluation were completed. A relaxed decrease in torque and pressure was identified [12]. In that study, regenerative ventilation was being evaluated for the possible arrangement of a new experimental record by mounting a turbine atop an oscillating water column. The DC systems were powered by the turbine, in which a NACA0015-based design profile was imposed on blades. The waft coefficient should be less than 0.1, and the turbine's Reynolds number should be high [13]. Wells turbines make a lot of noise and failed to work very well at high waft coefficients. These issues are resolved by skewing and reshaping the blades of the booster turbine and the rotor. The turbine was distorted by the Gurney flaps. The efficiency of blade profiles was enhanced [14].

The performance of Wells turbines with and without many planes was evaluated. Biplanes used contra-rotating rotors, while monoplanes relied on manually operated vanes. There were trials of a biplane turbine with two shafts. When compared to a monoplane turbine with guiding vanes, the contra-rotating turbine had a similar operational range, but it performed better post-stall and could withstand a pressure-drift ratio that was one mile higher. High viscous losses in the turbine rotor are optimal for recovery [15]. Turbine blades were imposed with symmetrical airfoils that revolve around a central hub. Energy generation can be maximized by using the rotational speed control loop. The turbine only spins in one direction, although the fluid can go in either direction. The goal of installing the stress sensors has been achieved. The first sensor tracked strain, while the second tracked pressure decrease from the faster generator [16]. Energy production is restricted by factors such as the predictability of waves, the efficiency of devices, etc. By optimizing the sweep angles of the blades, the performance and efficiency of the Wells turbine could be increased via numerical optimization. This investigation was finished using the RANS equations and determined the effectiveness in terms of strain, torque, and also waft. The authors' maximized the torque factor, and cross-validation errors were reduced with the use of Kriging. Increases in both turbine height and torque resulted from redesigned blade waft coefficients [17]. The notable observations are as follows: blades on monoplane turbines typically have a 90-degree angle, a Reynolds number of 8×10^5, a Mach number of 0.4 at the tip, and a hub-to-tip ratio of 0.6. Blades, tip clearance, and gliding efficiency are all distinct. Drift can be anticipated with the help of torque. Convergence of solutions was hampered by the discrepancy between axial velocity and tangential tip [18]. In this study, the authors compared static stall to dynamic stall and operational design settings. Next, they took take a theoretical and practical look at two different ratios of tip clearance to chord length, 0.0056 and 0.0111. Three-dimensional measurements of steady-state and pulsed flow were taken on Wells turbine prototypes. The rate at which a drift was occurring can modify the flow. Offshore utility REWEC3 seawalls were depicted as Wells turbines at a 1:10 scale. Gliding coefficients, torque, and strain drop were used to rate Wells turbines. Gliding became unsteady in the sinusoidal current. Performance of turbines was influenced by tip clearance [19]. Turbine performance was impacted by tip leaking glide. The investigation of the drift of turbine blades was executed. An NACA 0020 bladed Wells turbine was analyzed using CFD solver called FLUENT. The ratio of tip clearance to chord length grew, which negatively affected turbine performance and positively impacted glide coefficient. Turbine stall was brought on by boundary layer separation and tip leaking [20]. In the ANSYS simulations, the waft coefficients, angles, and rotation speed were all fixed. The SST turbulence model was imposed, and so the performance of the turbine improved as the adjustable vane angle and float path vortices were increased. The efficiency gained from a guide vane angle was significant, wherein an angle of 11.8 degrees and a glide ratio of 0.21 were optimal. Separating the floats and switching the vortex profiles both increased the efficiency of the guide vane attitude exchange [21].

In fragile systems, Wells turbine converters are typically used. Stronger lift forces and separation behaviors are generated by IFS profiles due to their concave shape in the submit-mid-chord region compared to NACA profiles. CFD analysis compared six different IFS and NACA blade designs. In comparison to NACA, IFS's operating spans and torque technology improved by 24.1%, and the IFS torque was also improved. The maximum torque coefficient was also increased by VTB20IFS, by 71 percent. VTB20IFS's superior separation region was validated by SS streamline contours [22]. Using NACA 0021 constant chord blades, the tool simulated the efficiency of a Wells turbine. Turbine performance and aerodynamics were predicted using the computer model. The model's prediction of the stall glide coefficient for the turbine was spot on. To better comprehend float physics and optimize the turbine's operating range and performance, both normal and stall scenarios were investigated. Turbine stalling due to a loss of tip clearance was explained by CFD data [23]. The aerodynamic forces, compressibility, and isolated and cascade float models of the blade profiles were analyzed using a CFD model. When taking into account prior data, the NACA 0015 blade profile was used and generated the strongest aerodynamic forces. Additionally, the 15% CA9 blade was favored based on the estimated slope of the cascade force and pressure curves. Once again, 15% CA9 performed better than NACA 0015 [24]. Power generation was improved by imitating the owl's and merganser's static trailing edges. The ideal amount of static for trailing chords was between 0 and 10%. Its glide coefficient was determined using the RANS equation with the help of ANSYS-CFX Float at a simulated 15.0. The 5%C torque is improved by prolonging the SETE. Above 5%C, the working range decreased, and the stall factor got close to FC. The SETE blade made it 23.4% of the way before stopping. The effectiveness of the SETE blades drops by 5.4% due to pressure loss. In comparison to blades with strong chord duration of the same size [25], SETE generated more torque. A lumped parameter model was used to simulate the interaction between the turbine and air mass in the chamber. The compressibility of the air in the chamber is the mechanism between air volume and the turbine duct, and it created hysteresis. Differences in performance do not matter for Wells turbines because they operate at non-dimensional frequencies [26]. The onset and severity of stall are the primary focus of this analysis of the turbine's performance. Highsolidity to simulate an oscillating water column wave power conversion device, a model utilizing Wells turbines, was required. The NACA 0021-based design profile was used. Spalart–Allmaras turbulence, and k-omega and Reynolds strain turbulence models were investigated using the RANS equations as a constant. The increment in nearby torque contribution has been achieved throughout the blade floor as the blade span is increased without blade separation. The torque dropped during a stall [27]. Testing and modeling were performed on a miniature version of the Wells turbine. A REWEC rotor with a NACA 0015 profile was implemented.

Non-dimensional torque, efficiency, and pressure drop were calculated, as well as the inlet and hub-nostril form of the turbine's tip. Principal drift features were related to stagnation pressure contours on iso-helicity surfaces and meridional tangential vorticity contours [28].

Based on the observations that are obtained from the literature survey [1–36], the process of designing a Wells turbine and optimizing its performance is identified out and so depicted in Figure 1.

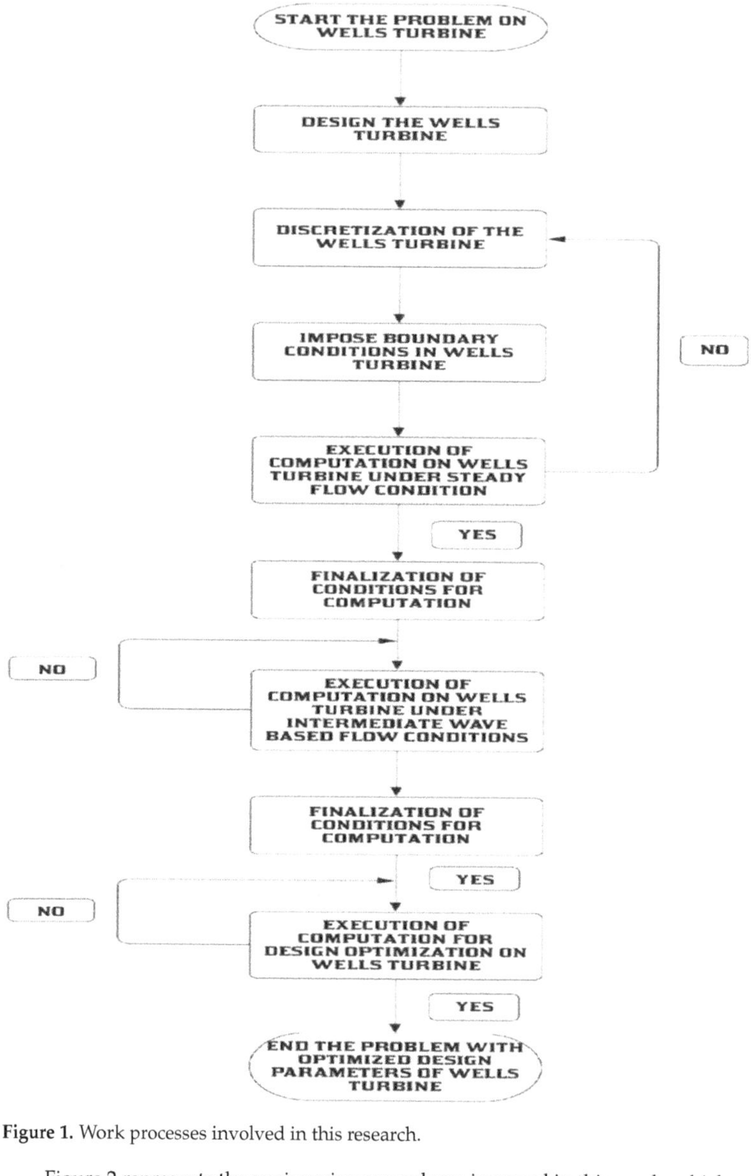

Figure 1. Work processes involved in this research.

Figure 2 represents the engineering procedures imposed in this work, which comprises CFD, computational structural, and modal analysis. CFD is used for the development hydrodynamic forces over the Wells turbine and for the estimation of torque produced by the Wells turbine. The computational structural analysis is used for the selection of suitable material that can effectively resist hydrodynamic loads and so provide a high lifetime to the blades. The modal analysis is used for the estimation of natural frequencies of the Wells turbine and so supports the development of energy extractions through the Wells turbine. Finally, the optimized Wells turbine with a high energy extraction rate and high structural lifetime has been adopted in the energy extraction process for real-time applications.

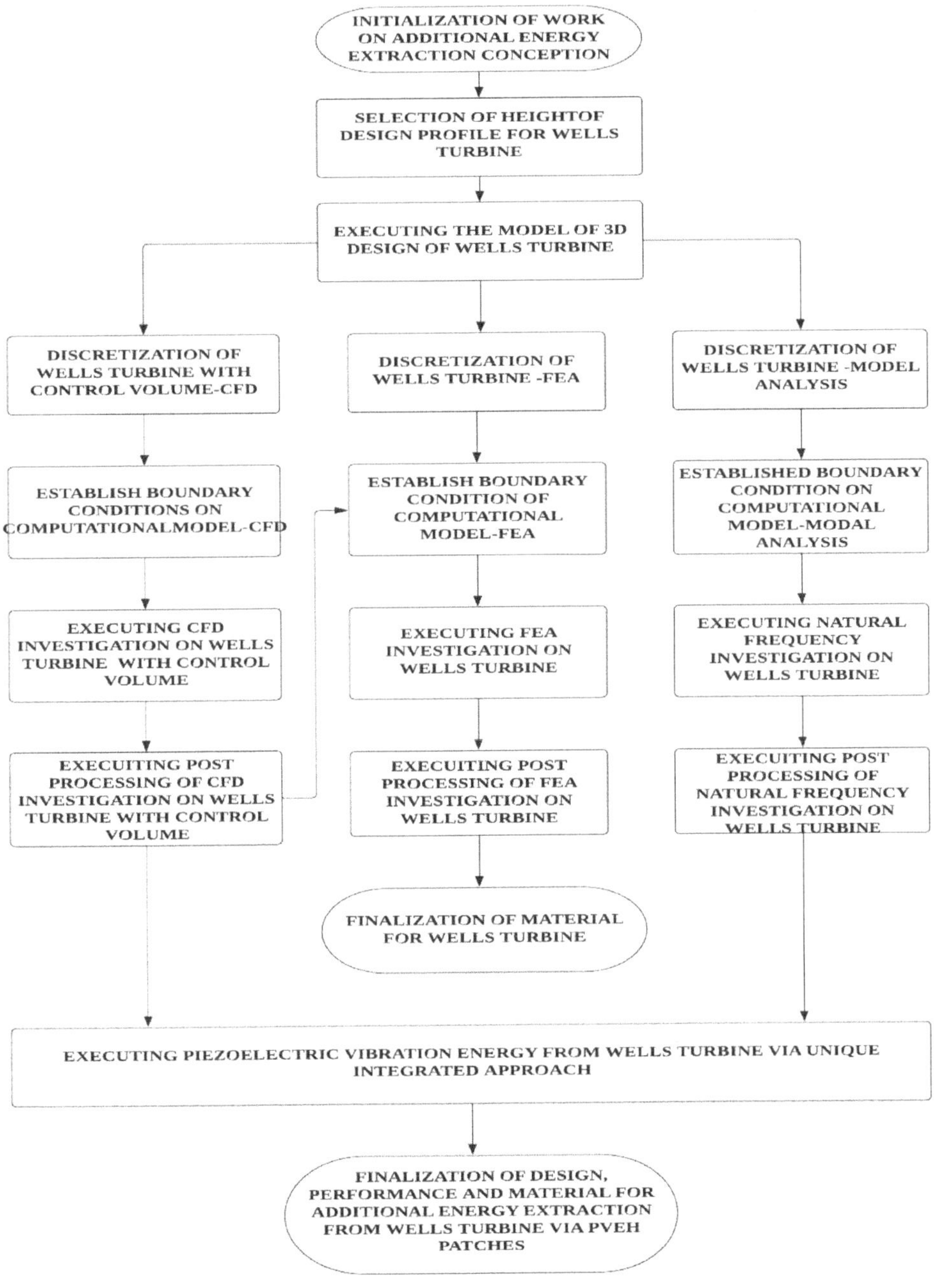

Figure 2. Work process involved.

2. Computational Methodologies

From the literature survey [1–36], it is observed that the previous works failed to impose the standard design calculations in the development of Wells turbine. Henceforth, this work is finalized to impose standard design procedures for the development of Wells turbine through the help of a bottom-up design approach. Since top-down approaches comprise assumption-based initialization, this work confirms the use of a bottom-up design approach. In a bottom-up approach, relevant design factors are chosen with previously obtained data, creating what are typically termed as historical relationships. The collected historical data and their typical relationships are revealed in Figures 3–5.

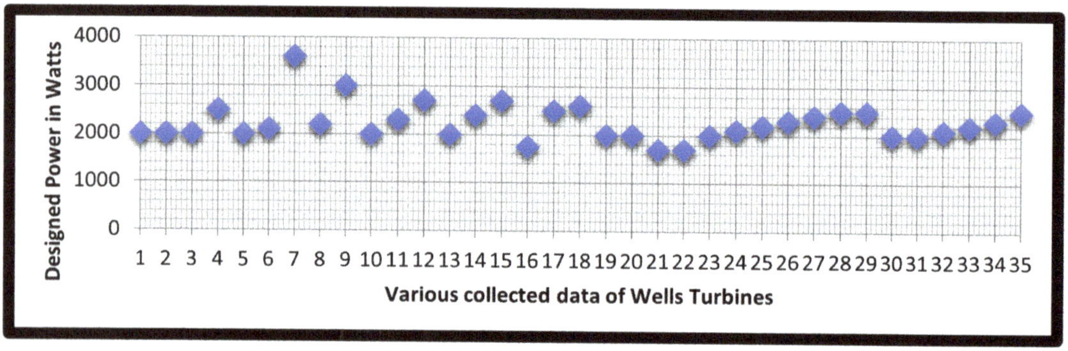

Figure 3. Various Wells turbines' data with respective power [1–36].

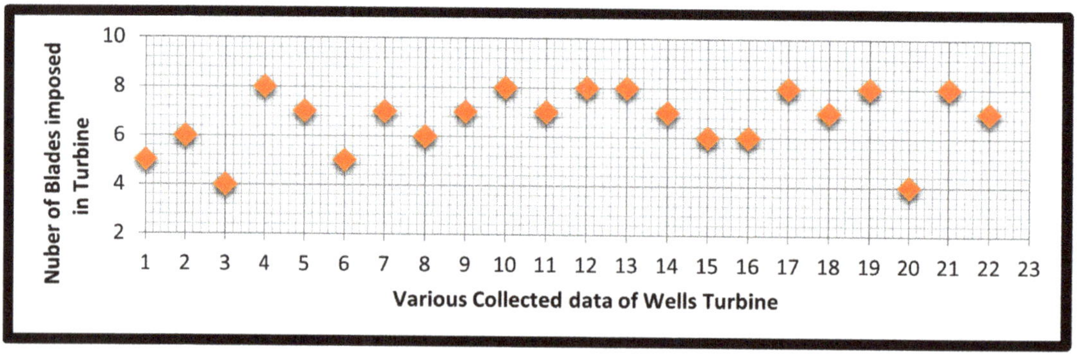

Figure 4. Various Wells turbines data with the respective number of blades [1–36].

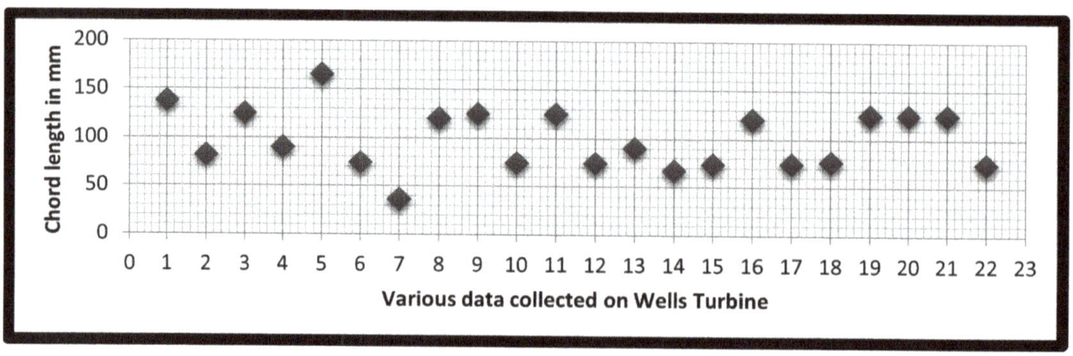

Figure 5. Various Wells turbine data with respective chord length [1–36].

2.1. Design Process

For this work, the imposed historical data and relationships concern the hydropower in watts, number of blades used, chord length in mm, and the radius of a hub to blade in mm. Figure 3 relates the various data of the Wells turbine with respective hydropower in watts, which is collected from the prior-discussed literature survey [1–36]. From Figure 3, it is observed that the maximum hydropower of 2250 W has been frequently imposed.

Figures 4 and 5 show the data of various Wells turbine with their corresponding number of blades and their chord length in mm. From Figure 4, the obtained number of blades is 7. Also, from Figure 5, the obtained chord length is 100 mm; i.e., $C^{B}_{WT} = 100$ mm.

$$A_{\text{Wells Turbine}} = \pi \times (r_{\text{blade}})^2 \tag{1}$$

$$\text{Circum}_{\text{Wells Turbine}} = 2\pi r_{\text{blade}} \tag{2}$$

$$\text{Circum}_{\text{Wells Turbine}} = 7 \times C^{B}_{WT} + 7 \times T^{B}_{WT} \tag{3}$$

From the shape of the conventional Wells turbine, the general relationships are derived, which are expressed in Equations (1)–(3). Equations (1)–(3) express the cross-sectional area of the Wells turbine and the circumference of the Wells turbine. To initiate the design process, the clearances between the blades play a major role. In this work, three different clearances are imposed, and so the different design data-based Wells turbines are framed. For sample calculation, the tolerance (clearance) between blades is assumed as 100 mm, and thus, $T^{B}_{WT} = 100$ mm.

$$\text{Circum}_{\text{blade}} = 1400\ \text{mm} \Rightarrow 2\pi r_{\text{blade}} = 1400 \Rightarrow r_{\text{blade}} = \frac{(1400 \times 7)}{(2 \times 22)} = 222.73\ \text{mm}$$

Additionally, another historical relationship has been organized between the hub's radius and the blade's radius. The comprehensive outcome of the recent historical relationship is revealed in Figure 6. Finally, from Figure 6, the unique relationship has been derived and is expressed in Equation (4).

$$\frac{r_{hub}}{r_{blade}} = 0.672 \tag{4}$$

$$\frac{r_{hub}}{334.1} = 0.672 \Rightarrow r_{hub} = 0.672 * 334.1 = 224.5152\ \text{mm}$$

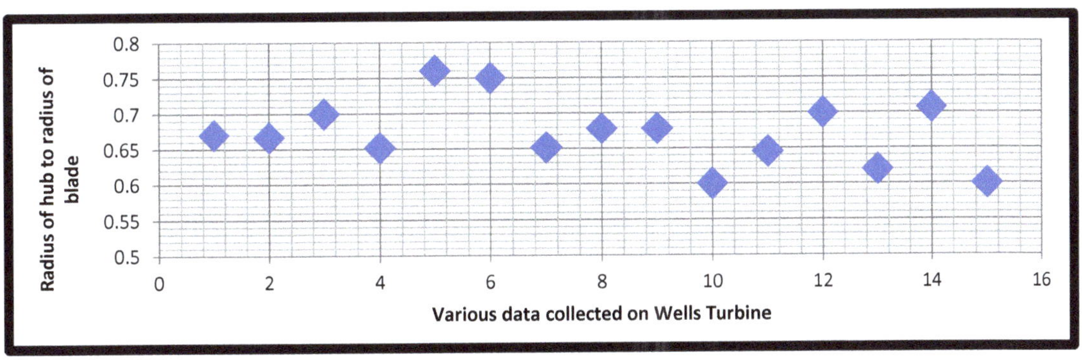

Figure 6. Various Wells turbines with respective radii of the hub to blade [1–36].

The shortlisted airfoil for this work from the literature survey [1–36] is NACA 0015. Table 1 depicts the final design data of the Wells turbine with different wingspan and main diameters in mm. Similarly, other cases are derived, and the seven cases' design data are listed in Table 2.

Table 1. Final design data of primary case.

Sl. No	Description	Value
1	Chord length	100 mm
2	Wingspan	219.1696 mm
3	Hub Diameter	449.0304 mm
4	Main Diameter	668.2 mm
5	Chosen airfoil	NACA 0015
6	Thickness for both hub and blades	15 mm (15% of the chord)
7	Number of blades	7

Table 2. Comprehensive design data of seven cases.

Design Cases	Dimensions of Various Design Factors
CASE I	Main diameter = 445.46 mm; Hub diameter = 299.35 mm; Wingspan = 146.11 mm; Tip clearance = 100 mm;
CASE II	Main diameter = 582.738 mm; Tip clearance = 100 mm
CASE III	Main diameter = 552.703 mm; Tip clearance = 150 mm
CASE IV	Main diameter = 732.036 mm; Tip clearance = 150 mm
CASE V	Main diameter = 668.2 mm; Tip clearance = 200 mm
CASE VI	Main diameter = 887.3696 mm; Tip clearance = 200 mm
CASE VII	Main diameter = 1106.5392 mm; Hub diameter = 449.0304mm; Wingspan = 438.3392 mm; Tip clearance = 200 mm

2.2. *Computational Fluid Dynamics*

Since this work primarily deals with the comprehensive investigations of various performance factors, an investigation based on the flexible and advanced engineering approach is mandatory for this work. Henceforth, the imposed Methodology for this work is computer-aided engineering, in which CFD provides a major contribution and also aims to determine the torque of the Wells turbine, hydrodynamic drag, hydrodynamic pressure on Wells turbine, and hydrodynamic velocities over the Wells turbine.

2.2.1. Computational Model

The Wells turbine, along with relevant control volume, is taken here as the computational model. Seven different design profiles of the Wells turbine were designed with the help of designed data. Each of the seven design profiles has a distinct hub diameter and blade length variation. The Wells turbine is designed using CATIA. The seven models have seven blades of NACA0015 airfoil with a 100 mm chord. These seven models were designed to obtain the result of maximum torque. In addition, a control volume is created around the Wells turbine. The typical fluid dynamic control volume and the Wells turbine's traces are revealed in Figure 7. For the development of control volume, the diameter of Wells turbine is picked as reference data, and so the entire volume is constructed. The rear position of the control volume is kept higher than the frontal position of the control volume. This imbalance in design data can permit the flow to settle after it impacts with Wells turbine.

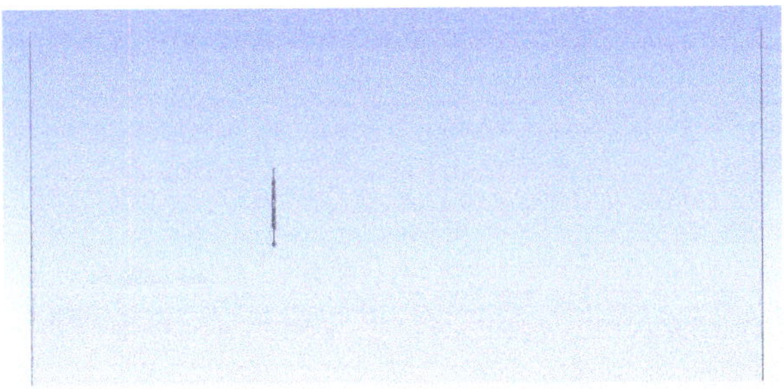

Figure 7. A systematic view of Wells turbine inside the control volume.

2.2.2. Discretization

Figure 8 represents the discretized mesh of the Wells turbine with control volume. The size function of the Wells turbine is "proximity and curvature" as the blade cross-section is of airfoil geometry. So, to captures the trailing edge in an effective manner, the above-said function is chosen. High smoothing is also chosen in addition to a fine relevance center. Table 3 provides the number of nodes and elements obtained for different mesh face sizes used for different cases of simulations that are imposed in grid independence study.

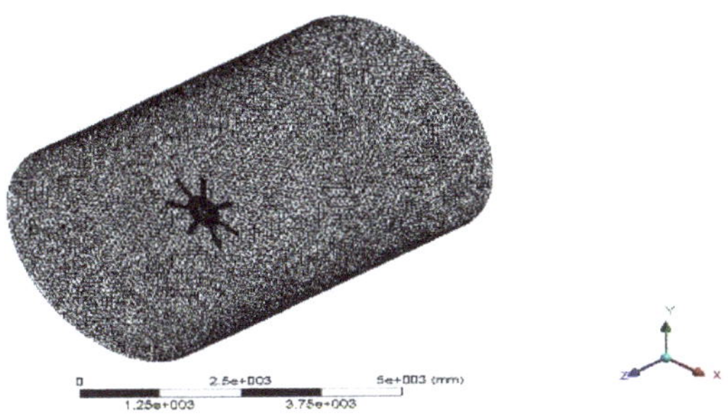

Figure 8. Discretized mesh of Wells turbine.

Table 3. Statistical data of nodes and elements for seven cases.

Various Mesh Cases	No. of Nodes	No. of Elements
CASE-1	170,185	934,788
CASE-2	196,670	1,083,869
CASE-3	157,498	865,671
CASE-4	681,382	3,889,953
CASE-5	176,180	972,195
CASE-6	238,700	1,324,759
CASE-7	693,333	3,640,060

2.2.3. Boundary Conditions

In this analysis, four inlet velocities are used such as average steady state velocity of water, shallow water velocity, wave celerity velocity, and the maximum wave velocity. The pressure outlet is zero for all the inlet velocities, and the operating pressure is 103,125 Pa. No slip condition is applied on the Wells turbine because of the friction. Instead, specified shear is applied on the wall. Four inlet velocities are the average velocity of water, shallow water velocity, intermediate water velocity, and deep water velocity. Average water velocity is obtained using surface water velocity estimated as 0.4 m/s through Equation (5) and the correction factor estimated as 0.85.

$$\text{Average velocity of water} = 0.4 \times 0.85 \Rightarrow 0.34\ \text{m/s} \quad (5)$$

$$\text{Shallow water velocity} = \sqrt{g \times \frac{\lambda}{20}} \Rightarrow \sqrt{g \times \frac{96.78}{20}} \Rightarrow 1.54\ \text{m/s} \quad (6)$$

The shallow water velocity is estimated as 1.54 m/s with the help of Equation (6),in which the imposed hydro-fluid properties density, dynamic viscosity, and kinematic viscosity are used. The values of these properties are 1030 kg/m^3, 1.793 $\times$ 10^{-3} kg/m-s, and 1.787 $\times$ 10^{-6} m^2 s^1.

2.2.4. Solver Data and Governing Equations

This study uses a pressure-based solver because of incompressible flow and constant density. The turbulence models used are Spalart–Allmaras and k-epsilon with enhanced wall treatment. The relevant shortlisted turbulence model is further imposed in all the models that give the best out-turn for boundary layers to acquire an adverse pressure gradient. Steady-state analysis is taken over for all seven cases. The turbulent viscosity ratio is fixed as 10% because of the high density of the imposed fluid. Concerning the new technology for renewable energy, the solver has to be setup with governing equations that are obtained from CFD fundamental concepts [37–39]. The RANS equations are certain to solve real problems using computational strategy, time continuity, and momentum relationships. The coupled algorithm-based pressure and velocity coupling has been implemented with higher-order approximations.

2.3. Computational Structural Analyses

In this part of the work, a structural analysis is carried out using the boundary conditions that will be discussed further down to determine the total deformation, equivalent stress, equivalent elastic strain, stress intensity, strain energy, shear stress, and normal stress for materials such as carbon fiber-reinforced polymer (CFRP) materials, glass fiber-reinforced polymer (GFRP) materials, and relevant lightweight alloys. This computational structural analysis aims to determine a suitable lightweight material for the Wells turbine that results in less deformation, stress, and strain with high lifetime.

The Wells turbine with the composition of seven blades is employed as a computational model for this structural analysis. This turbine is designed to obtain the maximum efficiency to extract the energy from ocean waves with the high-power design profile of NACA 0015. Figure 9 shows the computational model of a seven-bladed Wells turbine. The discretization is the next phase that is involved in this computational structural analysis. The discretized mesh of the computational model used in this investigation is seen in Figure 10. The "proximity and curvature" size function is implemented in this work, and the size function can be perfectly captures the Well turbine curvatures. In addition, high smoothing is selected, while a coarse relevant center is selected. Since all the seven cases are different in terms of design data, the meshed elements are obviously different for all the seven cases. For the sake of example, the statistical reports of case IV and case VI are provided that further enhance the reliability of this work. The total number of nodes and elements in case IV is 572,774, while the total number in case VI is 918,984.

In this study, the computational structural analysis has been computed with the help of a one-way coupling-based fluid–structure interaction approach. For hydrodynamic loading extraction purposes, two different inlet velocities are used for two different cases. One is the surface velocity of water, which is 0.34 m/s, and the second is deep water velocity, which is 23 m/s. The support used in this work is fixed support that is provided in all the interactions between the hub and blades of the Wells turbine. The material properties of the imposed lightweight materials have been extracted from the literature survey [40,41] and engineering data library from the imposed computational tool, i.e., ANSYS Workbench 17.2. The governing equations imposed in this work used are from force-and-displacement-based relationships, and stress-and-strain-based relationships [40,41]. For this analysis, each blade can be treated as a cantilever beam because of the production of fixed support to the hub. The uniformly distributed loads are imported from hydrodynamic pressure obtained through steady fluid flow analysis.

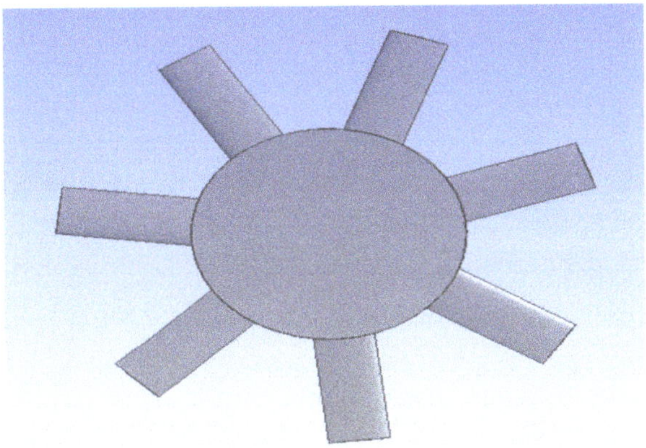

Figure 9. Typical representation of the computational model for structural analysis.

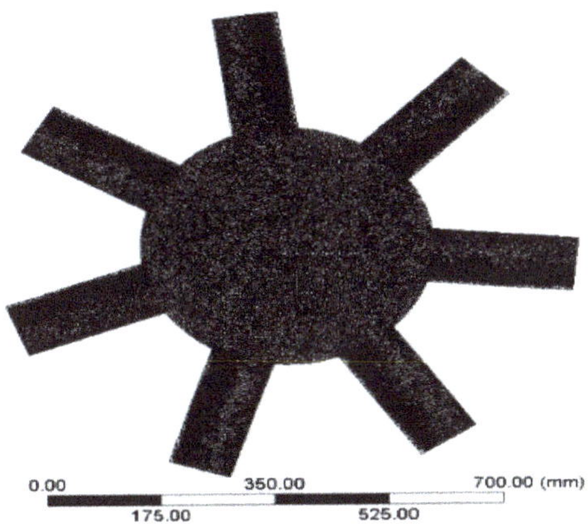

Figure 10. Discretized structure of Wells turbine.

2.4. Computational Analysis for PVEH

The Wells turbine, which has seven blades, can be disassembled into its component parts. To begin, the model is held in a stationary position while it is being supported. After that, there is no restriction on how it vibrates. The frequency of the required material is then determined via modal analysis, and this information is then employed in the power of piezoelectric vibrational energy harvester (PVEH) calculations. Figure 9 is a front-view illustration of a common computer model that is utilized in the process of vibrational analysis. The discretized structure of the Wells turbine is shown in Figure 10. The boundary condition that must be complied with is illustrated in Figure 11. For this modal analysis, the discretization has been executed, and the details are as follows: the number of nodes is 88,975, and the number of elements is 53,859. Also, the size function used in this computational vibrational analysis is curvature, and this smoothing is high.

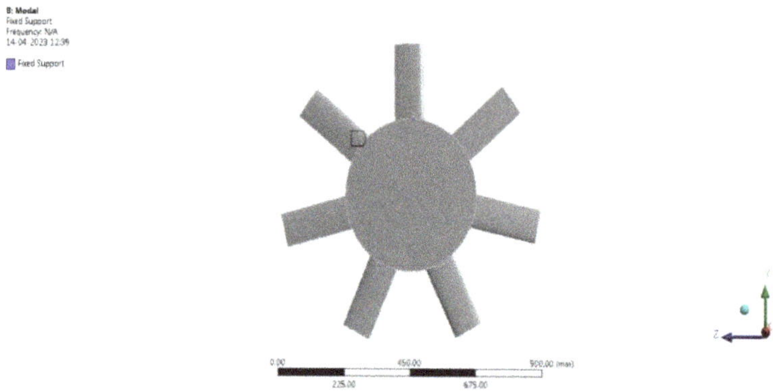

Figure 11. A typical isometric view-based representation of imposed boundary conditions on the Wells turbine (WT).

The inherent resonance frequencies and mode forms of a structure should be identified. A linear dynamical analysis is the process of studying modes. Since the overall acceleration, velocity, and location of the structure at each site are constantly unknown, the dynamic response of a structure must be calculated by first solving the fundamental equation of motion. Control of this system is defined by Equation (7) [42–47].

$$[m_{WT}]\left\{\frac{d^2u}{dx^2}\right\} + [C_{WT}]\left\{\frac{du}{dx}\right\} + [K_{WT}]\{u\} = \left\{F(t)_{Hydrodynamic}\right\} \tag{7}$$

Since the resonance frequency and mode forms are independent of the external load, they must be set to zero for modal analysis. However, the morphologies of natural frequencies and modes can only be expressed using complex numbers; therefore, damping effects are currently being disregarded. As seen in Equation (8), this first optimized governing equation is for modal simulation.

$$[m_{WT}]\left\{\frac{d^2u}{dx^2}\right\} + [K_{WT}]\{u\} = 0 \tag{8}$$

From the perspective of time domain dynamics, this problem can be thought of as a body either at rest or moving at a constant pace. Therefore, $\left\{\frac{d^2u}{dx^2}\right\}(t) = 0$; Equation (9), a condensed version of the governing equation, governs the modal computation.

$$[K_{WT}]\{u\} = 0 \tag{9}$$

2.4.1. Experimental Validation—1

A thorough investigation of experimental analysis is conducted to ascertain the efficacy of computational work. The MIDE quick pack QP 10W is utilized in the experimental configuration the authors suggested. A total of 2.5 mW of power was retrieved for this study. Therefore, in order to confirm the computational work and compare the experimental setup, a three-dimensional model of the experimental setup is produced with the assistance of CATIA, and then, it is analyzed in ANSYS Workbench 17.2. The authors suggest the boundary conditions and material characteristics. A modal analysis is then performed after the CFD simulation to ascertain the frequency of the relevant material. Figures 12 and 13 [42–47] vividly illustrate the well-established research of aerodynamic and modal analyses.

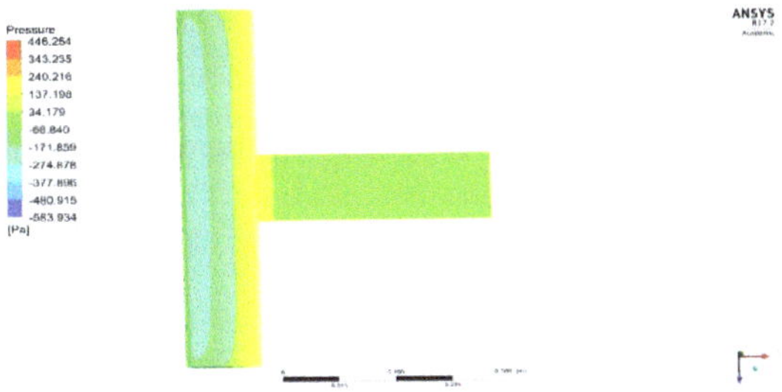

Figure 12. Pressure distribution on the proposed experimental setup.

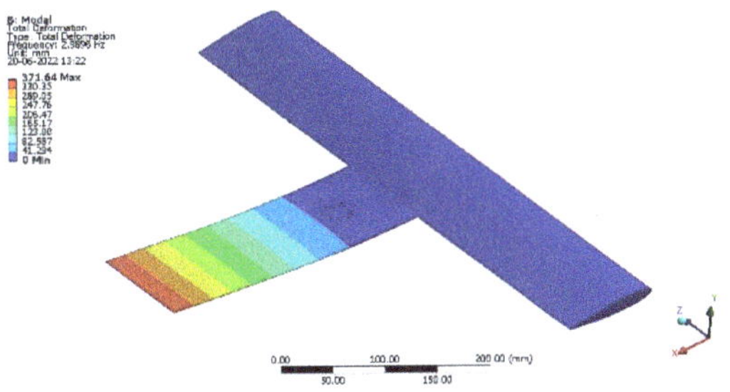

Figure 13. Total deformation on the experimental setup.

A computational simulation is used to determine the required values for the calculation. Quick pack QP 10W is used as a calculation sample. Equations (10) and (11) substitute the expected computational result-based data. The pressure on the piezoelectric patch is 34.179 N/m^2.

$$P^{Validation}_{Intermediate} = d^2_{lwm} w^2 f^2 \times \frac{18 \times T_{PL}}{\left[(W_P) \times (L_{PL})^2 \times ([t_P] + [T_{PL}])^4\right]} \times \frac{\rho_{lwm}}{[1 + f \times \varepsilon \times (\rho_{lwm})]} \quad (10)$$

$$P^{Validation}_{Final} = \left(P^{Validation}_{Intermediate}\right)\left(\frac{(0.44 * L)^5}{36} - \frac{L(0.44 \times L)^4}{6} + \frac{5L^2(0.44 \times L)^3}{12} - \frac{L^3(0.44 \times L)^2}{2} + \frac{L^4(0.44 \times L)}{4}\right) \quad (11)$$

Inputting the appropriate values into the equation yields an estimated power of 2.3 mW. Therefore, compared to experimental work, computational analysis error was 8%, which lies within the acceptable range. Table 4 contains a complete set of trial results and suggested data of this work. It is evident from Table 4 that the error percentage is within the allowed range, indicating that the PVEH patches can be used to extract electrical energy using the mathematical and analytical approach that is being presented for this work. With this imposed integrated strategy now verified, the authors are prepared to apply it to further real-time use cases. But first, nature-inspired Wells turbine uses the same methodology.

Table 4. Experimental and computational studies verify the legitimacy of electrical power extractions [42–47].

The Developed Electrical Power (mW) as Measured in Experimental Settings	The Resulting Electrical Power (in mW) from the Method Proposed Here	Error Percentage
2.5	2.3	8

2.4.2. Experimental Validation—2

In order to improve the reliability of the required procedure, a second research is required after the first one has confirmed the accuracy of the results of the energy extraction. This part is devoted to a validation study [42–47] because of the significance of the method that was implemented in this part of work. A separate and easier-to-understand validation design, which can be shown in Figure 14, has been developed. Figure 15 depicts the results of applying the models. Using ANSYS Fluent, we were able to come up with an estimate for the required pressure variation on the base model. Figure 16 illustrates the ultimate effect that the vibrations of the base object had.

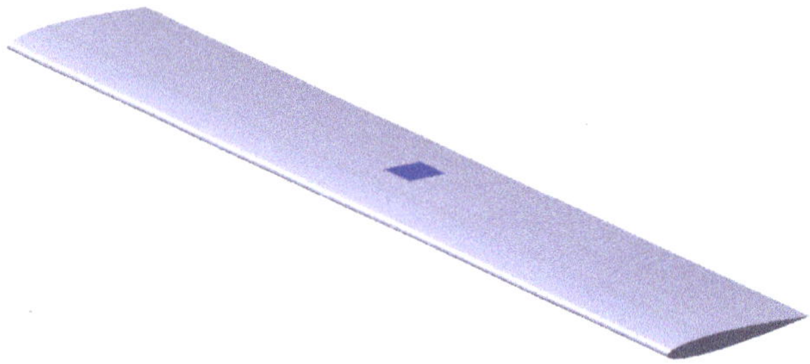

Figure 14. The typical isometric representations of secondly modified base model along with PVEH patches.

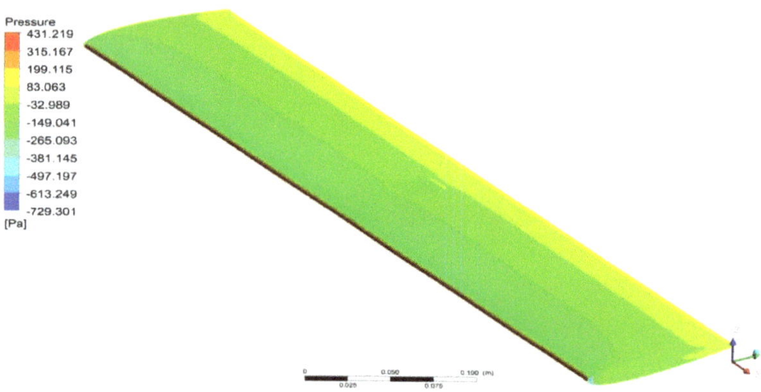

Figure 15. The typical isometric representations of pressure loads acting on the secondly modified base model along with PVEH patches.

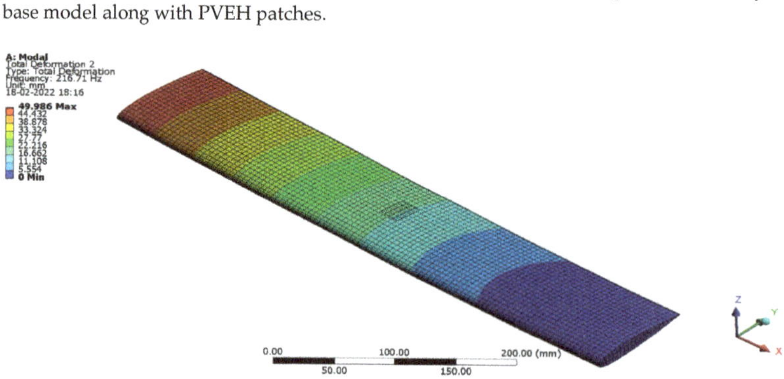

Figure 16. The typical isometric representations of freely vibrated structure of secondly modified base model along with PVEH patches.

Similarly to the first experimental validation test, the second experimental test is also conducted, and so the outcomes are listed in Table 5.

Table 5. Validation of electrical power extractions through experimental and computational outcomes [42–47].

The Developed Electrical Power (mW) as Measured in Experimental Settings	The Resulting Electrical Power (in mW) from the Method Proposed Here	Error PERCENTAGE
2.5	2.58	3.1

From Tables 4 and 5, it is clearly observed that the imposed energy extraction approach based on advanced computational method is reliable and suitable to impose in real-time applications. Also, it is learned that the additional approaches such as CFD and computational structural analysis are also validated with these experimental validations.

3. Results and Discussion

Firstly, the optimized Wells turbine is obtained with the help of CFD analysis for all seven cases with four different inlet velocities, such as 0.34 m/s, 1.54 m/s, 12.10 m/s, and 23 m/s. Secondly, the computational structural analysis is taken for the optimized Wells turbine using various lightweight materials.

3.1. CFD Outcomes

In CFD, the primarily focused outcomes have been finalized as torque and hydrodynamic drag; the secondary focused outcomes are hydrodynamic pressure acting on the Wells turbine and hydrodynamic velocity variations in and over the same Wells turbine.

3.1.1. Results and Discussion on Design Data-Based Modified Cases

Firstly, the CFD investigations are computed on first case of the Wells turbine. Figures 17–20 depict the hydrodynamic velocity and pressure variations that are acting in and over the first case of the Wells turbine for the inlet velocities of 0.34 m/s, 1.54 m/s, 12.10 m/s, and 23 m/s. The outcomes are compared, and so the comprehensive studies are carried out. Based on the integrative effect among the four inlet velocity outcomes, the velocity of 0.34 m/s performance is better than the others.

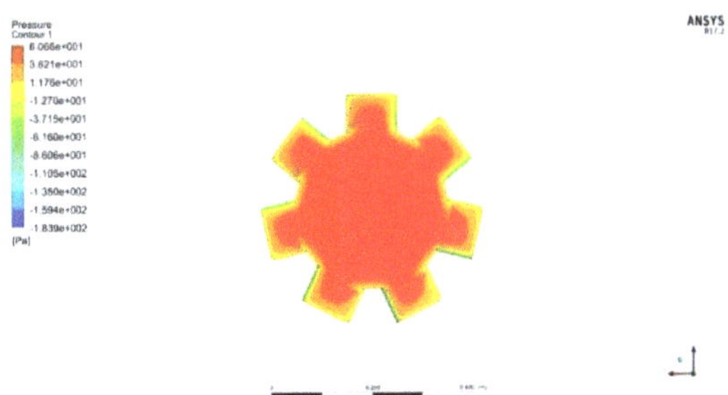

Figure 17. Pressure on first case based WT under hydrodynamic condition of 0.34 m/s.

Secondly, the CFD investigations are computed on the second case of the Wells turbine. Figures 21–24 depict the hydrodynamic velocity and pressure variations that are acting in and over the second case of the Wells turbine for the inlet velocities of 0.34 m/s, 1.54 m/s, 12.10 m/s, and 23 m/s. The outcomes are compared, and so the comprehensive studies are carried out. Based on the integrative effect among the four inlet velocity outcomes, the velocity of 23 m/s performance is better than the others.

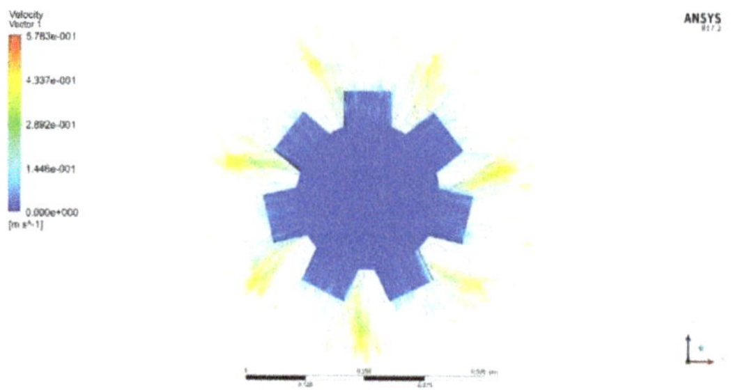

Figure 18. Hydrodynamic velocity on first case based Wells turbine—0.34 m/s speed– case-1.

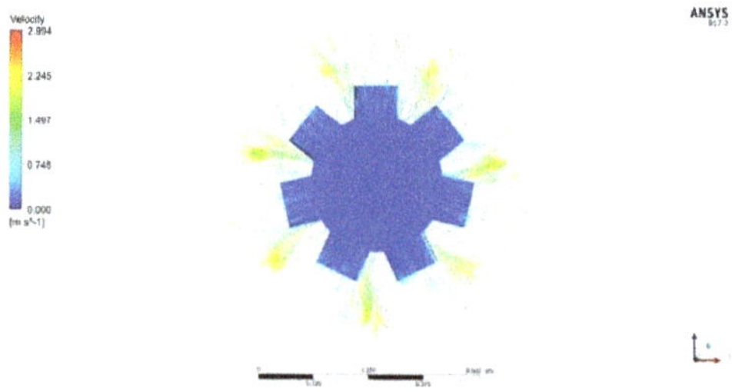

Figure 19. Hydrodynamic velocity on first case based Wells turbine—1.54 m/s speed.

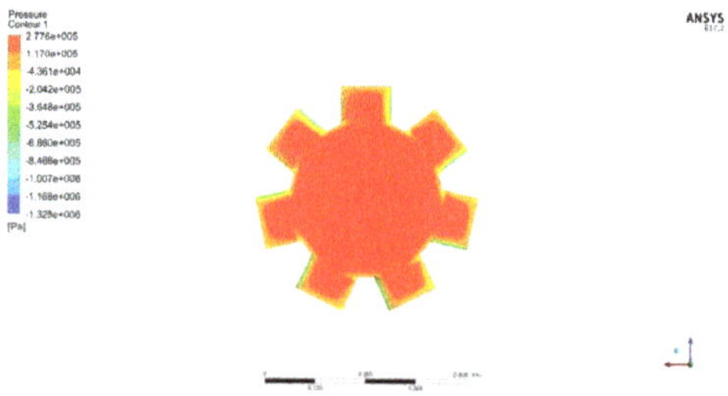

Figure 20. Pressure on first case based WT under hydrodynamic condition of 23 m/s.

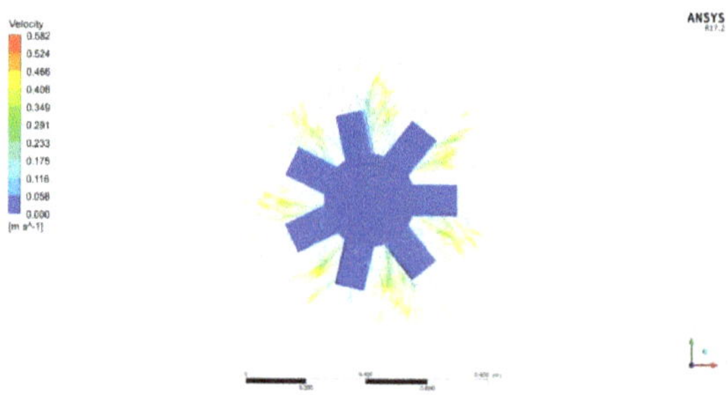

Figure 21. Hydrodynamic velocity on second case based Wells turbine—0.34 m/s speed.

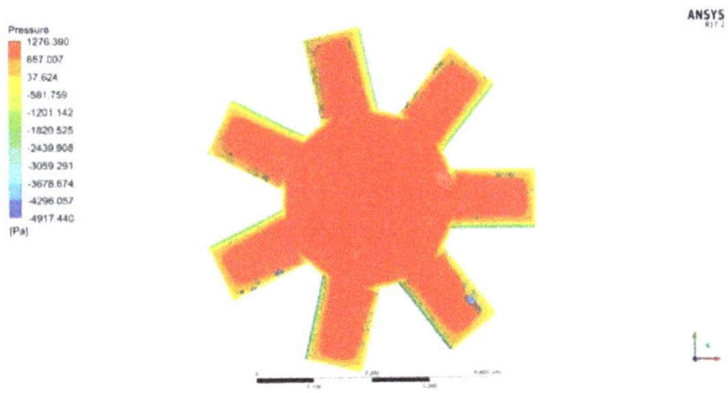

Figure 22. Pressure on second case based WT under hydrodynamic condition of 1.54 m/s.

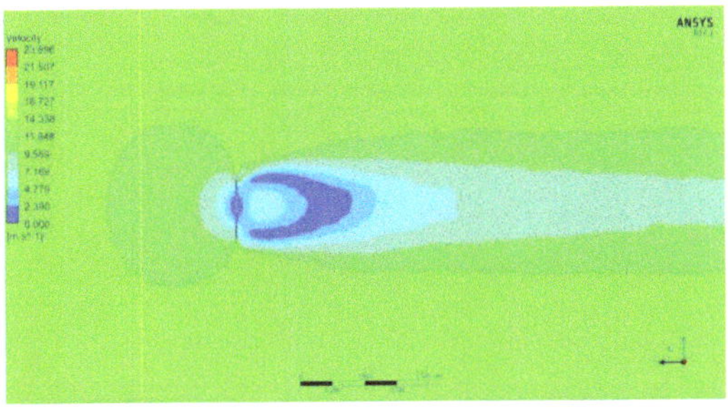

Figure 23. Hydrodynamic velocity on second case based Wells turbine—12.10 m/s speed.

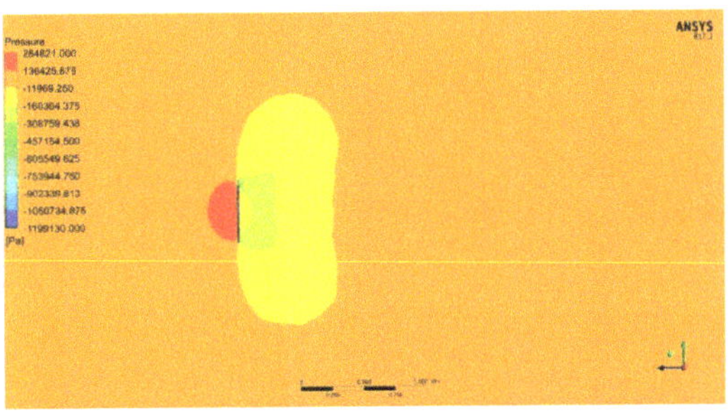

Figure 24. Pressure on second case based WT under hydrodynamic condition of 23 m/s.

Thirdly, the CFD investigations are computed on third case of the Wells turbine. Figures 25–28 depict the hydrodynamic velocity and pressure variations that are acting in and over the third case of the Wells turbine for the inlet velocities of 0.34 m/s, 1.54 m/s, 12.10 m/s, and 23 m/s. The outcomes are compared and so the comprehensive studies are carried out. Based on the integrative effect among the four inlet velocity outcomes, the velocity of 0.34 m/s performance is better than the others.

Fourthly, the CFD investigations are computed on the fourth case of the Wells turbine. Figures 29–32 depict the hydrodynamic velocity and pressure variations that are acting in and over the fourth case of the Wells turbine for the inlet velocities of 0.34 m/s, 1.54 m/s, 12.10 m/s, and 23 m/s, respectively. The other performance factors such as hydrodynamic forces acting on the Wells turbine and torque produced by the Wells turbine are also estimated through function calculator facility.

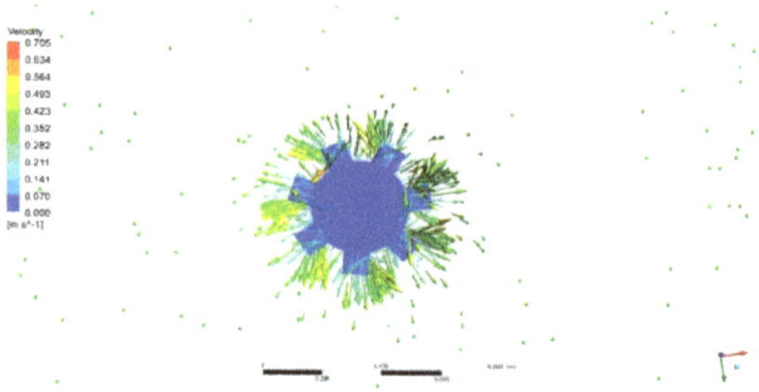

Figure 25. Hydrodynamic velocity on third case based Wells turbine—0.34 m/s speed.

Figure 26. Pressure on third case based WT under hydrodynamic condition of 1.54 m/s.

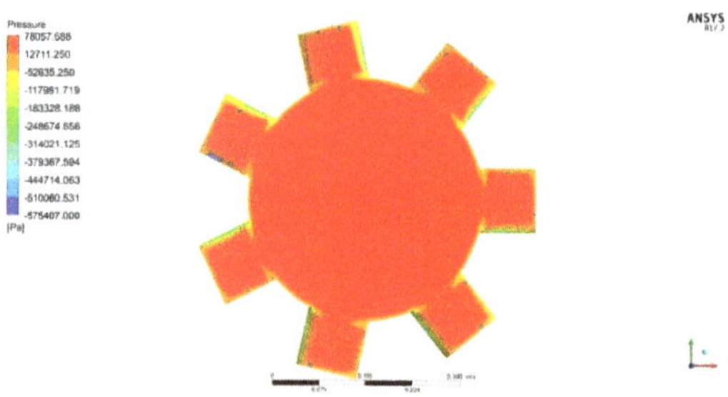

Figure 27. Pressure on third case based WT under hydrodynamic condition of 12.10 m/s.

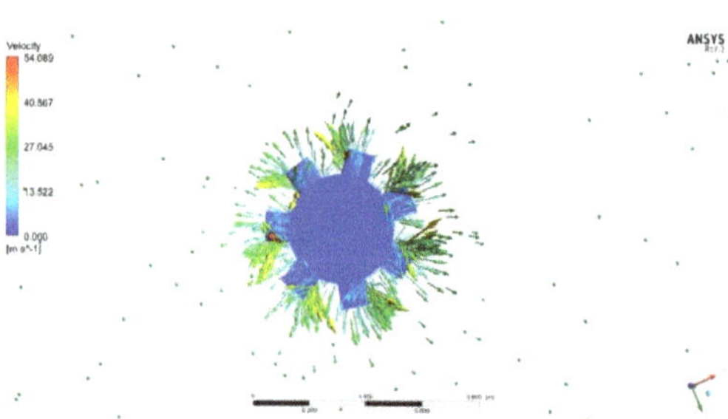

Figure 28. Hydrodynamic velocity on third case based Wells turbine—23 m/s speed.

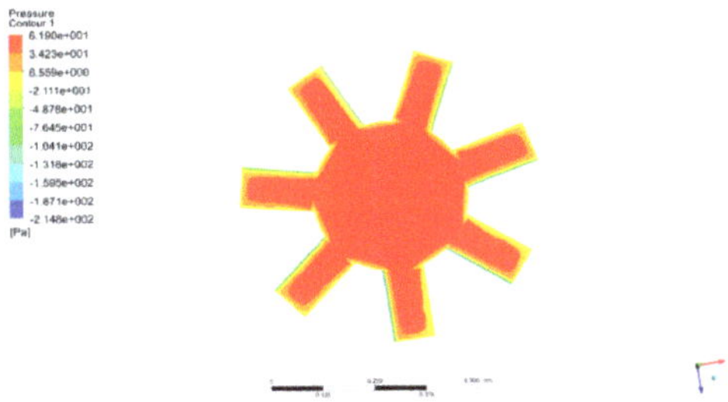

Figure 29. Pressure on fourth case based WT under hydrodynamic condition of 0.34 m/s.

Figure 30. Pressure on fourth case based WT under hydrodynamic condition of 1.54 m/s.

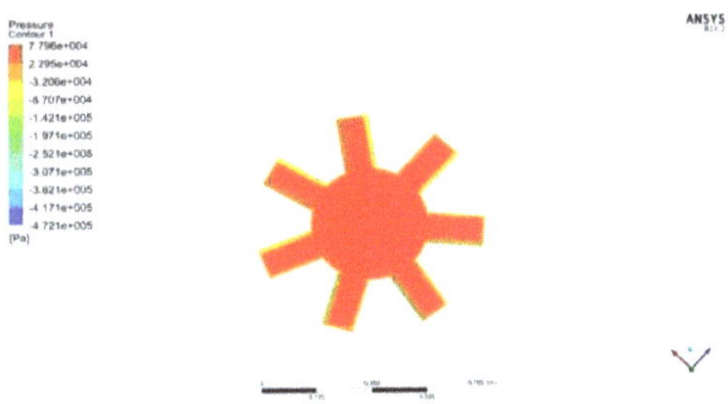

Figure 31. Pressure on fourth case based WT under hydrodynamic condition of 12.10 m/s.

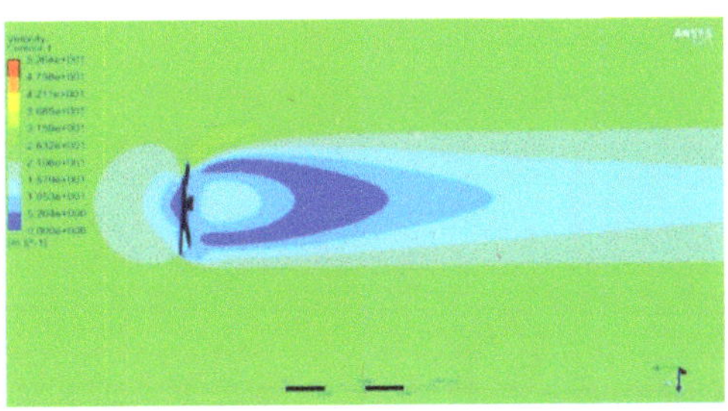

Figure 32. Hydrodynamic velocity on fourth case based Wells turbine—23 m/s speed.

Fifthly, the CFD investigations are computed on the fifth case of the Wells turbine. Figures 33–36 depict the hydrodynamic velocity and pressure variations that are acting in and over the fifth case of the Wells turbine for the inlet velocities of 0.34 m/s, 1.54 m/s, 12.10 m/s, and 23 m/s, respectively. As similar as other cases, the performance factors are also determined for this fifth model.

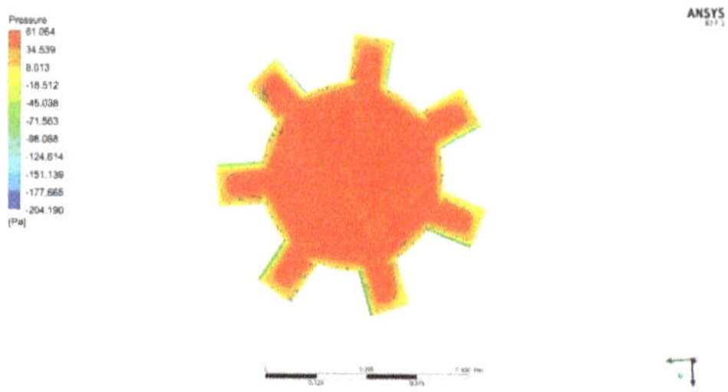

Figure 33. Pressure on fifth case based WT under hydrodynamic condition of 0.34 m/s.

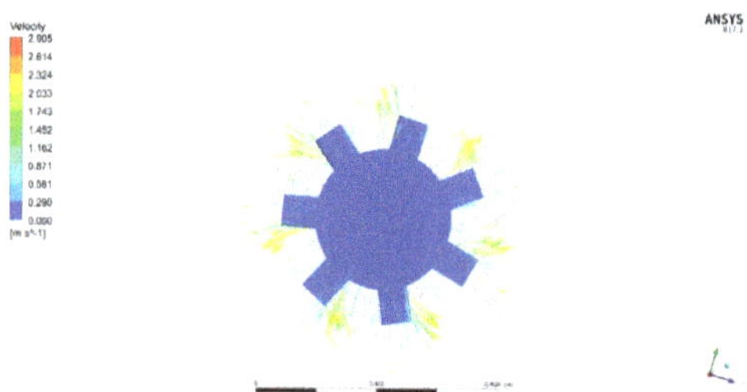

Figure 34. Hydrodynamic velocity on fifth case based Wells turbine—1.54 m/s speed.

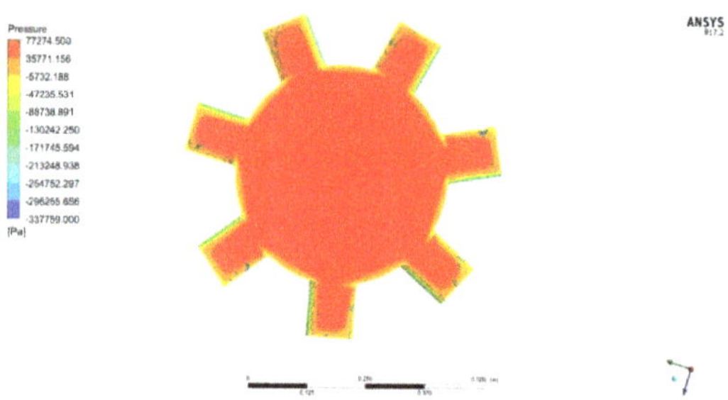

Figure 35. Pressure on fifth case based WT under hydrodynamic condition of 12.10 m/s.

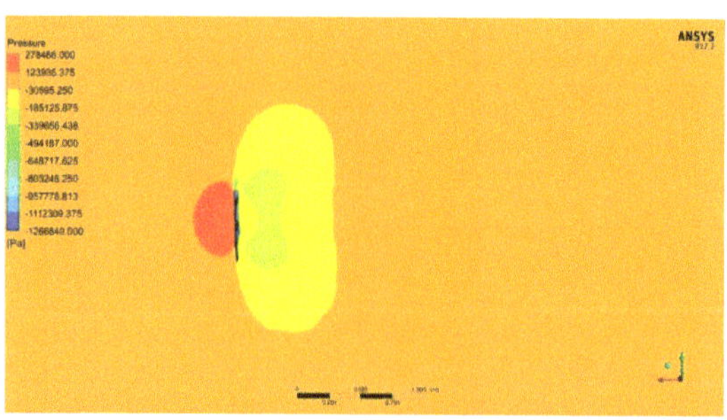

Figure 36. Pressure on fifth case based WT under hydrodynamic condition of 23 m/s.

Sixthly, the CFD investigations are computed on sixth case of the Wells turbine. Figures 37–40 depict the hydrodynamic velocity and pressure variations that are acting in and over the sixth case of the Wells turbine for the inlet velocities of 0.34 m/s, 1.54 m/s, 12.10 m/s, and 23 m/s. The outcomes are compared, and so the comprehensive studies are carried out. Based on the integrative effect, the 0.34 m/s inlet velocity result is the most effective of the four possible inlet velocities.

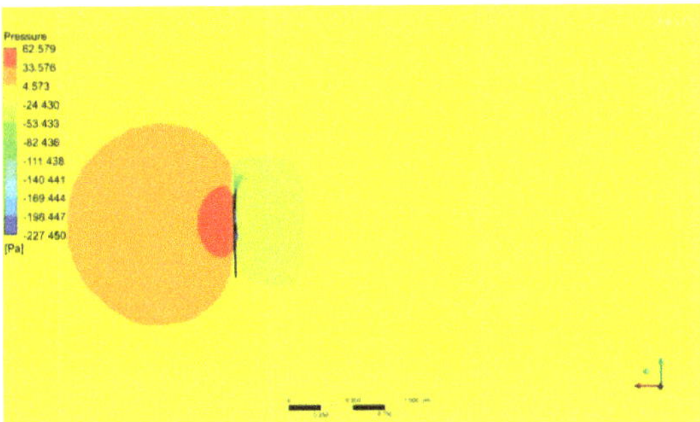

Figure 37. Pressure on sixth case based WT under hydrodynamic condition of 0.34 m/s.

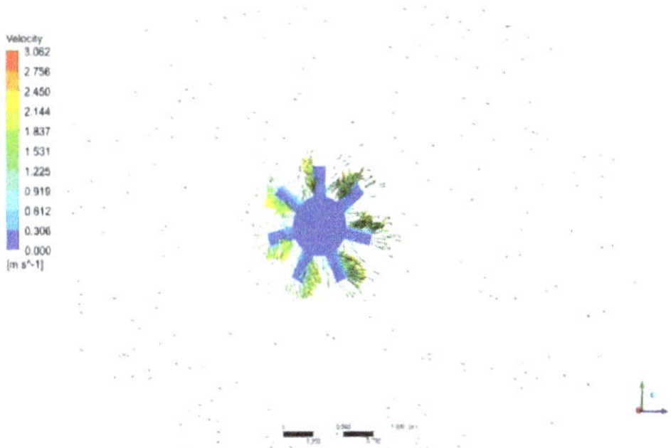

Figure 38. Hydrodynamic velocity on sixth case based Wells turbine—1.54 m/s speed.

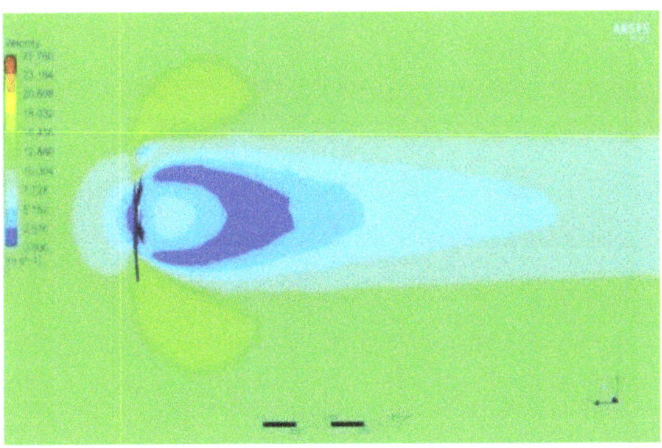

Figure 39. Hydrodynamic velocity on sixth case based Wells turbine—12.10 m/s speed.

Seventhly, the CFD investigations are computed on seventh case of the Wells turbine. Figures 41–44 depict the hydrodynamic velocity and pressure variations that are acting in and over the seventh case of the Wells turbine for the inlet velocities of 0.34 m/s, 1.54 m/s, 12.10 m/s, and 23 m/s. The outcomes are compared, and so the comprehensive studies are carried out. Based on the integrative effect, the performance of the 0.34 m/s inlet velocity is superior to the other four possible outcomes.

Figure 40. Pressure on sixth case based WT under hydrodynamic condition of 23 m/s.

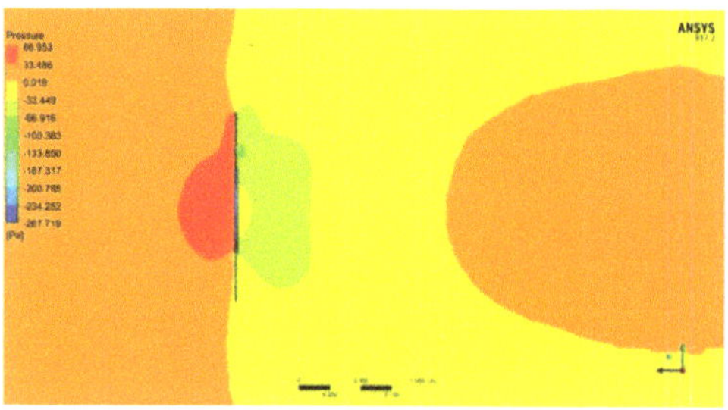

Figure 41. Pressure on seventh case based WT under hydrodynamic condition of 0.34 m/s.

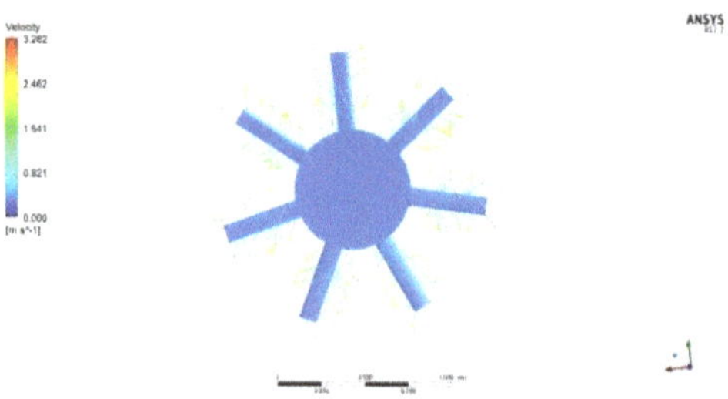

Figure 42. Hydrodynamic velocity on seventh case based Wells turbine—1.54 m/s speed.

Figure 43. Pressure on seventh case based WT under hydrodynamic condition of 12.10 m/s.

Figure 44. Pressure on seventh case based Wells turbine under hydrodynamic condition of 23 m/s.

Finally, the comprehensive outcomes of all four cases are listed in Tables 6–8. Table 6 comprises data on aerodynamic forces-based outcomes, Table 7 comprises comprehensive data on torque, and Table 8 comprises comprehensive data on power extraction.

Table 6. Torque for various imposed Wells turbine cases.

Velocity (m/s)	Torque (Nm)						
	Case 1	Case 2	Case 3	Case 4	Case 5	Case 6	Case 7
0.34	0.055215	0.154057	0.142265	0.246834	0.158302	0.418652	1.27369
1.54	1.49314	3.73196	3.9291	6.83129	3.84614	10.0088	30.0695
12.10	103.202	243.903	260.64	471.9	251.783	656.869	1919.9
23	380.127	884.07	957.654	1279.2	925.113	2393.45	6713.63

Table 7. Drag force for various imposed Wells turbine cases.

Velocity (m/s)	Drag (N)						
	Case 1	Case 2	Case 3	Case 4	Case 5	Case 6	Case 7
0.34	9.09072	13.4306	13.2765	18.8315	19.0282	26.3446	59.7228
1.54	185.191	272.619	271.242	383.472	388.983	537.146	1219.24
12.10	11,454.7	16,795.5	16,702.1	23,630.1	24,011	33,159.2	73,799.7
23	40,891.8	59,962.9	58,965.2	82,569.6	81,523.9	113,105	245,682

Table 8. Power for various imposed Wells turbine cases.

Velocity (m/s)	Power (W)						
	Case 1	Case 2	Case 3	Case 4	Case 5	Case 6	Case 7
0.34	0.143368	0.30774	0.362951	0.417528	0.28004	0.581275	1.540195
1.54	20.07226	37.19742	44.78828	60.95917	33.44393	69.07728	178.3818
12.10	11,463.88	20,004.22	26,920.69	35,328.13	18,015.37	38,139.29	92,229.6
23	81,805.03	134,024.3	98,913.09	183,986.5	127,204.8	263,356.2	594,207.1

The high power extraction rate and low hydrodynamic drag force are the predominant selection factors involved in this work. From CFD outcomes, the torque and drag forces are calculated, and they are listed in Tables 6–8. From Tables 6–8, based on integrative effect, the reasonable drag developer with torque producer-based design model is shortlisted for further investigation. From the outcomes listed in Tables 6–8, it is clearly observed that in case-4,there is better extraction of a high amount of energy from the hydro-fluid sources and also less drag produced than in other enhanced cases. Based on the integrative effect, this decision has been made, and so case-4 is finalized as the best performer.

3.1.2. Results and Discussion on Design Profile Modified Cases

Because of the high imposition of modernization, the hydro-fluid source is reducing drastically. Thus, the authors finalized further design profile modifications on case-4. Three more additional design profile modifications are imposed on the case-4-based Wells turbine. The typical representations of all three profile-enhanced Wells turbines are revealed in Figures 45–47.

In three advanced cases, two cases belong to stepped-back airfoil-based modified design profiles, and one case belongs to a zigzagged airfoil-based modified design profile. As per the aforesaid same boundary conditions and control volume dimensions, the computational investigations are computed. The typical hydrodynamic velocity variations in and over of all three advanced cases are revealed in Figures 48–50. Additionally, to obtain a further view about the turbulence development and its effect that supports the development of high hydropower, studies on turbulence behavior are also executed. The major outcomes of the turbulence prediction such as turbulence dissipation rate and turbulence kinetic energy are determined. The variations in turbulence dissipation rate and turbulence kinetic energy in and over the Wells turbine are systematically revealed in Figures 51–56.

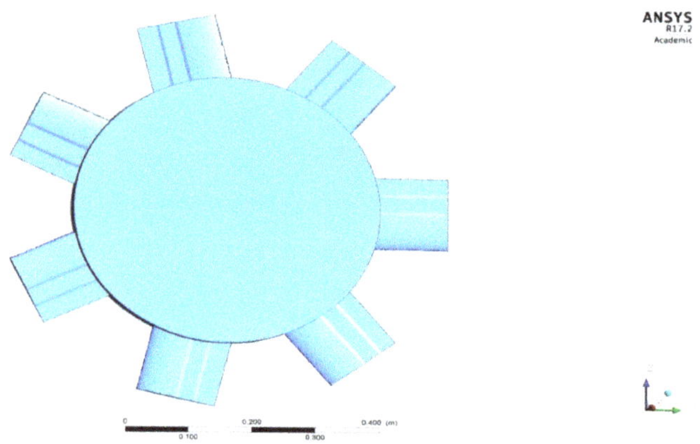

Figure 45. A typical frontal view-based representation of first stepped-back airfoil-based Wells turbine (advanced case-1).

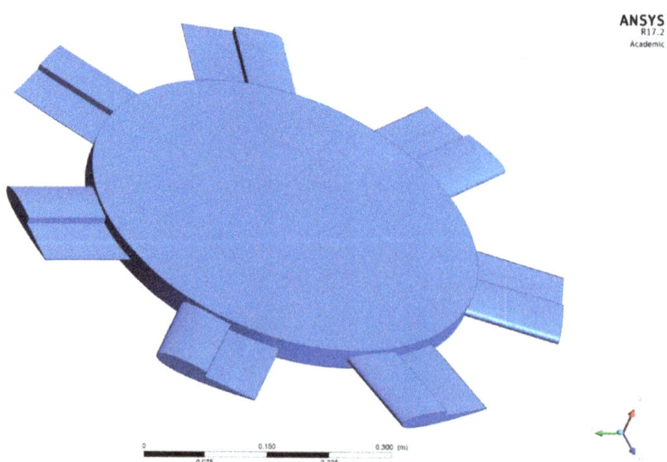

Figure 46. A typical frontal view-based representation of second stepped-back airfoil-based Wells turbine (advanced case-2).

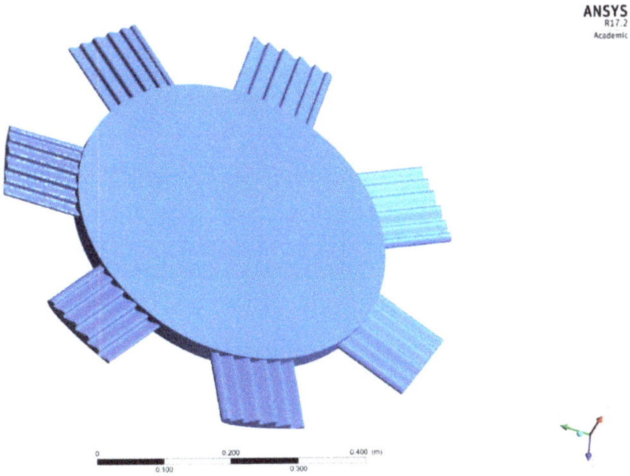

Figure 47. A typical frontal view-based representation of third zigzagged-cum-stepped-back airfoil-based Wells turbine (advanced Case-3).

After the successful completions of computations on all three additional advanced cases, the torques are captured and listed in Table 8.

From Table 9, it is clearly understood that advanced case-2 is capable of enhancing the hydro-power by 15.19 percent as compared to the conventional case-4.

Table 9. Comprehensive hydro-power outcomes of all the three advanced cases.

Number of Special Cases	Description of Special Cases	Torque (Nm)	Observations
Base	Case-IV normal cases	1279.2341	Not applicable
Case-1	Step-1	172.946	Torque is decreased by the percentage of 86.48
Case-2	Step-2	1473.54	Torque is increased by the percentage of 15.19
Case-3	Serration at top-up approach	1137.64	Torque is decreased by the percentage of 11.07

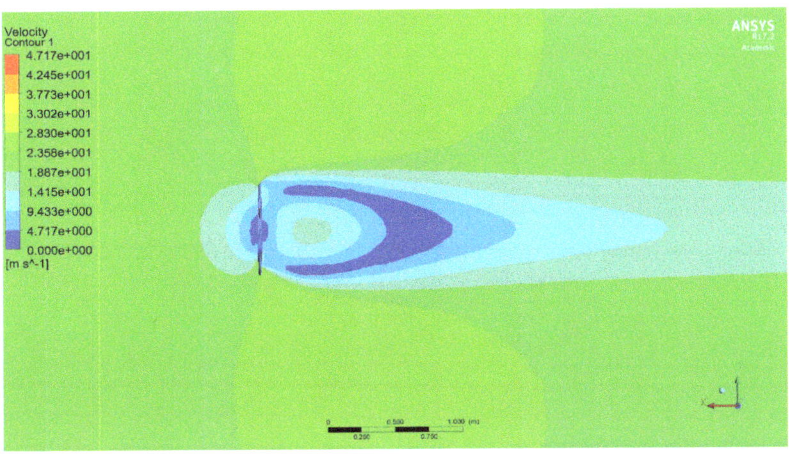

Figure 48. Hydrodynamic velocity variations in and over the Case-3-based Wells turbine.

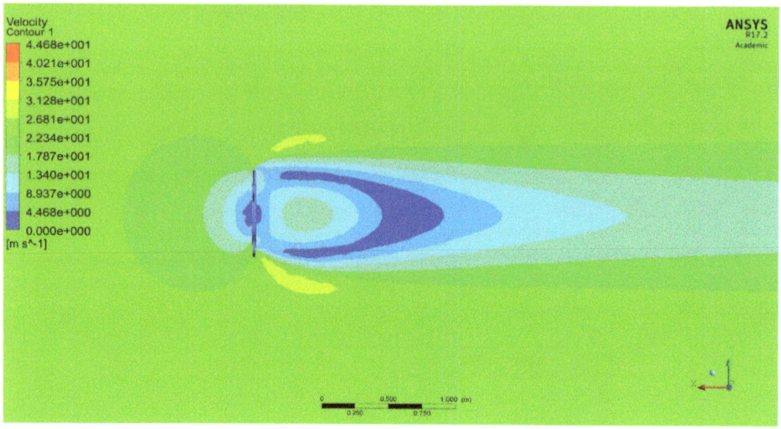

Figure 49. Hydrodynamic velocity variations in and over the Case-2-based Wells turbine.

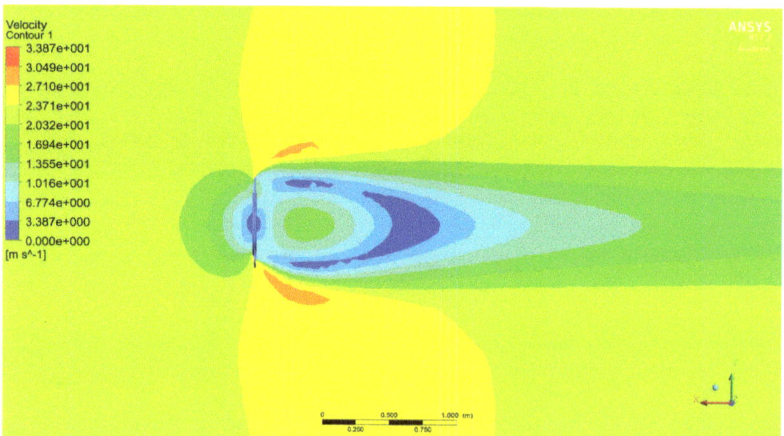

Figure 50. Hydrodynamic velocity variations in and over the Case-1-based Wells turbine.

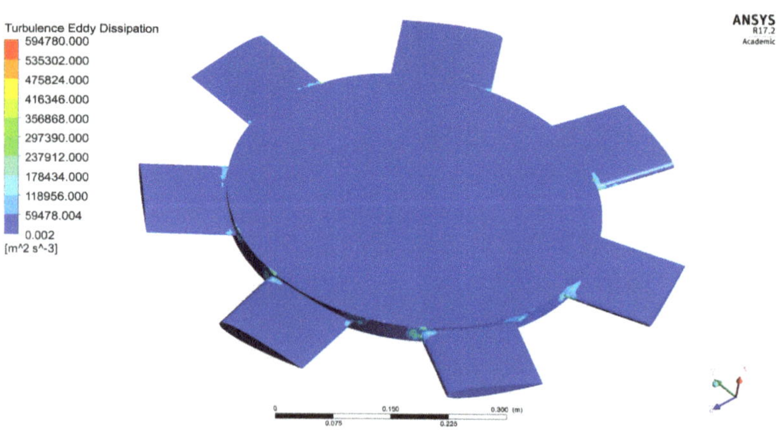

Figure 51. Turbulence eddy dissipation variations in base case.

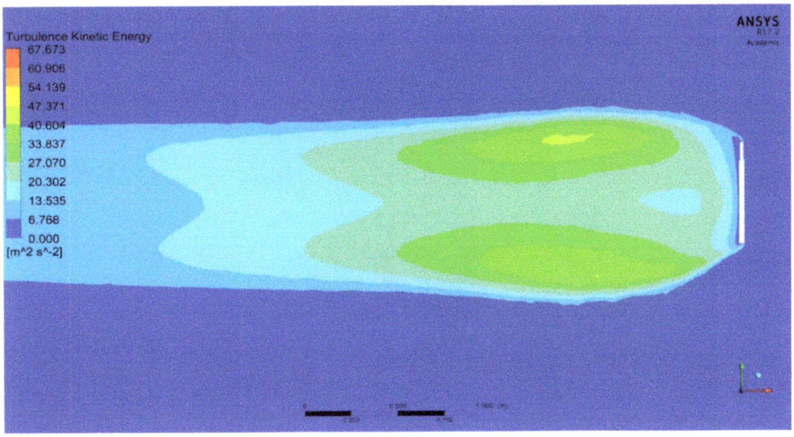

Figure 52. Turbulence kinetic energy variations in base case.

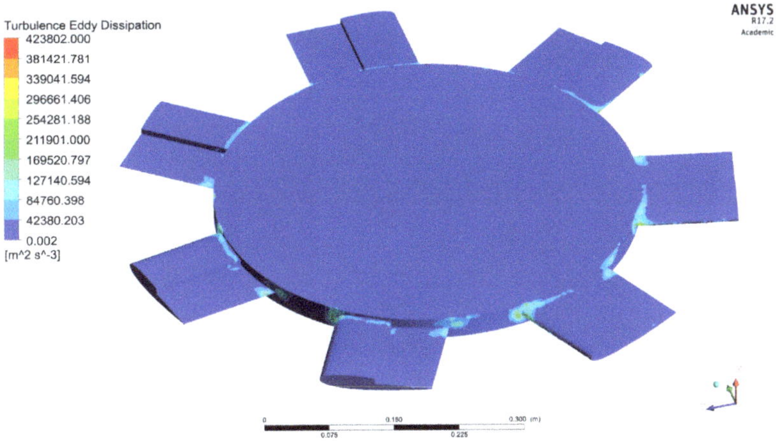

Figure 53. Turbulence Eddy Dissipation variations in stepped-2 case.

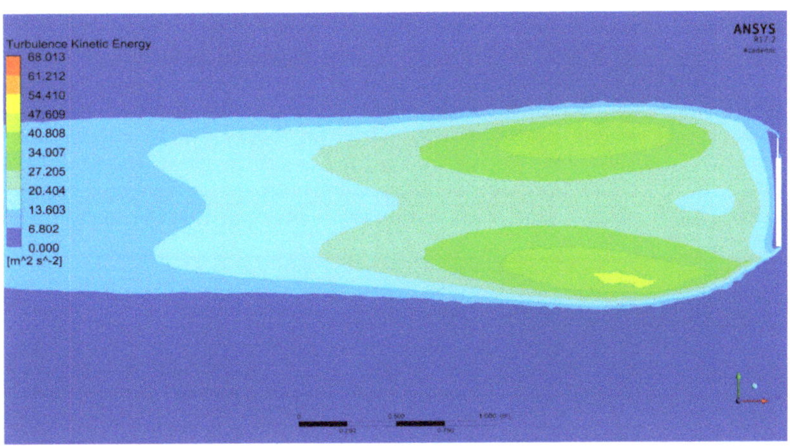

Figure 54. Turbulence kinetic energy variations in stepped-2 case.

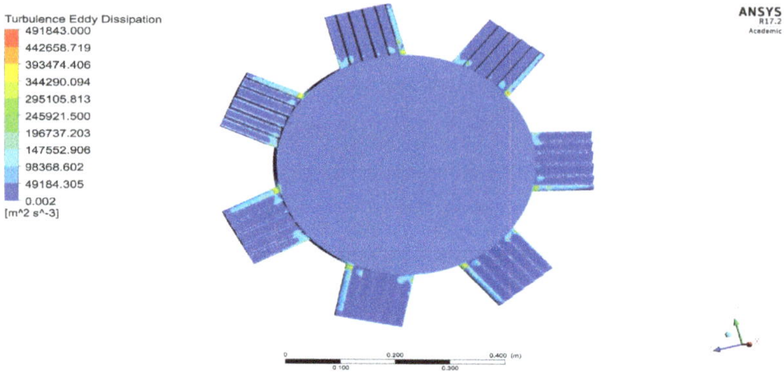

Figure 55. Turbulence eddy dissipation variations in serration at top-up approach-based case.

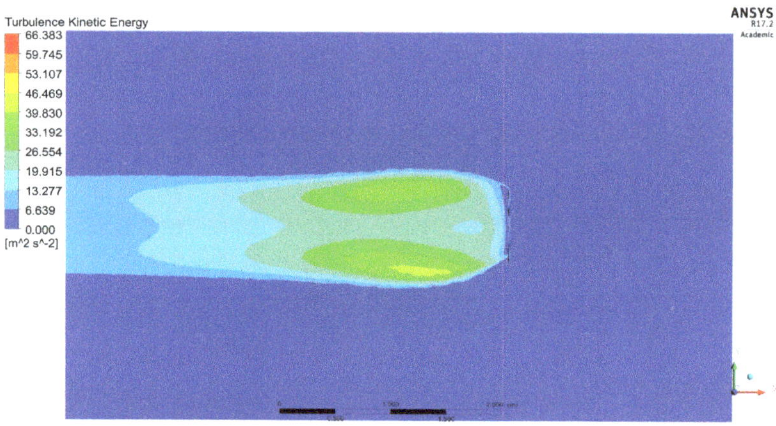

Figure 56. Turbulence kinetic energy variations in serration at top-up approach-based case.

3.2. Computational Structural Analyses

Computational structural analyses are performed using the boundary conditions that are mentioned above. In this analysis, total deformation, equivalent elastic strain, equivalent stress, shear stress, normal stress, stress intensity, and strain energy for various materials are obtained to validate the material which can withstand stress, strain, and deformation. The CASE-IV-based Wells turbine model is shortlisted as base computational model for this structural analysis, in which the inlet velocity of hydrodynamic flow simulation is analyzed at 23 m/s. Through the help of one-way coupling-based fluid–structure interaction-based advanced approaches, the aforesaid computational structural analyses are carried out. The typical front view-based structural outcomes such as stress intensity, shear stress, strain energy, normal stress, equivalent stress, equivalent elastic strain, and total deformation are revealed in Figures 57–63.

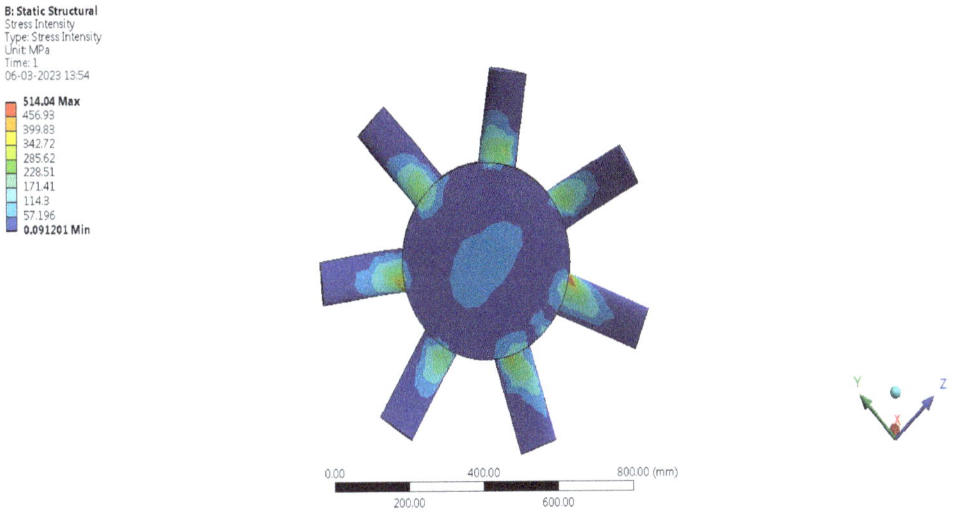

Figure 57. Stress intensity of Wells turbine of the material epoxy carbon woven pre-230.

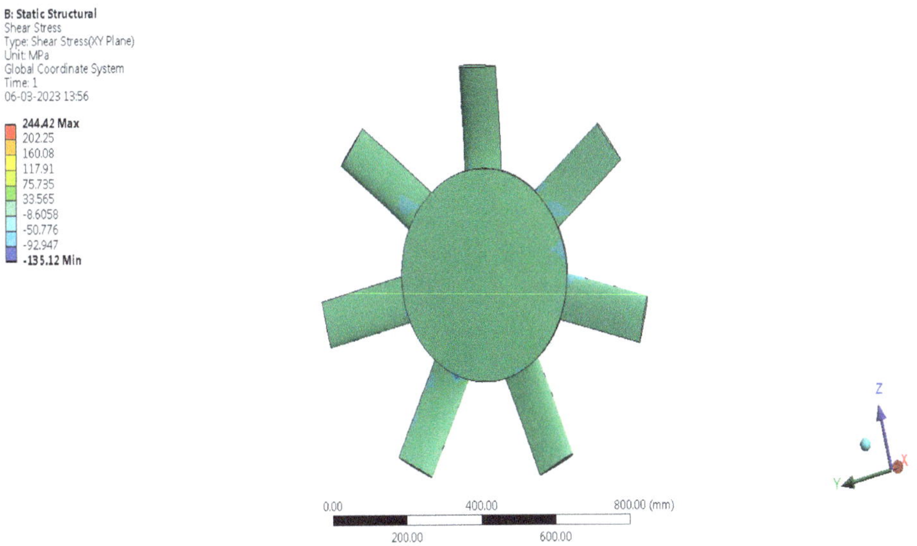

Figure 58. Shear stress of Wells turbine of the material epoxy carbon woven pre-230.

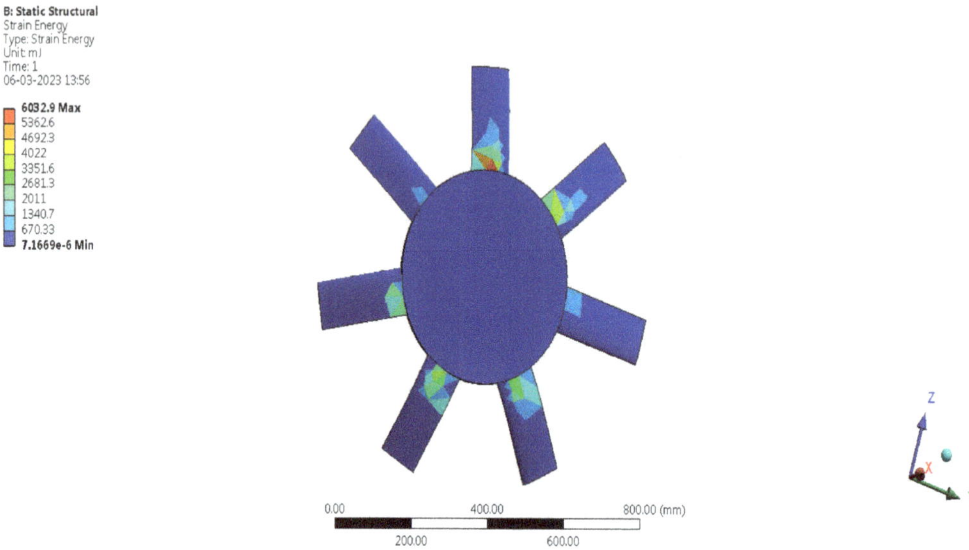

Figure 59. Strain energy of Wells turbine of the material epoxy carbon woven pre-230.

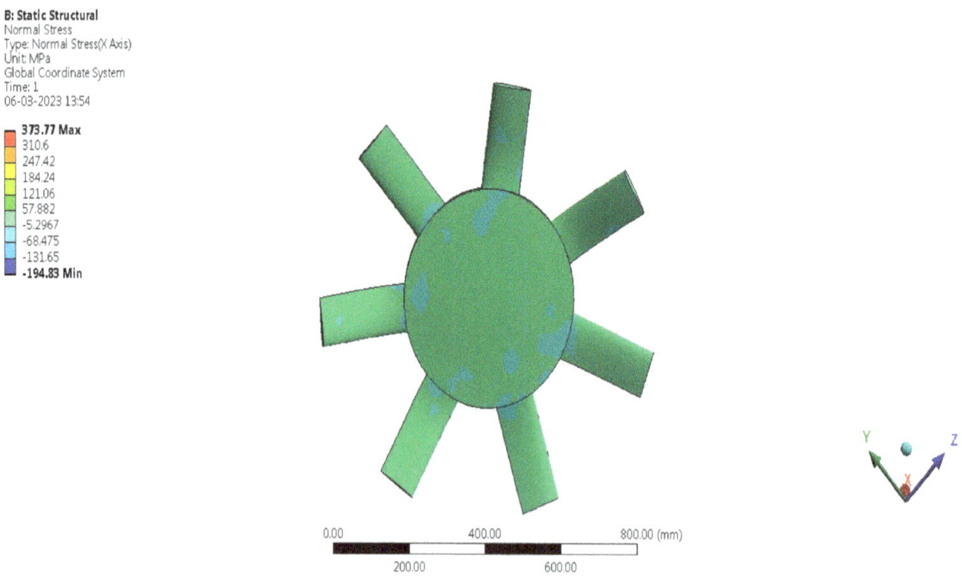

Figure 60. Normal stress of Wells turbine of the material epoxy carbon woven pre-230.

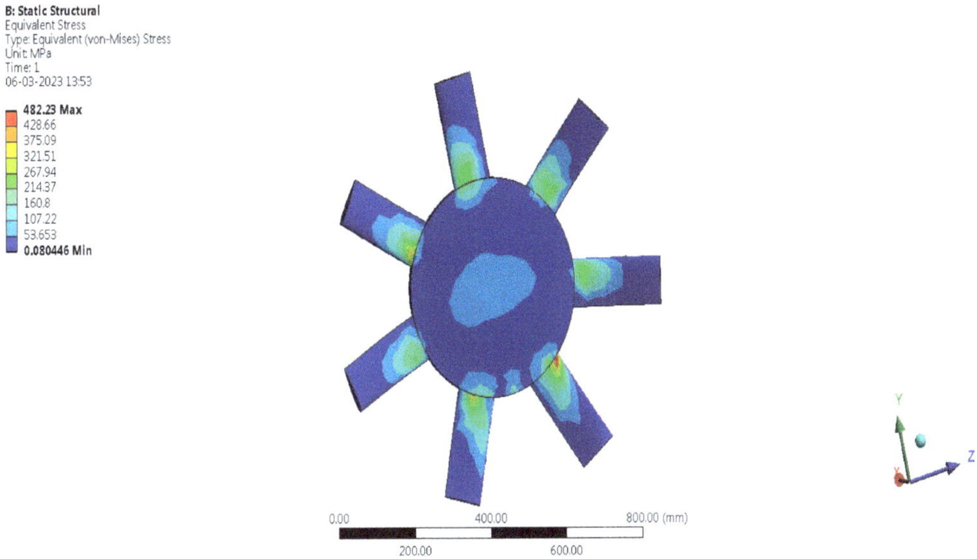

Figure 61. Equivalent stress of Wells turbine of the material epoxy carbon woven pre-230.

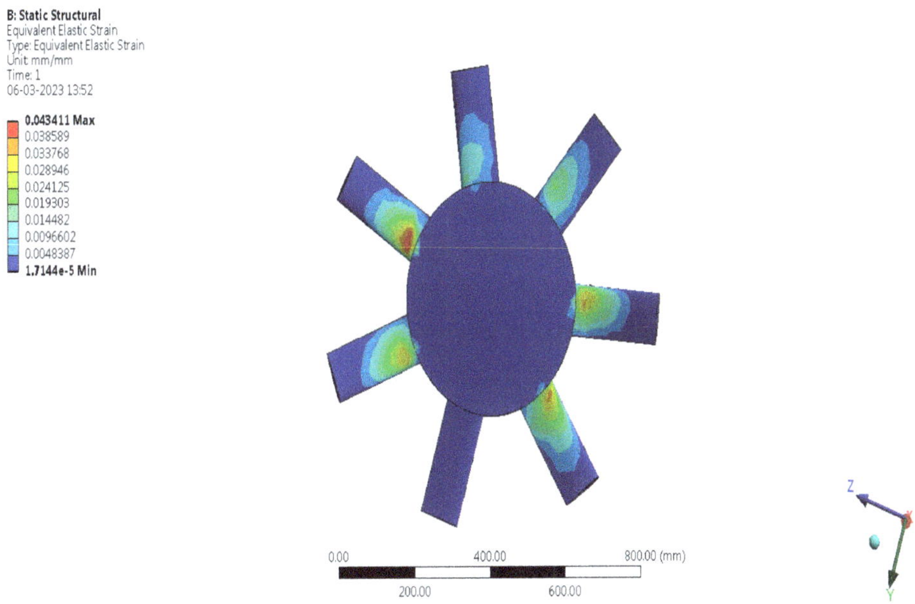

Figure 62. Equivalent elastic strain of Wells turbine of the material epoxy carbon woven pre-230.

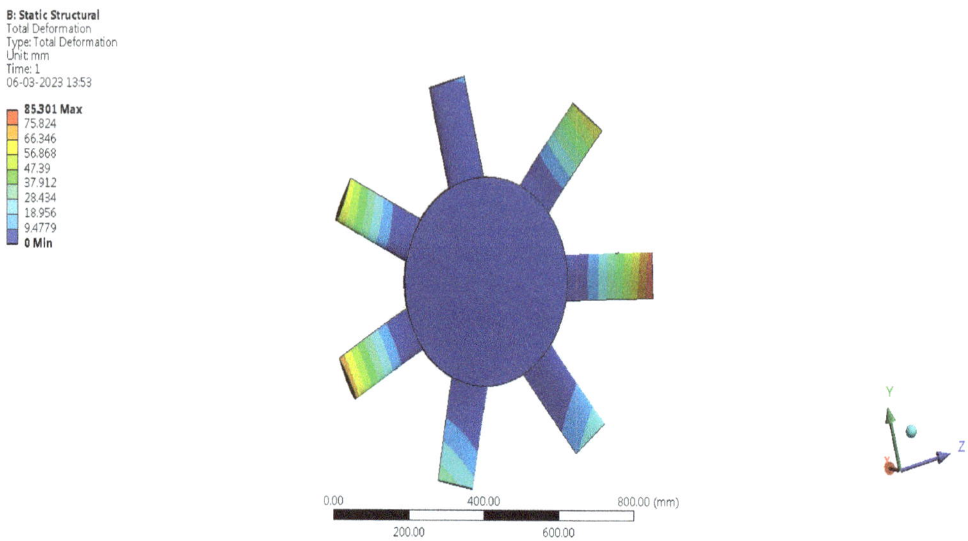

Figure 63. Total deformation of Wells turbine of the material epoxy carbon woven pre-230.

The structural outcomes are carefully monitored, wherein the maximum and minimum induced regions are noted. The attained outcomes of maximum and minimum occurred regions are exactly fulfilled with the theoretical concepts of structural physics. In particular, the stress factors are reacted maximally at the fixed end, and so the deformation is maximally reacted at the free end of the Wells turbine. Since the basic structural physics is confirmed, the same computational structural analysis is extended for other lightweight materials such as alloys, CFRP composites, and GFRP composites. The lightweight alloys-based structural outcomes are computed, and so the comprehensive data are listed and shown in Figures 64–70.

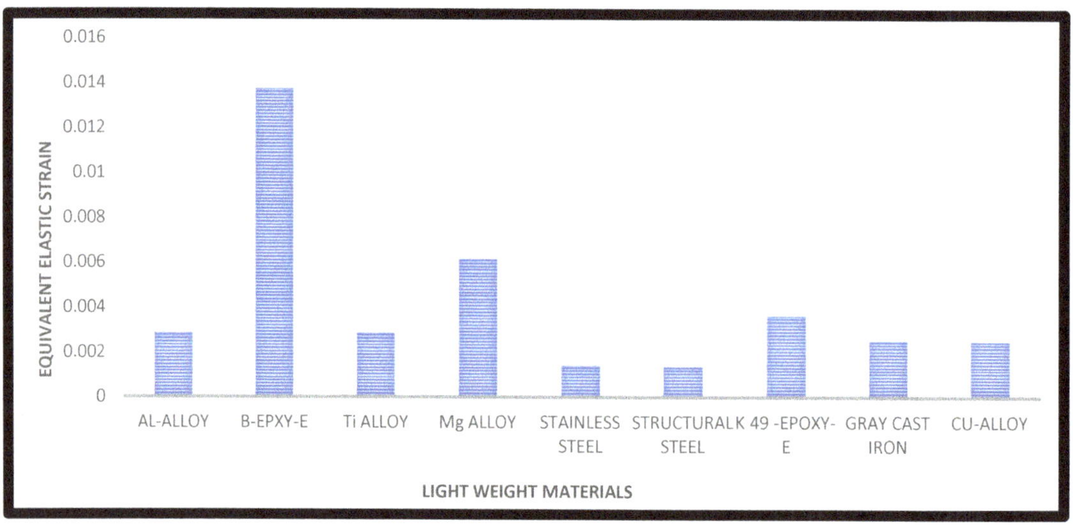

Figure 64. Comprehensive outcomes of equivalent elastic strain for various lightweight alloys and unique composites.

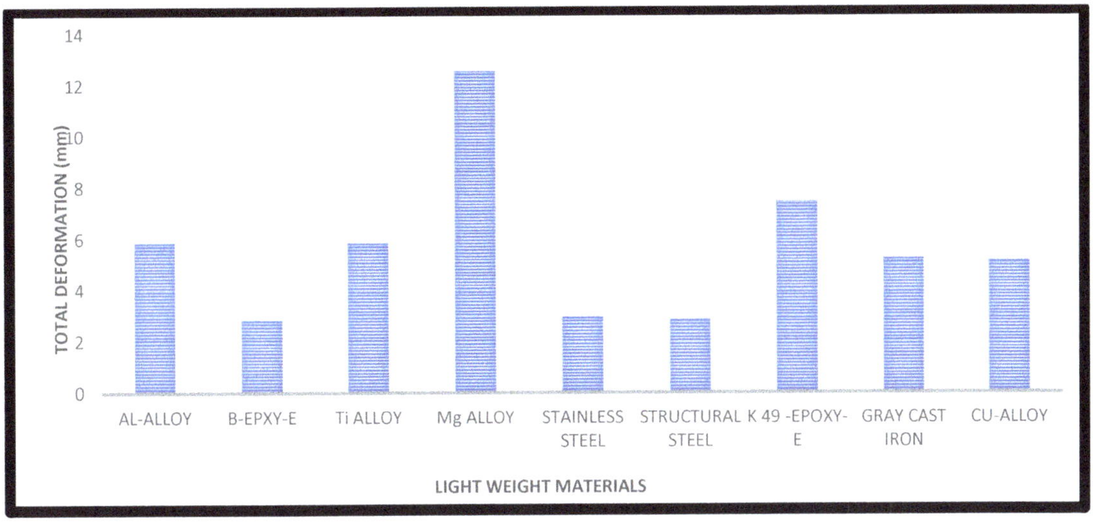

Figure 65. Comprehensive outcomes of total deformation for various lightweight alloys and unique composites.

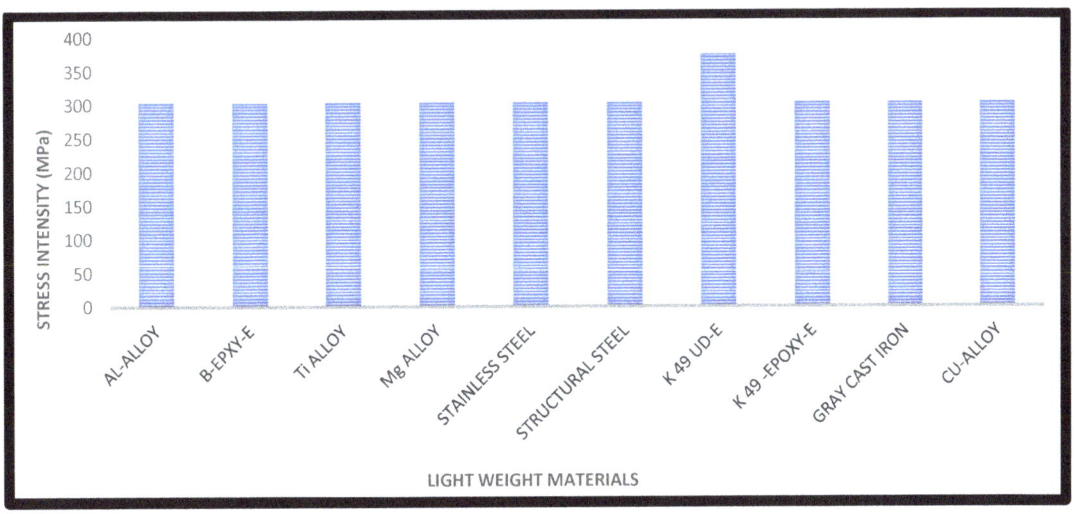

Figure 66. Comprehensive outcomes of stress intensity for various lightweight alloys and unique composites.

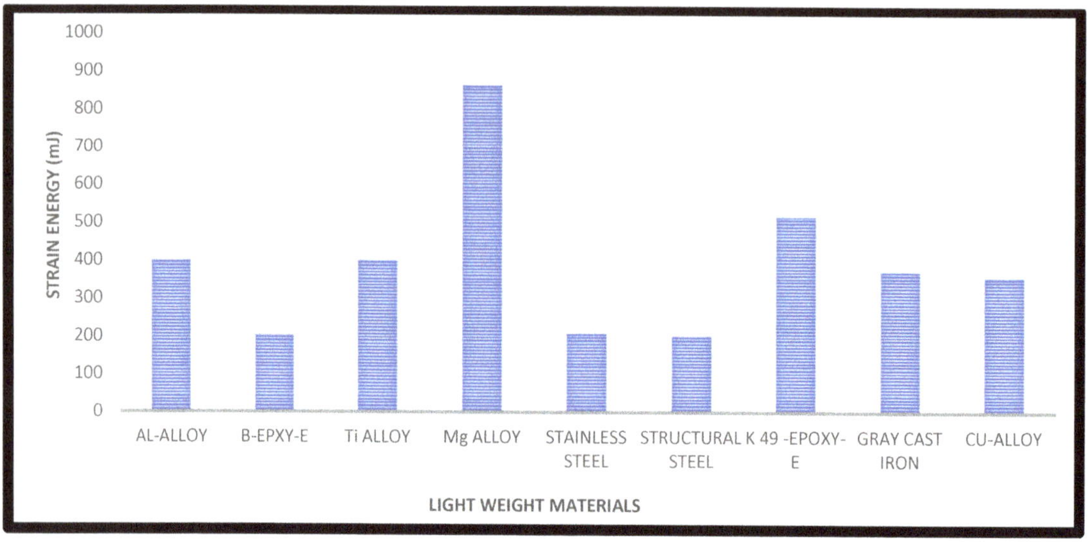

Figure 67. Comprehensive outcomes of strain energy for various lightweight alloys and unique composites.

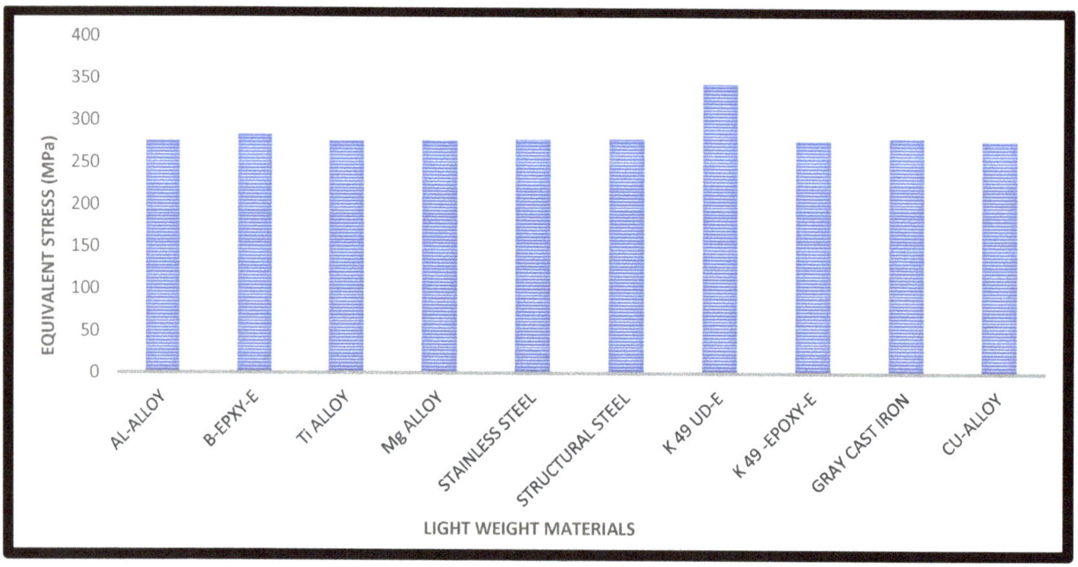

Figure 68. Comprehensive outcomes of equivalent stress for various lightweight alloys and unique composites.

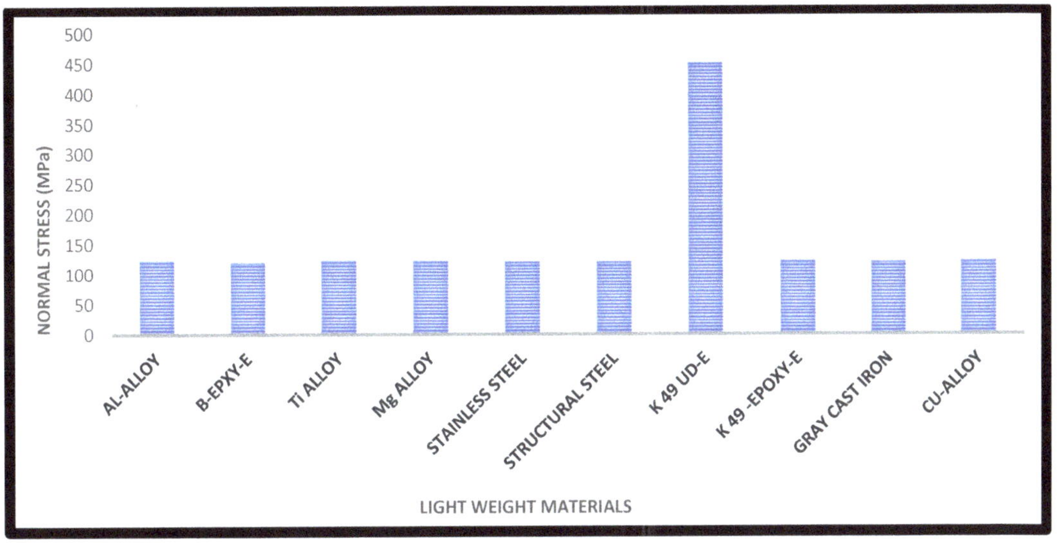

Figure 69. Comprehensive outcomes of normal stress for various lightweight alloys and unique composites.

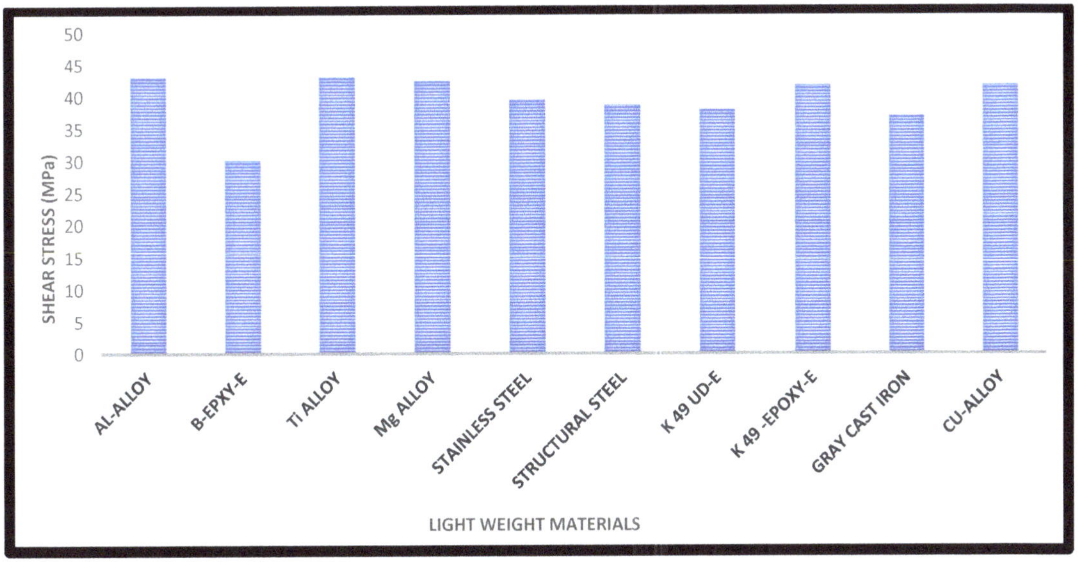

Figure 70. Comprehensive outcomes of shear stress for various lightweight alloys and unique composites.

Figure 64 shows the equivalent elastic strain results of all lightweight materials except Kevlar-49-Epoxy because it has more equivalent elastic strain when compared to other materials, which is shown in the above bar chart. The equivalent elastic strain obtained from Kevlar-49-Epoxy is 0.055941 at the inlet velocity of 23 m/s.

Figure 65 depicts the total deformation results of all lightweight materials except Kevlar-49-Epoxy because it is highly deformed when compared to other materials, which is shown in the above bar chart. The total deformation obtained from Kevlar-49-Epoxy is 105.82 mm at the inlet velocity of 23 m/s.

The stress intensity outcomes of all the lightweight alloys and few unique composites are systematically revealed in Figure 66. With the consideration of low-stress-induced level, the boron fiber and epoxy resin-associated composite performs well.

Figure 67 shows the strain energy results of all lightweight materials except Kevlar-49-Epoxy because it has more strain energy when compared to other materials, which is shown in the above graph. The strain energy obtained from Kevlar-49-Epoxy is 7402.7 mJ at the inlet velocity of 23 m/s.

The equivalent stress-, normal stress-, and shear stress-based structural outcomes are computed for the shortlisted lightweight alloys and few unique composites. The respective comprehensive outcomes are revealed in Figures 68–70. Additionally, the deformed magnitudes of the imposed lightweight materials based wells turbine are systematically revealed in Figure 71. Since the load and supports are common for all the materials, the low-reacting material is superior to provide a high lifetime. Henceforth, the boron fiber and epoxy resin-associated composite reacts in a less effective manner, and so the same lightweight material is chosen as the best material.

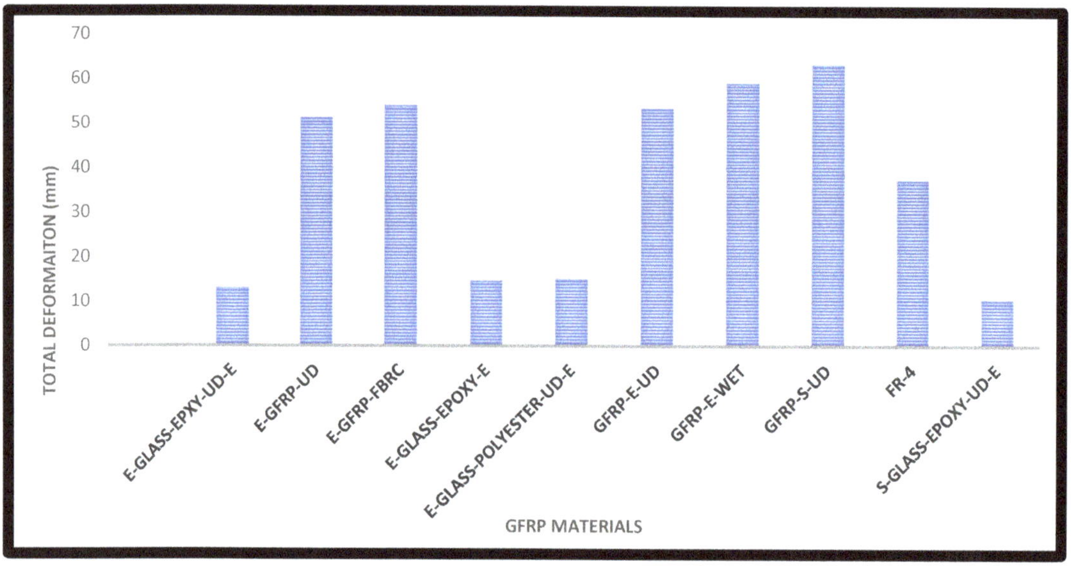

Figure 71. Comprehensive outcomes of total deformation for various GFRP composites.

After the successful completion of first case of comprehensive structural studies, the second comprehensive structural investigations are computed for various lightweight GFRP composites. The imposed GFRP composites are E-GFRP-UD-, E-GFRP-Fabric-, E-GFRP-UD-Polyester-, E-GFRP-Wet-, S-GFRP-UD-, and FR-4-GFRP-based composites and their associates' productions. The comprehensive structural outcomes of the third case that contains GFRP-based outcomes are revealed in Figures 72–77.

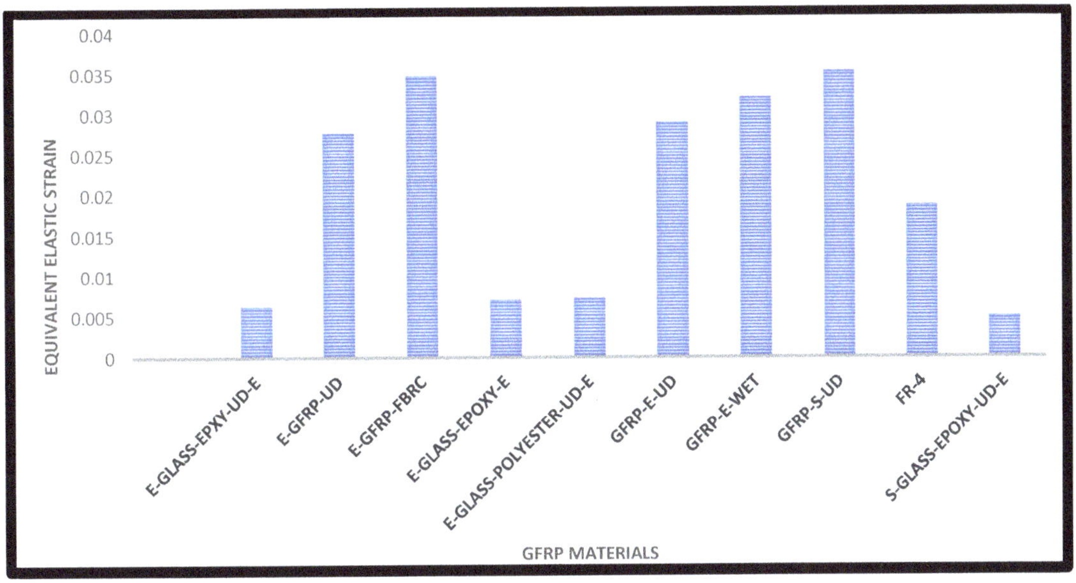

Figure 72. Comprehensive outcomes of equivalent elastic strain for various GFRP composites.

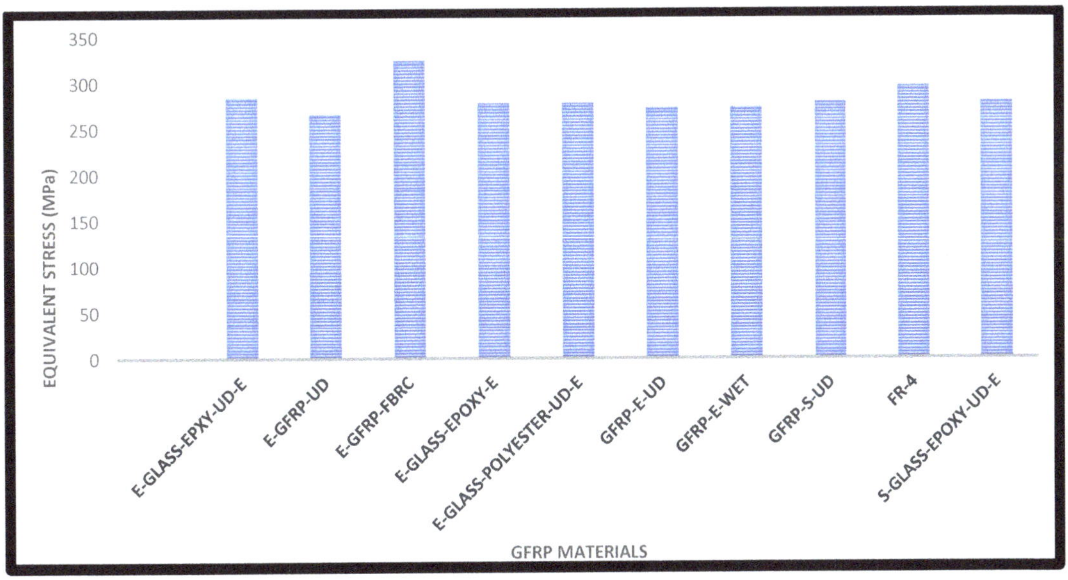

Figure 73. Comprehensive outcomes of equivalent stress for various GFRP composites.

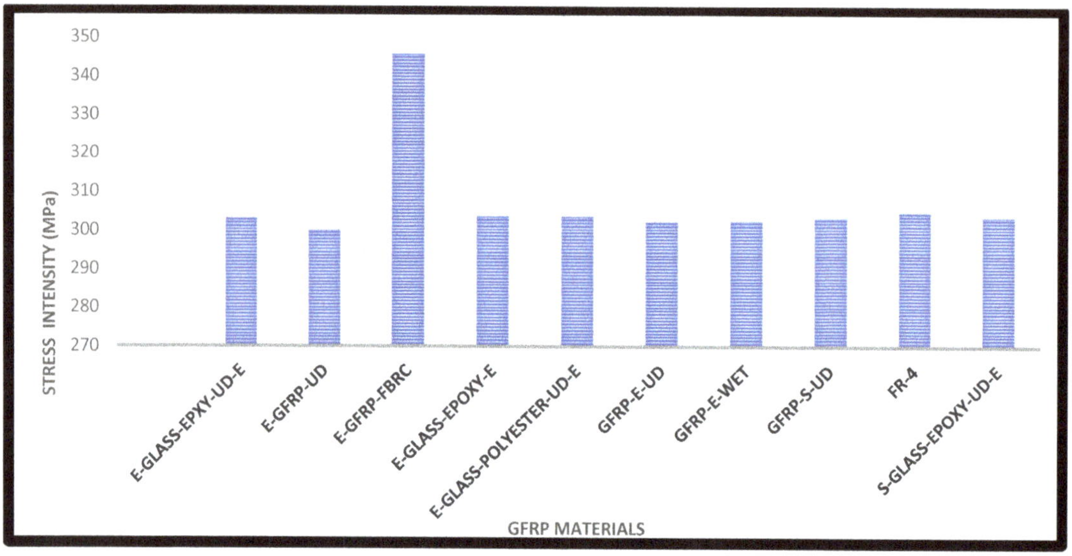

Figure 74. Comprehensive outcomes of stress intensity for various GFRP composites.

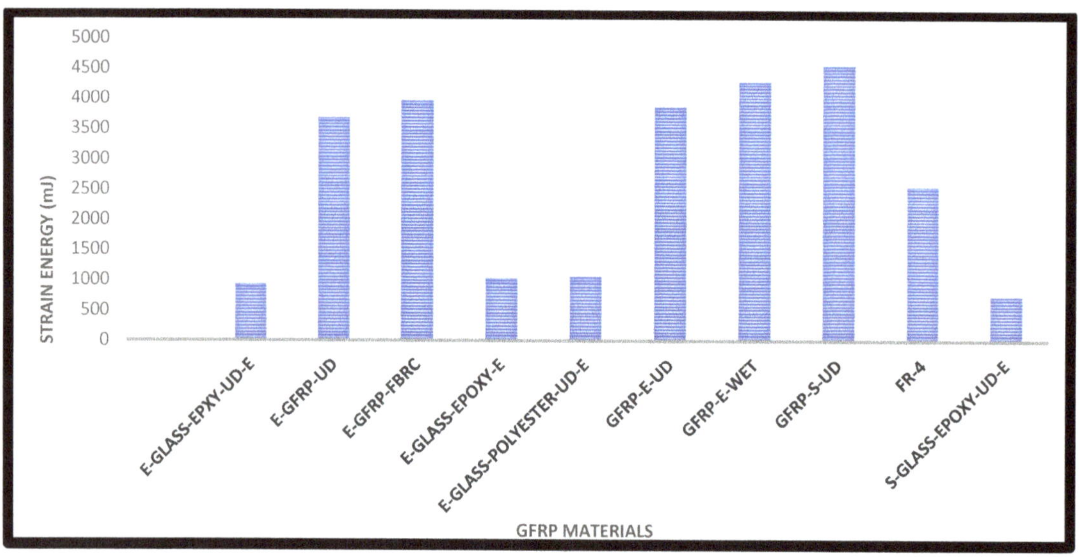

Figure 75. Comprehensive outcomes of strain energy for various GFRP composites.

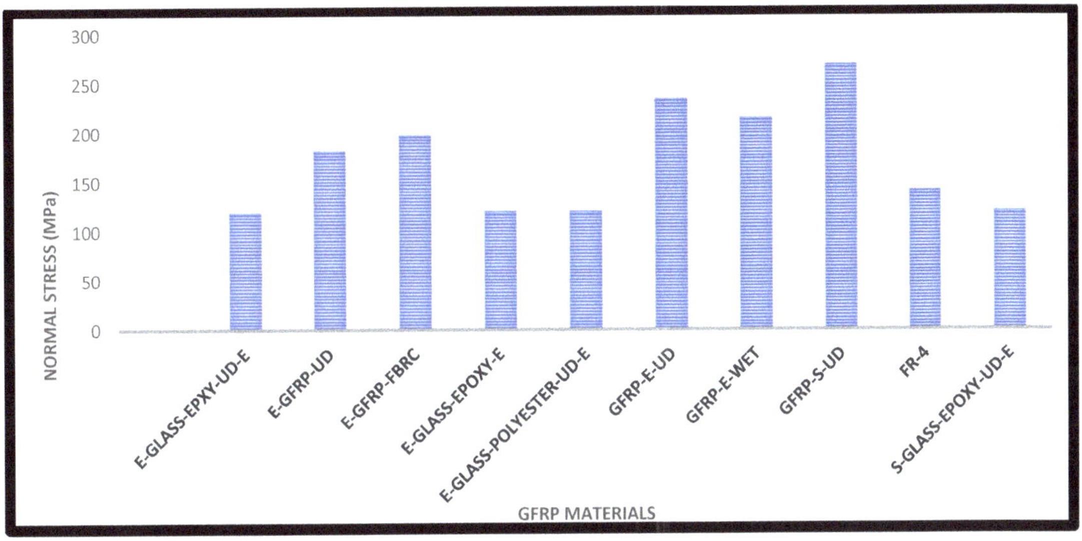

Figure 76. Comprehensive outcomes of normal stress for various GFRP composites.

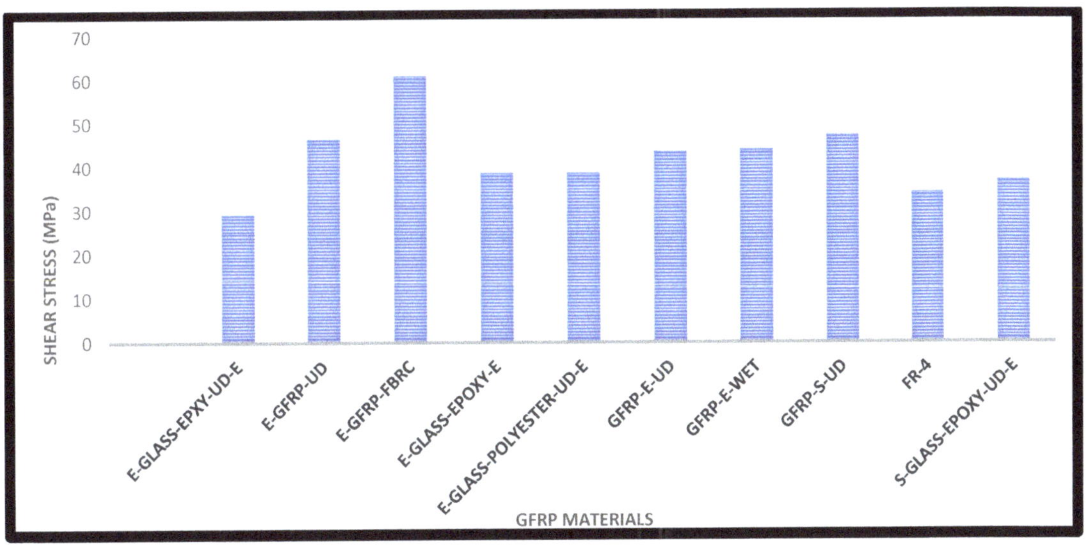

Figure 77. Comprehensive outcomes of shear stress for various GFRP composites.

From Figures 72–77, it is observed that the S-GFRP-UD-Epoxy-based composite reacts in a less effective manner than the other GFRP composites. Henceforth, the same S-GFRP-UD-Epoxy-based composite is chosen as the best performer. After the successful completion of the second case of comprehensive structural studies on GFRP composites, the third comprehensive structural investigations are computed for various lightweight CFRP composites. The comprehensive structural outcomes of the third case that contains CFRP-based outcomes are revealed in Figures 78–84.

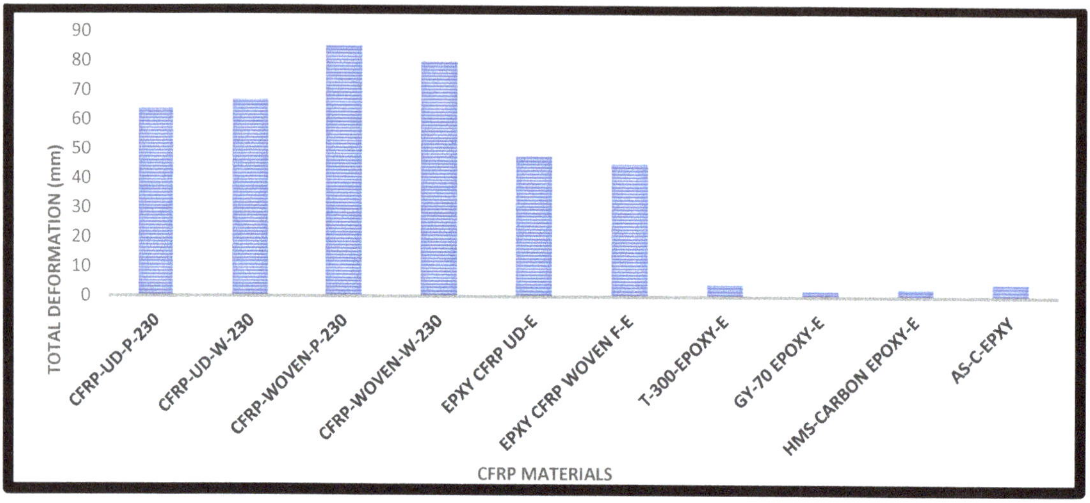

Figure 78. Comprehensive outcomes of total deformation for various CFRP composites.

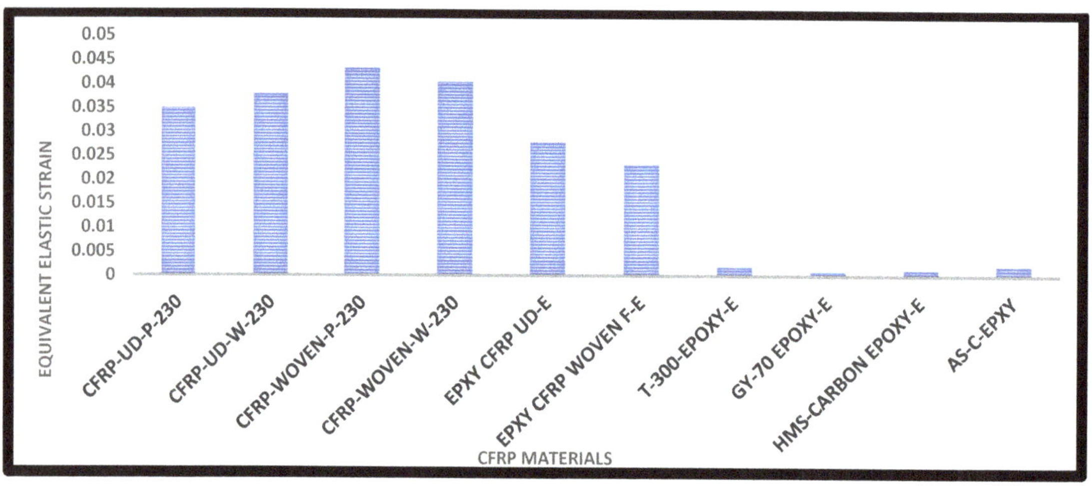

Figure 79. Comprehensive outcomes of equivalent elastic strain for various CFRP composites.

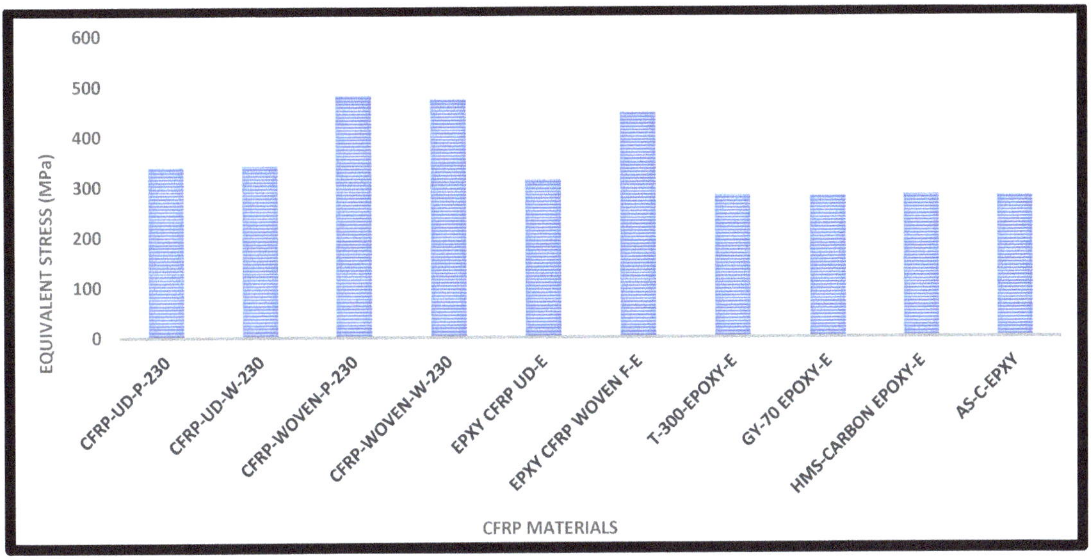

Figure 80. Comprehensive outcomes of equivalent stress for various CFRP composites.

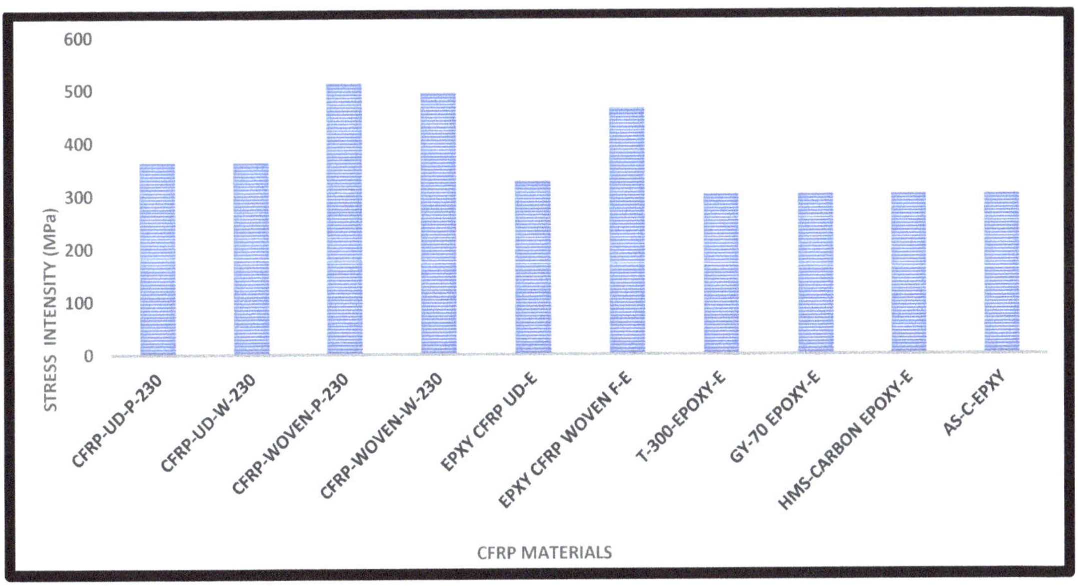

Figure 81. Comprehensive outcomes of stress intensity for various CFRP composites.

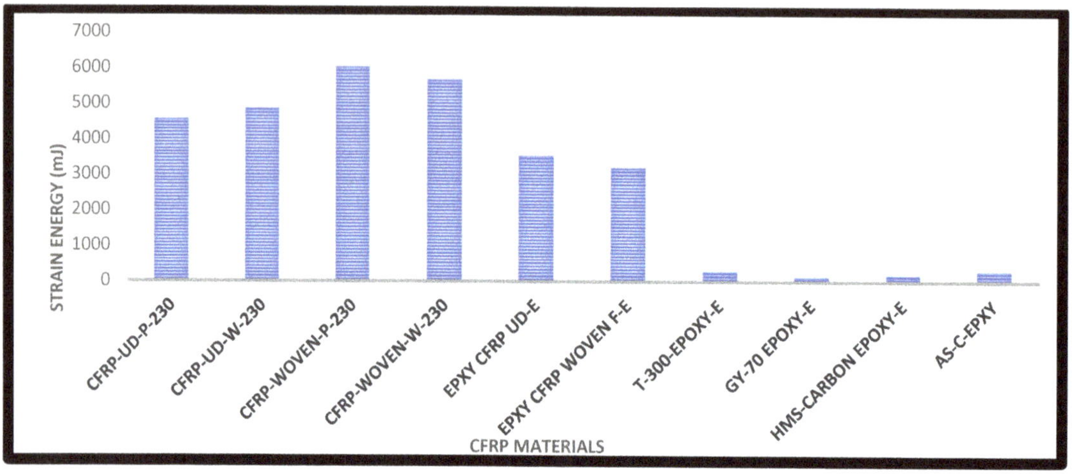

Figure 82. Comprehensive outcomes of strain energy for various CFRP composites.

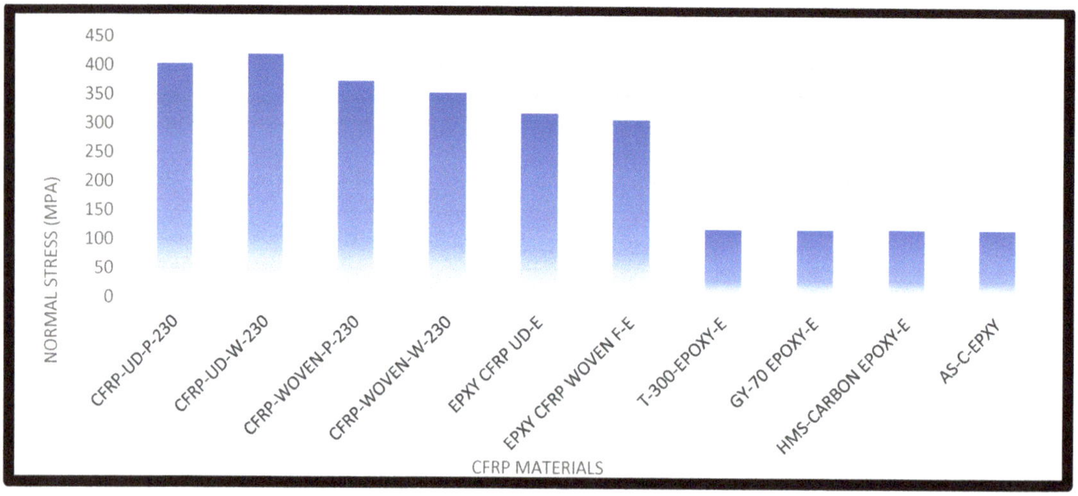

Figure 83. Comprehensive outcomes of normal stress for various CFRP composites.

Figures 78–84 depict the graphical representation of structural analyses of the Wells turbine at the inlet velocity of 23 m/s (deep water velocity), which comprises total deformation, equivalent elastic strain, equivalent stress, stress intensity, strain energy, normal stress, and shear stress with the respective CFRP materials. By comparing the above-shown CFRP materials, T-300 epoxy- and GY-70 epoxy-based composites in particular performed better than other materials. Since the load-resisting behavior of these shortlisted lightweight materials is high, the reaction is generated in a less effective manner than other materials.

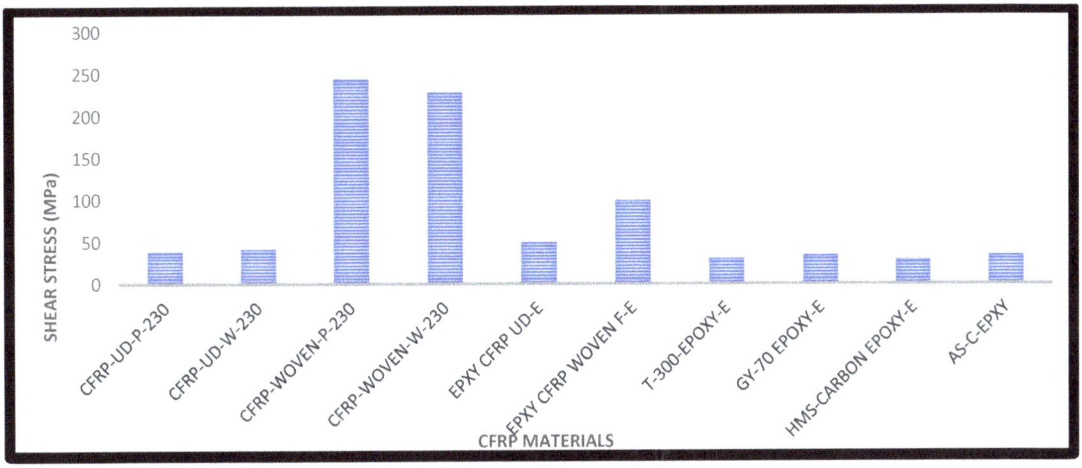

Figure 84. Comprehensive outcomes of shear stress for various CFRP composites.

3.3. Estimation of Power from Energy Harvester

Important steps in this evaluation include picking and calculating PVEH patches. Several common and hybrid materials were analyzed in ANSYS. In the first step, the authors calculated the power of various materials. Then, the computations were carried out in two stages: an intermediate power estimate and a final power estimate. Equation (12) necessitates the first iteration of intermediate power extraction.

$$P_{\text{Intermediate}}^{\text{Hybrid Wells Turbine}} = d_{\text{lwm}}^2 w^2 f^2 \frac{18 * T_{PL}}{\left[(W_{L-WT}) * (L_{PL})^2 * ([t_{WT}] + [T_{PL}])^4\right]} \frac{\rho_{\text{lwm}}}{[1 + f * \varepsilon * (\rho_{\text{lwm}})]} \tag{12}$$

$$P_{\text{Final}}^{\text{Hybrid Wells Turbine}} = \left(P_{\text{Intermediate}}^{\text{Hybrid Wells Turbine}}\right)\left(\frac{(0.44 * L)^5}{36} - \frac{L(0.44 * L)^4}{6} + \frac{5L^2(0.44 * L)^3}{12} - \frac{L^3(0.44 * L)^2}{2}\frac{L^4(0.44 * L)}{4}\right) \tag{13}$$

For the purpose of determining the power output of the PVEH patches, the previously described mathematical procedures (Equations (12) and (13)) must be utilized. Table 10 provides a summary of the findings that were acquired through the utilization of a variety of lightweight materials to harvest energy from the rotor span.

Table 10. Estimated energy outcome on 100% of the blade span.

Materials	Energy Extracted (W)	Materials	Energy Extracted (W)
KFRP-49-UD	3375.524253	GFRP-E-Wn	1705.544864
CFRP-Wn-wet	3849.998201	GFRP-E-UD	2813.393605
CFRP-wet-UD	5148.895683	GFRP-E-wet	2250.131784
CFRP-UD-Pg	5294.984336	GFRP-S-UD	2960.020325
CFRP-Wn-Pg	3978.721358	GFRP-FR-4-Wn	1545.139361

Table 11 shows the total output power of CASE-IV for the inlet velocity of 12.10 m/s using different lightweight materials by placing PVEH patches on the seven blades of the Wells turbine.

Table 11. All-inclusive forecast of the Wells turbine's total output power from the hybrid energy scheme.

Lightweight Materials	Energy Extracted Through PVEH Patches (W)							CFD 12.10 m/s	Resultant Renewable Power (kW)
	Blade-1	Blade-2	Blade-3	Blade-4	Blade-5	Blade-6	Blade-7		
GFRP-E-Wn	1705	1705	1705	1705	1705	1705	1705	38,139.29	50.07429
CFRP-wet-UD	5148	5148	5148	5148	5148	5148	5148		74.17529
CFRP-UD-Pg	5295	5295	5295	5295	5295	5295	5295		75.20429
CFRP-Wn-Pg	3978	3978	3978	3978	3978	3978	3978		65.98529
CFRP-Wn-wet	3849	3849	3849	3849	3849	3849	3849		65.08229
GFRP-E-UD	2813	2813	2813	2813	2813	2813	2813		57.83029
GFRP-E-wet	2250	2250	2250	2250	2250	2250	2250		53.88929
GFRP-S-UD	2960	2960	2960	2960	2960	2960	2960		58.58929
GFRP-FR-4-Wn	1545	1545	1545	1545	1545	1545	1545		48.95429
KFRP-49-UD	3375	3375	3375	3375	3375	3375	3375		61.76429

4. Conclusions

From this analysis, seven different seven-bladed Wells turbines are designed. Then, CFD analysis for seven different Wells turbine is performed with four different velocities such as 0.34 m/s, 1.54 m/s, 12.10 m/s, and 23 m/s. Through CFD investigation, pressure, velocity, force, and torque are obtained. Using the obtained torque, CASE-IV has a torque of 1729.2 Nm at 23 m/s, and CASE-VI has a torque of 2393.45 at 23 m/s, which are concluded as the optimized Wells turbines. Then, CASE-IV is adopted for the comprehensive structural analyses. The computational structural analyses on the Wells turbine use different lightweight alloys, CFRP, GFRP materials, and few unique composites. From the help of comprehensive structural outcomes, the following observations are obtained: Boron Epoxy is the strongest element because it shows the minimum deformation, stress, and strain compared to other lightweight materials. Compared to CFRP materials, GY-70-Epoxy is considered the strongest element as it has low deformation, stress, and strain. By comparing the GFRP materials, S-Glass-Epoxy-UD is considered the strongest material because it has low deformation, stress, and strain. The shortlisted lightweight materials are used for the hybrid energy calculation. The PVEH-based hybrid energy extraction concept has emerged and has been imposed in this work for various shortlisted lightweight materials. The hydrodynamic medium-based energy extraction from the Wells turbine is listed in Table 10. The hybrid energy concept-based important outcomes are listed in Table 11. From Tables 10 and 11, CASE-VI is a potential case for the enhancement power from hydrodynamic medium as compared to other cases. Finally, a suitable material for high-energy extractions through a hybrid energy mechanism is also found.

Author Contributions: Conceptualization, V.R., J.T. and H.P.J.; methodology, V.R., J.T. and H.P.J.; software, V.R. and P.R.; validation, S.K.M. and H.A.Z.A.-b.; formal analysis, S.K.M., V.R. and P.R.; investigation, S.K.M., R.R., V.R. and P.R.; resources, V.R., R.R., P.R., and S.Z.; data curation, S.K.M., V.R. and P.R.; writing—original draft preparation, S.K.M., P.R. and H.A.Z.A.-b.; writing—review and editing, V.R., P.R., J.T. and S.Z.; visualization, B.S.A., R.R. and S.Z.; supervision, S.K.M., V.R., P.R. and S.Z.; project administration, S.K.M., V.R. and P.R.; funding acquisition, S.Z. and P.R. All authors have read and agreed to the published version of the manuscript.

Funding: This work was supported by a Universiti Sains Malaysia Bridging GRANT under Project No: 304/PAERO/6316608.

Data Availability Statement: The authors confirm that the data supporting the findings of this study are available within the article.

Acknowledgments: This study has been performed based on the authors' interests. It uses computational and experimental resources from the design and simulation laboratory and the Kumaraguru College of Technology (KCT) aerodynamics laboratory.

Conflicts of Interest: The authors declare no conflict of interest.

Nomenclatures

Design Variables	Explanations
C^{B}_{WT}	Chord length of blade in Wells turbine
$A_{\text{Wells Turbine}}$	Frontal area of the Wells turbine
r_{blade}	Radius of blade in the Wells turbine
$\text{Circum}_{\text{Wells Turbine}}$	Circumference of the Wells turbine
T^{B}_{WT}	Thickness of blade in the Wells turbine
r_{hub}	Radius of hub in the Wells turbine
g	Gravitational force
m_{WT}	Mass of the Wells turbine
C_{WT}	Damping coefficient of the Wells turbine
K_{WT}	Stiffness of the Wells turbine
$F(t)_{\text{Hydrodynamic}}$	Hydrodynamic forces acting on the Wells turbine
u	Displacement
$P^{\text{Hybrid Wells Turbine}}_{\text{Intermediate}}$	Intermediate generated theoretical power from the Wells turbine
$P^{\text{Hybrid Wells Turbine}}_{\text{Final}}$	Final generated theoretical power from the Wells turbine
d_{lwm}	The piezoelectric material constant for PZT is 120.
w	Force acting on the blade of the Wells turbine
f	Natural frequency of the blade of the Wells turbine
W_{L-WT}	Width of the Wells turbine's blade
L_{PL}	Length of the piezoelectric patches
t_{WT}	Thickness of the Wells turbine's blade
T_{PL}	Thickness of the piezoelectric patches
ε	Dielectric constant of the material
ρ_{lwm}	Density of the base light materials

References

1. Das, T.K.; Kerikous, E.; Venkatesan, N.; Janiga, G.; Thévenin, D.; Samad, A. Performance improvement of a Wells turbine through an automated optimization technique. *Energy Convers. Manag. X* **2022**, *16*, 100285. [CrossRef]
2. Shehata, A.S.; Saqr, K.M.; Xiao, Q.; Shehadeh, M.F.; Day, A. Performance analysis of wells turbine blades using the entropy generation minimization method. *Renew. Energy* **2016**, *86*, 1123–1133. [CrossRef]
3. Barnabei, V.F.; Castorrini, A.; Corsini, A.; Rispoli, F. FSI analysis and simulation of flexible blades in a Wells turbine for wave energy conversion. *E3S Web Conf.* **2020**, *197*, 11008. [CrossRef]
4. Kincaid, K.; MacPhee, D.W. CFD Analysis of stall in a wells turbine. In Proceedings of the ASME 2018 Power Conference POWER 2018, Lake Buena Vista, FL, USA, 24–28 June 2018.
5. Harby, A.H.; Shehata, A.S.; Afify, R.S.; Hanafy, A.A. Experimental investigation for suction slots of wells turbine and shapes of point absorber. *Energy Rep.* **2022**, *8*, 1275–1287. [CrossRef]
6. Gareev, A.; Cooper, P.; Kosasih, P.B. *CFD Analysis of Air Turbines as Power Take-Off Systems in Oscillating Water Column Wave Energy Conversion Plant*; University of Wollongong: Wollongong, NSW, Australia, 2009.
7. Shehata, A.S.; Xiao, Q.; Saqr, K.M.; Alexander, D. Wells turbine for wave energy conversion: A review. *Int. J. Energy Res.* **2016**, *41*, 6–38. [CrossRef]
8. Karthick, N.; Sanjai, M. Optimization of the Wells Turbine by using Slotted blade and Guide Vanes. *Int. J. Creat. Res. Thoughts* **2018**, *6*, 969–978.
9. Suzuki, M.; Arakawa, C. Flowon Blade of Wells Turbine for Wave Power Generation. *J. Vis.* **2006**, *9*, 83–90. [CrossRef]
10. Thakker, A.; Abdulhadi, R. The performance of Wells turbine under bi-directional airflow. *Renew. Energy* **2008**, *33*, 2467–2474. [CrossRef]
11. Torresi, M.; Camporeale, S.M.; Pascazio, G.; Fortunato, B. Fluid Dynamic Analysis of a Low Solidity Wells Turbine. In Proceedings of the 59 Congresso ATI, Genova, Italy, 14–17 October 2004.
12. Torresi, M.; Camporeale, S.M.; Pascazio, G. Performance of a small prototype of a highsolidity wells turbine. In Proceedings of the 7th European Conference on Turbomachinery: Fluid Dynamics and Thermodynamics (ETC, 2007), Athens, Greece, 5–9 March 2007.

13. Dorrell, G.; Hsieh, M.-F. Performance of Wells Turbines For Use In Small-Scale Oscillating Water Columns. Available online: https://onepetro.org/ISOPEIOPEC/proceedings/ISOPE08/All-ISOPE08/ISOPE-I-08-165/10667 (accessed on 6 July 2008).
14. Sharma, S.; Peng, S.-L.; Agrawal, J.; Shukla, R.K.; Le, D.-N. *Data, Engineering and Applications*; Springer Nature: Berlin/Heidelberg, Germany, 2022.
15. Curran, R.; Gato, L.M.C. The energy conversion performance of several types of Wells turbine designs. *Proc. Inst. Mech. Eng. Part A J. Power Energy* **1997**, *211*, 133–145. [CrossRef]
16. Lekube, J.; Garrido, A.; Garrido, I.; Otaola, E.; Maseda, J. Flow Control in Wells Turbines for Harnessing Maximum Wave Power. *Sensors* **2018**, *18*, 535. [CrossRef]
17. Halder, P.; Samad, A.; Thévenin, D. Improved design of a Wells turbine for higher operating range. *Renew. Energy* **2017**, *106*, 122–134. [CrossRef]
18. Watterson, J.K.; Raghunathan, S. Numerical Analysis of Wells Turbine Aerodynamics. *J. Propuls. Power* **1999**, *15*, 253–259. [CrossRef]
19. Torresi, M.; Stefanizzi, M.; Fornarelli, F.; Gurnari, L.; Filianoti, P.G.F.; Camporeale, S.M. Performance characterization of a wells turbine under unsteady flow conditions. In Proceedings of the Second International Conference on Material Science, Smart Structures and Applications: ICMSS-2019, Erode, India, 20–21 November 2019. [CrossRef]
20. Taha, Z.; Sugiyono; Sawada, T. A comparison of computational and experimental results of Wells turbine performance for wave energy conversion. *Appl. Ocean Res.* **2010**, *32*, 83–90. [CrossRef]
21. Samad, A. Effect of guide vane angle on wells turbine performance. In Proceedings of the ASME 2014 Gas Turbine India Conference, GTINDIA, New Delhi, India, 15–17 December 2014. [CrossRef]
22. Valizadeh, R.; Abbaspour, M.; TaeibiRahni, M.; Pour, M.S.; Hulme-Smith, C. Improved Wells Turbine Using a Concave Sectional Profile. *Preprints* **2020**, 2020100045. [CrossRef]
23. Cruz, J. *Ocean Wave Energy: Current Status and Future Prespectives*; Springer: Berlin/Heidelberg, Germany, 2008.
24. Oaly, J.; Frawley, P.; Thakker, A. A 2-D computational fluid dynamics analysis of wells turbine blade profiles in isolated and cascade flow. In Proceedings of the OMAEí02 21st International Conference on Offshore Mechanics and Arctic Engineering, Oslo, Norway, 23–28 June 2002.
25. Kumar, P.M.; Samad, A. Nature-inspired design of a turbine blade harnessing wave energy. *Proc. Inst. Mech. Eng. Part A J. Power Energy* **2019**, *234*, 670–689. [CrossRef]
26. Cambuli, F.; Ghisu, T.; Virdis, I.; Puddu, P. Dynamic interaction between OWC system and Wells turbine: A comparison between CFD and lumped parameter model approaches. *Ocean Eng.* **2019**, *191*, 106459. [CrossRef]
27. Torresi, M.; Camporeale, S.M.; Pascazio, G. Detailed CFD Analysis of the Steady Flow in a Wells Turbine Under Incipient and Deep Stall Conditions. *J. Fluids Eng.* **2009**, *131*, 071103. [CrossRef]
28. Torresi, M.; Camporeale, S.M.; Pascazio, G. Experimental and numerical investigation on the performance of a wells turbine prototype. In Proceedings of the 7th European Wave and Tidal Energy Conference, Porto, Portugal, 11–13 September 2007.
29. Ghisu, T.; Puddu, P.; Cambuli, F. A detailed analysis of the unsteady flow within a Wells turbine. *Proc. Inst. Mech. Eng. Part A J. Power Energy* **2017**, *231*, 197–214. [CrossRef]
30. Torresi, M.; Stefanizzi, M.; Gurnari, L.; Filianoti, P.G.F.; Camporeale, S.M. Experimental characterization of the unsteady performance behavior of a Wells turbine operating at high flow rate coefficients. *E3S Web Conf.* **2020**, *197*, 08009. [CrossRef]
31. Kotb, A.T.M.; Nawar, M.A.A.; Attai, Y.A.; Mohamed, M.H. Performance optimization of a modified Wells turbine for wave energy conversion. *Ocean Eng.* **2023**, *280*, 114849. [CrossRef]
32. Morais, F.J.F.; Carrelhas, A.A.D.; Gato, L.M.C. Biplane-rotor Wells turbine: The influence of solidity, presence of guide vanes and comparison with other configurations. *Energy* **2023**, *276*, 127514. [CrossRef]
33. Stefanizzi, M.; Camporeale, S.M.; Torresi, M. Experimental investigation of a Wells turbine under dynamic stall conditions for wave energy conversion. *Renew. Energy* **2023**, *214*, 369–382. [CrossRef]
34. Zarketa-Astigarraga, A.; Martin-Mayor, A.; Maeso, A.; de Miguel, B.; Martinez-Agirre, M.; Penalba, M. A holistic optimization tool for the design of power take-off systems in realistic wave climates: The Wells turbine case. *Ocean Eng.* **2023**, *285*, 115332. [CrossRef]
35. Geng, K.; Yang, C.; Zhao, B.; Zhao, W.; Gao, J.; Hu, C.; Zhang, H.; Wu, W. Residual circulation budget analysis in a Wells turbine with leading-edge micro-cylinders. *Renew. Energy* **2023**, *216*, 119049. [CrossRef]
36. Kotb, A.T.M.; Nawar, M.A.A.; Attai, Y.A.; Mohamed, M.H. Performance assessment of a modified wells turbine using an integrated casing groove and Gurney flap design for wave energy conversion. *Renew. Energy* **2022**, *197*, 627–642. [CrossRef]
37. Raja, V.; Prakash, R.A.; Kumar, A.; de Pacheco, D.A.J. Multi-disciplinary engineering design of a high-speed nature-inspired unmanned aquatic vehicle. *Ocean Eng.* **2023**, *270*, 113455. [CrossRef]
38. Raja, V.; Murugesan, R.; Rajendran, P.; Palaniappan, S.; Al-Bonsrulah, H.A.Z.; Jayaram, D.K.; Al-Bahrani, M. Multi-Domain Based Computational Investigations on Advanced Unmanned Amphibious System for Surveillances in International Marine Borders. *Aerospace* **2022**, *9*, 652. [CrossRef]
39. Raja, V.; Madasamy, S.K.; Rajendran, P.; Ganesan, S.; Murugan, D.; Al-Bonsrulah, H.A.Z.; Al-Bahrani, M. Nature-Inspired Design and Advanced Multi-Computational Investigations on the Mission Profile of a Highly Manoeuvrable Unmanned Amphibious Vehicle for Ravage Removals in Various Oceanic Environments. *J. Mar. Sci. Eng.* **2022**, *10*, 1568. [CrossRef]

40. Kumar, K.N.; Meenakshi, S.; Deviparameswari, K.; Vaidegi, R.; Nandhagopal, R.; Ramesh, M.; Vijayanandh, R. Investigation of energy generation on large rotary wing unmanned aerial vehicle's propeller using coupled engineering approaches. In *Advances in Environment Engineering and Management*; Springer: Cham, Switzerland, 2021; pp. 209–224.
41. Gnanasekaran, R.K.; Shanmugam, B.; Raja, V.; Al-Bonsrulah, H.A.Z.; Rajendran, P.; Radhakrishnan, J.; Eldin, S.M.; Narayanan, V. Comprehensive computational investigations on various aerospace materials under complicated loading conditions through conventional and advanced analyses: A verified examination. *Front. Mater.* **2023**, *10*, 1147310. [CrossRef]
42. Wang, Y.; Kumar, L.; Raja, V.; AL-bonsrulah, H.A.Z.; Kulandaiyappan, N.K.; Amirtharaj Tharmendra, A.; Marimuthu, N.; Al-Bahrani, M. Design and Innovative Integrated Engineering Approaches Based Investigation of Hybrid Renewable Energized Drone for Long Endurance Applications. *Sustainability* **2022**, *14*, 16173. [CrossRef]
43. Kulandaiyappan, N.K.; John, B.R.R.; Raja, V. Investigation of Aeroelastic Energy Extraction from Cantilever Structures under Sustained Oscillations. *Processes* **2023**, *11*, 830. [CrossRef]
44. Raja, V.; Al-Bonsrulah, H.A.Z.; Raji, A.P.; Madasamy, S.K.; Ramaiah, M.; Gnanasekaran, R.K.; Kulandaiyappan, N.K.; Al-Bahrani, M. Design and multi-disciplinary computational investigations on PVEH patches attached horizontal axis hybrid wind turbine system for additional energy extraction in HALE UAVs. *IET Renew. Power Gener.* **2023**, *17*, 617–643. [CrossRef]
45. Morada, A.M.; Elzahaby, A.; Abdallah, S.; Kameld, M.; Khalile, M.K. Application of Piezoelectric Materialsfor Aircraft Propeller Blades Vibration Damping. *Int. J. Sci. Eng. Res.* **2015**, *6*, 513–520.
46. Du, S.; Jia, Y.; Seshia, A. Maximizing Output Powerina Cantilevered Piezoelectric Vibration Energy Harvester by Electrode Design. *J. Phys. Conf. Ser.* **2015**, *660*, 012114. [CrossRef]
47. Dunnmon, J.; Stanton, S.; Mann, B.; Dowell, E. Power extraction from aeroelastic limit cycle oscillations. *J. Fluids Struct.* **2011**, *27*, 1182–1198. [CrossRef]

processes

MDPI

Article

A Database Extension for a Safety Evaluation of a Hydrogen Refueling Station with a Barrier Using a CFD Analysis and a Machine Learning Method

Hyung-Seok Kang [1,*], Ji-Won Hwang [2] and Chul-Hee Yu [3]

1 Korea Atomic Energy Research Institute, Daejeon 34057, Republic of Korea
2 Dahan Tech Inc., Seoul 06106, Republic of Korea; wonny@dahan.co.kr
3 Korea Gas Safety Corporation, Yeongwol-gun 26203, Republic of Korea; gusari@kgs.or.kr
* Correspondence: hskang3@kaeri.re.kr; Tel.: +82-42-868-8948

Abstract: A methodology is proposed to extend datasets in a database suitable for use as a reference tool to support an evaluation of damage mitigation by a barrier wall in a hydrogen refueling station (HRS) during a vapor cloud explosion (VCE) accident. This is realized with a computational fluid dynamic (CFD) analysis and machine learning (ML) technology because measured data from hydrogen explosion tests with various installed barrier models usually require considerable amounts of time, a secured space, and precise measurements. A CFD sensitivity calculation was conducted using the radXiFoam v1.0 code and the established analysis methodology with an error range of approximately ±30% while changing the barrier height from that was used in an experiment conducted by the Stanford Research Institute (SRI) to investigate the effect of the barrier height on the reduction in peak overpressures from an explosion site to far fields in an open space. The radXiFoam code was developed based on the open-source CFD software OpenFOAM-v2112 to simulate a VCE accident in a humid air environment at a compressed gaseous or liquefied HRS. We attempted to extend the number of datasets in the VCE database through the use of the ML method on the basis of pressure data predicted by a CFD sensitivity calculation, also uncovering the possibility of utilizing the ML method to extend the VCE database. The data produced by the CFD sensitivity calculation and the ML method will be examined to confirm their validity and applicability to hypothetical VCE accident simulations if the related test data can be produced during experimental research. The database constructed using core data from the experiment and extended data from the CFD analysis and the ML method will be used to increase the credibility of radXiFoam analysis results for VCE accident scenarios at HRSs, ultimately contributing to safety assurances of HRSs in Republic of Korea.

Keywords: hydrogen energy; hydrogen refueling station; barrier; vapor cloud explosion; peak overpressure; CFD; OpenFOAM; machine learning; safety evaluation

1. Introduction

To build a stronger economic society using hydrogen energy in Republic of Korea, a roadmap for the construction of 1200 hydrogen refueling stations (HRSs) for transportation vehicles by 2040 was proposed by the Korean government [1,2]. However, this plan is facing delays when compared to the original timeline, as residents around the HRS sites to be built in a large city are opposing its construction. They claim that HRSs are not safe, referring to an explosion accident that occurred at a compressed hydrogen storage tank connected to a water electrolysis facility in Gangwon technopark in Republic of Korea [3]. In addition, the memory of a hydrogen explosion accident at a nuclear power plant in Japan and its detrimental effects on the environment continually enhances residents' anxiety over the use of hydrogen energy in Republic of Korea [4–6]. From the end of 2023, liquefied hydrogen (LH2) will be produced in large quantities at several facilities and supplied to consumers through LH2 hydrogen stations in Republic of Korea [7]. Thus, the residents' concerns over the safety of HRSs are considerably greater because stations with LH2 contain more hydrogen compared to typical refueling stations.

The Korea Gas Safety Corporation (KGS), the organization responsible for the safety of hydrogen energy facilities in Republic of Korea, decided to reinforce the specific technical regulations pertaining to HRSs to increase safety on the basis of experimental research and numerical simulations [8]. One such effort is specifically to present a method by which to devise a standard model for a damage-mitigation wall, also referred to as a barrier, around the storage tanks at a HRS or the boundary of a HRS on the basis of the results of a quantitative risk assessment (QRA) and based on the distance between the storage tank at the HRS and the facilities surrounding the HRS [8–10]. To perform the QRA, various accident scenarios should be taken into account after investigating the possibility of a hydrogen gas release from certain components and the corresponding frequencies based on datasets and the design of the HRS [11]. The extent of such damage resulting from each accident scenario, including individual risk and social risk levels, must be predicted using a verified computational code [12].

The most frequent and damaging type of accident in resident areas around a HRS, transferring the damaging effects over long distances from the explosion site considering various accident scenarios, may be a vapor cloud explosion (VCE), which occurs due to hydrogen released from a storage tank into an air environment at a HRS [13]. This likely stems from how the pressure wave propagates at the speed of sound to distant locations in all directions in an air environment from the explosion site at the HRS as soon as the pressure builds due to the combustion energy released from the hydrogen–air chemical reaction. If a pressure wave with a magnitude exceeding a certain value, as shown in Table 1, arrives at structures or where people are, harmful effects, such as broken glass or even ear injuries, may occur [14,15]. Thus, a three-dimensional computational fluid dynamic (CFD) code is useful for quantitatively evaluating the propagation of pressure waves due to a VCE at an HRS because pressure waves generally tend to lose strength as well as change direction when colliding with structures around an HRS as they propagate through air [16]. In addition, applying the CFD codes, including the open-source software OpenFOAM, for a VCE analysis has been conducted wildly in the areas of process safety and loss prevention [17].

Table 1. Damage criteria according to overpressure values [14,15].

Overpressure (kPa)	Damage Effect
0.2	Partially broken windows
1	Breaking glass
3	Minor damage to structures and injuries to people
5	Broken structures in houses
7	Partially damaged houses and much harm to people
14	Starting value for lethal effects on humans
15	Slightly damage to roofs and walls of structure
18	Approximately 50% of houses damaged per block
20	Steel structure of building is damaged and pulling away from the foundation
35–50	Structure destroyed

We developed a CFD code, radXiFoam, using OpenFOAM-v2112 to simulate a VCE accident in a humid air environment at an HRS after validation through several test results, which included the overpressure reduction effect by a barrier [18,19]. We used open-source software because we can freely distribute radXiFoam to engineers who work on the safety of HRSs. The released radXiFoam will be used to predict harmful effects due to hypothetical VCE accidents at HRSs, and the QRA, on the basis of the predicted CFD results, will be reviewed by the KGS to approve the safety evaluation results for HRSs before the granting of a construction permit from the licensing institute.

To increase the confidence of hypothetical VCE accident analysis results, using radXiFoam as a reference tool, we used a database consisting of experimental data and validated analysis data fed into radXiFoam with an error range to utilize quantitative data to describe the barrier effect of overpressure reduction on several VCE types. Herein, radXiFoam analysis results validated against test data can provide overpressure data at various locations where measurement devices are not installed, as well as visualized pressure contours from the explosion site to a far-field in an air environment. The datasets in the database can provide criteria to safety engineers, who must judge the credibility of radXiFoam results before using the calculated overpressures through the barrier as the QRA inputs, as the CFD results for hydrogen explosions occurring around structures in open spaces tend to vary according to the geometrical model, the mesh distribution in the grid model, the numerical and empirical constants used in the CFD models, and the time step size in the transient calculation [20]. Such efforts to use databases for safety evaluations of gas explosion accidents are also underway in other countries [21–23].

However, it is difficult to obtain sufficient VCE test data with various barrier models in a short period before starting to construct HRSs actively around protected facilities because gas explosion tests usually require safe experimentation at secured locations [24,25]. One feasible means of overcoming this difficulty in obtaining the data is to extend the database using CFD sensitivity calculations on the basis of validated radXiFoam results for VCE test data with the installation of the barrier. In addition, we are now considering the use of a machine learning (ML) method [26] to produce new datasets based on existing data in the VCE database when no available data for a certain HRS exists in the database. However, this idea of using the ML method to generate datasets for a safety evaluation should be examined if the related data can be produced in our experimental research [8], despite the fact that this type of trial using the ML method to reinforce a database is currently used in other industries [27].

2. Use of the VCE Database for HRS' Safety Evaluations

To build a database to be used as a reference tool for a hypothetical VCE scenario analysis at an HRS with a barrier, first we collected published test data produced by a foreign institute, in this case, Stanford Research Institute (SRI) [24,25,28]. The plan is to add Republic of Korea's own hydrogen explosion test results by installing various barrier models to the VCE database after analyzing the data, with this enacted by the KGS and other institutes [8]. According to the test plan, a hydrogen–air mixture volume of 40 m^3 with the barrier installed will be exploded to measure the overpressure reduction through the barrier. This may create unique data which can extend the range of the database because previous VCE tests with the installation of such a barrier were generally gained when using a hydrogen–air mixture volume of 5.2 m^3 under a stoichiometric condition [24,25,28]. The datasets of the overpressure in the VCE database will be categorized according to the hydrogen–air mixture volume, hydrogen concentration, ignition method, existence of obstacle geometries, weather conditions, and the barrier model. Such a categorized database will contain the criteria to be used to anticipate the hydrogen explosion strength, the harmful effects, and the degree of damage mitigation by the barrier.

Safety engineers can compare their VCE accident analysis results, which will be used as the input for the QRA, with the suitable datasets in the database if conditions similar to those of the accident scenarios, such as the hydrogen concentration, relative humidity of the air, and the distributed range of the hydrogen–air mixture, exist in the database, as shown in Figure 1. This comparison can increase the credibility of the VCE accident analysis results by radXiFoam because CFD results basically have uncertainty arising from the modeling process of the VCE accident scenarios [20]. Thus, the database should contain a broad range of data related to VCE accident scenarios at HRSs, including several barrier models, to the greatest extent possible, to increase the credibility of safety evaluations of HRSs to be constructed with a barrier around protected facilities so as to meet KGS regulations. Therefore, the data produced by radXiFoam sensitivity calculations for various barrier models can be efficiently used to increase the utilization of the VCE database, as such data compensate for the lack of test data stemming from the limited gas explosion experiments. If the ML can properly produce reference data for HRSs, which will be constructed in various areas in large cities, using a published theory based on existing datasets in a database [26,29], safety engineers can use the produced data as reference materials for comparisons with calculated radXiFoam results for a VCE accident analysis.

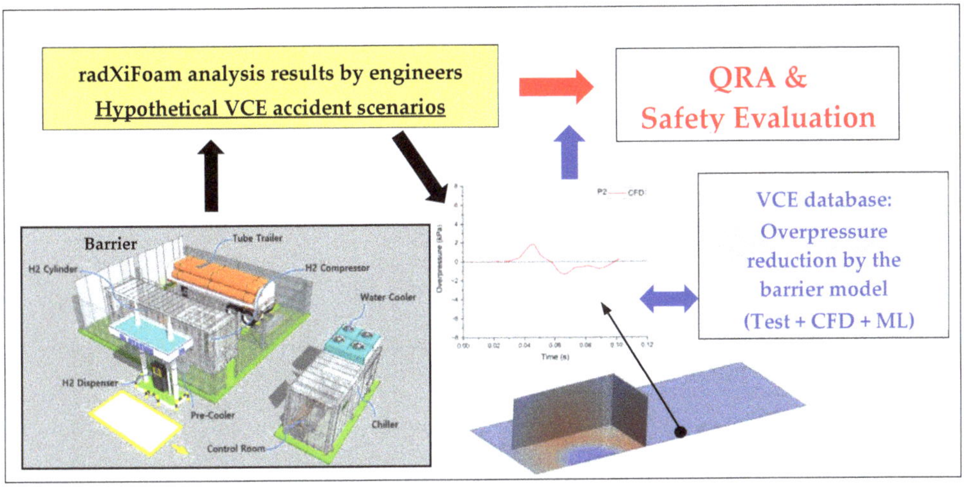

Figure 1. Plan for the utilization of the VCE database.

3. Producing Datasets Using the CFD Analysis

3.1. RadXiFoam Solver

We developed an open-source CFD solver, radXiFoam, to simulate a VCE accident in a humid air environment at a HRS and its damaging effects on residents around a HRS on the basis of XiFoam implemented in OpenFOAM-v2112 [18,19]. The combustion solver XiFoam, using a flamelet progress variable model, is usually applied to simulate the flame acceleration phenomena in a premixed combustion area considering an effective computational cost when compared to a solver based on a detailed chemical kinetic model [30]. However, XiFoam does not currently simulate humid air conditions, which may affect the VCE analysis results, as XiFoam only considers the transport equations of species for the fuel mixture fraction (ft) and regress variable (b) [19]. From the calculated results of ft and b in the fuel-oxygen mixture, XiFoam can calculate only the mass fraction of the fresh fuel and the burnt fuel in an inhomogeneous mixture model in OpenFOAM-v2112.

Therefore, we added species equations for the water vapor (wv) and nitrogen gas (n_2) to the existing species equation in XiFoam (Equation (1)) to consider a humid air condition, setting the relative humidity to 40%, and developed what is termed the vaporInnhomogeneous mixture model, which can calculate the variations in the gas concentrations in a hydrogen–air mixture as the combustion proceeds [18]. In Equation (1), the variable i indicates a certain species, such as water vapor or nitrogen gas. In addition, we implemented a radiative heat transfer term to the energy equation into XiFoam (Equation (2)) to calculate the precise enthalpy variation in the mixture of hydrogen, air, and water vapor, as the existence of water vapor may affect the heat transfer phenomenon through the absorption and emittance of thermal photons discharged from the combustion region.

To simulate thermal photon absorption by water vapor accurately according to the relevant concentration in the air, we modified "GreyMeanAbsorptionEmission.H" in XiFoam to include the partial pressure of the water vapor (p_{wv}) in the gas mixture to calculate the absorption coefficient (a_{wv}) and then proposed it as "VaporGreyMeanAbsorptionEmission.H" in radXiFoam. In Equation (3), k_{wv} is the absorption coefficient depending on the gas temperature [31]. For a radiative heat transfer model, first we apply the P-1 radiation model (Equation (4)) to calculate the radiative heat transfer through the water vapor in the air, which is described in detail in the literature [32,33]. We also proposed an effective correlation model (Equation (5)) for the laminar flame speed based on the fuel equivalent ratio in the hydrogen–air mixture considering the initial water vapor, which was modified from a complicated form implemented in XiFoam [18,33–35]. In Equation (5), T_u is the unburnt gas temperature, T_o is the reference temperature (300 K), p is the pressure, p_o is the reference pressure (1 bar), S_{uo} is the laminar flame speed in the reference condition, and φ is the fuel equivalence ratio.

$$\frac{\partial(\rho Y_i)}{\partial t} + \nabla\cdot(\rho U Y_i) - \nabla\cdot\left[\left(\frac{\mu_t}{Sc_t}\right)\nabla Y_i\right] = S_i \tag{1}$$

$$\frac{\partial}{\partial t}(\rho h) + \nabla\cdot(\rho U h) + \frac{\partial}{\partial t}(\rho K) + \nabla\cdot(\rho U K) - \frac{\partial p}{\partial t} + \nabla\cdot\left(\frac{\mu_t}{Sc_t}\nabla h\right) = -\nabla\cdot q_r \tag{2}$$

$$a_{wv} = k_{wv} \times p_{wv} \tag{3}$$

$$\nabla\cdot q_r = aG - 4\left(e\sigma_{SB}T^4 + E\right) \tag{4}$$

$$S_u = S_{uo}\left(\frac{T_u}{T_o}\right)^{\alpha}\left(\frac{p}{p_o}\right)^{\beta} \tag{5}$$

$$\alpha = 2.18 - 0.8(\varphi - 1) \tag{6}$$

$$\beta = -0.16 + 0.22(\varphi - 1) \tag{7}$$

3.2. *Analysis of SRI Test 4-02 Using radXiFoam*

In order to determine whether radXiFoam can accurately simulate the VCE phenomena and the overpressure decrease due to the use of a barrier in an air environment, a validation analysis was conducted on the basis of measured data from SRI Test 4-02 [24,25]. SRI conducted a gas explosion test using a hydrogen–air mixture volume of 5.2 m^3 under a stoichiometric condition of approximately 30 vol.% with the installation of a barrier, as shown in Figure 2 and Table 2 [24,25]. The barrier, which had a height of 2 m, width of 10 m, and thickness of 0.1 m, was located 5.1 m from the ignition point in a tent to reduce the propagation of the damaging effects due to the VCE. The gas explosion in the tent was started using an electric spark device, which emitted a spark for 2 ms, located 2.5 cm above the center of the bottom plate, with the removal of the plastic film that covered the boundary of the tent. An obstacle type of geometry to induce flame acceleration was not used, except for the supporting bars of the plastic film and the measurement devices in the tent. Pressure sensors to measure the overpressure reduction due to the barrier were located 2 m behind (P2) and 2 m in front (P4) of the barrier, as shown in Figure 2, also measuring the overpressure increase owing to the VCE in the tent (P3). The overpressures at 11 m, 21 m, and 41 m from the ignition point were also measured to record the characteristics of the pressure wave propagation behind the barrier.

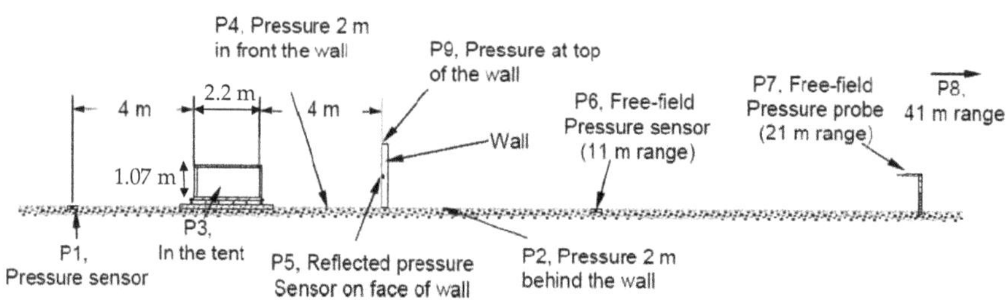

Figure 2. Schematic diagram of the SRI test facility [24].

Table 2. Conditions of the SRI hydrogen explosion tests [24,25].

Test No.	H_2–Air Mixture Volume (m^3)	H_2 Con. (vol.%)	Ambient Temp. (K)	Wind (m/s)	Ignition Method	Barrier Existence
4-02	5.2	29.9	283.45	2.0	Electric spark	O

Barrier dimensions: Height × Width × Thickness = 2 m × 10 m × 0.1 m.

A three-dimensional grid model utilizing a half-symmetric condition to simulate the SRI test facility was developed using the blockMesh utility in OpenFOAM-v2112 [19], as shown in Figure 3. To determine the horizontal length along the x-direction in the grid model, we considered a required separation distance of 30 m between the HRS and the first protected facilities located around the HRS in the KGS codes [9,10]. Thus, we did not include pressure sensor P8, located 41 m from the ignition point in the grid model, to reduce the computational time. We also decreased the width of the grid model to 5 m from the distance of 10 m used in a previous analysis [36] to ensure effective sensitivity calculations of the various barrier heights during a suitable computational time. A total of 2,569,440 hexahedral cells with an approximate cell length of 1.5 cm around the tent region and 15 to 25 cm at the far field region from the tent were generated in the grid model to resolve the rapid flame acceleration due to the spark ignition and the propagation of the pressure wave in the air environment. The cell length sizes in the grid model were determined based on previous mesh sensitivity calculation results [36]. The boundary conditions for the grid model are shown in Figure 3a. The opening condition allows for the inflow and outflow of air through the applied surfaces and simultaneously uses a wave transmissive outflow condition that can be expressed by Equations (8) and (9) [19,36]. In Equation (9), W_p is the patch wave speed, ϕ_p is the patch face flux, S_f is the patch face area vector, φ_p is the patch compressibility, and γ is the ratio of specific heat. A patch refers to a surface on which the opening condition is applied. An emissivity value of 0.072 was assigned to the bottom wall region in the tent on the basis of radiative property literature [32] on the simulation of the radiative heat transfer from a combusted gas to a steel wall.

$$\left.\frac{DW_p}{Dt}\right|_{outflow} = 0 \tag{8}$$

$$W_p = \frac{\varnothing_p}{\left|S_f\right|} + \sqrt{\frac{\gamma}{\varphi_p}} \tag{9}$$

The initial conditions used to simulate the test conditions in Table 2, including the hydrogen concentration in the tent, were given as the corresponding hydrogen mass fraction (ft), as shown in Figure 3b, using the utility program setFields in OpenFOAM-v2112 [18,19]. The spark ignition model, previously developed on the basis of other SRI test cases [24,25], used to initiate the hydrogen combustion in the tent region, was applied to simulate electric spark operation for 2 ms with an equivalent energy of 40 J [20]. To capture the pressure wave propagation and reflection accurately after collision with the barrier in the air environment, we used a small time step of approximately 1.0×10^{-6} s to 1.0×10^{-4} s in the transient calculation time of 0.1 s.

The calculated temperature and pressure distribution on the symmetric plane using radXiFoam with the established methodology (Table 3) over time are correspondingly shown in Figures 4 and 5, which indicate that the results reasonably predict the progress of the VCE from the ignition in the tent region to the pressure wave propagation to the air environment with the formation of a semi-spherical pressure band. The pressure buildup in the combustion region results from the combustion energy due to the hydrogen–air chemical reaction, which progresses continually as the flame propagates in the tent region. Figure 5 shows that the barrier effectively prevents pressure wave propagation along the horizontal x-direction, i.e., from the tent region to the air environment. As a result of this prevention, the pressure in the region below the barrier height is lower than those of other regions in the rear area of the barrier. Thus, the calculated peak overpressure of 5.27 kPa at P4, located 2 m in front of the barrier, is reduced to 1.67 kPa at P2 at 2 m behind the barrier. The comparison of pressure waves at the local positions of P2, P4, P6, and P7 shows that the CFD results accurately predict the measured peak overpressures with an error range of approximately ±30%, as shown in Figure 6. However, the predicted pressure waves at each measurement location pass approximately 28 ms faster than in the test data. This can be explained by the fact that we neglected the transition period from the laminar flame to the turbulent flame by assuming fully turbulent mixing and combustion in the spark ignition model.

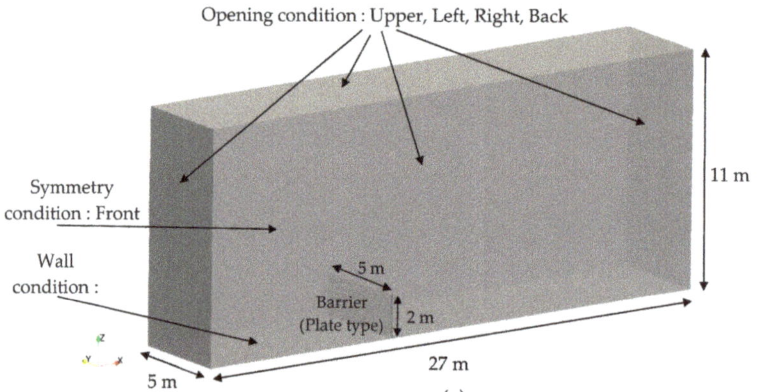

(a)

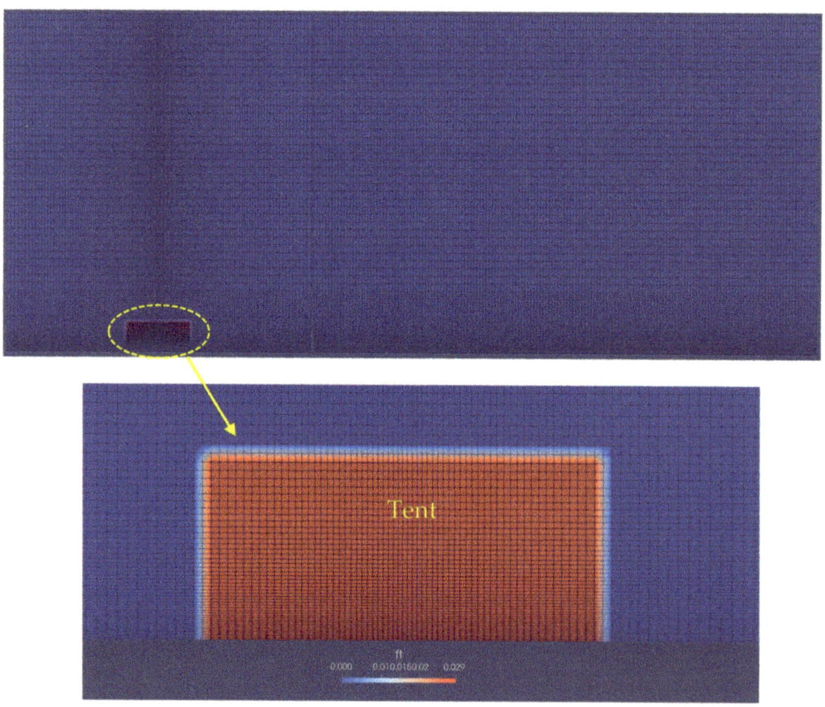

(b)

Figure 3. Grid model of the SRI Test 4-02 facility: (**a**) Geometric model and boundary conditions; (**b**) Mesh distribution and hydrogen mass fraction (ft) in the grid model.

Table 3. Analysis methodology for the VCE explosion test with a barrier.

Parameter	Models
• Open-source software	radXiFoam v1.0 based on OpenFOAM-v2112
• Thermal–hydraulic solver algorithm	PIMPLE [19]
• Combustion model	Flamelet progress variable
• Turbulence model	k-ω SST
• Wall function	kqR/omega
• CFL number	<0.8
• Mesh type	Hexahedral
• Mesh size at the far field	~25 cm
• Ignition model	Spark ignition model

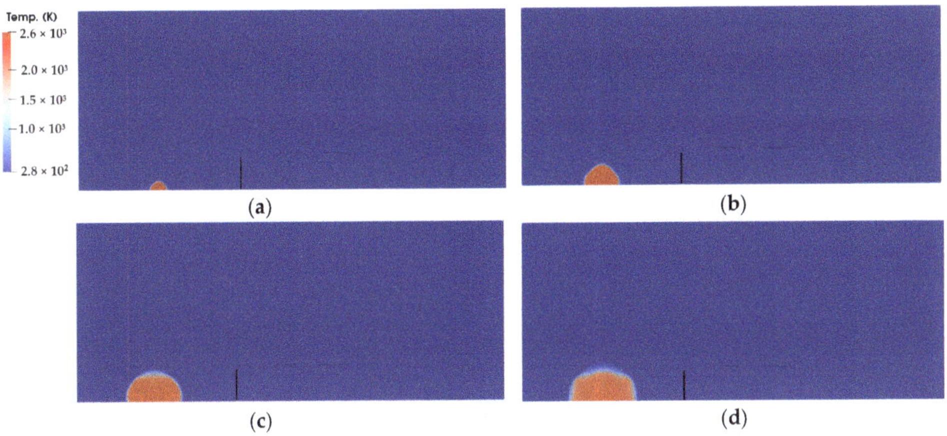

Figure 4. Temperature distribution over time (front view): (**a**) t = 0.01 s; (**b**) t = 0.02 s; (**c**) t = 0.03 s; (**d**) t = 0.04 s.

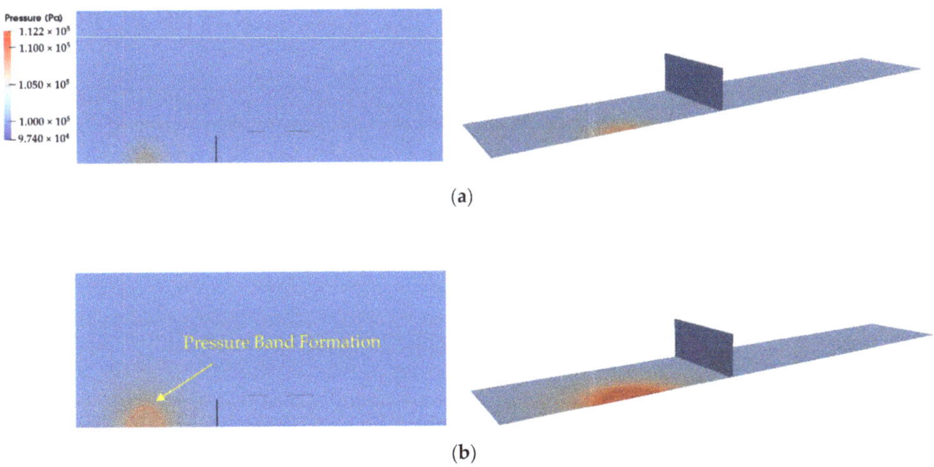

Figure 5. *Cont.*

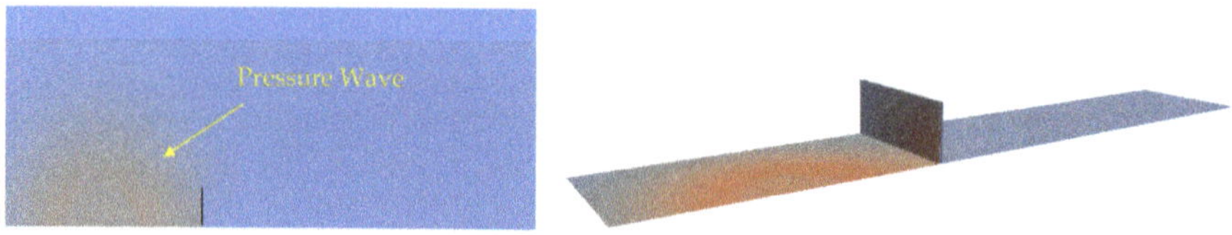

(c)

(d)

Figure 5. Pressure distribution on symmetry and wall planes: (**a**) t = 0.01 s; (**b**) t = 0.02 s; (**c**) t = 0.03 s; (**d**) t = 0.04 s.

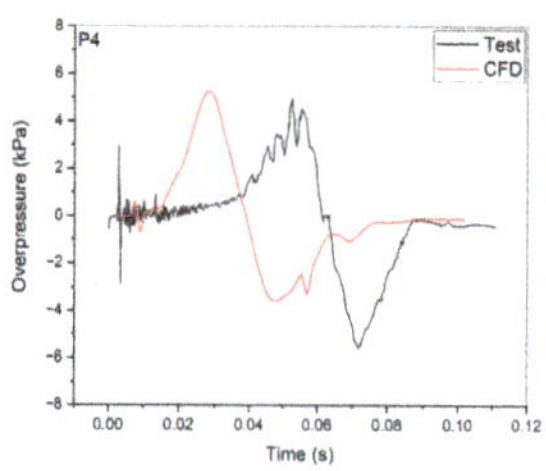

(a)

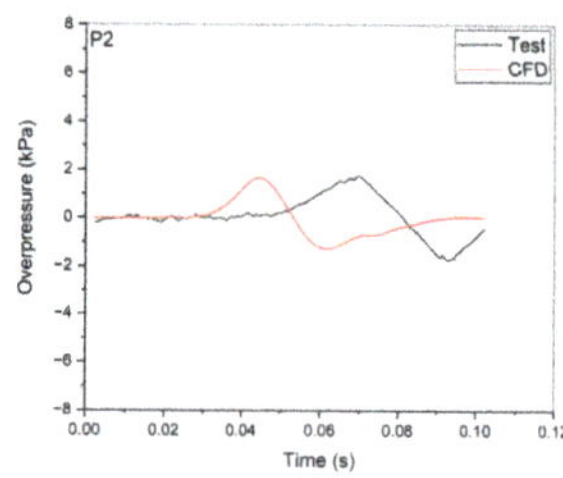

(b)

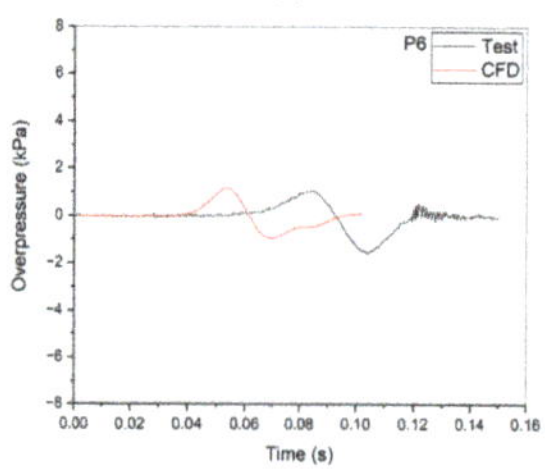

(c)

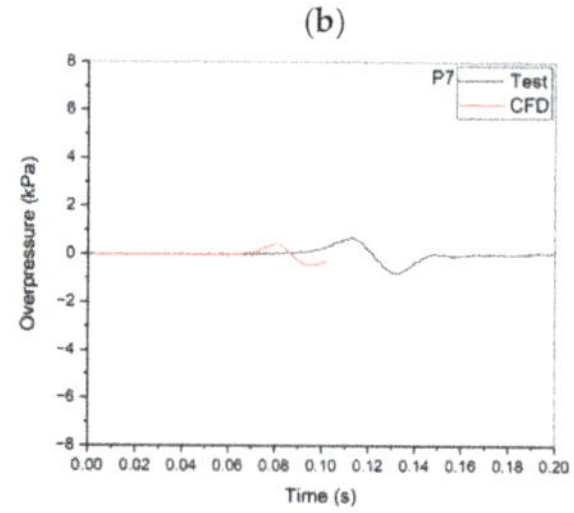

(d)

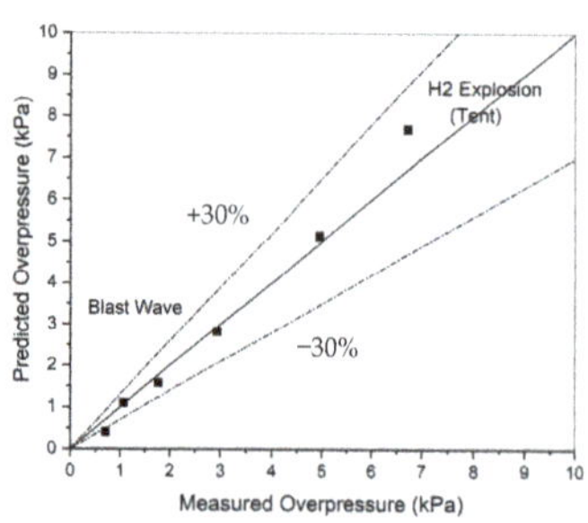

(e)

Figure 6. Comparison of overpressures at local positions between the test data and the CFD results: (**a**) P4 (2 m front of the barrier); (**b**) P2 (2 m behind the barrier); (**c**) P6 (range: 11 m); (**d**) P7 (range: 21 m); (**e**) Prediction uncertainty.

3.3. CFD Sensitivity Calculation Results and Discussion

To assess the effect of the barrier height on the reduction in the overpressure due to the VCE, we conducted a sensitivity calculation while changing only the barrier height from 2 m to 3 m and 4 m in the input files and with the established analysis methodology used for the SRI Test 4-02 calculation (Table 4), as we judged that the height variation in the barrier had an obvious effect on the overpressure reduction on the basis of simulation results from an overseas institute [28]. We carefully controlled the input parameters for the blockMesh utility to produce a nearly identical mesh distribution around the barrier between the grid models for Cases−1 to −4 to eliminate the mesh dependency on the calculated pressure distribution. In addition, we calculated Case−1 again using a new grid model, in which the width only extended from 5 m to 10 m, to investigate the effect of the location of the outflow condition along the *y*-direction in the grid model on the pressure field calculation (Case−4). It is expected that the comparison results of the local overpressures between Case−1 and Case−4 will produce a good guideline for the development of a grid model to analyze hypothetical VCE accident scenarios at HRSs to be performed by safety engineers.

Table 4. Sensitivity calculation on the basis of a CFD simulation for SRI Test 4-02.

Case No.	Barrier Height (m)/Width (m)/R (m)	Grid Model Length (m)/Width (m)/Height (m)	Number of Cells (Grid Model)
Case−1	2/0.1/5.1	27/5/11	2,569,440
Case−2	3/0.1/5.1	27/5/11	2,568,800
Case−3	4/0.1/5.1	27/5/11	2,568,160
Case−4	2/0.1/5.1	27/10/11	3,478,390

Range (m): horizontal distance from the ignition point.

The CFD sensitivity calculation results for various barrier heights of 2 m, 3 m, and 4 m (Figure 7a–c) show that the height of the pressure wave that propagates from the tent region to the air environment blocked by the barrier is increased as the barrier height increases. This feature indicates that the affected area on the ground in the rear region of the barrier from the propagating pressure wave is also increased as the barrier height increases. As a result of these effects according to the barrier height, the predicted peak overpressures at the P2 location are decreased from approximately 1.7 kPa to 0.85 kPa as the barrier height is increased from 2 m to 4 m (Figure 8a). This distinct difference may stem from the fact that the magnitude of the pressure wave and the elevation of the pressure band formation around the tent region are nearly identical regardless of the barrier height because the exploded hydrogen–air volume in the tent region is equal to 5.2 m^3 in Cases−1 to 3.

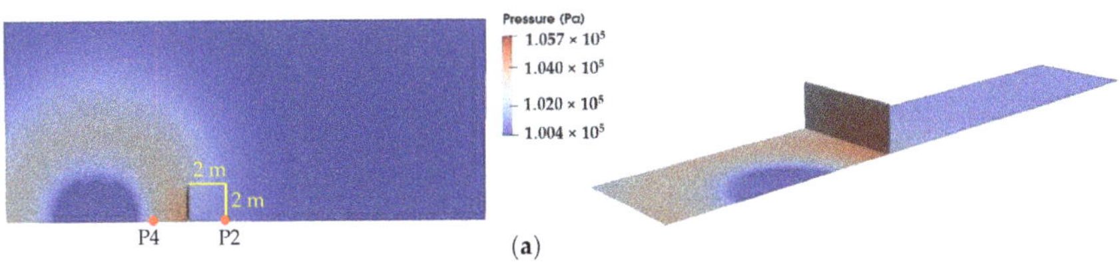

(**a**)

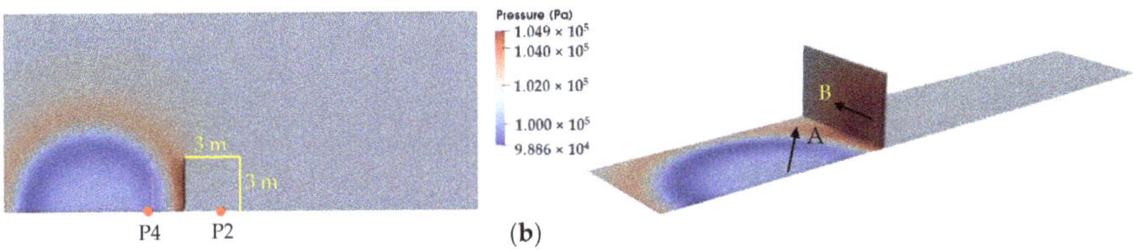

(**b**)

Figure 7. *Cont.*

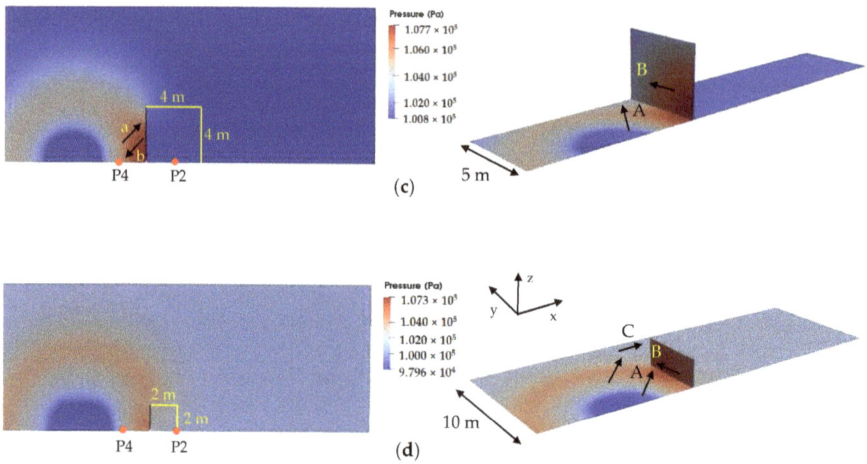

Figure 7. Pressure distribution on symmetry and wall planes according to the case: (**a**) t = 0.037 s (Case−1); (**b**) t = 0.030 s (Case−2); (**c**) t = 0.031 s (Case−3); (**d**) t = 0.034 s (Case−4); A (Pressure wave propagating to the outer region from the tent region); B (Reflected pressure wave propagating to the outer region along the surface of the barrier); C (Pressure wave propagating to the outer region after bypassing the barrier); a (Pressure wave propagating to the barrier from the ten tent region); b (Reflected pressure wave propagating to the location of P4 from the barrier).

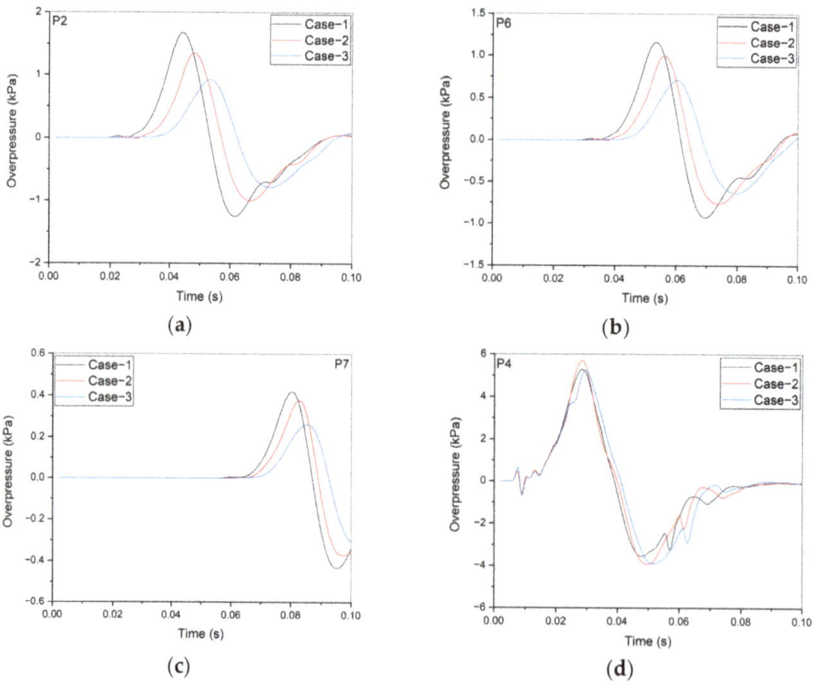

Figure 8. Comparison of overpressures at local positions between Cases−1, 2 and −3: (**a**) P2 (2 m behind the barrier); (**b**) P6 (range: 11 m); (**c**) P7 (range: 21 m); (**d**) P4 (2 m front of the barrier).

However, the calculated overpressures at P4, of which the location is 2 m in front of the barrier, show no significant differences between the cases tested here, as shown in Figure 8d. This result can be explained by the fact that the propagation of the pressure wave in front of the barrier proceeds continually as hydrogen combustion takes place from the tent region to the outward region along the y-direction while maintaining a hemi-spherical-type shape on the ground (Figure 7b–d, "A"). The reflected pressure wave after the collision with the barrier (Figure 7b–d, "B") mainly flows to the outward region along the y-direction on the surface of the barrier because the strength of the pressure wave from the tent region exceeds that of the pressure wave reflected from the barrier.

As a result of this propagation pattern of the pressure wave in front of the barrier, the location of the wave transmissive outflow condition, for which the Neumann condition for the wave velocity applies (Equation (8)), may affect the calculation of the pressure field in the area in front of the barrier by radXiFoam using the PIMPLE algorithm (Figure 9a) [19]. In addition, this effect appears to exist in the calculated overpressure at the P2 location in the rear region of the barrier (Figure 9b). This can be explained by the fact that the direction of the pressure wave passing the barrier along the y-direction turns to the P2 location despite the fact that the corresponding wave's strength is considerably decreased. Therefore, we can conclude that the wave outflow condition in the VCE simulation should be located at some distance away from the explosion site to calculate the pressure field accurately when assuming proper pressure wave decay while the wave propagates to the outflow region.

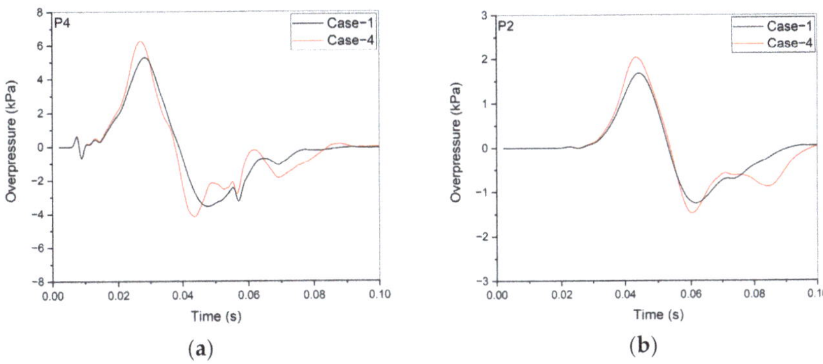

Figure 9. Comparison of overpressures at local positions between Cases−1 and 4: (**a**) P4 (2 m front of the barrier); (**b**) P2 (2 m behind the barrier).

4. Producing Datasets Using Machine Learning

4.1. Development of the Neural Network Model

To develop a neural network model, using the open-source software of python 3.7 [37], TensorFlow 2.3.0 [38], and Keras 2.3.1 [39], capable of predicting the overpressures in the rear region of the barrier according to the barrier height and the distance from the explosion site during the VCE of the hydrogen–air mixture volume 5.2 m^3, we used the calculated overpressures for 0.1 s at the locations of P2, P6, and P7 with barrier heights of 2 m and 4 m (Figure 8) as the training data for the neural network model. This development of the neural network model started with a conceptual design model, as shown in Figure 10, which was configured with hidden layers, neurons, activation functions, and loss functions using the commercial software Mathematica [29,40,41]. The completed model consisted of eight hidden layers with different numbers of neurons and activation functions separately applied to each layer to minimize the loss value, as shown in Table 5. The loss value calculated by the mean squared error function in the developed neural network model was approximately ±1.24%.

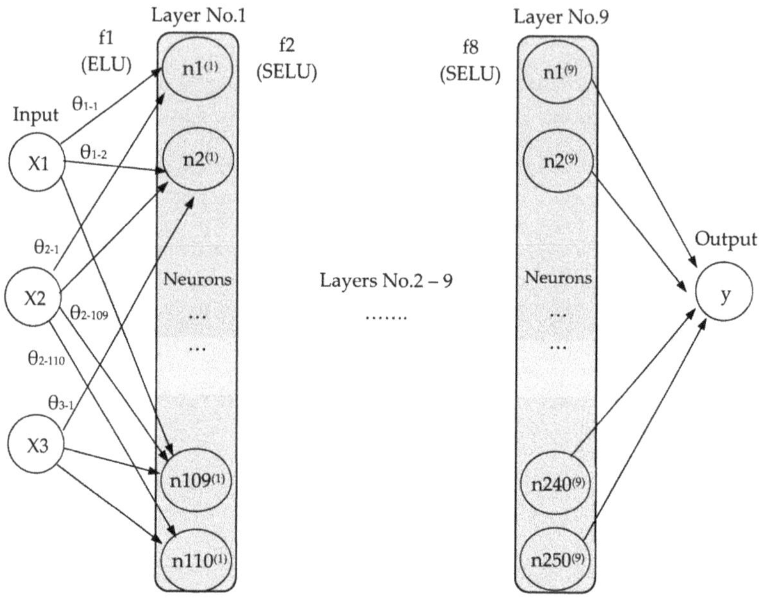

(**a**)

Function Name	Exponential Linear Unit (ELU)	Rectified Linear Unit (ReLU)	Self-Gated Activation Function (Swish)	Scaled Exponential Linear Unit (SELU)
Plot	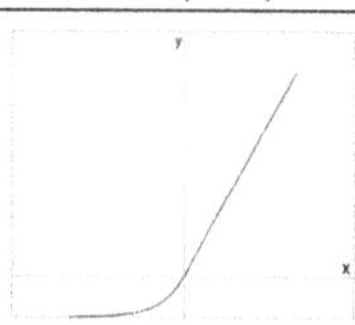	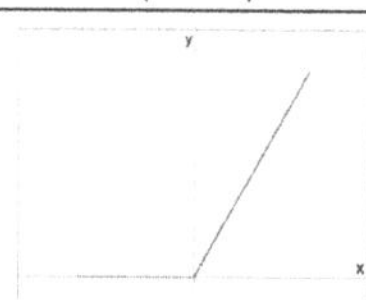	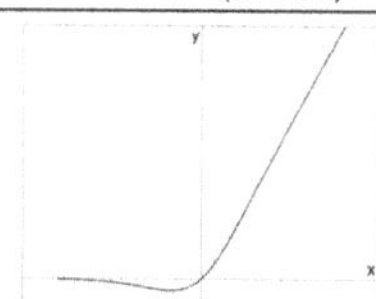	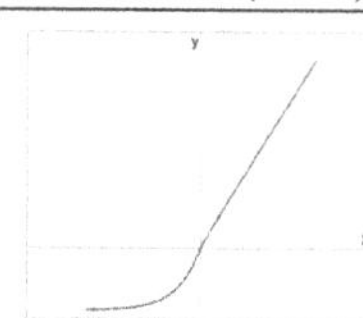
Numerical Expression	$\begin{cases} x \geq 0 & x \\ x < 0 & e^x - 1 \end{cases}$	$\begin{cases} x \geq 0 & x \\ x < 0 & 0 \end{cases}$	$\frac{x}{e^{\beta x} + 1}, \quad \beta = -1$	$\begin{cases} x \geq 0 & 1.0507x \\ x < 0 & 1.7581(e^x - 1) \end{cases}$

(**b**)

Figure 10. Conceptual design of hidden layers, neurons, and activation functions for the neural network model: (**a**) Conceptual design model; (**b**) Plots and numerical expressions of the activation functions.

Table 5. Number of neurons and activation functions according to the hidden layers.

Hidden Layers	Number of Neurons	Name of Activation Function
1	110	Exponential Linear Unit (ELU)
2	280	Scaled Exponential Linear Unit (SELU)
3	190	Rectified Linear Unit (ReLU)
4	170	Exponential Linear Unit (ELU)
5	100	Self-Gated Activation Function (Swish)
6	140	Scaled Exponential Linear Unit (SELU)
7	50	Exponential Linear Unit (ELU)
8	70	Scaled Exponential Linear Unit (SELU)
9	250	Rectified Linear Unit (ReLU)

Loss Function and Value: Mean Squared Error and 0.0124; Learning Rate: 0.0005.

4.2. Overpressure Data Produced by the ML Method and Discussion

The overpressure data produced at the P2, P6, and P7 locations by the developed neural network model for barrier heights of 2.5 m and 3.5 m during VCE with a hydrogen–air mixture volume of 5.2 m^3 are shown in Figure 11. From a comparison of the overpressures between the data produced by the ML method and the training data for the neural network model, we can judge that the data generated by the ML method accurately represent the characteristics of the pressure wave in the rear region of the barrier. The peak overpressures at P2, P6, and P7 locations decrease as the barrier height increases and the distance from the ignition point increases.

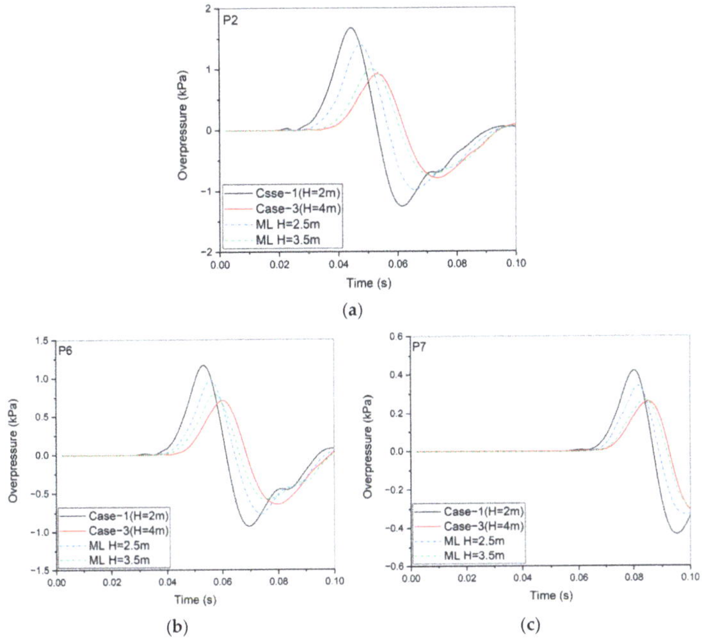

Figure 11. Overpressure data produced in the rear region of the barrier using the ML method: (**a**) P2 (2 m behind the barrier); (**b**) P6 (range: 11 m); (**c**) P7 (range: 21 m).

Therefore, we can expect that the ML method will be effective when used to extend the datasets using credible data in a VCE database when proper datasets for safety evaluations of HRSs to be constructed around protected facilities do not exist. However, through this study, which generated overpressures using the ML method, we also found that a neural network model that considers the features of pressure wave propagation, which arose in the area in front of the barrier and the region behind the barrier, should be developed separately.

5. Conclusions

A methodology was developed to configure a database to provide criteria to safety engineers who will conduct CFD analyses of hypothetical VCE accidents to evaluate the safety of HRSs to be constructed around protected facilities in a large city. The proposed methodology is to extend the datasets in the VCE database using the CFD sensitivity calculations and the ML method on the basis of measured data obtained from VCE experimentation, assuming the installation of a barrier wall. CFD sensitivity calculations using an established analysis methodology for the VCE phenomena will not only generate new datasets but will also provide visualized results to support evaluations by safety engineers. In addition, datasets based on various grid models in CFD sensitivity calculations can provide a guideline for the development of a grid model for VCE accident analyses in HRSs. We also uncovered the possibility of using the ML method to produce new datasets quickly when proper test datasets for comparisons with CFD analysis results for hypothetical VCE accidents do not exist. However, from this study, we realized that the training dataset to be used should be carefully chosen, considering the application purpose when developing a neural network model. Finally, we conclude such an extended VCE database, as discussed here, can sufficiently compensate for the lack of gas explosion test results with a barrier installed and can be effectively used as a reference tool to increase the safety of HRSs in Republic of Korea.

6. Future Studies

The datasets produced by the CFD sensitivity calculation and the ML method will be examined to confirm their validity and applicability after related test data are produced experimentally. VCE accident simulations with the installation of various barrier models in HRSs will be conducted, and their results will be confirmed on the basis of the datasets in the VCE database.

Author Contributions: Conceptualization, C.-H.Y.; methodology, H.-S.K. and J.-W.H.; software, H.-S.K. and J.-W.H.; validation, H.-S.K. and J.-W.H.; formal analysis, H.-S.K. and J.-W.H.; investigation, H.-S.K. and C.-H.Y.; resources, H.-S.K. and J.-W.H.; data curation, H.-S.K. and J.-W.H.; writing—original draft preparation, H.-S.K.; writing—review and editing, J.-W.H. and H.-S.K.; visualization, H.-S.K.; supervision, H.-S.K.; project administration, C.-H.Y.; funding acquisition, C.-H.Y. All authors have read and agreed to the published version of the manuscript.

Funding: This work was supported by a grant from the Korea Institute of Energy Technology Evaluation and Planning (KETEP) funded by the Korean government (Ministry of Trade, Industry and Energy) (No. 20215810100020).

Data Availability Statement: Research data will be provided through the official website that will be operated by KGS and Dahan Tech Inc. after the completion of the research project.

Conflicts of Interest: The authors declare no conflict of interest.

Nomenclature

Variable	Definition	Unit
a_{wv}	Absorption coefficient of water vapor	[−]
E	emission contribution	[−]
ft	Fuel mixture fraction	[−]
G	radiation intensity	[W/m^2]
h	Enthalpy	[J/kg]
K	Thermal conductivity	[W/m·K]
k	Absorption coefficient depending on the gas temp.	[−]
p	Pressure	[Pa]
q	Heat flux	[W/m^2]
S_i	Source/Sink of species-i	[kg/m^2·s]
S_u	Laminar flame speed	[m/s]
S_{ct}	Turbulent Schmidt number	[−]
T	Temperature	[K]
U	Velocity	[m/s]
Y_i	Species-i mass fraction	[−]
μ_t	Turbulence effective viscosity	[kg/m·s]
ρ	Density	[kg/m^3]
σ_{SB}	Stefan-Boltzmann coefficient	[−]
φ	Fuel equivalent ratio	[−]
Subscripts		
o	Reference condition	
u	Unburned	
wv	Water vapor	

References

1. Ministry of Trade, Industry, and Energy. *Hydrogen Economy Fostering and Hydrogen Safety Management Act*; No. 16942; Ministry of Trade, Industry, and Energy: Sejong, Republic of Korea, 2021.
2. National Library of Korea. Hydrogen Economy Revitalization Roadmap, Policy Information. 2019. Available online: https://policy.ni.go.kr (accessed on 23 September 2023).
3. Chuncheon District Prosecutors' Office, Gangneung Branch. *Investigation Results for the Hydrogen Tank Explosion in Gangwon Techno Park*; Press Release Report; Chuncheon District Prosecutors' Office, Gangneung Branch: Gangneung, Republic of Korea, 2019.
4. International Atomic Energy Agency. *The Fukushiama Daiichi Accident, Vol. 1, Description and Context of the Accident*; Technical Report; International Atomic Energy Agency: Vienna, Austria, 2015.
5. International Atomic Energy Agency. *IAEA Review of Safety Related Aspects of Handling ALPS Treated Water at TEPCO's Fukushima Daiichi Nuclear Power Station*; Technical Report; International Atomic Energy Agency: Vienna, Austria, 2023.
6. Yoon, B.H. Hydrogen refueling station deployment can't keep up with hydrogen vehicles. In *Electronic Times*; Korea Electricity Newpaper: Seoul, Republic of Korea, 2023.
7. Kang, S.G.; Kim, H. Liquid hydrogen technology trends and safety standards development status. In Proceedings of the Korean Institute of GAS Spring Meeting, Jeju, Republic of Korea, 26–27 May 2022.
8. Kim, K.S. *Development of Design Technology and Safety Standard in the Protection Wall for Blast Mitigations in a Hydrogen Station*; Research Plan Report, No. 20215810100020; Korea Gas Safety Corporation: Yeongwol, Republic of Korea, 2021.
9. *KGS FP216*; Facility/Technical/Inspection Code for Fuel Vehicles Refueling by Type of On-Site Hydrogen Production. Korea Gas Safety Corporation: Eumseong, Republic of Korea, 2020.
10. *KGS FP217*; Facility/Technical/Inspection Code for Fuel Vehicles Refueling by Type of Compressed Hydrogen Delivery. Korea Gas Safety Corporation: Eumseong, Republic of Korea, 2020.
11. You, E.G.; Kim, M.I.; Lee, Y.S.; Kang, S.G. A study in the risk assessment of LH2 refueling station. In Proceedings of the Korean Institute of GAS Spring Meeting, Jeju, Republic of Korea, 25–26 May 2023.
12. Kwon, D.; Choi, S.K.; Yu, C. Improved Safety by Crossanalyzing quantitative risk assessment of hydrogen refueling stations. *Int. J. Hydrogen Energy* **2022**, *47*, 10788–10798. [CrossRef]
13. Li, Z.; Pan, X.; Meng, X.; Ma, J. Study on the harm effects of releases from liquid hydrogen tank by consequence modeling. *Int. J. Hydrogen Energy* **2012**, *37*, 10624–17629. [CrossRef]
14. *KOSHA P-31-2001*; Predictive Methods for Accident Damage. Korea Occupational Safety and Health Agency: Ulsan, Republic of Korea, 2001.
15. Sochet, I.; Viossat, A.-L.; Rouyer, J.-L.; Hemmerich, P. Safe hydrogen generation by nuclear HTR. In Proceedings of the ICAPP-2004, Pittsburg, PA, USA, 13–17 June 2004.
16. Kang, H.S.; NO, H.C.; Kim, S.B. Application of the developed CFD analysis methodology to H_2 explosion accidents in an open space. *Int. J. Hydrogen Energy* **2017**, *42*, 1306–1317. [CrossRef]
17. Shen, R.; Jiao, Z.; Parker, T.; Sun, Y.; Wang, Q. Recent application of computational fluid dynamics (CFD) in process safety and loss prevention: A review. *J. Loss Prev. Process Ind.* **2020**, *67*, 104252. [CrossRef]
18. Kim, S.M.; Kang, H.S.; Choi, K.S. *radXiFoam v1.0*; VCE Solver; KAERI: Daejeon, Republic of Korea. Available online: https://github.com/ksm0226/radXiFoam (accessed on 23 September 2023).
19. OpenCFD Ltd. *OpenFOAM-v2112 User Guide*; Technical Report; ESI: Rungis, France, 2021. Available online: https://www.openfoam.com (accessed on 23 September 2023).
20. Kang, H.S.; No, H.C.; Kim, S.B.; Kim, M.H. Methodology of CFD analysis for evaluating H_2 explosion accidents in an open space. *Int. J. Hydrogen Energy* **2015**, *40*, 3075–3090. [CrossRef]
21. Wen, J.X.; Marnon, M.; Moretto, P.; Reinceke, E.-A.; Sathiah, P.; Studer, E.; Vyazmina, E.; Melideo, D. Statistics, lessons learned and recommendations from analysis of HIAD 2.0 database. *Int. J. Hydrogen Energy* **2022**, *47*, 3075–3090. [CrossRef]
22. ARIA. Available online: https://www.aria.developpement-durable.gouv.fr (accessed on 23 September 2023).
23. RISCAD. Available online: https://riss.aist.go.jp/sanpo/riscad (accessed on 23 September 2023).
24. Groethe, M.A.; Colton, J.D. *FY01 Annual Report on Hydrogen Safety in the World Energy Network*; Technical Report; SRI: Menlo Park, CA, USA, 2002.
25. Groethe, M.A. *FY02 Annual Report on Hydrogen Safety in the World Energy Network*; Technical Report; SRI: Menlo Park, CA, USA, 2002.
26. Bangert, P. *Machine Learning and Data Science in the Power Generation Industry*, 1st ed.; Elsevier: Amsterdam, The Netherlands, 2021.
27. Krishnan, S.; Yang, Z.; Hellerstein, J.; Stoica, I. What Is the Role of Machine Learning in Database? Available online: https://rise.cs.berkeley.edu/blog (accessed on 23 September 2023).
28. Nozu, T.; Tanaka, R.; Ogawa, T.; Hibi, K.; Sakai, Y. Numerical simulation of hydrogen explosion tests with a barrier wall for blast mitigation. In Proceedings of the ICHS-2005, Pisa, Italy, 8–10 September 2005.
29. Hwang, J.; Park, S.; Lee, Y.; Min, H. Research on predicting liquid-hydrogen explosion pressure using machine learning. In Proceedings of the Korean Institute of GAS Spring Meeting, Jeju, Republic of Korea, 26–27 May 2022.
30. Choi, J.Y.; Jeung, I.S.; Yoon, Y. Computational fluid dynamics algorithms for unsteady shock-induced combustion, part 1: Validation. *AIAA J.* **2000**, *38*, 1179–1187. [CrossRef]

31. Mossi, A.; Galarca, M.M.; Brittes, R.; Vielmo, H.A.; Franca, F.H.R. Comparison of spectral models in the computation of radiative heat transfer in participating media composed of gases and soot. *ABCM J.* **2012**, *34*, 112–119. [CrossRef]
32. Modest, M.F. *Radiative Heat Transfer, International Editions*; McGraw-Hill, Inc.: Singapore, 1993.
33. Kim, S.; Kim, J. Effect of radiation model on simulation of water vapor—Hydrogen premixed flame using flamelet combustion model in OpenFOAM. *Nuclear Eng. Tech.* **2022**, *54*, 1321–1335. [CrossRef]
34. Metghalchi, M.; Keck, J.C. Burning velocities of mixtures of air with methanol, isooctane, and indolence at high pressure and temperature. *Combust. Flame* **1982**, *48*, 191–210. [CrossRef]
35. Koroll, G.W.; Kumar, R.K.; Bowles, E.M. Burning velocities of hydrogen-air mixtures. *Combust. Flame* **1993**, *94*, 330–340. [CrossRef]
36. Kang, H.-S.; Kim, S.-M.; Kim, J. Safety issues of a hydrogen refueling station and a prediction for an overpressure reduction by a barrier using OpenFOAM software for an SRI explosion test in an open space. *Energies* **2022**, *15*, 7556. [CrossRef]
37. Python 3.7.0. Available online: https://python.org (accessed on 23 September 2023).
38. TensorFlow 2.3.0. Available online: https://www.tensorflow.org (accessed on 23 September 2023).
39. Keras 2.3.1. Available online: https://pypi.org/project/keras (accessed on 23 September 2023).
40. Wolfram. Neural Networks. Available online: https://reference.wolf.com/language/guide/NeuralNetworks.html (accessed on 23 September 2023).
41. Cho, J. Risk Assessment for a Severe Accident Using a Deep Learning Technology. In Proceedings of the KAERI, Daejeon, Republic of Korea, 13 December 2021.

processes

Article

Influence of Surface Roughness Modeling on the Aerodynamics of an Iced Wind Turbine S809 Airfoil

Leidy Tatiana Contreras Montoya [1,2], Adrian Ilinca [1,*] and Santiago Lain [2,*]

1 Mechanical Engineering Department, Faculty of Engineering, École de Technologie Supérieure, University of Québec, Montreal, QC H3C 1K3, Canada; leidy-tatiana.contreras-montoya.1@ens.etsmtl.ca
2 PAI+ Group, Energetics & Mechanics Department, Faculty of Engineering, Universidad Autonoma de Occidente, Cali 760030, Valle del Cauca, Colombia
* Correspondence: adrian.ilinca@etsmtl.ca (A.I.); slain@uao.edu.co (S.L.)

Abstract: Ice formation on structures like wind turbine blade airfoils significantly reduces their aerodynamic efficiency. The presence of ice on airfoils causes deformation in their geometry and an increase in their surface roughness, enhancing turbulence, particularly on the suction side of the airfoil at high angles of attack. An approach for understanding this phenomenon and assessing its impact on wind turbine operation is modeling and simulation. In this contribution, a computational fluid dynamics (CFD) study is conducted using FENSAP-ICE 2022 R1 software available in the ANSYS package. The objective was to evaluate the influence of surface roughness modeling (Shin et al. and beading models) in combination with different turbulence models (Spalart–Allmaras and k-ω shear stress transport) on the estimation of the aerodynamic performance losses of wind turbine airfoils not only under rime ice conditions but also considering the less studied case of glaze ice. Moreover, the behavior of the commonly less explored pressure and skin friction coefficients is examined in the clean and iced airfoil scenarios. As a result, the iced profile experiences higher drag and lower lift than in the no-ice conditions, which is explained by modifying skin friction and pressure coefficients by ice. Overall, the outcomes of both turbulence models are similar, showing maximum differences not higher than 10% in the simulations for both ice regimes. However, it is demonstrated that the influence of blade roughness was critical and cannot be disregarded in ice accretion simulations on wind turbine blades. In this context, the beading model has demonstrated an excellent ability to manage changes in roughness throughout the ice accretion process. On the other hand, the widely used roughness model of Shin et al. could underestimate the lift and overestimate the drag coefficients of the wind turbine airfoil in icy conditions.

Keywords: turbulent flow; rime and glaze ice; wind turbine rough surface; CFD simulation; aerodynamic loss

1. Introduction

Simulation and modeling techniques have become increasingly valuable in studying and analyzing ice accretion on wind turbines. These methods offer numerous advantages compared to experimental approaches, including cost-effectiveness, efficiency, and the ability to explore diverse icing events. Nonetheless, achieving accurate ice accretion modeling on wind turbines demands a comprehensive approach involving multiple disciplines, encompassing thermodynamics, aerodynamics, heat transfer, and mass transfer.

Computer-aided engineering techniques are commonly utilized to conduct these analyses, employing various tools and approaches for numerically solving coupled differential equations through finite element and finite volume methods [1–3]. Additionally, emerging methodologies such as deep learning, genetic algorithms, and artificial intelligence have been utilized in some instances, as evidenced by a few examples in recent years [4–6]. In real terms, during the last twenty years, several investigations have been performed to study the influence of icing on wind turbines using simulation methods. These studies aim to enhance our understanding of the phenomenon and provide valuable insights for designing more resilient and efficient wind turbine systems [7–10].

Computational fluid dynamics (CFD) is a common tool used to simulate various phenomena in turbomachinery [11], including the accumulation of ice on specific blade sections of the wind turbine [12]. The amount of ice accretion is predicted by numerical models, which can affect the airfoil geometry and decrease aerodynamic performance. Some examples of such studies include Etemaddar et al. [13]; Villalpando et al. [14]; Han, Kim and Kim [2]; Jin and Virk [15,16]; Yirtici et al. [17], Yirtici et al. [18].

In most computational fluid dynamics (CFD) studies focused on icing events in wind turbines, the flow along the blade span is often neglected due to its complexity and high computational costs. Instead, these studies usually focus on two-dimensional airfoil profiles placed at specific sections of the blade span. The decision to simplify the analysis using 2D airfoil models and neglecting the 3D rotating effect is driven by practical considerations. Accounting for the full 3D rotational flow requires significantly more computational resources and time. By focusing on specific sections of the blade span, researchers can still gain valuable insights into the aerodynamic behavior of the wind turbine under icing conditions while keeping the computational requirements within manageable limits. While this approach may introduce some limitations in capturing the complete 3D flow physics, it still provides valuable information for understanding the impact of ice accretion on the performance of wind turbines. Researchers continue to explore advancements in modeling techniques and computational capabilities to better address the complexities of 3D rotating flows in future studies [19–21].

After conducting a review of the literature, it was noticed that popular icing programs were initially designed for simulation icing in aeronautics [3,22–24]. The icing programs primarily developed for simulating aircraft icing are not specifically tailored to account for wind turbines' distinct operational and weather conditions. Wind turbines and aircraft exhibit distinct icing dynamics attributed to various factors. These factors encompass differences in operational altitude, angle of attack (AoA), airfoil positioning concerning the ground, the contrast between fixed-wing and rotating blades, and the impact of air compressibility at varying airspeeds [24].

Indeed, a few programs for in-flight icing have undergone adaptations, testing, and validation to replicate the formation of ice on the blades of wind turbines. The main emphasis of these adjustments lies in integrating the distinctive atmospheric conditions during operation and accounting for the geometry and rotation of the wind turbine blades. Consequently, models originally crafted for simulating icing on aircraft may display incongruent behaviors when employed in simulations for wind turbines. The distinctions in operational conditions and blade characteristics necessitate adjustments to the existing models. For instance, wind turbine blades experience different airspeeds, air temperatures, and humidity levels compared to aircraft wings. The rotation of wind turbine blades introduces a time-varying flow field, leading to variations in the ice accretion patterns.

Additionally, the complex geometry of wind turbine blades, including their twist and taper along the span, requires specific considerations for accurate simulation. By adapting, testing, and validating these in-flight icing programs for wind turbines, researchers and engineers aim to address these incompatibilities and develop models that effectively capture the icing behavior unique to wind turbines. These modifications enable more reliable assessments of ice accretion effects and aid in designing and optimizing wind turbine systems for safe and efficient operation in icy conditions [12,14,24]. For instance, Etemaddar et al. [13] concluded that wind turbines could operate under icy conditions by reducing their cut-off speed to minimize the risk of damage to components. The authors used LEWICE 1.6 software and the Blade Element Momentum (BEM) code WT-Perf to reach this conclusion. Another example is the study of Han, Kim and Kim [2], in which a CFD model was used to suggest that modifying the pitch angle is necessary for maintaining steady power in wind turbines during icing periods. However, the study concluded that maximum turbine efficiency under these conditions can be achieved below the rated speed by operating at a variable speed via generator torque control.

While software packages like NASA's LEWICE and FENSAP-ICE (currently part of ANSYS) are commonly employed for investigating icing, they have predominantly been utilized in aeronautical applications. LEWICE and FENSAP-ICE have been extensively used and validated for simulating ice accretion on aircraft surfaces. These packages incorporate specialized algorithms and models tailored to aircraft icing phenomena, considering droplet impingement, ice shape evolution, and ice shedding. However, their application to wind turbine icing simulations requires careful evaluation and adaptation to account for wind turbines' unique characteristics and operational conditions.

To effectively utilize these software packages for wind turbine icing studies, validating their performance against experimental data and real-world observations specific to wind turbines is essential. This process involves assessing their ability to capture ice accretion patterns, evaluating the effects on aerodynamic performance, and accounting for the complex interactions between rotating blades, atmospheric conditions, and ice-shedding dynamics. By conducting thorough testing and validation, researchers and engineers can enhance the reliability and accuracy of these software packages for wind turbine icing simulations, enabling a better understanding and mitigation of icing-related challenges in the wind energy industry [19,25].

However, using these two programs, various investigations have devised approaches for replicating ice accumulation on wind turbine blades [15,26–28]. Homola et al. [29] employed FENSAP-ICE to anticipate the formation of ice on the airfoils of the NREL 5MW benchmark wind turbine. In the study of Etemaddar et al. [13], FLUENT was utilized for aerodynamic computations and ice accretion was simulated using LEWICE. The resultant lift and drag coefficients, C_L and C_D, were computed through ANSYS FLUENT and then corroborated against experimental measurements from the wind tunnel at "LM Wind Power".

The simulation of ice accretion usually comprises a sequence of four primary modules [24,30,31]. The first module involves aerodynamic calculations to determine the flow characteristics around the wind turbine blade. This is achieved by solving the Navier–Stokes equations, including continuity (Equation (1)), momentum (Equation (2)), and energy (Equation (3)).

$$\frac{\partial \rho_a}{\partial t} + \vec{\nabla} \cdot \left(\rho_a \vec{U} \right) = 0 \tag{1}$$

$$\frac{\partial \rho_a \vec{U}}{\partial t} + \vec{\nabla} \cdot \left(\rho_a \vec{U} \vec{U} \right) = \vec{\nabla} \cdot \sigma^{ij} + \rho_a \vec{g} \tag{2}$$

$$\frac{\partial (\rho_a E_a)}{\partial t} + \vec{\nabla} \cdot \left(\rho_a \vec{U} H_a \right) = \vec{\nabla} \cdot \left[k_a \left(\vec{\nabla} T_a \right) + U_i \tau^{ij} \right] + \rho_a \vec{g} \cdot \vec{U} \tag{3}$$

In this context ρ_a denotes air density and $\vec{U}$ represents the fluid velocity vector. The subscript a pertains to the air solution, T denotes the static air temperature measured in Kelvin, σ^{ij} signifies the stress tensor, k_a is the thermal conduction coefficient, and E and H are the total initial energy and enthalpy, respectively. Usually, such equations are supplemented with turbulence models aimed at describing the fluctuating characteristics of the flow around the wind turbine blades.

The second module focuses on calculating the trajectory of water droplets in the airflow. This can be performed using either a Lagrangian approach, which tracks individual droplets, or an Eulerian approach, which considers the behavior of droplets as a continuous phase. Simulating the droplet trajectories, the module determines where the droplets impinge on the blade surface. The third module entails thermodynamic computations to evaluate the rate of ice accumulation in a specific location over a designated period. These calculations consider air temperature, humidity, and droplet properties to estimate how ice accumulates on the wind turbine blade surface. The fourth module deals with the geometry of the wind turbine blade and enables updating the blade shape as the ice grows. This is important because ice accretion alters the blade's geometry and surface roughness, affecting its aerodynamic performance. The module ensures that the evolving geometry due to ice growth is accurately represented in the simulation. By integrating these four modules into a comprehensive simulation framework, researchers and engineers can obtain insights into the aerodynamic effects of ice accretion on wind turbine blades and make informed decisions regarding turbine design, operation, and ice protection strategies.

This layer of ice enhances the likelihood of boundary layer detachment on the suction side of the airfoil, known as the extrados, resulting in an aerodynamic stall occurring at a lower AoA compared to a clean scenario without ice. Given the intricate nature of this phenomenon, the fluctuations in flow velocity and the formation of eddies become exceedingly challenging to predict. Addressing these disturbances at a small scale necessitates employing an unsteady Navier–Stokes equation and incorporating a high level of detail. However, these vortices and fluctuations diminish in size at higher Reynolds numbers. Hence, the Reynolds averaged Navier–Stokes (RANS) approach becomes more appropriate where the velocity fluctuations in the flow field are averaged over time [24].

Various Reynolds-averaged Navier–Stokes (RANS) models and adaptations are accessible, contingent upon the specific application, each relying on distinct methodologies for computing turbulent eddy viscosity. Addressing this diversity, numerous turbulence models, including but not limited to Spalart–Allmaras, k-epsilon, k-omega, shear-stress transport (k-omega SST model), and large eddy simulation (LES), are usually employed. These models are extensively applied in studying flows surrounding airfoils and wind turbines [24].

The Spalart–Allmaras model (SA) is a one-equation turbulence model. This additional equation models the turbulent viscosity transport. This model is employed in aerodynamics because of its compromise between computational cost and accuracy [24]. The k-epsilon model stands as a turbulence model with two equations, addressing both the turbulent kinetic energy (k) and the rate at which it dissipates (epsilon). It has demonstrated adequate performance in simulating ice accretion, gaining widespread popularity owing to its numerical stability, efficiency in computational resources, and rapid convergence rate [24,29,32]. The k-omega model is also a two-equation model that solves the turbulent kinetic energy (k) and the specific dissipation rate of kinetic energy (omega). It is used in cases where k-epsilon is insufficient but has a lower convergence rate as it is more non-linear [33]. Finally, the k-omega SST model, introduced by Menter [34], represents a two-equation model that amalgamates features from the k-omega and k-epsilon models. It employs these models in distinct flow regions, activating the k-omega model close to the wall and resorting to the k-epsilon model when situated away from the surface [24]. Such a turbulence model is more accurate than others because of its ability to handle flow recirculation zones, offering a satisfactory approximation of flow separation and elucidating the creation of distinct vortices at both the trailing and leading edges [24,34,35].

Reference [24] addressed the various modeling approaches and simulation techniques available for wind turbine icing. This article specifically highlighted the distinct characteristics of wind turbine icing simulations, including the unique operational conditions and software capabilities required. Furthermore, the potential and suitability of various software tools for wind turbine icing simulations were thoroughly discussed. In particular, the adaptability of software FENSAP-ICE for wind turbine icing simulations, its integration with ANSYS, and the insights provided by the previously published articles collectively contribute to the growing understanding and progress in the modeling and simulating of icing on wind turbines. These developments facilitate more accurate assessments of the impact of icing on wind turbine performance and aid in designing mitigation strategies to ensure safe and efficient wind turbine operations.

Recently, Martini, Ibrahim, Contreras M, Rizk and Ilinca [12] performed a computational analysis on a NACA 64-618 airfoil of the iced NREL 5MW benchmark wind turbine, utilizing ANSYS FLUENT and FENSAP-ICE. The investigation focused on evaluating the precision of aerodynamic loss estimation for airfoils affected by icing using two turbulence models (Spalart–Allmaras and k-ω SST) and the effect of surface roughness distribution using the Shin et al. [36] model available in the ICE3D module in FENSAP-ICE. Etemaddar et al. [13] and Homola et al. [29] used published research, numerical investigations and experimental studies in the existing literature for comparison. These authors found that neglecting surface roughness from calculations underestimates the effect of ice on aerodynamic performance. On the other hand, they concluded that the choice of turbulence model had a limited influence on the resulting aerodynamic losses caused by icing, compared to the impact of considering roughness. A similar conclusion was attained by the authors of [37], where they found that the extent of the ice-induced roughness and its height drove the decrease in the aerodynamic performance of the studied airfoils as the angle of attack increased.

The previous discussion demonstrates that the accuracy of icing modeling depends heavily on roughness, which has been emphasized by various authors in the field [24,38–40]. Roughness plays a significant role in airfoil performance as it affects the boundary layer transition and flow separation, critical factors in aerodynamic efficiency [39–41]. Even small amounts of ice can significantly affect an airfoil's performance. Therefore, it is crucial to consider the effects of roughness at every step of the ice growth calculation [37,41]. Considering the roughness height as a parameter in heat transfer analysis ensures that the effects of surface roughness on convective heat transfer are adequately accounted for, leading to more reliable and precise thermal assessments and design considerations [42,43]. For example, modelers studying aircraft icing extensively utilized computational fluid dynamics (CFD) to analyze the local heat transfer coefficient across various aircraft components. The findings suggested that surface roughness could substantially amplify local heat transmission, even in the presence of a thin layer of ice [44].

However, determining the roughness height of structures such as wind turbine blades necessitates conducting experiments, as detailed in the study of Blasco, Palacios and Schmitz [40]. These experiments can be tedious and costly, and their applicability is limited as they depend on the airfoil type and wind tunnel configuration. In response to the above, different research centers have various ways of describing roughness [24]. For instance, the traditional NACA roughness model originated through distributing typical grain sizes uniformly from the leading edge downstream on both the pressure and suction surfaces [45]. The sand-grain roughness model of Shin et al. [36], initially designed for aeronautics, is the most frequently employed correlation for predicting ice surface roughness along wind turbine blades. This model was specifically designed to match the ice shapes predicted by the LEWICE code to experimental ones and for the atmospheric conditions typical of aviation, which makes the model lack generality [37]. This empirical correlation relies on the Shin and Bond formula, which computes the non-dimensional height of small-scale surface roughness: k_s/c as a function of static temperature, airfoil chord length c, median volume diameter (MVD), liquid water content (LWC), and the relative wind speed (see Equation (4)).

$$k_s = 0.6839\left[\frac{k_s/c}{(k_s/c)_{base}}\right]_{LWC}\cdot\left[\frac{k_s/c}{(k_s/c)_{base}}\right]_{T}\cdot\left[\frac{k_s/c}{(k_s/c)_{base}}\right]_{MVD}\cdot\left(\frac{k_s}{c}\right)_{base}\cdot c \quad (4)$$

where each sand-grain roughness parameter is provided by Shin et al. [36]:

$$\left[\frac{k_s/c}{(k_s/c)_{base}}\right]_{LWC} = 0.5714 + 0.2457(LWC) + 1.2571(LWC)^2 \text{ for LWC } > 1\text{ g/m}^3 \quad (5)$$

$$\left[\frac{k_s/c}{(k_s/c)_{base}}\right]_{T} = 0.047(T) - 11.27 \quad (6)$$

$$\left(\frac{k_s}{c}\right)_{base} = 0.001177 \quad (7)$$

$$\left[\frac{k_s/c}{(k_s/c)_{base}}\right]_{MVD} = \begin{cases} 1 & MVD \leq 20 \\ 1.667 - 0.0333(MVD) & MVD > 20 \end{cases} \tag{8}$$

An alternative to the previous roughness model is the so-called beading model available in ANSYS-FENSAP-ICE. This model considers both constant and varying distributions of sand-grain roughness and integrates them into the existing turbulence models. After activation of the beading model, the prediction of sand-grain roughness height on the surface caused by moving and freezing beds is enabled [38], i.e., there is an automatic transfer of the spatially and temporally evolving roughness data to the airflow module at the end of each shot. As a result, the roughness height changes dynamically and depends on the contaminated area.

As a matter of fact, the sand-grain roughness model of Shin et al. [36] is used in nearly every CFD investigation focused on icing, where it is employed to gauge the roughness of the ice-covered surfaces on wind turbine blades [24,37]. However, as far as we know, there is a scarcity of icing studies in the literature comparing the influence of the surface roughness model, particularly the beading model, on the aerodynamic performance of wind turbine airfoils.

Therefore, this paper presents a numerical study conducted in the iced S809 airfoil where the impact of the surface roughness model on its aerodynamic coefficients is addressed. Moreover, the influence of the employed turbulence model is also evaluated. The results for the clean airfoil were validated with the experiments in [46] and simulations of [47]. The predicted rime ice shape was validated with the experimental studies of [48]. Two methodologies for estimating roughness (the Shin et al. [36] model and beading model) are compared, as well as the effect of icing type, rime ice, and the much less studied case of glaze ice on aerodynamic performance. Such influence of icing conditions and roughness is evaluated and discussed by examining the lift and drag coefficients and skin friction and pressure coefficients.

2. Geometrical Configuration and Mesh Generation

The geometrical model was based on an airfoil used by the National Renewable Energy Laboratory (NREL). The S809 airfoil is used in the two-bladed NREL Phase VI turbine [49], which is present along the entire length of the blade. This study will focus on the section at 95% radius of the blade, primarily because it corresponds to the area closest to the blade tip, where ice accumulation is most prevalent. Furthermore, this region of the blade is the main region responsible for turbine performance [50,51]. The chord of the S809 at this radius is 0.358 m, and the airfoil has the following specifications [52]:

Max thickness: 21% at 39.5% of the chord from the leading edge.

Max camber: 1% at 82.3% of the chord from the leading edge.

The geometric model was constructed using the DesignModeler 2022 R1 software, which is part of the ANSYS suite. Figure 1 illustrates the model, showcasing a sizable computational domain encompassing crucial flow disturbances, particularly downstream. The positioning of boundaries and the size of the domain were determined based on prior studies that demonstrated satisfactory agreement with experimental and numerical data. For example, the computational domain used by Villalpando, Reggio and Ilinca [14] and Hildebrandt and Sun [39] was extended by 12.5 chords in front of the airfoil and 20 chords behind; in the case of Zanon et al. [53], the domain boundaries were located at 20 chord lengths. Considering the above, the chosen dimensions of the computational domain were the same as the ones used by [53]. This guarantees that the boundary conditions applied to the external domain do not disrupt the flow in the neighboring region or compromise the accuracy of the results.

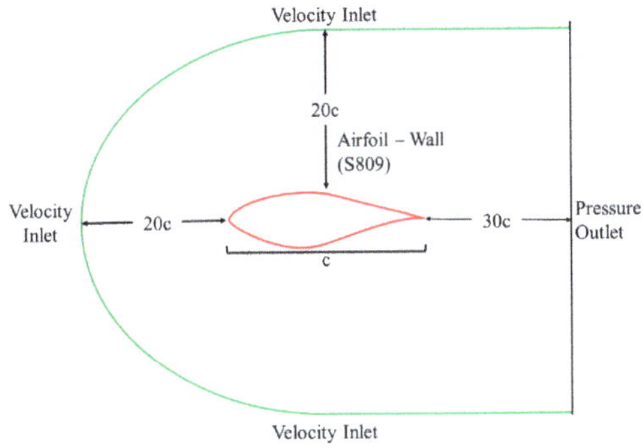

Figure 1. Illustration of the computational domain geometry.

The computational domains have been discretized using a non-structured grid generated with ANSYS Meshing 2022 R1 software. Figure 2 provides an overview of the utilized meshes. To adequately describe the boundary layer development, the mesh around the airfoil has an O-grid topology, including 25 prism layers (see Figure 3). Beyond the prisms zone, a non-structured mesh based on tetrahedra was created, ensuring their aspect ratio is close to the prisms to guarantee a smooth transition between both mesh regions.

The following sections present the calculations for ice accretion, including an estimation of the aerodynamic characteristics of clean and iced airfoils. Additionally, the performance losses occurring under specific icing scenarios are evaluated.

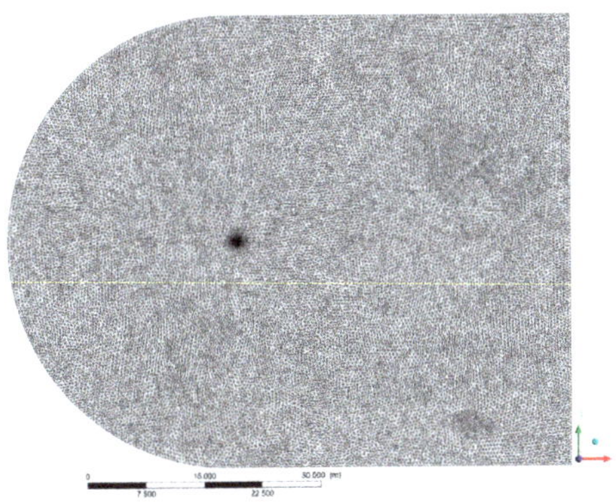

Figure 2. Overview of the mesh in the computational domain.

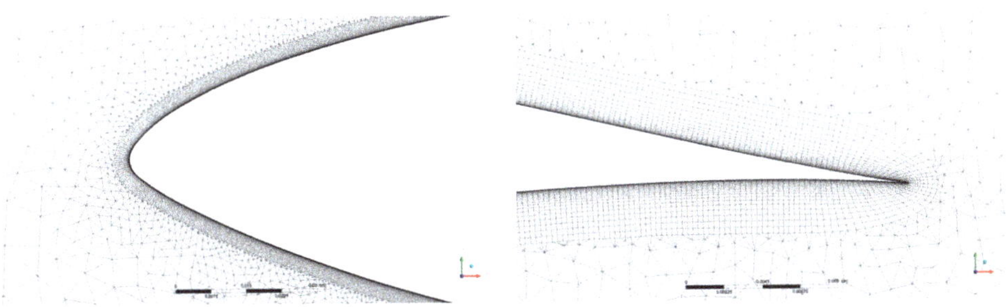

(a) (b)

Figure 3. Detail of the prism layers around the airfoil. (**a**) Leading edge. (**b**) Trailing edge.

3. Numerical Simulation Setup and Verification Study

Simulations of ice accretion on the S089 airfoil were carried out with the software FENSAP-ICE integrated into the ANSYS platform. They were performed on a desktop computer with a Windows 11 operating system, Intel Core i5 10th generation processor @ 2.90 GHz, and 24 GB RAM.

FENSAP-ICE operates in a modular system (see Figure 4) with three components: the FENSAP module is used for aerodynamic calculations, DROP3D is used for droplet impingement, and ICE3D is used for ice growth calculations. Moreover, FENSAP-ICE has two methods to estimate the roughness due to ice accretion: the beading model and the Shin et al. [36] model; both methods were used to compare their effect on the airfoil aerodynamic coefficients. Data regarding the simulation setup can be found in Table 1 for two ice conditions: rime (dry regime) and glaze (wet regime).

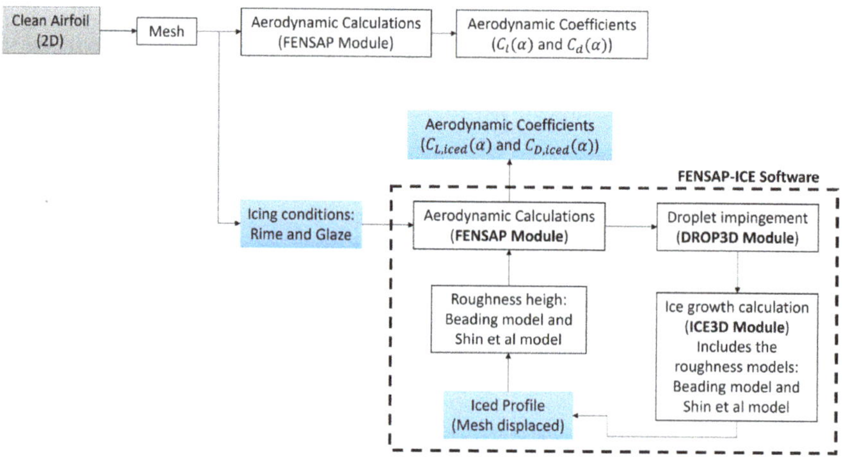

Figure 4. Flowchart of the CFD simulation using FENSAP-ICE.

Table 1. Overview of the simulation parameters.

Parameters	Rime	Glaze
Air speed U [m/s]	42	42
Angle of attack, AoA [°]	4	4
Reynolds number	3.4×10^6	3.4×10^6
MVD [μm]	20	27
LWC [g/m^3]	0.05	0.12
Ice density [kg/m^3]	750	917
Accretion time (min)	30	30
Temperature (°C)	−10	−5
Turbulence model	Spalart–Allmaras/k-ω SST	Spalart–Allmaras/k-ω SST
Convergence criterion	1×10^{-5}	1×10^{-5}

Note: the parameters for the rime condition were taken from Han, Palacios and Schmitz [48] to validate and compare the results. The glaze ice condition parameters were estimated based on in situ measurements from the non-profit applied research center "Nergica" in Québec, Canada.

The validation of the flow around the clean airfoil was performed with the two considered turbulence models under the same conditions described in Table 1. The Spalart–Allmaras (SA) model was chosen because it is widely used in aerodynamics and due to its good computational cost/accuracy performance ratio and the k-ω SST model was chosen because it is more robust and performs better in flows with strong adverse pressure gradients.

A spatial verification test or mesh independence study was performed in the no-ice conditions. Three distinct grids were created to verify mesh convergence, adjusting the number of elements. Localized mesh refinement was opted for, indicating that variations in spatial discretization are concentrated around the airfoil. The procedure commenced with a coarse discretization at the outer limit of the boundary layer and progressed to a finer discretization along the profile wall (see Table 2).

Table 2. Number of elements in the considered meshes.

Mesh	Number of Elements
1	870,726
2	1,235,620
3	1,655,379

Intending to resolve the boundary layer development, all the grids use a thin layer around the airfoil, which is discretized by 25 layers of prisms. The initial height of these layers was set at 7.6×10^{-6} m, a value determined to maintain $y^+ < 2$. This value of the variable y^+ is suggested by the numerical requirements imposed by the k-ω SST turbulence model.

The CFD grid independency study involved calculations in successively refined meshes, evaluating the convergence of the most relevant variables. The lift and drag coefficients, C_L and C_D, were chosen in this case. They are defined as:

$$C_L = \frac{L}{\frac{1}{2}\rho c U^2}; \; C_D = \frac{D}{\frac{1}{2}\rho c U^2} \quad (9)$$

where L and D represent the dimensional lift and drag forces, and c is the airfoil chord length.

Table 3 showcases the verification outcomes for the generated meshes. The convergence error, expressed as a percentage, was computed by assessing the difference in the aerodynamic coefficient for each mesh compared to the most refined mesh, i.e., mesh no. 3.

Table 3. Convergence errors for each mesh.

	K-ω SST Model				Spalart–Allmaras Model			
Mesh No.	C_L	C_L Error	C_D	C_D Error	C_L	C_L Error	C_D	C_D Error
1	0.5405	1.66%	0.014877	5.81%	0.5617	3.81%	0.013611	3.55%
2	0.5506	0.18%	0.015824	0.19%	0.5459	0.89%	0.014540	2.84%
3	0.5496		0.015794		0.5411		0.014058	

As can be seen in Table 3, the maximum difference between mesh no. 2 and 3 is less than 3%. Either mesh could be used for the study, so the computational cost was also used to decide. Although mesh no. 3 has 33% more elements, the difference in computational time was minimal, so it was decided to work with it. For example, the complete simulation (i.e., of the three modules FENSAP, DROP3D, and ICE3D) took 24 h for mesh #2 and 26 h for mesh no. 3. The convergence study was performed with the two turbulence models, the k-ω SST being the one with the lowest error percentage. In addition, as depicted in Figure 5, the lift and drag coefficients attain a nearly constant value, demonstrating that the achieved solution is mesh-independent.

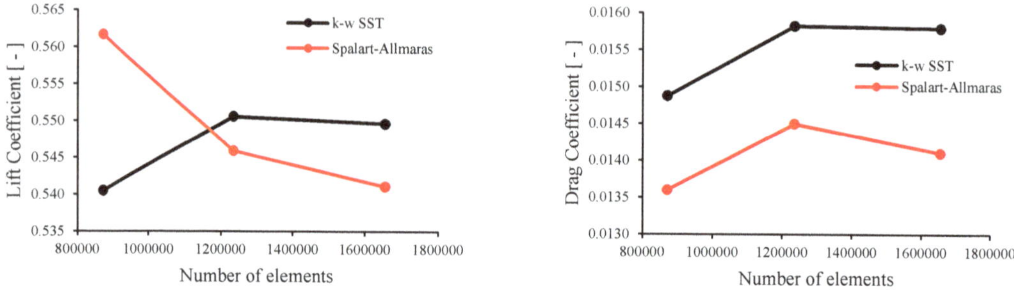

Figure 5. Aerodynamic coefficients vs. the number of elements. Lift coefficient (**left**) and drag coefficient (**right**).

As previously mentioned, FENSAP-ICE has three modules: FENSAP, DROP3D, and ICE3D; for each of them, the values of the relevant variables are given in Table 4.

Table 4. Parameter values in each ANSYS FENSAP-ICE module.

Parameters	FENSAP	DROP3D	ICE3D
Characteristic length		0.358 m	
Airspeed U		42 m/s	
Air static pressure		101325 Pa	
Air static temperature		−10 °C (rime) −5 °C (glaze)	
Velocity angle of attack		4°	
Liquid water content (LWC)	n. a	0.05 g/m^3 (rime) 0.12 g/m^3 (glaze)	0.05 g/m^3 (rime) 0.12 g/m^3 (glaze)
Droplet diameter (MVD)	n. a	20 μm (rime) 27 μm (glaze)	n. a
Droplet distribution	n. a	Monodisperse	n. a
Total time of ice accretion	n. a		1800 s
Roughness model	n. a		Beading model/ Shin et al. [36] model

Upon completing the simulation, ICE3D produced the geometry of the iced airfoil, generating the computational mesh, which was subsequently employed to estimate aerodynamic lift and drag coefficients under rime and glaze ice conditions.

4. Results and Discussion

The results presented here demonstrate the validation of the obtained aerodynamic parameters for the S809, comparing them with experimental data and findings from prior numerical studies on this airfoil [46–48]. The performed CFD simulations consider two turbulence models in conjunction with two roughness scenarios and two different types of icing (rime and glaze).

In computational fluid dynamics (CFD) simulations, several models are used to account for turbulence development, including the models chosen in this work: Spalart–Allmaras (SA) and k-ω SST, which belong to the Reynolds-averaged Navier–Stokes (RANS) family. These models have been widely studied in the literature for wind turbines [35]. The k-ω SST model is advantageous for modeling flows around wind turbine blades as it allows analyzing the turbulent boundary layer at large angles of attack [54]. On the other hand, the Spalart–Allmaras model is often used due to its low computational cost and acceptable precision in turbulent flow modeling, particularly in aerospace applications [29, 55]. However, it has a disadvantage in describing the boundary layer behavior with adverse pressure gradients compared to the k-ω SST model [12].

4.1. Clean Airfoil

Clean airfoil drag and lift coefficients were estimated to assess the aerodynamic loss due to icing. Then, the results were compared with the measured coefficients on the same airfoil under varying angles of attack, serving as a validation for the simulation. The CFD simulation results of the S809 airfoil were validated with experimental data from Somers [46]. The numerical flow coefficients were also compared with the CFD study by Tan [47], using the same turbulence model (Spalart–Allmaras).

The experimental study carried out by Somers [46] took place at the low-turbulence wind tunnel of the Delft University of Technology Low-Speed Laboratory in the Netherlands. The study's primary objectives were to achieve a maximum lift coefficient insensitive to roughness and obtain low-profile drag coefficients within a specific range of lift coefficients and Reynolds numbers. The airfoil model used in the experimental study was made of aluminum, with a chord length of 600 mm and a span of 1248 mm. The model was tested at various Reynolds numbers based on the airfoil chord, ranging from 1.0×10^6 to 3.0×10^6. The tests were conducted with the airfoil surface both smooth (without roughness) and with roughness introduced at specific locations (0.02c on the upper surface and 0.05c on the lower surface) to promote the transition of the boundary layer from laminar to turbulent.

Regarding Reynolds number effects, at the design Reynolds number (2.0×10^6), the maximum lift coefficient achieved was approximately 1.01, which met the design objective of 1.0 [46]. The trailing-edge stall behavior was gentle and occurred at approximately 20° angle of attack. Minimal hysteresis was observed at angles of attack beyond the maximum lift and below the minimum lift. Concerning the effect of roughness, Somers [46] found that the maximum lift coefficient was not significantly affected by the addition of roughness at any of the Reynolds numbers tested. However, when using fixed transition, the drag coefficients tended to be excessively high at both low and high lift coefficients. This fact was primarily attributed to the roughness height being comparable to the boundary-layer thickness on the upper surface at low lift coefficients and on the lower surface at high lift coefficients. This resulted in an additional contribution to drag known as pressure drag, caused by the roughness itself. The impact of roughness on drag was more pronounced at higher Reynolds number.

The simulations by Tan [47] assessed the aerodynamic performance of the NREL S809 airfoil with a trailing-edge flap. The length of the airfoil chord was 1 m, and a structured mesh of 150,000 cells around the airfoil was used. The simulations were performed using the commercial CFD software ANSYS Fluent 2022 R1 to solve the RANS equations in conjunction with three turbulence models (Spalart–Allmaras, SST k-ω, and Wray–Agarwal) at Reynolds number 10^6 (free stream velocity equal to 15 m/s) at angles of attack from 0° to 20°. The computations were performed with double precision, with a second-order upwind scheme for the convection terms and a second-order central difference scheme for the diffusion terms, and the SIMPLE algorithm was used for the pressure-velocity coupling. Only the results of the unmodified profile with the Spalart–Allmaras turbulence model were used to validate this study.

As shown in Figure 6, the lift coefficients estimated with the two turbulence models are close to those obtained experimentally by [46]; however, at an AoA of 15°, the CFD results slightly overpredict the experimental values. The disagreement between computations and experimental data at this AoA can be explained by the stall experienced by the airfoil, which is a phenomenon difficult to compute.

In the case of the drag coefficient (see Figure 7), both turbulence models presented a slight overestimation for angles of attack of 0° and 5° and an underestimation for angles of attack of 10° and 15°. This behavior is similar to that found by [12] and is associated with the stall phenomenon. The difference between the present SA computations and those of [47] are only significant at an AoA of 15°, and they are attributed to the known drawbacks of the SA turbulence model in regions characterized by high-pressure gradients [1]. Nevertheless, the disparity noted at elevated angles of attack between the experimental data and numerical drag outcomes for the no-iced airfoil was deemed less critical because it appears beyond the operational ranges of the considered wind turbine [9,49].

Figures 8 and 9 depict the streamlines around the S809 airfoil, showing the flow recirculation zone (enclosed in the red circle) at an angle of attack of 15° for both turbulence models; the presence of such zones generates a decrease in lift and an increase in drag in the airfoil. The maximum error obtained with the k-ω SST model in C_L was below 10%, while the same error obtained with the Spalart–Allmaras model was 15%. This shows that the k-ω SST model better estimates the lift coefficient. On the other hand, when comparing the present C_L curve obtained with SA turbulence model with the results of [47], a maximum difference of 6% is observed even though the same turbulence model was used; this may be associated with the fact that this author used a much coarser structured mesh and the ANSYS Fluent software.

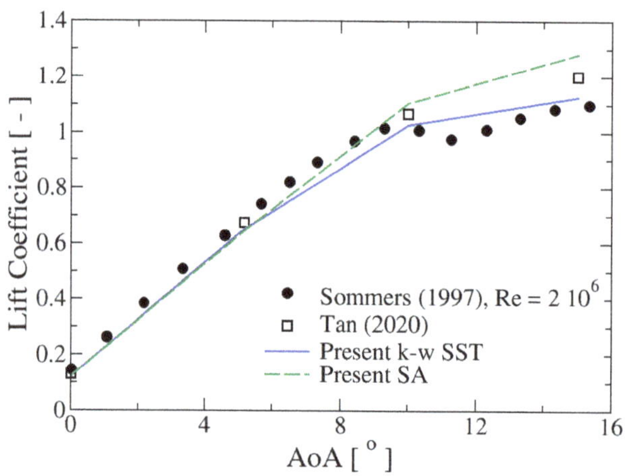

Figure 6. Lift coefficient versus AoA for the S809 airfoil, Sommers's data from [46], Tan's data from [47].

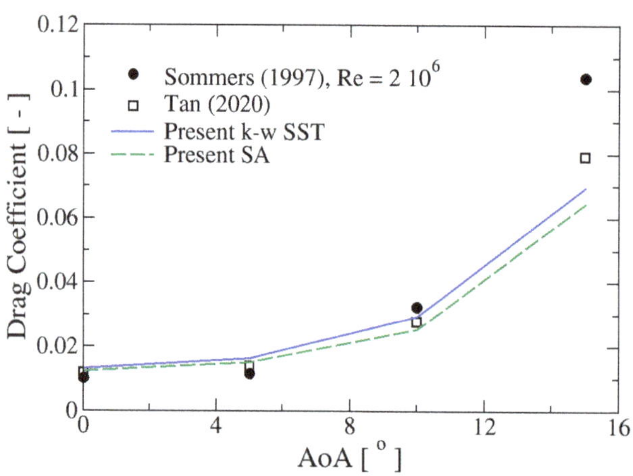

Figure 7. Drag coefficient versus AoA for the S809 airfoil, Sommers's data from [46], Tan's data from [47].

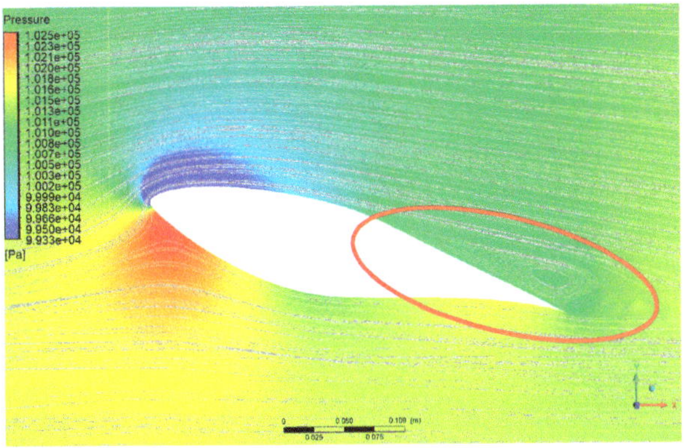

Figure 8. Pressure contours and streamlines at AoA of 15° of the clean profile using Spalart–Allmaras turbulence model.

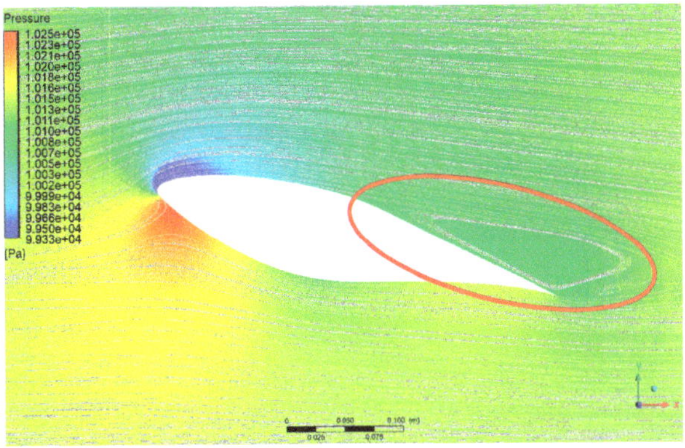

Figure 9. Pressure contours and streamlines at AoA of 15° of the clean profile using k-ω SST turbulence model.

Comparing the outcomes of both turbulence models, the lift coefficient estimate obtained with the Spalart–Allmaras model was approximately 5% higher on average than that obtained with k-ω SST. Consistently, the drag coefficient for the Spalart–Allmaras estimate was about 10% lower than the estimate obtained with the k-ω SST model.

Further insights about the flow behavior around the airfoil under separated conditions can be devised by analyzing the skin friction, C_f, and pressure, C_p, coefficients. They are defined as:

$$C_f = \frac{\tau_w}{\frac{1}{2}\rho U^2};\ C_p = \frac{p_g}{\frac{1}{2}\rho U^2} \tag{10}$$

$$\tau_w = \rho {u^*}^2 \tag{11}$$

where τ_w is the wall shear stress and p_g is the local gauge pressure.

Figure 10 shows the skin friction coefficient, C_f at AoA of 15°. The trends of both turbulence models are very similar, showing overlapping curves along the intrados (pressure side) and small differences in the extrados (suction side). In the figure, the stagnation and separation points can be identified as those with $C_f = 0$; the first ones are located close to the leading edge, while the second ones are approximately placed at half of the chord. It is seen that the Spalart–Allmaras model predicts a later boundary layer separation than the k-ω SST model, which is consistent with the streamlines depicted in Figures 8 and 9.

Figure 11 displays the pressure coefficients C_p also at an AoA of 15°. The upper side curves correspond to the suction side (extrados) and those of the lower side to the pressure side (intrados). Both turbulence models predict very similar values of C_p in the intrados but the SA model provides more negative pressure coefficients than the k-ω SST model along the first half of the chord in the extrados. The two models provide similar values in the second half of the chord. The higher suction forecasted by the SA model reflects a higher lift coefficient than that obtained with the k-ω SST model, as confirmed in Figure 6; consequently, the SA also yields a lower drag coefficient than the two-equation model.

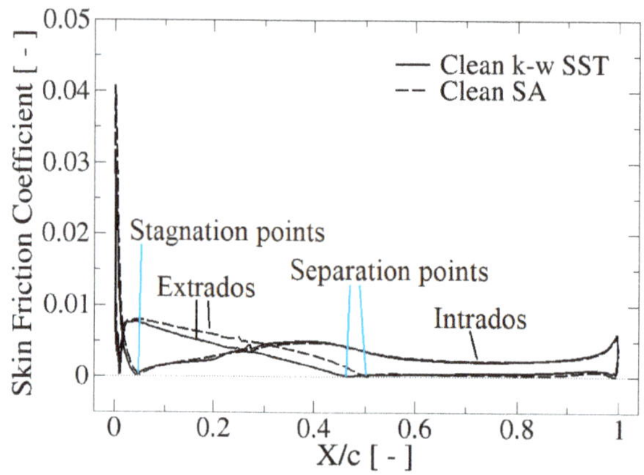

Figure 10. Skin friction coefficient at AoA of 15° of the clean profile with both turbulence models.

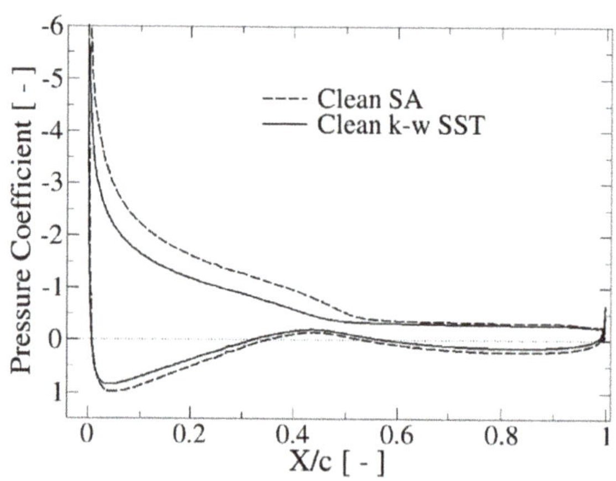

Figure 11. Pressure coefficient at AoA of 15° of the clean profile with both turbulence models.

In the subsequent analysis, considering that the likelihood of flow separation is increased by icing, the performance of both turbulence models in the presence of rime and glaze ice was investigated by comparing their results with those of the clean airfoil. Moreover, the effect of the roughness modeling on the aerodynamic coefficients is also studied.

4.2. Iced Airfoil

In Figure 12, the accumulation of ice at the leading edge is illustrated. A minor portion of the collected supercooled water does not instantly freeze upon contact with the airfoil's leading edge, leading the water to subsequently flow back and freeze at the trailing edge beyond the impingement zone, a phenomenon observed in the study by Fortin and Perron [42].

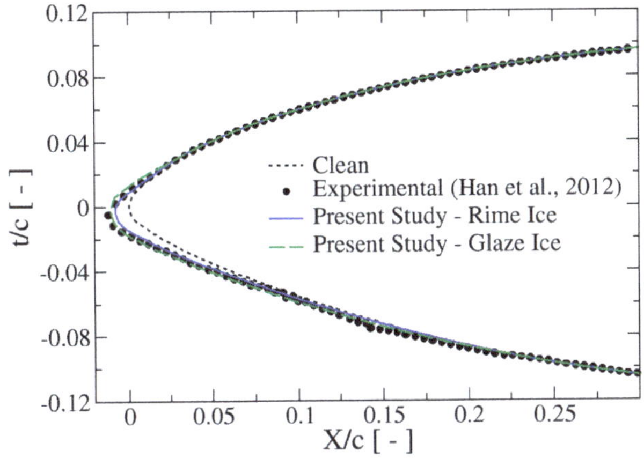

Figure 12. Comparison of the predicted ice shape (non-dimensional thickness versus non-dimensional chord) with the experimental results of Han et al. [48]. The clean shape, i.e., without ice, is also shown for comparison.

Concerning the leading edge (see Figure 12), the ice shape closely resembled the one depicted by Han et al. [48]. This similarity enables us to assert that the model and the simulation tools employed are adequate. However, the aerodynamic coefficients could not be validated because this study only showed the shape of the ice generated at the wind tunnel airfoil. However, since the alteration in the airfoil aerodynamic coefficients depends on the airfoil shape, it was decided that simulating a shape similar to that obtained experimentally would be sufficient. The validation was performed for both turbulence models.

It can be observed in Figure 12 that the ice shape for the two types of ice is quite similar, agreeing with [48], which is remarkable. Despite a 50% increase in LWC and a 35% increase in MVD in the case of the glaze ice, the difference in the accumulated ice shape at the leading edge is minimal and even comparable to the results obtained experimentally under rime ice conditions.

Since FENSAP-ICE performs the remeshing for the iced airfoil, it was decided to verify if the condition imposed for the y^+ was still satisfied. This parameter is essential since the performance of the k-ω SST model depends on its ability to describe the boundary layer development, so this value should be below 5 (or even less than 1) for better accuracy [56]. Although not shown, the values of y^+ for the clean airfoil are less than 1 throughout the profile, except for the leading-edge zone (30% of the profile), where it reaches a maximum value of 2.15. However, in the iced airfoil cases, the y^+ moderately increases, reaching a maximum value of 3.14, mainly at the airfoil's leading edge. This is to be expected since it is the area where the ice accumulates. Therefore, the corresponding elements experience a rearrangement or deformation.

Despite the increase in y^+, it can be stated that the remeshing process for the conditions studied is acceptable, and the meshes generated by the software satisfy the primary conditions for the turbulence model to provide accurate results. This, in turn, highlights the importance and attention that should be paid to the initial meshing process.

The results obtained for each simulated ice regime (dry and wet) are presented below. For each case, the effect of the turbulence and roughness models on the aerodynamic coefficients of the iced airfoil was evaluated.

4.2.1. Dry Regime—Rime Ice

Rime ice forms when supercooled water droplets, which are liquid droplets at temperatures well below freezing, i.e., lower than −10 °C, come into contact with the cold surfaces of the turbine blades. These supercooled droplets freeze quickly upon contact, adhering to the blade surfaces. Following [45], rime ice formation is favored at low temperatures, low liquid water content, small mean volume diameter, and low air velocities.

Regarding the results for the lift coefficient (see Figure 13), the obtained C_L was slightly higher than in the case of the no-ice airfoil at an angle of attack of 5°, 1.04% with the Spalart–Allmaras model and less than 2% with the k-ω SST model. This fact happens at low angles of attack, and it has been observed previously [37,40,46], probably because the ice-induced roughness at this AoA has a positive effect of triggering the transition from laminar to turbulent boundary layer, which is less prone to separation than the laminar one. There is a noticeable increase in drag (Figure 14) at this incidence angle, promoted by both the ice shape and the induced roughness [42,45]. However, for larger angles of attack, C_L in the iced configurations is notably reduced, and C_D is largely increased regarding the clean case; in fact, drag augments faster with growing AoA. This is because apparent roughness causes the stall phenomenon to be anticipated up to an angle of around 10°, as indicated in Figure 13. The most significant differences in lift between the different models are observed at an AoA of 10°. This fact can be explained by the sudden expansion of the separation zone, the uncertainty of the transition position, and the inaccuracy of the turbulence and roughness models in predicting them.

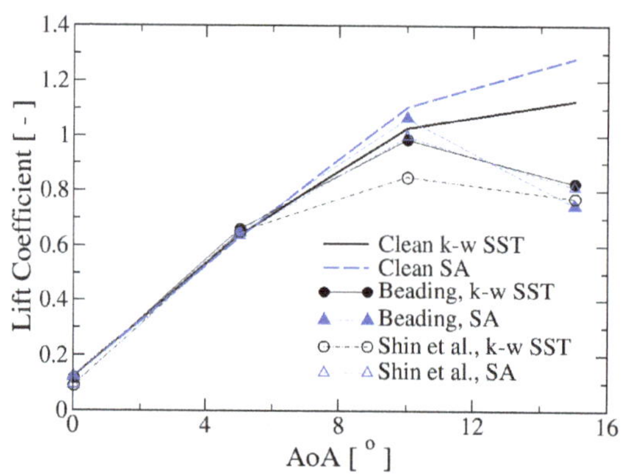

Figure 13. Lift coefficient for the rime ice, Shin's data from [36].

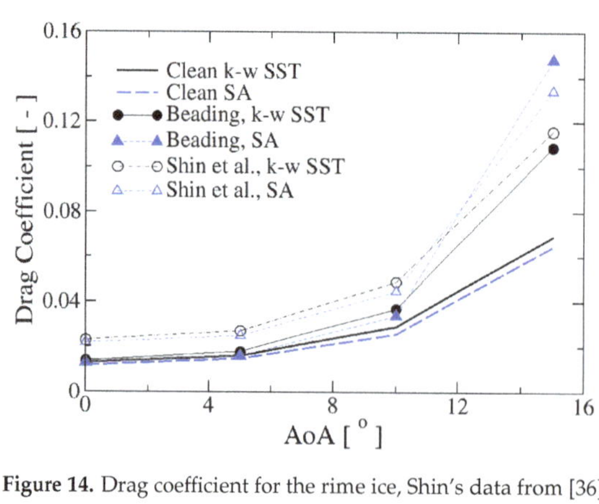

Figure 14. Drag coefficient for the rime ice, Shin's data from [36].

Interestingly, the Shin et al. [36] roughness model predicts a larger lift decrease than the beading model at high angles of attack. However, in the case of the SA turbulence model at 15°, such a trend is inverted, and the C_L provided by the beading model is lower than for the Shin et al. [36] model, a fact that does not happen for the k-ω SST formulation.

Figures 15 and 16 show the behavior of the skin friction (C_f) and pressure (C_p) coefficients in the iced configurations with rime ice at an AoA of 15°. Comparing with Figures 10 and 11, it can be seen that the peak of skin friction at the leading edge is higher in the case of ice than in the clean configuration; conversely, the maximum pressure coefficient magnitude at the same location is lower in the iced scenario than in the clean situation. Moreover, the overpressure in intrados is higher in the clean profile than in the case with ice; at the same time, in the first half of the chord, the suction is also higher than in the iced configuration. As a result, the lift coefficient decreases with the presence of ice. Regarding the friction coefficient, rime ice occurrence increases C_f in the intrados and promotes an earlier boundary layer detachment than in the clean profile. As a consequence, the separated flow region is the largest in the presence of rime ice, aggravating the stall phenomenon and implying lower lift and higher drag than in the no-ice configuration. Looking at Figures 15 and 16, the performance of both turbulence models in the iced scenario is very similar, showing some differences only in the location of the flow separation point in the extrados, which is predicted earlier by the SA than by the k-ω SST model.

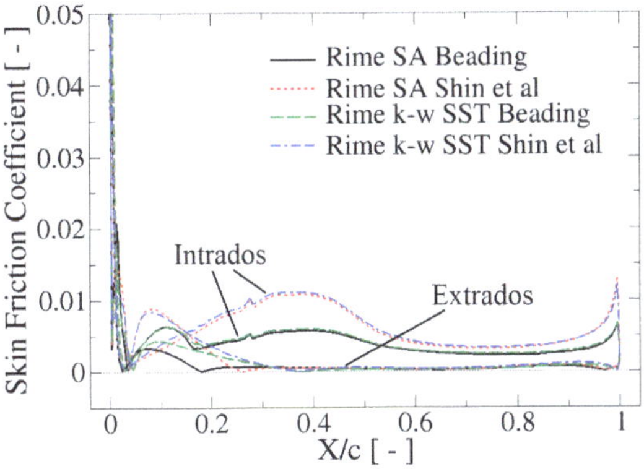

Figure 15. Rime ice: skin friction coefficient at an AoA of 15° of the iced profile with both turbulence and roughness models, Shin's data from [36].

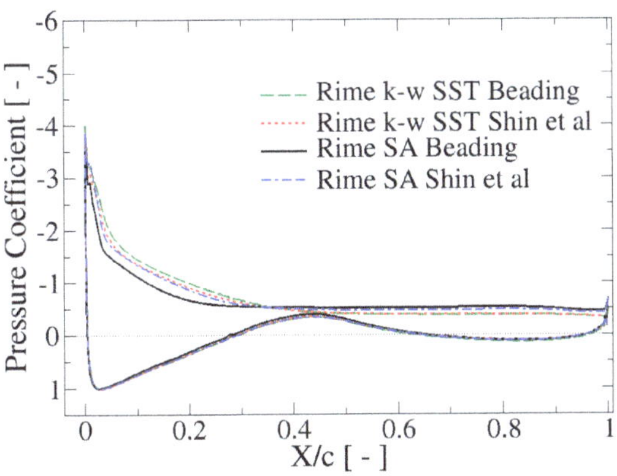

Figure 16. Rime ice: pressure coefficient at AoA of 15° of the clean profile with both turbulence and roughness models, Shin's data from [36].

Regarding the influence of the roughness model, it is found that its larger influence happens in the skin friction coefficient (Figure 15), showing only a moderate effect on C_p distribution along the leading half chord of the profile extrados (Figure 16). In the intrados, the Shin et al. [36] model predicts a noticeably higher C_f than the beading model, which is attributed to the extent of the rough region: the roughness height is uniform in the case of Shin et al. [36], while in the beading model, it is higher close to the leading edge, where the ice layer develops. Therefore, the larger the rough surface, the larger the wall shear stress, as should be expected. A similar situation happens in the extrados; logically, the friction coefficient is close to zero behind the point of flow separation. Here, the beading model predicts an earlier flow separation than in the Shin et al. [36] model, which implies a larger recirculation bubble on the extrados; this is illustrated in Figures 17 and 18 by the streamlines in the case of the SA turbulence model. Interestingly, in the intrados, the beading model predicts a secondary maximum of C_f close to the leading edge, around 10% of the chord length, precisely where the ice generates the higher roughness height. On the other hand, owing to the regular roughness distribution, the Shin et al. model displays a maximum much later, around 40% of the chord.

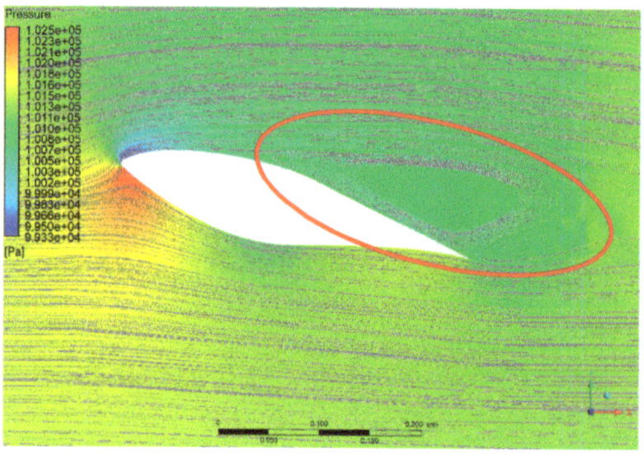

Figure 17. Rime ice: pressure contours and streamlines at AoA of 15°; beading model and Spalart–Allmaras model.

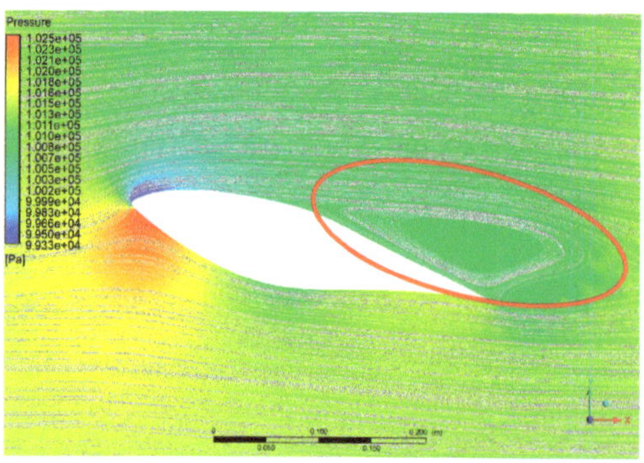

Figure 18. Rime ice: pressure contours and streamlines at AoA of 15°; Shin et al. [36] model and Spalart–Allmaras model.

As shown in Figure 19, the roughness estimated by the beading model and caused by ice affects mainly the airfoil leading edge (17% of the chord in this case). In comparison, the roughness estimated by Shin et al. is a fixed value along the 100% of the airfoil. This may imply an underestimation of the lift coefficient since the presence of roughness may accelerate the separation of the boundary layer. As can be seen from Figure 19, the roughness estimated with the Spalart–Allmaras model is somewhat lower than that estimated by the k-ω SST model. From that figure, it can be seen that in the beading model, the ice roughness extends much more on the intrados than on the extrados.

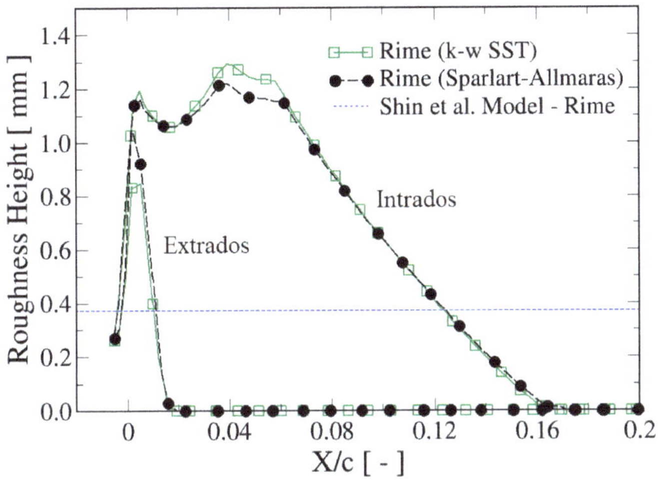

Figure 19. Distribution of the roughness height along the S809 airfoil for rime ice at an AoA of 15°, Shin's data from [36].

Using the beading model (see Table 5), it was found that for angles of attack of 0° and 10°, the loss in lift coefficient using both turbulence models was less than 5%. For an angle of attack of 5°, around a 3% increase in the lift coefficient was obtained with the two turbulence models. However, for 15°, there was a loss of 41.74% with the Spalart–Allmaras model and 26.78% with the k-ω SST model. The average loss in the lift coefficient was 9.10% using the k-ω SST model and 12.59% using the Spalart–Allmaras model. Regarding the drag coefficient, the behavior was as expected, and as the angle of attack increased, so did the drag coefficient, reaching its maximum at 15°. Again, the highest increase was estimated by the Spalart–Allmaras model, with an average value of 43.93%.

Table 5. Aerodynamic loss estimated using the beading model with rime ice.

	Clean				Iced (Beading Model)				% Difference (Iced vs. Clean)			
	Spalart–Allmaras		k-ω SST		Spalart–Allmaras		k-ω SST		Spalart–Allmaras		k-ω SST	
AOA (°)	C_L	C_D	C_L	C_D	C_L	C_D	C_L	C_D	C_L	C_D	C_L	C_D
0	0.124	0.012	0.123	0.013	0.124	0.013	0.120	0.014	−1.88%	6.73%	−2.16%	8.68%
5	0.632	0.015	0.639	0.016	0.651	0.016	0.662	0.018	3.13%	8.81%	3.43%	10.61%
10	1.106	0.026	1.029	0.029	1.066	0.034	0.988	0.037	−3.61%	31.67%	−4.04%	26.37%
15	1.284	0.065	1.129	0.069	0.748	0.148	0.827	0.109	−41.74%	128.51%	−26.78%	57.79%
				Average:					−12.59%	43.93%	−9.10%	25.86%

The Shin et al. model, shown in Table 6, predicts a more pronounced drop in C_L regarding the beading model: an average value of 17% for the SA model and around 19% for the k-ω SST model. However, the trend at an AoA of 5° is maintained with a forecasted slight increment of around 1% by both turbulence models. Likewise, a considerable augmentation in the drag coefficient is predicted by the Shin et al. [36] model, with mean increments of more than 80% and 68% for the SA and k-ω SST models, respectively.

Table 6. Aerodynamic loss estimated using the Shin et al. [36] model with rime ice.

	Clean				Iced (Shin et al. Model)				% Difference (Iced vs. Clean)			
	Spalart–Allmaras		k-ω SST		Spalart–Allmaras		k-ω SST		Spalart–Allmaras		k-ω SST	
AOA (°)	C_L	C_D	C_L	C_D	C_L	C_D	C_L	C_D	C_L	C_D	C_L	C_D
0	0.124	0.012	0.123	0.013	0.095	0.022	0.089	0.023	−22.96%	72.35%	−27.33%	72.70%
5	0.632	0.015	0.639	0.016	0.638	0.025	0.652	0.027	1.04%	66.96%	1.87%	68.00%
10	1.106	0.026	1.029	0.029	0.999	0.045	0.851	0.049	−9.61%	77.00%	−17.33%	65.94%
15	1.284	0.065	1.129	0.069	0.815	0.134	0.774	0.116	−36.50%	106.89%	−31.48%	66.70%
				Average:					−17.53%	80.80%	−19.50%	68.34%

Therefore, looking at the numbers in Tables 5 and 6, which are shown in Figures 13 and 14, it can be stated the Shin et al. [36] model tends to predict lower C_L and higher C_D in the iced airfoil than the beading model. This fact is attributed to the fact that the Shin et al. [36] model assumes that the entire profile is covered by ice. Therefore, the whole profile is considered to be a rough surface. On the contrary, the beading model assumes that the increase in roughness occurs only in areas contaminated by ice. In the case of the S809 profile, only the leading edge is contaminated (see Figures 12 and 19).

Additionally, looking at Figures 15 and 16, it can be concluded that the impact of turbulence modeling on the pressure and skin friction coefficients of the iced airfoil is lower than that of the roughness model.

4.2.2. Wet Regime—Glaze Ice

Glaze ice is formed in conditions of temperatures between −10 and 0 °C where the droplet, after impact, runs along the surface in the flow direction before freezing completely, occupying a wider surface than in the case of rime ice. The frozen liquid forms a smooth, transparent, and translucent layer of ice on the blade surface.

Regarding the results for the lift coefficient (see Figure 20), similarly to what happened for rime ice, the obtained C_L was a little bit over that of the clean airfoil at an angle of attack of 5°, between 2% and 3% depending on the turbulence and roughness models. However, for larger angles of attack, the lift coefficient decreases noticeably, even more than in the case of rime ice. Moreover, the drag coefficient increases at the same time, reaching higher values than in the case of rime ice (see Figure 21). Of course, this is again due to the anticipation of the stall phenomenon, which is illustrated by the streamline's behavior shown in Figures 22 and 23 in the case of the SA turbulence model and for the two considered roughness models.

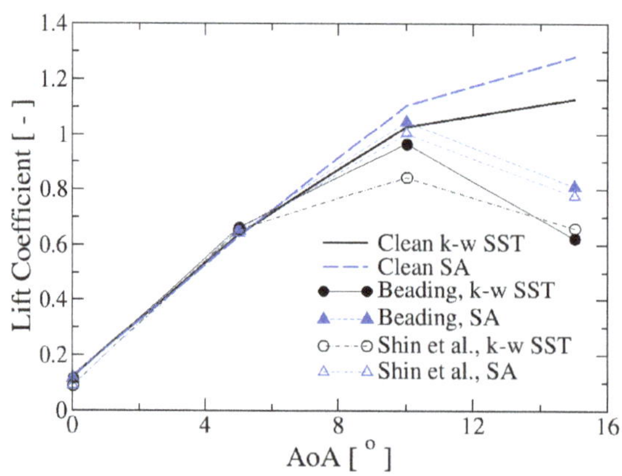

Figure 20. Lift coefficient for the glaze ice, Shin's data from [36].

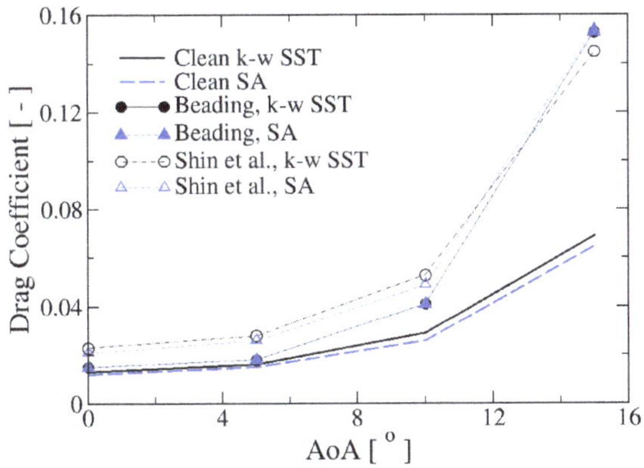

Figure 21. Drag coefficient for the glaze ice, Shin's data from [36].

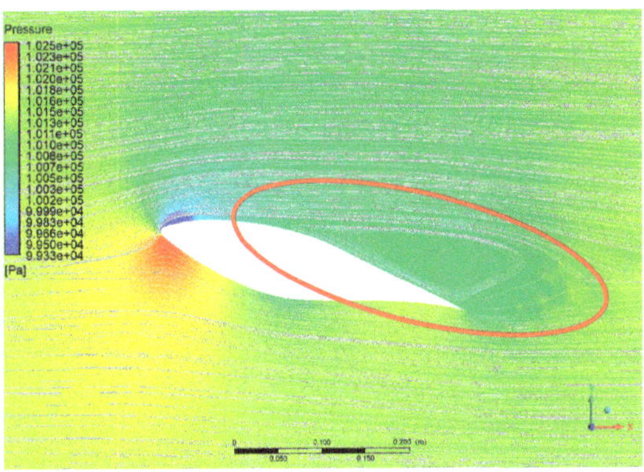

Figure 22. Glaze ice: pressure contours and streamlines at an AoA of 15°; beading model and Spalart–Allmaras model.

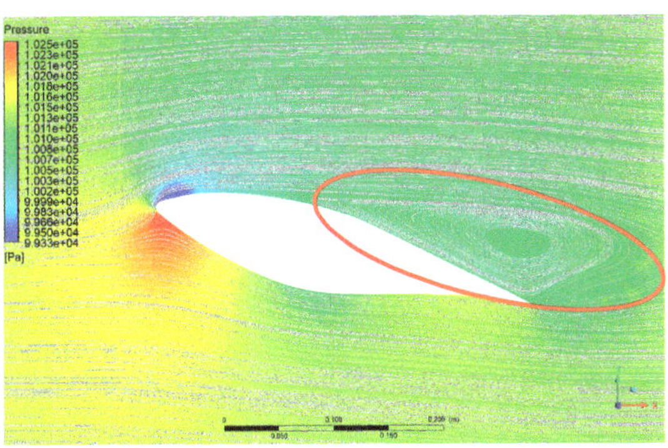

Figure 23. Glaze ice: pressure contours and streamlines at an AoA of 15°; Shin et al. [36] model and Spalart–Allmaras model.

Figures 24 and 25 show the behavior of the skin friction (C_f) and pressure (C_p) coefficients, respectively, in the iced configurations with glaze ice at an AoA of 15°. When comparing such coefficients with the clean profile scenario, the trends are similar to those observed with rime ice: the C_f peak at the leading edge is higher than in clean conditions and C_p magnitude in the same position is lower than in the no-ice scenario, being even lower than in the case of rime ice (Figure 25). This observation is consistent with the lower C_L coefficients attained under glaze ice (Figure 20). In the case of the friction coefficient, shown in Figure 24, the separation points in the extrados are not so clearly identified as in the clean profile; the C_f curves show, for all the cases, a smooth decreasing shape reaching very low values already at 15% of the chord, suggesting a detached flow, as observed in Figures 22 and 23. As for rime ice, both turbulence models provide very similar results for both coefficients, with the differences mainly located in the area between the leading edge and 20% of the chord. Here, the C_p and C_f predicted by the SA model are somewhat higher than those provided by the k-ω SST model.

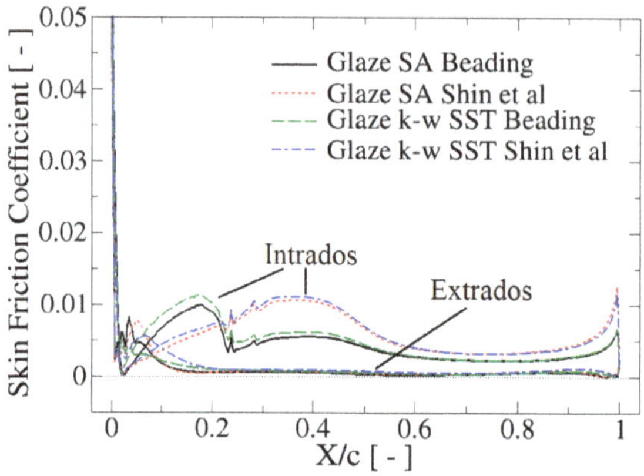

Figure 24. Glaze ice: skin friction coefficient at an AoA of 15° of the iced profile with both turbulence and roughness models, Shin's data from [36].

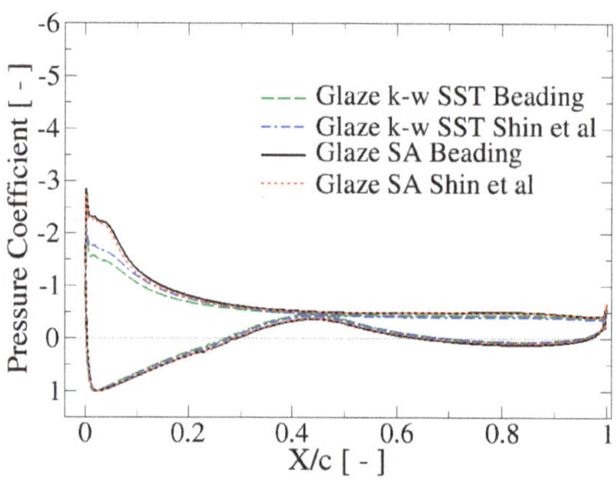

Figure 25. Glaze ice: pressure coefficient at an AoA of 15° of the clean profile with both turbulence and roughness models, Shin's data from [36].

The influence on the same coefficients of the roughness model can also be seen in Figures 24 and 25. Apparently, C_p is only slightly modified by the roughness model, being noticeable only in the leading 20% of the chord in the profile extrados for the k-ω SST model. However, differences are more obvious in the skin friction coefficient. In the intrados, similar to what happened with rime ice, the secondary maximum of C_f predicted by the Shin et al. model is close to 40% of the chord, while the beading model predicts a bump near the leading edge with a maximum of around 20% of the chord; the extent of such a bump coincides with the region of the profile contaminated with ice (see Figure 26). In the extrados, the Shin et al. [36] model predicts higher values of C_f than the beading model, except perhaps in the area very close to the leading edge. In summary, it can be stated that the Shin et al. model provides higher values of the skin friction coefficient than the beading model; however, the last one can concentrate the friction effects on the area where the ice is deposited, while the first model distributes such an influence along the whole profile. In this sense, it can be argued that the beading model responds better to the physics of the problem than the Shin et al. [36] model, at least for the present application to wind turbine profiles.

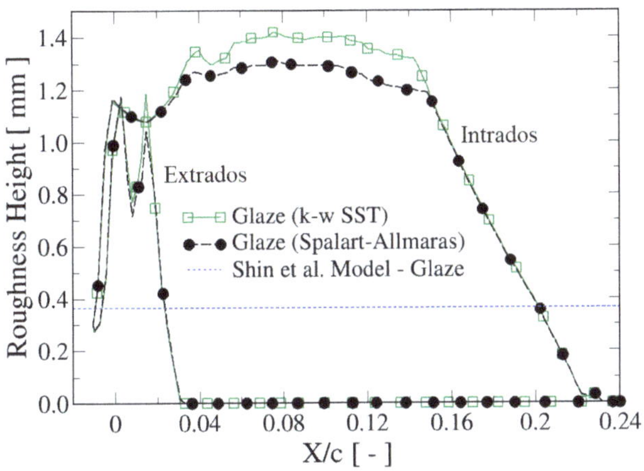

Figure 26. Distribution of the roughness height along the S809 airfoil for glaze ice at an AoA of 15°, Shin's data from [36].

As shown in Figure 26, the roughness estimated by the beading model in the case of glaze ice affects approximately 24% of the airfoil. As seen from that figure, the roughness estimated with the Spalart–Allmaras model is lower than that estimated by the k-ω SST model (around 8%). On the contrary, the Shin et al. model predicts a uniform roughness along the profile with a very similar value to the case of rime ice.

Using the beading model (see Table 7), it was found that for angles of attack of 0° and 10°, the loss in lift coefficient using both turbulence models was less than 7%. For the angle of attack of 5°, a 3.5% increase in the lift coefficient was obtained with the two turbulence models. However, at 15°, there was a loss of 36.71% with the Spalart–Allmaras model and 44.65% with the k-ω SST model. The average loss in the lift coefficient was 15.10% using the k-ω SST model and 12.80% using the Spalart–Allmaras model. As in the rime ice conditions, the drag coefficient followed the expected behavior, increasing monotonically with the angle of attack, reaching its maximum at 15°. Again, the highest increase was estimated by the Spalart–Allmaras model, with an average value of 59.34%.

Table 7. Aerodynamic loss estimated using the beading model with glaze ice.

	Clean				Iced (Beading Model)				% Difference (Iced vs. Clean)			
	Spalart–Allmaras		k-ω SST		Spalart–Allmaras		k-ω SST		Spalart–Allmaras		k-ω SST	
AOA (°)	C_L	C_D	C_L	C_D	C_L	C_D	C_L	C_D	C_L	C_D	C_L	C_D
0	0.124	0.012	0.123	0.013	0.117	0.015	0.116	0.015	−5.80%	18.25%	−6.04%	19.46%
5	0.632	0.015	0.639	0.016	0.653	0.018	0.663	0.018	3.32%	20.08%	3.55%	20.68%
10	1.106	0.026	1.029	0.029	1.047	0.041	0.966	0.041	−5.39%	61.91%	−6.15%	48.20%
15	1.284	0.065	1.129	0.069	0.812	0.153	0.625	0.153	−36.71%	137.12%	−44.65%	119.05%
				Average:					−12.80%	59.34%	−15.10%	51.85%

The results obtained with the Shin et al. model in the wet regime are presented in Table 8 and follow similar trends as in rime ice. In this case, the average drop in the lift coefficient in the ice scenario is around 19% with the SA model and 23% in the k-ω SST model. A slight increase of about 2% in C_L is predicted for an AoA of 5°, similar to what happened with the beading model. At the same time, C_D increases with mean increments of around 94% and 83% for the one and two equation turbulence models.

Table 8. Aerodynamic loss estimated using the Shin et al. [36] model with glaze ice.

	Clean				Iced (Shin et al. Model)				% Difference (Iced vs. Clean)			
	Spalart–Allmaras		k-ω SST		Spalart–Allmaras		k-ω SST		Spalart–Allmaras		k-ω SST	
AOA (°)	C_L	C_D	C_L	C_D	C_L	C_D	C_L	C_D	C_L	C_D	C_L	C_D
0	0.124	0.012	0.123	0.013	0.092	0.021	0.089	0.023	−25.39%	72.10%	−29.98%	72.81%
5	0.632	0.015	0.639	0.016	0.643	0.026	0.656	0.028	1.85%	70.89%	2.47%	71.71%
10	1.106	0.026	1.029	0.029	1.005	0.049	0.844	0.053	−9.12%	93.16%	−18.05%	78.95%
15	1.284	0.065	1.129	0.069	0.778	0.154	0.661	0.145	−39.37%	138.91%	−41.44%	109.14%
				Average:					−18.93%	93.77%	−22.98%	83.15%

In general, the Shin et al. model predicts lower C_L and higher C_D than the beading model, which is the same trend as in rime ice. Again, this difference is mainly attributed to the airfoil area covered by ice (see Figures 12 and 26). On the other hand, the trends of the influence of the ice roughness model on the aerodynamic performance of the S809 airfoil are very similar for both turbulence models, SA and k-ω SST (see Tables 7 and 8). This fact, together with the curves of C_f and C_p shown in Figures 24 and 25, means that, in terms of the simulation outcomes, the modeling of ice-induced roughness is more critical than the election of turbulence model.

4.3. *Analysis of the Aerodynamic Performance of the Iced Airfoils*

As anticipated, ice on the S809 profile reduced the lift coefficient under both regimes, rime and glaze (Figures 13 and 20), while C_D increased (Figures 14 and 21). This agrees with the results in the literature [12,29,48,57,58]. The following tables compare the aerodynamic coefficients between the different ice scenarios considering the roughness and turbulence models employed.

The average difference in the lift coefficient (see Table 9) between the dry regime and the wet regime using the beading model was 7.67% with the k-ω SST model and 3.66% with the Spalart–Allmaras model, indicating a variation between the two turbulence models of less than 5%. However, for an angle of attack of 15°, the Spalart–Allmaras model estimates a gain in the lift coefficient of 8.63%, while the k-ω SST model estimates a loss of 24.41%. This difference can be attributed to the limitations of the Spalart–Allmaras model under adverse pressure gradient conditions.

Table 9. Lift coefficient comparison of regimes using the beading model.

AoA (°)	k-ω SST Model			Spalart–Allmaras Model		
	C_L Rime	C_L Glaze	% Diff	C_L Rime	C_L Glaze	% Diff
0	0.120	0.116	−3.97%	0.121	0.117	−3.99%
5	0.662	0.663	0.11%	0.652	0.653	0.18%
10	0.988	0.966	−2.20%	1.066	1.047	−1.84%
15	0.827	0.625	−24.41%	0.748	0.812	8.63%
Average			7.67%			3.66%

The average difference in the drag coefficient (see Table 10) was higher, 18.78% with the k-ω SST and 11.97% with the Spalart–Allmaras models, indicating a difference between the two turbulence models of less than 7%. This increase in C_D is expected since the surface of the airfoil covered by ice was 5% greater in the wet regime than in the dry regime.

Table 10. Drag coefficient comparison of regimes using the beading model.

AoA (°)	k-ω SST Model			Spalart–Allmaras Model		
	C_D Rime	C_D Glaze	% Diff	C_D Rime	C_D Glaze	% Diff
0	0.015	0.016	9.92%	0.013	0.015	10.80%
5	0.018	0.020	9.10%	0.017	0.018	10.36%
10	0.037	0.044	17.27%	0.034	0.041	22.96%
15	0.110	0.152	38.83%	0.148	0.153	3.77%
Average			18.78%			11.97%

The average difference in the lift coefficient (Table 11) between the dry and wet regimes using the Shin et al. [36] model was 4.91% with the k-ω SST model and 2.25% with the Spalart–Allmaras model. The average difference in the drag coefficient (see Table 12) was slightly higher, being 8.89% with the k-ω SST model and 6.78% with the Spalart–Allmaras model. In these cases, the difference between the turbulence models was less than 3%. Similar findings were reported by Martini et al. [12] for a different airfoil (NACA 64-618).

Table 11. Lift coefficient comparison of regimes using the Shin et al. model.

AoA (°)	k-ω SST Model			Spalart–Allmaras Model		
	C_L Rime	C_L Glaze	% Diff	C_L Rime	C_L Glaze	% Diff
0	0.089	0.086	−3.65%	0.095	0.092	−3.15%
5	0.652	0.656	0.60%	0.638	0.643	0.80%
10	0.851	0.844	−0.87%	1.000	1.005	0.54%
15	0.774	0.661	−14.53%	0.815	0.778	−4.52%
Average			4.91%			2.25%

Table 12. Drag coefficient comparison of regimes using the Shin et al. [36] model.

AoA (°)	k-ω SST Model			Spalart–Allmaras Model		
	C_D Rime	C_D Glaze	% Diff	C_D Rime	C_D Glaze	% Diff
0	0.023	0.023	0.06%	0.022	0.022	−0.14%
5	0.028	0.028	2.21%	0.025	0.026	2.36%
10	0.049	0.053	7.84%	0.045	0.049	9.13%
15	0.116	0.146	25.46%	0.134	0.155	15.48%
Average			8.89%			6.78%

As can be observed from the previous values, the predictions of aerodynamic characteristics by the Shin et al. [36] model are less sensitive to the type of ice regime than those by the beading model, presenting lower variations in both the lift and drag coefficients.

Regarding the ice type, the present results indicate that the wet regime negatively impacts the aerodynamic performance of the airfoil. Comparing the aerodynamic coefficients of the airfoil evaluated under glaze ice versus the clean airfoil, the lift loss using the k-ω SST model and the beading model was 6% higher than that obtained with the rime ice (compare values in Tables 5 and 7). In contrast, with the Shin et al. model, the increase was 3.48% (compare values in Tables 6 and 8). However, when evaluating the Spalart–Allmaras model, the increase in lift loss between the two regimes using either roughness model was less than 2%. In the case of the drag coefficient, the difference between the regimes was more evident.

Regarding the combination of k-ω SST and the beading model, the increase was 25.99% (compare values in Tables 5 and 7), while with the Shin et al. [36] model, it was 14.82% (compare values in Tables 6 and 8). As with the lift coefficient, the increase was lower using the Spalart–Allmaras model, 15.41% combined with the beading model and 12.97%

with the Shin et al. [36] model. The above affirms that the airfoil performance will be more affected under conditions of glaze ice.

The combination of the k-ω SST turbulence model and the Shin et al. [36] model for the roughness modeling generated the highest loss of lift coefficient for both regimes (19.50% for rime ice and 22.98% for glaze ice). Similar findings were reported by Martini et al. [12] for a NACA 64-618 airfoil where the C_L loss was 22.54% under rime ice conditions. The combination of the Spalart–Allmaras turbulence model and the Shin et al. model generated the highest drag coefficient increase for both regimes (80.80% for rime ice and 93.77% for glaze ice). The findings reported by Martini et al. [12] under rime ice using the same models are notably higher, with some exceeding 120%. Differences in the airfoil design can account for these variations. For instance, the S809 profile has been specifically designed to obtain lower drag coefficients and maximum lift and to be insensitive to roughness [46]. The clean airfoil yielded comparable outcomes with a slight discrepancy in drag estimation at a 15° angle of attack when employing both turbulence models. This variation is attributed to the proximity to conditions of flow separation on the airfoil's suction side. As previously explained, the Spalart–Allmaras turbulence model exhibits limitations in boundary layer separation scenarios, which are accentuated when ice accumulates on the airfoil.

A related study was carried out by Caccia and Guardone [37], in which the authors searched to verify whether the negative effect on the aerodynamic performance of the profiles was due to the ice shape or the roughness. The roughness model of Shin et al. was employed in [37], and the model known as Wright's relation, which was determined from experimental measurements of the roughness height as a function of the freezing fraction at the stagnation point. The analysis was performed only for the dry regime (freezing fraction equal to 1) in different sections of the NREL 5MW turbine blade [59] composed at the base by DU profiles and, at the tip, by NACA 643-618 profiles. Using the Spalart–Allmaras turbulence model, the roughness values were evaluated in two ways: the first one only in the area where ice accumulated (similar to what the FENSAP-ICE beading model does) and the second one at 0.44 m along the leading edge of the profile on both the suction and pressure sides. Caccia and Guardone [37] found that ice caused a degradation in profile aerodynamic performance in all the evaluated cases due to both ice shape and roughness. Regarding the ice shape, the icing condition evaluated generated a "horn" on the leading edge of the NACA 643-618 profiles. In contrast, in the DU profiles, the ice followed the shape of the profile. This horn shape produced a significant decrease in the performance of the NACA profiles. In the case of roughness, the effect of the extent of roughness and the value of its height became more critical as the angle of attack increased.

From the results presented in the previous sections and the performed discussions, it is seen that there is a close agreement between the attained conclusions in the present study and those of Caccia and Guardone [37], i.e., the stall was anticipated, the lift coefficient slope decreased, and the drag coefficient increased in the case of iced airfoil. Furthermore, both studies found that the most significant difference occurred when the roughness height was higher and covered a larger airfoil area. The present results extend the same conclusions for the glaze ice scenario.

5. Conclusions

This study examines the impact of two surface roughness models, combined with two turbulence models, on the aerodynamic performance of the S809 wind turbine airfoil during rime and glaze ice via CFD simulation with FENSAP-ICE software. It has been shown that computational results of the aerodynamic parameters and ice shape were in reasonable agreement with the experimental and numerical data for clean and iced airfoils. The main conclusions of the study are summarized as follows:

1. Iced airfoil analysis revealed increased drag and reduced lift, consistent with the literature. Under the rime ice condition, C_D increased by around 50%, and C_L decreased by approximately 15%. These aerodynamic coefficients showed a higher

impact in the glaze ice condition, increasing to 70% for C_D and decreasing to 20% for C_L, emphasizing the need for further investigation into glaze ice conditions.

2. The modification of the airfoil lift and drag characteristics in the presence of ice is explained by the behavior of the skin friction and pressure coefficients. The occurrence of the ice layer and its induced roughness promotes an alteration in the pressure and wall shear stress around the profile, which increases skin friction and decreases the pressure coefficient. Also, it has been shown that the airfoil experiences an earlier flow separation with accreted ice compared to clean conditions, with the effect of increasing drag and reducing lift.
3. On the other hand, the turbulence model selection did not significantly affect outcomes for aerodynamic coefficients. Average differences were around 3% for rime ice conditions and 2% for glaze ice conditions.
4. However, surface roughness was crucial, requiring consideration at each growth stage during ice accretion simulations. Validation of data obtained with experiments is recommended to determine which roughness model yields better results.
5. The beading model has shown excellent handling of the roughness variation along the ice accretion process. However, further validation of the model is needed to evaluate accuracy fully.
6. The widely used Shin et al. model may underestimate the lift coefficient and overestimate the drag coefficient compared to the beading model, suggesting that the first, which assumes that roughness is uniformly distributed along the entire airfoil, does not accurately reflect the actual physical nature of roughness distribution in the wind turbine simulated conditions.
7. Accurate predictions cannot rely solely on empirical correlations from the aeronautics domain. In this context, on-site roughness measurements are imperative to eliminate uncertainties.

The validated modeling methods and simulation utilities presented in this research would benefit scholars and engineers conducting reliable simulations for icing on wind turbine blades. However, it is essential to emphasize that the analysis was limited to two-dimensional aerodynamic performance; future analyses should be carried out to understand the effect of icing from a three-dimensional point of view that includes other phenomena, such as the influence of airflow along the blade span induced by its rotation.

Author Contributions: Conceptualization, L.T.C.M. and A.I.; methodology, L.T.C.M. and S.L.; software, L.T.C.M.; validation, L.T.C.M.; formal analysis, L.T.C.M., A.I. and S.L.; data curation, L.T.C.M.; writing—original draft preparation, L.T.C.M.; writing—review and editing, A.I. and S.L.; supervision, A.I. and S.L. All authors have read and agreed to the published version of the manuscript.

Funding: This research received no external funding.

Data Availability Statement: Data are contained within the article.

Acknowledgments: Special acknowledgment to Nergica for the data provided.

Conflicts of Interest: The authors declare no conflict of interest.

References

1. O'Brien, J.M.; Young, T.M.; O'Mahoney, D.C.; Griffin, P.C. Horizontal axis wind turbine research: A review of commercial CFD, FE codes and experimental practices. *Prog. Aerosp. Sci.* **2017**, *92*, 1–24. [CrossRef]
2. Han, W.; Kim, J.; Kim, B. Study on correlation between wind turbine performance and ice accretion along a blade tip airfoil using CFD. *J. Renew. Sustain. Energy* **2018**, *10*, 023306. [CrossRef]
3. Pedersen, M.C.; Sørensen, H. Towards a CFD Model for Prediction of Wind Turbine Power Losses due to Icing in Cold Climate. In Proceedings of the 16th International Symposium on Transport Phenomena and Dynamics of Rotating Machinery, Honolulu, HI, USA, 10–15 April 2016; p. 7.
4. Fei, C.-W.; Han, Y.-J.; Wen, J.-R.; Li, C.; Han, L.; Choy, Y.-S. Deep learning-based modeling method for probabilistic LCF life prediction of turbine blisk. *Propuls. Power Res.* **2023**. [CrossRef]
5. Chen, J.-Y.; Feng, Y.-W.; Teng, D.; Lu, C.; Fei, C.-W. Support vector machine-based similarity selection method for structural transient reliability analysis. *Reliab. Eng. Syst. Saf.* **2022**, *223*, 108513. [CrossRef]

6. Hacıefendioğlu, K.; Başağa, H.B.; Yavuz, Z.; Karimi, M.T. Intelligent ice detection on wind turbine blades using semantic segmentation and class activation map approaches based on deep learning method. *Renew. Energy* **2022**, *182*, 1–16. [CrossRef]
7. Fakorede, O.; Feger, Z.; Ibrahim, H.; Ilinca, A.; Perron, J.; Masson, C. Ice protection systems for wind turbines in cold climate: Characteristics, comparisons and analysis. *Renew. Sustain. Energy Rev.* **2016**, *65*, 662–675. [CrossRef]
8. Hu, L.; Zhu, X.; Hu, C.; Chen, J.; Du, Z. Wind turbines ice distribution and load response under icing conditions. *Renew. Energy* **2017**, *113*, 608–619. [CrossRef]
9. Hu, L.; Zhu, X.; Chen, J.; Shen, X.; Du, Z. Numerical simulation of rime ice on NREL Phase VI blade. *J. Wind Eng. Ind. Aerodyn.* **2018**, *178*, 57–68. [CrossRef]
10. Rizk, P.; Younes, R.; Ilinca, A.; Khoder, J. Wind turbine ice detection using hyperspectral imaging. *Remote Sens. Appl. Soc. Environ.* **2022**, *26*, 100711. [CrossRef]
11. Laín, S.; Contreras, L.T.; López, O. A review on computational fluid dynamics modeling and simulation of horizontal axis hydrokinetic turbines. *J. Braz. Soc. Mech. Sci. Eng.* **2019**, *41*, 375. [CrossRef]
12. Martini, F.; Ibrahim, H.; Contreras Montoya, L.T.; Rizk, P.; Ilinca, A. Turbulence Modeling of Iced Wind Turbine Airfoils. *Energies* **2022**, *15*, 8325. [CrossRef]
13. Etemaddar, M.; Hansen, M.O.L.; Moan, T. Wind turbine aerodynamic response under atmospheric icing conditions. *Wind Energy* **2014**, *17*, 241–265. [CrossRef]
14. Villalpando, F.; Reggio, M.; Ilinca, A. Prediction of ice accretion and anti-icing heating power on wind turbine blades using standard commercial software. *Energy* **2016**, *114*, 1041–1052. [CrossRef]
15. Jin, J.Y.; Virk, M.S. Study of ice accretion along symmetric and asymmetric airfoils. *J. Wind Eng. Ind. Aerodyn.* **2018**, *179*, 240–249. [CrossRef]
16. Jin, J.Y.; Virk, M.S. Study of ice accretion and icing effects on aerodynamic characteristics of DU96 wind turbine blade profile. *Cold Reg. Sci. Technol.* **2019**, *160*, 119–127. [CrossRef]
17. Yirtici, O.; Cengiz, K.; Ozgen, S.; Tuncer, I.H. Aerodynamic validation studies on the performance analysis of iced wind turbine blades. *Comput. Fluids* **2019**, *192*, 104271. [CrossRef]
18. Yirtici, O.; Ozgen, S.; Tuncer, I.H. Predictions of ice formations on wind turbine blades and power production losses due to icing. *Wind Energy* **2019**, *22*, 945–958. [CrossRef]
19. Pedersen, M.C. Modelling Icing on Structures for Wind Power Applications. Ph.D. Dissertation, Aalborg University, Aalborg, Denmark, 2018.
20. Shu, L.; Li, H.; Hu, Q.; Jiang, X.; Qiu, G.; He, G.; Liu, Y. 3D numerical simulation of aerodynamic performance of iced contaminated wind turbine rotors. *Cold Reg. Sci. Technol.* **2018**, *148*, 50–62. [CrossRef]
21. Jin, J.Y.; Virk, M.S.; Hu, Q.; Jiang, X. Study of Ice Accretion on Horizontal Axis Wind Turbine Blade Using 2D and 3D Numerical Approach. *IEEE Access* **2020**, *8*, 166236–166245. [CrossRef]
22. IEA Wind. *Wind Energy in Cold Climates Available Technologies—Report*; VTT Technical Research Centre on Finland Ltd.: Tampere, Finland, 2016; p. 120.
23. Pedersen, M.C.; Yin, C. Preliminary Modelling Study of Ice Accretion on Wind Turbines. *Energy Procedia* **2014**, *61*, 258–261. [CrossRef]
24. Contreras Montoya, L.T.; Lain, S.; Ilinca, A. A Review on the Estimation of Power Loss Due to Icing in Wind Turbines. *Energies* **2022**, *15*, 1083. [CrossRef]
25. ANSYS Inc. In-Flight Icing Simulation. Available online: https://www.ansys.com/products/fluids/ansys-fensap-ice (accessed on 10 November 2022).
26. Hann, R.; Hearst, R.J.; Sætran, L.R.; Bracchi, T. Experimental and Numerical Icing Penalties of an S826 Airfoil at Low Reynolds Numbers. *Aerospace* **2020**, *7*, 46. [CrossRef]
27. Knobbe-Eschen, H.; Stemberg, J.; Abdellaoui, K.; Altmikus, A.; Knop, I.; Bansmer, S.; Balaresque, N.; Suhr, J. Numerical and experimental investigations of wind-turbine blade aerodynamics in the presence of ice accretion. In Proceedings of the AIAA Scitech 2019 Forum, San Diego, CA, USA, 7–11 January 2019.
28. Li, Y.; Wang, S.; Sun, C.; Yi, X.; Guo, W.; Zhou, Z.; Feng, F. Icing distribution of rotating blade of horizontal axis wind turbine based on Quasi-3D numerical simulation. *Therm. Sci.* **2018**, *22*, 681–691. [CrossRef]
29. Homola, M.C.; Virk, M.S.; Nicklasson, P.J.; Sundsbø, P.A. Performance losses due to ice accretion for a 5 MW wind turbine. *Wind Energy* **2012**, *15*, 379–389. [CrossRef]
30. Makkonen, L.; Laakso, T.; Marjaniemi, M.; Finstad, K.J. Modelling and Prevention of Ice Accretion on Wind Turbines. *Wind Eng.* **2001**, *25*, 3–21. [CrossRef]
31. Virk, M.; Mughal, U.; Hu, Q.; Jiang, X. Multiphysics Based Numerical Study of Atmospheric Ice Accretion on a Full Scale Horizontal Axis Wind Turbine Blade. *Int. J. Multiphys.* **2016**, *10*, 237–246. [CrossRef]
32. Fu, P.; Farzaneh, M. A CFD approach for modeling the rime-ice accretion process on a horizontal-axis wind turbine. *J. Wind Eng. Ind. Aerodyn.* **2010**, *98*, 181–188. [CrossRef]
33. Bai, C.-J.; Wang, W.-C. Review of computational and experimental approaches to analysis of aerodynamic performance in horizontal-axis wind turbines (HAWTs). *Renew. Sustain. Energy Rev.* **2016**, *63*, 506–519. [CrossRef]
34. Menter, F. Zonal Two Equation k-w Turbulence Models For Aerodynamic Flows. In Proceedings of the 23rd Fluid Dynamics, Plasmadynamics, and Lasers Conference, Orlando, FL, USA, 6–9 July 1993.

35. Sagol, E.; Reggio, M.; Ilinca, A. Assessment of Two-Equation Turbulence Models and Validation of the Performance Characteristics of an Experimental Wind Turbine by CFD. *ISRN Mech. Eng.* **2012**, *2012*, 428671. [CrossRef]
36. Shin, J.; Berkowitz, B.; Chen, H.; Cebeci, T. Prediction of ice shapes and their effect on airfoil performance. In Proceedings of the 29th Aerospace Sciences Meeting, Reno, NV, USA, 7–10 January 1991; p. 264.
37. Caccia, F.; Guardone, A. Numerical simulation of ice accretion on wind turbine blades: Are performance losses due to ice shape or surface roughness? *Wind Energy Sci.* **2023**, *8*, 341–362. [CrossRef]
38. ANSYS Inc. *ANSYS FENSAP-ICE 18.1 User Manual*; ANSYS Inc.: Canonsburg, PA, USA, 2017.
39. Hildebrandt, S.; Sun, Q. Evaluation of operational strategies on wind turbine power production during short icing events. *J. Wind Eng. Ind. Aerodyn.* **2021**, *219*, 104795. [CrossRef]
40. Blasco, P.; Palacios, J.; Schmitz, S. Effect of icing roughness on wind turbine power production. *Wind Energy* **2017**, *20*, 601–617. [CrossRef]
41. Turkia, V.; Huttunen, S.; Wallenius, T. Method for estimating wind turbine production losses due to icing. *VTT Tech. Res. Cent. Finl.* **2013**, *114*, 44.
42. Fortin, G.; Perron, J. Wind Turbine Icing and De-Icing. In Proceedings of the 47th AIAA Aerospace Sciences Meeting Including the New Horizons Forum and Aerospace Exposition, Orlando, FL, USA, 5–8 January 2009; Aerospace Sciences Meetings. American Institute of Aeronautics and Astronautics: Reston, VA, USA, 2009.
43. Sagol, E. Three Dimensional Numerical Prediction of Icing Related Power and Energy Losses on a Wind Turbine. Ph.D. Thesis, École Polytechnique de Montréal, PolyPublie Polytechnique Montréal, Montréal, QC, Canada, 2014.
44. Farzaneh, M. *Atmospheric Icing of Power Networks*, 1st ed.; Springer: Berlin/Heidelberg, Germany, 2008; p. 381.
45. Battisti, L. Aerodynamic Performances on Ice Contaminated Rotors. In *Wind Turbines in Cold Climates*, 1st ed.; Springer International Publishing: Cham, Switzerland, 2015; p. 341.
46. Somers, D.M. *Design and Experimental Results for the S809 Airfoil*; National Renewable Energy Laboratory (NREL): Golden, CO, USA; U.S. Department of Energy, Office of Scientific and Technical Information: Oak Ridge, TN, USA, 1997.
47. Tan, H. CFD Analysis of a Wind Turbine Airfoil with Flap. Master's Thesis, Washinton University, St. Louis, MO, USA, 2020.
48. Han, Y.; Palacios, J.; Schmitz, S. Scaled ice accretion experiments on a rotating wind turbine blade. *J. Wind Eng. Ind. Aerodyn.* **2012**, *109*, 55–67. [CrossRef]
49. Hand, M.M.; Simms, D.A.; Fingersh, L.J.; Jager, D.W.; Cotrell, J.R.; Schreck, S.; Larwood, S.M. *Unsteady Aerodynamics Experiment Phase VI WindTunnel Test Configurations and Available Data Campaigns*; National Renewable Energy Laboratory—NREL: Golden, CO, USA, 2001; p. 310.
50. Li, Y.; Sun, C.; Jiang, Y.; Yi, X.; Zhang, Y. Effect of liquid water content on blade icing shape of horizontal axis wind turbine by numerical simulation. *Therm. Sci.* **2019**, *23*, 1637–1645. [CrossRef]
51. Burton, T.; Jenkins, N.; Sharpe, D.; Bossanyi, E. Aerodynamics of Horizontal Axis Wind Turbines. In *Wind Energy Handbook*; John Wiley & Sons, Inc.: Hoboken, NJ, USA, 2011; pp. 39–136.
52. Airfoil Tools. NREL's S809 Airfoil (s809-nr). Available online: http://airfoiltools.com/airfoil/details?airfoil=s809-nr (accessed on 5 February 2022).
53. Zanon, A.; De Gennaro, M.; Kühnelt, H. Wind energy harnessing of the NREL 5 MW reference wind turbine in icing conditions under different operational strategies. *Renew. Energy* **2018**, *115*, 760–772. [CrossRef]
54. Tardif d'Hamonville, T. Modélisation et Analyse des Phénomènes Aéroélastiques Pour Une Pale D'éolienne. Master's Thesis, Université du Québec à Chicoutimi, Chicoutimi, QC, Canada, 2009.
55. Hudecz, A.; Koss, H.; Laver, M.O. Ice Accretion on Wind Turbine Blades. In Proceedings of the 15th International Workshop on Atmospheric Icing of Structures (IWAIS XV), St. John's, NL, Canada, 8–13 September 2013; p. 8.
56. ANSYS Inc. *ANSYS Fluent User's Guide Release 15.0*; ANSYS Inc.: Canonsburg, PA, USA, 2013; p. 2692.
57. Bodenlle-Toral, D.; García-Regodeseves, P.; Pandal-Blanco, A. Performance evaluation of an airfoil under ice accretion using CFD simulations. *J. Phys. Conf. Ser.* **2022**, *2217*, 012088. [CrossRef]
58. Jasinski, W.J.; Noe, S.C.; Selig, M.S.; Bragg, M.B. Wind Turbine Performance Under Icing Conditions. *J. Sol. Energy Eng.* **1998**, *120*, 60–65. [CrossRef]
59. Jonkman, J.; Butterfield, S.; Musial, W.; Scott, G. *Definition of a 5-MW Reference Wind Turbine for Offshore System Development*; NREL/TP-500-38060; TRN: US200906%%69 United States 10.2172/947422 TRN: US200906%%69 NREL English; National Renewable Energy Lab. (NREL): Golden, CO, USA, 2009; p. 75.

processes

MDPI

Article

Study on Oxidation Activity of Hydrogenated Biodiesel–Ethanol–Diesel Blends

Jianbo Zhou [1], Lyu Chen [1], Rui Zhang [1,2] and Weidong Zhao [3,*]

1 School of Aeronautical Electromechanical Equipment Maintenance, Airforce Aviation Repair Institute of Technology, Changsha 410073, China; nchu_zhou@163.com (J.Z.); hnu_chen@163.com (L.C.); nudtzhang@163.com (R.Z.)
2 School of Aeronautics and Astronautics, Central South University, Changsha 410073, China
3 School of Automotive and Traffic Engineering, Jiangsu University, Zhenjiang 212013, China
* Correspondence: zhaowd@ujs.edu.cn

Abstract: In the pursuit of understanding the oxidation mechanisms of hydrogenated biodiesel fuels and elucidating the combustion behavior of biomass fuels when blended with diesel, this study presents a comprehensive investigation into the reaction mechanism of hydrogenated biodiesel–ethanol–diesel mixtures. We develop a comprehensive reaction mechanism encompassing 187 components and 735 reactions for hydrogenated biodiesel–ethanol–diesel mixtures. Through kinetics analysis under varied conditions, including 1.0 MPa pressure, an equivalence ratio of 1.0, and temperatures of 900 K and 1400 K, we explore the impact of cross-reactions and changing fuel blend ratios on low- and high-temperature oxidation. Our findings indicate that oleic and stearic acid methyl esters serve as better substitutes for representing hydrogenated biodiesel kinetics than methyl decanoate. At lower temperatures, increased hydrogenated biodiesel and ethanol content leads to reduced OH generation, impacting reactivity. Conversely, higher temperatures result in enhanced OH production with increased hydrogenated biodiesel and ethanol concentrations, promoting reactivity. A cross-reaction analysis reveals CH_2O as a prominent product, with the $CH_2O \rightarrow HCO \rightarrow CO$ pathway playing a pivotal role. In summary, our research unveils the intricate oxidation mechanisms of hydrogenated biodiesel–ethanol–diesel mixtures, providing insights into their combustion characteristics and offering implications for optimizing fuel blends for cleaner and more efficient energy solutions.

Keywords: hydrogenated biodiesel; ethanol; diesel; blended fuels; oxidation mechanism; ternary fuel blend; combustion; reaction kinetics mechanism; numerical simulation

1. Introduction

The rapid escalation in global energy demand, propelled by technological advancement and economic growth, underscores energy as a pivotal pillar in socio-economic development and a principal contributor to carbon emissions [1–4]. To mitigate the impact of carbon dioxide emissions on the global climate, prioritizing the development of renewable, clean fuels is paramount in the internal combustion engine industry, with alternative fuels such as biodiesel garnering significant attention from scholars [5–8]. The inception of biodiesel traces back to Rudolf (1859–1913) in 1895, with Niel Company achieving a breakthrough in 1988 by refining biodiesel from rapeseed oil for diesel combustion. The oxidation process of biodiesel, primarily governed by $C_5H_8O_3$, alongside smaller molecules like C_2H_2 and C_2H_4, initiates under high-temperature conditions. Despite the widespread sources of biodiesel, its low saturation results in poor oxidative stability, a drawback that can be effectively ameliorated by hydrogenation, which also enhances its cetane number. However, hydrogenated biodiesel suffers from deteriorated low-temperature flow properties. Ethanol, characterized by hydroxyl and alkyl groups, exhibits unique polarity, rendering it miscible with various organic and inorganic solvents, including hydrocarbons, ethers, benzene, and esters. Consequently, alcohols find widespread applications across industrial domains. The ethanol oxidation reaction predominantly involves small molecules like CH_2O and HCO. With its lower kinematic viscosity, it can moderate the high viscosity of hydrogenated biodiesel, and exhibits commendable miscibility with the diesel–hydrogenated biodiesel blend [9–13]. Given the complexity of multi-component fuel components, representative constituents are selected to construct an oxidation mechanism that characterizes fuel combustion properties, with chemical reaction kinetics emerging as a primary research methodology. Chemical reaction kinetics elucidate the combustion characteristics of substances through an exploration of their molecular composition and structural attributes. The combustion process within an engine cylinder represents a complex physicochemical frontier, where myriad chemical reactions occur simultaneously, including carbon chain scission, small-molecule polymerization, and molecular isomerization. In the case of mixed fuels, the combustion process engenders a plethora of intersecting reactions, involving the further decomposition of large-molecular products and polymerization reactions among various small molecules, thereby altering the combustion reactivity, product distribution, and properties within the cylinder. The current research into biodiesel and alcohol blended fuels' combustion predominantly revolves around experimentally exploring their combustion and emission traits within internal combustion engines. However, there exists a dearth of in-depth examinations into the combustion mechanism and dynamic simulation of such blended fuels through mixed-fuel chemistry. This paper delves into a microscopic analysis of the hydrogenated biodiesel–ethanol–diesel system.

The initial research on biodiesel primarily focused on employing methyl caprate as a single-component substitute [14–17], aiming to reflect its physicochemical properties more accurately. However, multi-component substitution has become the mainstream approach. Chang Yachao et al. [18] utilized a mixture of n-decane, methyl decanoate, and 5-decenoic acid methyl ester to represent biodiesel in their studies. Their results indicated a proficient prediction of the fuel's oxidation characteristics in real engines, albeit neglecting the cross-reactions between fuel molecules. Zhai Yitong et al. [19] selected widely recognized C4 fatty acid methyl esters (butyric acid methyl ester and butenoic acid methyl ester) to characterize biodiesel. Their model efficiently simulated experimental data from the previous literature. Regarding hydrogenated biodiesel, given the changes in its physicochemical properties and cetane number post-hydrogenation, research on its microscopic aspects is scarce, with most studies focusing on experimental aspects. Mei Deqing et al. [20] conducted engine performance experiments using B20 and HB20 on a 186FA diesel engine. Their findings showed that for B20, the emissions of HC, CO, and smoke decreased by 9.9%, 9.3%, and 15.2%, respectively, while NOx emissions increased by 8.5%. For PHB20, the reductions in HC, CO, and smoke emissions were 12.4%, 13.5%, and 17.1%, respectively, with a 6.7% increase in NO_x emissions, indicating that PHB20 enhances combustion and emission performance in diesel engines more effectively than B20. Leng Xianyin et al. [21] conducted European Steady Cycle (ESC) emission tests on engines using hydrogenated catalytic biodiesel blended in ratios of 5%, 10%, 20%, and 30% with national IV standard diesel. The results showed that with increasing proportions of hydrogenated catalytic biodiesel, emissions of THC, NO_x, PM, and CO decreased to varying degrees. Notably, CO emissions showed a linear decrease, with the most significant reduction in NO_x emissions occurring at high loads and low velocity. Hence, it is crucial to investigate ester alcohol diesel with varying mixing ratios to enhance the combustion efficiency and emission performance of diesel engines. At present, research on the chemical reaction kinetics of small-molecule ester fuels has reached a certain level of maturity, while investigations into the chemical reaction kinetics of large-molecule oxygen-containing fuels have also shown progress. Nevertheless, further advancements are necessary to develop mechanisms and characterize large-molecule, oxygen-containing mixed fuels comprehensively. With the widespread adoption of the "carbon peak and carbon neutrality" concept, the consumption of biodiesel and alcohol fuels has been steadily increasing. In light of this, a numerical simulation study was conducted, coupling a chemical reaction kinetics analysis with computational fluid dynamics (CFD), based on a constructed mechanism model to characterize the hybrid fuel comprising hydrogenated biodiesel, ethanol, and diesel. The study delved into the impact of ester alcohol fuel blending ratios on the combustion reaction activity of the hybrid fuel, as well as the generation and consumption of major active free radicals.

Given the limited research on the microscopic aspects of hydrogenated biodiesel, this paper utilizes a constructed dual-component hydrogenated biodiesel mechanism as a foundation, incorporating a simplified ethanol and n-heptane mechanism to form a comprehensive chemical kinetics mechanism for the hydrogenated biodiesel–n-heptane–ethanol fuel blend. This study meticulously analyzes the effects of varying ethanol blend ratios with hydrogenated biodiesel under different temperature conditions. The aim is to enrich the theoretical understanding of high-cetane-number fuel blends. The analysis will contrast and examine the impacts of cross-reactions on fuel oxidation at high and low temperatures changes in the formation of major components, and reveal the influence of the combustion reactivity of these blended fuels on the entire combustion system. This provides a theoretical basis for the practical application of biodiesel as an alternative fuel.

2. Selection of Hydrogenated Biodiesel Substitution Mechanism and Validation of Ternary Fuel Mechanism Simplification

2.1. Selection of Hydrogenated Biodiesel Substitution Mechanism

Methyl decanoate (MD), a long-chain fatty acid methyl ester with a decyl chain comprising ten carbon atoms, aptly simulates the long-chain fatty acid methyl esters (C16-C22) present in actual biodiesel. Analyzing MD's physical properties, such as density and flash point, reveals a close resemblance to real biodiesel. Chemically, MD possesses a cetane number similar to actual biodiesel and exhibits comparable ignition characteristics and combustion reaction properties. Thus, this paper selects methyl oleate and stearic acid methyl ester as substitutes for hydrogenated biodiesel for three main reasons [22]. Firstly, as shown in Table 1, the cetane number of hydrogenated biodiesel is 71.2, whereas MD's cetane number is 47, making MD an unsuitable substitute. Secondly, Table 2 showcases the comparison of the proportion of the main substance components between hydrogenated biodiesel and biodiesel. The peak area represents the proportion of each substance in the total substance. This comparison is based on the predominant presence of methyl oleate in hydrogenated biodiesel and its notable deviation in cetane number compared to hydrogenated biodiesel. Thus, the addition of 20% stearic acid methyl ester to methyl oleate, resulting in a cetane number of 70.66, closely matches that of hydrogenated biodiesel. Diesel, being a complex mixture, is typically characterized by one or two components when studying its combustion mechanism. Given that the cetane number of n-heptane closely mirrors that of diesel, n-heptane is commonly employed to represent the actual ignition characteristics of diesel. The third reason pertains to the reaction pathways of the blended fuels, which will be elaborated in the subsequent sections.

Table 1. Cetane numbers of cottonseed methyl ester (CME) and partially hydrogenated cottonseed methyl ester assisted by ultrasonic (UPHCME) [23].

Sample	Cetane Number
CME	59.3
UPHCME	71.2
MD	47
Methyl Oleate	62.3
Stearic Acid Methyl Ester	104.1

Table 2. Composition and contents of CME and UPHCME [24].

Composition	Molecular Formula	Peak Area/%	
		CME	UPHCME
Methyl Myristic Acid Ester (C14:0)	$C_{15}H_{30}O_2$	0.13	0.15
Methyl Palmitate (C16:0)	$C_{17}H_{34}O_2$	23.05	23.11
Methyl Stearate (C18:0)	$C_{19}H_{38}O_2$	3.17	10.54
Trans-Methyl Oleate (trans-C18:1)	$C_{19}H_{36}O_2$	0	12.18
Cis-Methyl Oleate (cis-C18:1)	$C_{19}H_{36}O_2$	21.47	31.62
Methyl Linoleate (C18:2)	$C_{19}H_{34}O_2$	49.94	20.67

2.2. Chemical Reaction Kinetics of Methyl Oleate and Stearic Acid Methyl Ester

Researchers at Lawrence Berkeley National Laboratory, including Westbrook and others [25], constructed a comprehensive mechanism for methyl oleate and stearic acid methyl ester under a wide range of conditions: temperatures of 800–1500 K, equivalence ratios of 0.5–2, and pressures of 0.1–6 MPa. This detailed mechanism encompasses 402 species with 16,188 reactions for methyl oleate, and 423 species with 17,436 reactions for stearic acid methyl ester, validated through experimentation. Based on this detailed mechanism, our study simplifies it using the direct relation graph method and the error-propagation-incorporated direct relation graph method. Figure 1 illustrates an example of the DRG method's simplification mechanism, which identifies strong contributions from component A to component B, and reciprocal contributions between component B and component D, thereby establishing a close interrelationship among components A, B, and D. The direct relation graph method facilitates rapid simplification with minimal error, typically serving as the initial step in mechanism development simplification. As depicted in Figure 2, a comparison of the ignition delay predictions between the detailed, semi-detailed, and simplified mechanisms of methyl oleate shows minimal errors in the simplified mechanism. By combining a sensitivity analysis and the error propagation direct relation graph method, and a full-species sensitivity analysis, the less sensitive reaction components are removed, resulting in a simplified mechanism for methyl oleate comprising 71 species with 273 reactions, and for stearic acid methyl ester, 67 species with 246 reactions.

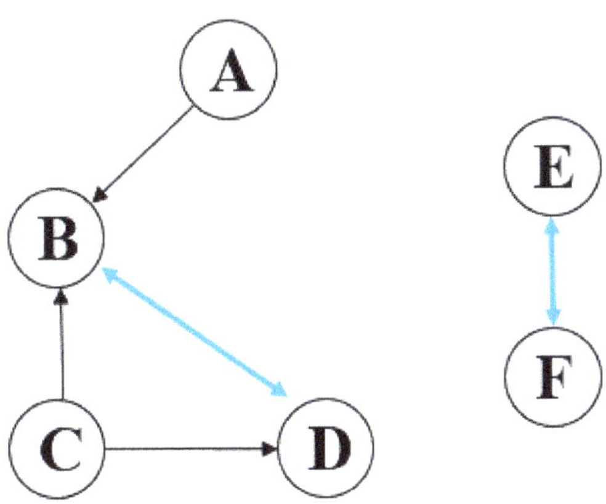

Figure 1. Example of simplifying mechanism with DRG method.

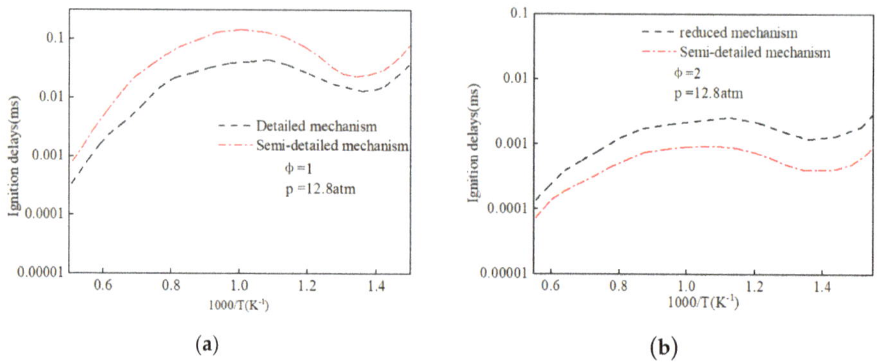

Figure 2. Comparison of the predicted ignition delay periods between the detailed and reduced mechanisms of methyl oleate. (**a**) Comparison of detailed and semi-detailed mechanisms; (**b**) Comparison of semi-detailed and reduced mechanisms.

Figures 3 and 4 compare the predicted and experimental values of the ignition delay periods for the simplified mechanisms of methyl oleate and stearic acid methyl ester under various conditions. The close fit between the experimental and simulated values validates the accuracy of the mechanism simplification and paves the way for subsequent simulation work.

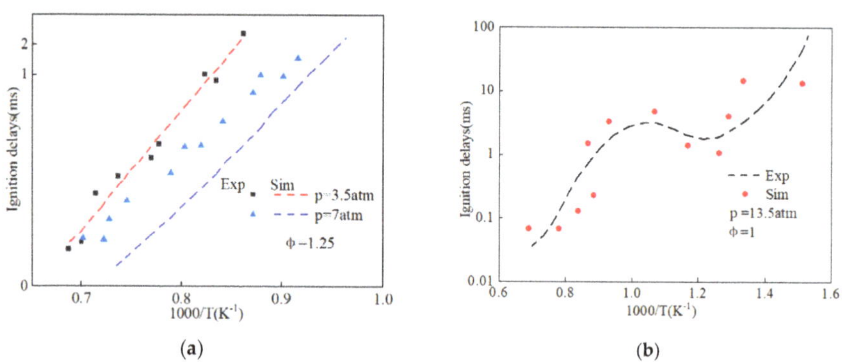

Figure 3. Comparison of the predicted ignition delay periods of the simplified mechanism of methyl oleate with experimental values. (**a**) Comparison of predicted and experimental values at pressures of 3.5 atm and 7 atm, and an equivalence ratio of 1.25; (**b**) Comparison of predicted and experimental values at a pressure of 13.5 atm and an equivalence ratio of 1.

2.3. Chemical Kinetics Mechanism of Hydrogenated Biodiesel–Ethanol–Diesel

The detailed mechanism for n-heptane was adopted from the comprehensive chemical reaction mechanism of Currant et al. [26], which initially included 531 components and 2539 reactions, and was ultimately simplified to 42 components with 168 reactions. The detailed mechanism for ethanol was sourced from Metcalfe W.K. et al. [27], comprising 254 components and 7568 reactions, and was eventually reduced to 48 components and 163 reactions. The simplification process for n-heptane and ethanol follows the same methodology discussed earlier and is not reiterated here.

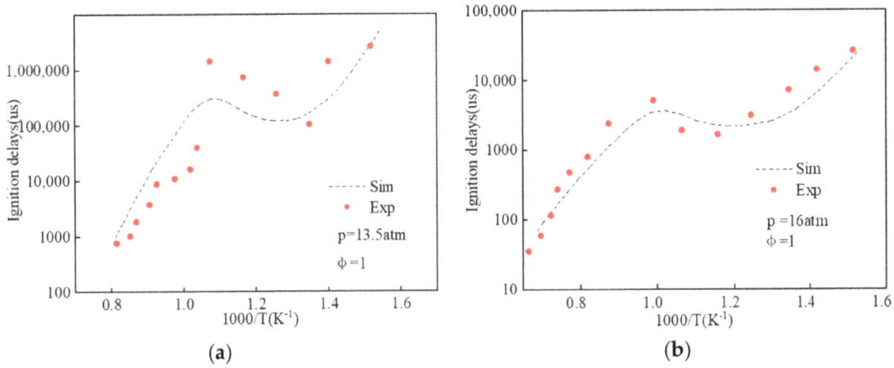

Figure 4. Comparison of the predicted ignition delay periods of the simplified mechanism of stearic acid methyl ester with experimental values. (**a**) Comparison of predicted and experimental values at a pressure of 13.5 atm and an equivalence ratio of 1; (**b**) Comparison of predicted and experimental values at a pressure of 16 atm and an equivalence ratio of 1.

The mixed mechanism for n-heptane–methyl oleate–ethanol constructed in this study was validated by comparison with the individual fuel mechanisms of methyl oleate and ethanol, which remained unchanged and hence do not require further validation. Figure 5 presents the comparison of ignition delay periods calculated using the n-heptane–methyl oleate–stearic acid methyl ester mixed mechanism against experimental values measured by Fiewger [28], Shen [29], Herzler [30], Hartm [31], and others, resulting in a final count of 187 components and 735 reactions. As seen in Figure 5, the mechanism's prediction of the ignition delay period for n-heptane closely matches experimental values, indicating its suitability for further research.

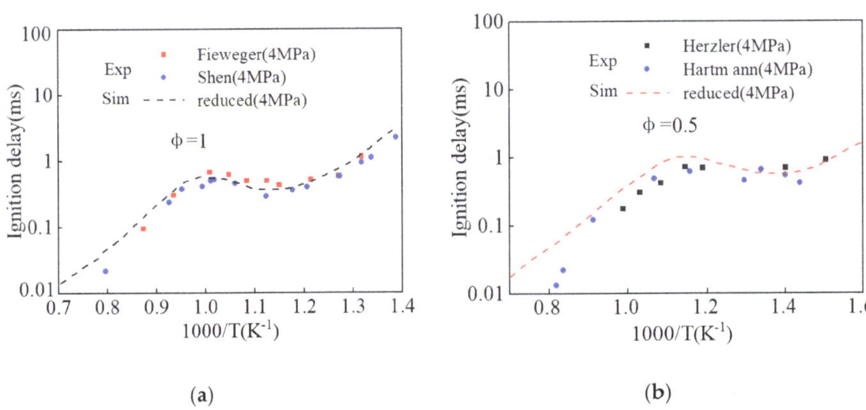

Figure 5. Comparison between the predicted ignition delay periods of the mixed mechanism for n-heptane–methyl oleate–ethanol and the experimental values for n-heptane. (**a**) Comparison of predicted and experimental values at a pressure of 4 MPa and an equivalence ratio of 1; (**b**) Comparison of predicted and experimental values at a pressure of 4 MPa and an equivalence ratio of 0.5.

2.4. Theoretical Basis of Multi-Component Fuel Cross-Reaction

Utilizing the closed homogeneous reaction model within the CHEMKIN 2016 software, we scrutinize the impact of varying temperatures and components on the fuel oxidation process. Through an examination of the cross-reaction mechanism among components with diverse mixing ratios, we unveil the influence of small-molecule active free radical OH on the combustion reaction system. The combustion dynamics within a diesel engine are governed by both the low-temperature and high-temperature oxidation processes of individual fuel components, alongside the cross-reactions occurring among different components. To delve deeper into the ramifications of cross-reactions on the combustion reaction of blended fuels, we employ a simplified mechanism model of hydrogenated biodiesel–ethanol–diesel mixed fuel. This analysis dissects the effects of cross-reactions on both the low-temperature and high-temperature oxidation of ester alcohol diesel, along with the differential consumption of free radicals, viewed from the vantage point of chemical reaction kinetics.

Currently, within the realm of chemical reaction kinetics, cross-reactions are primarily categorized into broad and narrow senses. The overarching concept of "cross-reaction" pertains to interactions occurring between small-molecular free radicals within mixed fuels and the fuel molecules themselves, as well as with other small-molecular free radicals. On the other hand, the more narrowly defined "cross-reaction" involves the generation of macromolecular free radicals subsequent to the dehydrogenation of macromolecules within multi-component fuel systems. These macromolecular free radicals then actively partake in the elementary reactions of various components and engage in interactions with other elements. In terms of chemical kinetics, the accurate prediction of ignition delay periods is significantly influenced by narrow cross-reactions, whereas the reaction dynamics among different components of mixed fuels are predominantly impacted by generalized cross-reactions. We have opted to employ highly reactive free radicals, such as OH, as conduits for facilitating cross-reactions among diverse constituents in fuels. Through the lens of generalized "cross-reactions," we aim to elucidate the mechanisms underlying the interplay between diesel's single-component and multi-component substitutes.

OH serves as a catalyst in fuel cracking and energy liberation, exerting a pivotal influence on the oxidation reaction activity and reaction rate of the entire fuel combustion system. Its significance extends to the realm of free radicals within the cylinder, where it assumes a vital role. Consequently, leveraging the generalized "cross-reaction" theory, we have opted to scrutinize the substitution mechanism of hydrogenated biodiesel–ethanol–diesel hybrid fuel. This endeavor aims to delve into the cross-reaction pathways during both low-temperature and high-temperature oxidation reactions of the hybrid fuel. Furthermore, it seeks to unveil the role and impact of reactive groups on the combustion reaction system throughout the combustion process of ester alcohol diesel.

3. Results and Analysis

3.1. Cross-Reaction Flux Analysis of Methyl Decanoate and n-Heptane at 900 K Involving OH Radicals

From the perspective of chemical reaction kinetics, the oxidation behavior of fuels is significantly influenced by the intermediate product, the OH radical. As depicted in Figure 6, under conditions of an initial temperature of 900 K, initial pressure of 6 MPa, and an equivalence ratio of 1.0, the analysis reveals that in the low-temperature combustion reaction system of MD10 fuel, the dominant pathway consuming OH radicals is MD + OH = MD_xJ + H_2O, whereas the pathway generating OH is MD + O = MD_xJ + OH. The consumption of OH exceeds its generation, resulting in a net OH consumption rate of 13.48%. The processes of MD_xJ undergoing oxygenated isomerization to form MD_xO_2 and the cracking of MD_xJ to produce MP2D do not generate OH [32]. Subsequently, $MD_xOOH_yO_2$ undergoes intramolecular hydrogen transfer followed by cracking into MDKET24 and OH. MDKET24 further reacts with oxygen, releasing OH, contributing 10.44% to the OH pathway. The dehydrogenation initiation reaction for diesel at low temperatures is C_7H_{16} + OH = C_7H_{15} + H_2O [14], with an OH consumption rate of 68.93%. Analyzing the dehydrogenation initiation reactions of MD, it is evident that the reactions of n-heptane and methyl decanoate are competitive, which is detrimental to the combustion initiation of the mixed fuel. The cracking process of n-heptane also does not generate OH in its deoxygenation reactions. MD, after undergoing dehydrogenation initiation, forms MD_xJ, which has many isomers, leading to numerous pathways for OH consumption. In the methyl decanoate-n-heptane mixture, the concentration of OH decreases, inhibiting the decomposition of n-heptane. Furthermore, the consumption rate of OH by methyl decanoate exceeds its generation rate, with a total consumption rate of 13.48%. Therefore, n-heptane must provide some of the OH needed for its decomposition to react with methyl decanoate, which impacts its own reaction. Consequently, adding 10% methyl decanoate to n-heptane at 900 K is disadvantageous for enhancing the reactivity of the reaction system.

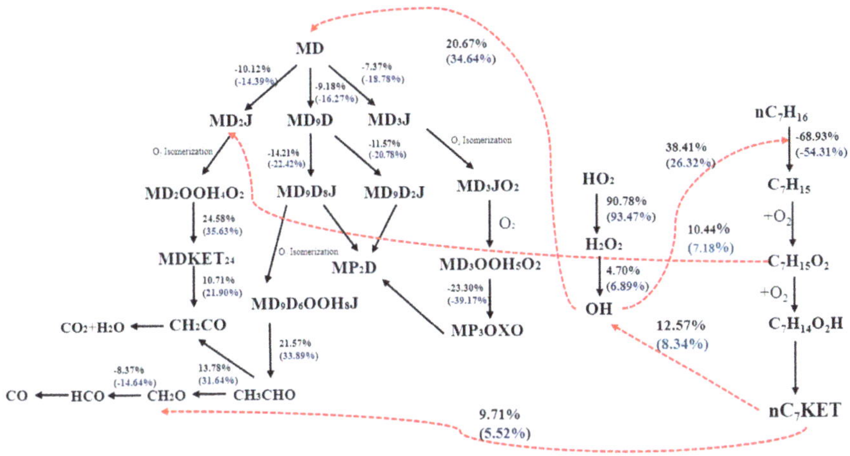

Figure 6. Reaction flux concerning OH in the combustion reaction system of MD10 (and MD20) fuels at 900 K.

Compared to the MD10 fuel, with the increased proportion of methyl decanoate (MD) and decreased content of n-heptane in the MD20 fuel, the OH consumption ratio by MD in the combustion reaction system increases, with a total consumption rate of 23.48%. This rise accentuates the competitive interaction between MD and n-heptane for OH during the dehydrogenation initiation reaction. As evident from the illustrations, in comparison to MD10, the production and consumption of OH by n-heptane during cracking and dehydrogenation reactions in MD20 are progressively diminishing. This indicates that at 900 K, an increased concentration of methyl decanoate leads to a reduction in the decomposition rate of n-heptane, which is not conducive to promoting the reaction. Therefore, it is not advisable to add a large proportion of MD to n-heptane.

3.2. Cross-Reaction Flux Analysis of Methyl Oleate and n-Heptane at 900 K under Low Temperature Conditions

Given the high cetane number and oxygenated nature of hydrogenated biodiesel, which can generate a significant number of OH radicals, the cross-reaction flux analysis of OH in the HB10 (and HB20) fuel combustion reaction system at an initial temperature of 900 K, initial pressure of 6 MPa, and an equivalence ratio of 1.0 is illustrated in Figure 7. Methyl oleate, a mono-unsaturated methyl ester produced by substituting a –(CH)=(CH)– double bond for –(CH_2)–(CH_2)– in stearic acid methyl ester, plays a crucial role in the reaction system. Under low-temperature conditions, the dehydrogenation initiation reaction of methyl oleate in the HB10 fuel combustion system primarily consumes OH through the dehydrogenation of stearic acid methyl ester, producing methyl oleate and a small amount of H_2O. However, in this environment, H_2O contributes minimally to OH generation, resulting in an OH consumption rate of 4.70%. The total OH consumption rate for the dehydrogenation of methyl oleate is 48.43%, with its cracking producing an overall OH contribution of 29.35%. This indicates a higher consumption than generation of OH.

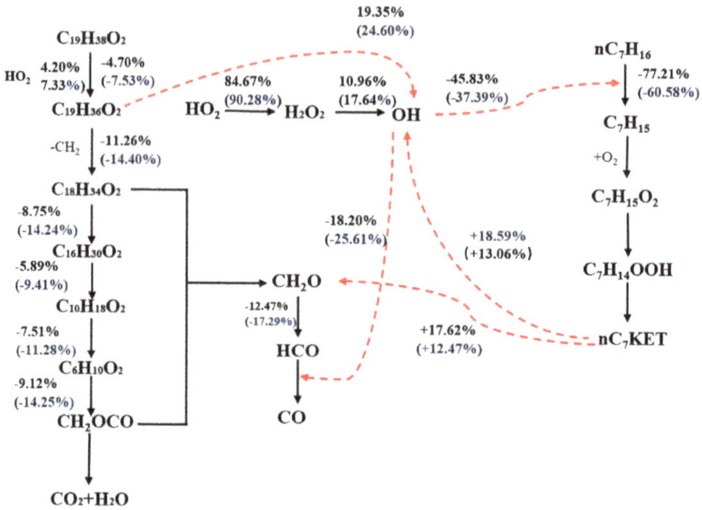

Figure 7. Reaction flux concerning OH in the combustion reaction system of HB10 (and HB20) fuels at 900 K.

In comparison to the MD10 mixed-fuel system under similar conditions, which has a total OH consumption rate of 13.48%, the OH consumption rate in HB10 is higher at 19.08%. This increase is due to the fact that, apart from the initial cracking reaction producing OH, the subsequent dehydrogenation of methyl oleate generates fewer olefinic compounds, producing almost no additional OH for its reaction. The dehydrogenation initiation reaction for diesel at low temperatures is $C_7H_{16} + OH = C_7H_{15} + H_2O$, with an OH consumption rate of 77.21%. Analyzing the dehydrogenation initiation reaction of methyl oleate reveals that OH flows from methyl oleate to n-heptane. The dehydrogenation initiation reactions of n-heptane and methyl oleate are synergistic, facilitating the combustion initiation of the mixed fuel. In n-heptane reactions, the two oxygenation reactions also do not generate OH, with $C_7H_{14}O_2HO_2$ cracking into nC_7KET and OH, contributing 18.59% to the OH pathway. The mixed-fuel reaction system supplies 50.83% of the OH needed for the cracking of n-heptane, whereas the OH produced from n-heptane's own cracking accounts for 36.21%. Therefore, it can be inferred that the OH produced by methyl oleate supports the further cracking of n-heptane, accelerating its reaction rate and leading to more thorough decomposition. Consequently, in the HB10 reaction system, the addition of methyl oleate promotes the overall reaction process. Hence, adding 10% hydrogenated biodiesel to diesel is beneficial for enhancing the reactivity of the reaction system.

Compared to the HB10 fuel, it can be observed from the analysis that with the increased proportion of methyl oleate in the HB20 fuel combustion reaction system, the consumption ratio of OH by n-heptane decreases. This is attributed to the high cetane number and oxygenated nature of methyl oleate, which can generate more OH radicals. In the HB20 mixed fuel, the increased amount of methyl oleate results in a higher supply of OH for n-heptane. The proportion of OH produced by the cracking of methyl oleate and subsequently utilized for the cracking of n-heptane further increases, which is conducive to promoting the reaction.

Additionally, the overall contribution rate of OH in the reaction system at an initial temperature of 900 K, initial pressure of 6 MPa, and an equivalence ratio of 1.0 is higher for the HB20 (and HB10) fuel compared to MD10 (and MD20). This indicates that the reaction system of methyl oleate and n-heptane is more effective than that of methyl decanoate and n-heptane. The presence of methyl oleate enhances the overall reactivity of the fuel blend, making it a more favorable choice in terms of boosting the combustion efficiency and reactivity, particularly under the specified conditions.

In summary, when comparing the same conditions as depicted in Figures 6 and 7, it is evident that both the n-heptane–methyl oleate and n-heptane–methyl decanoate systems share a common reaction pathway of $CH_2O \rightarrow HCO \rightarrow CO$. In the n-heptane–methyl decanoate system, the OH consumption for MD10 (and MD20) is 8.37% and 14.64%, respectively, whereas in the n-heptane–methyl oleate system, the OH consumption for HB10 (and HB20) along this pathway is 12.47% and 17.29%, respectively. This difference is primarily attributed to the variance in cetane numbers, as in the HB10 (and HB20) mixed fuels, methyl oleate initiates cracking reactions to provide OH for n-heptane reactions. In contrast, in the MD10 (and MD20) mixed fuels, methyl decanoate undergoes cracking reactions after n-heptane, with n-heptane supplying OH for methyl decanoate reactions, thereby impacting the cracking and dehydrogenation reactions of n-heptane itself. Consequently, the OH consumption in the HB10 (and HB20) mixed fuels is greater than that in the MD10 (and MD20) mixed fuels.

In the HB10 (and HB20) mixed fuels, the total OH consumption rates for n-heptane are 77.21% and 60.58%, with OH generation rates of 36.21% and 25.53% respectively. In contrast, for the MD10 (and MD20) mixed fuels, the total OH consumption rates for n-heptane are 68.93% and 54.31%, with OH generation rates of 32.72% and 21.04% respectively. Considering the overall generation and consumption rates of OH in both groups of mixed fuels, the flow of OH from n-heptane to methyl decanoate in the n-heptane–methyl decanoate system affects the cracking and dehydrogenation reactions of n-heptane. Therefore, considering these factors, selecting methyl oleate as a representative for the substitution mechanism of hydrogenated biodiesel aligns more closely with real combustion scenarios.

3.3. Cross-Reaction Flux Analysis of n-Heptane–Methyl Oleate–Ethanol at High and Low Temperatures

3.3.1. Low-Temperature OH Cross-Reaction Flux Analysis for n-Heptane–Methyl Oleate–Ethanol

Under the conditions of an initial temperature of 900 K, initial pressure of 6 MPa, and an equivalence ratio of 1.0, the cross-reaction flux analysis concerning OH in the HB5E5 (HB10E10) (HB15E15) fuel combustion reaction system is depicted in Figure 8. The analysis shows that in the 900 K environment, stearic acid methyl ester in HB5E5 initially undergoes a dehydrogenation reaction to form methyl oleate, with an OH consumption rate of 4.90%. This rate increases with a higher concentration of methyl oleate and stearic acid methyl ester in the system. Moreover, a small amount of H_2O is generated, but at 900 K, most water transforms into more stable H_2O_2, rendering the OH produced by this pathway negligible.

In the HB5E5 mixed-fuel system, the OH contribution from the cracking of methyl oleate is 18.15%. Of this, 4.35% of OH is supplied to ethanol and 9.80% to n-heptane for reactions. Given that ethanol accounts for only 5% and n-heptane for 90% of the system, it implies that ethanol's competition for OH is greater than that of n-heptane at this temperature. The OH generated in the cracking process of n-heptane ($C_7H_{14}O_2HO_2$ to nC_7KET) and the OH produced in the process of nC_7KET undergoing a series of reactions to form CH_2O, as well as the OH consumed by n-heptane itself, decreases with the reduction in n-heptane content. This suggests that the increase in methyl oleate and ethanol content adversely affects the cracking and heat release of n-heptane.

Considering the supply of OH to ethanol and the system's provision of OH to n-heptane, the changes in OH contribution rates are not significantly pronounced in a low-temperature environment due to the incomplete cracking and dehydrogenation reactions of the mixed fuel. At 900 K, ethanol initially undergoes cracking reactions to form smaller molecules. Concurrently, in the presence of OH and H_2O, ethanol undergoes dehydrogenation reactions, mainly producing CH_3CHOH, CH_3CH_2O, and C_2H_4OH, which further crack and oxidize into smaller molecules.

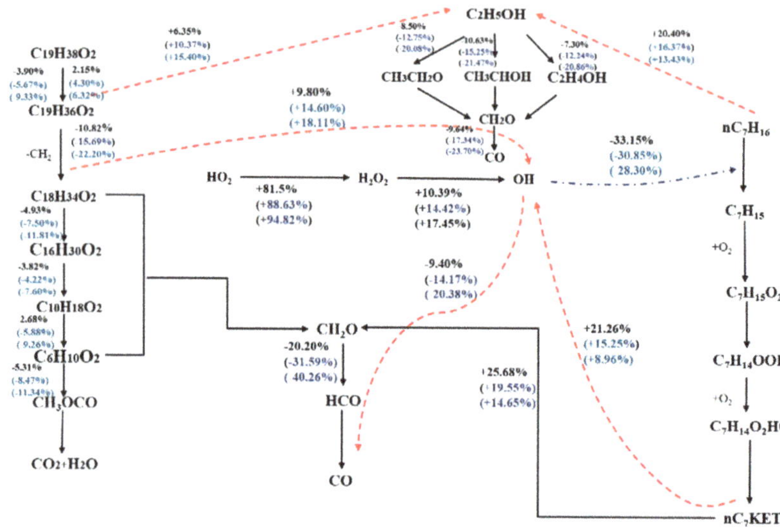

Figure 8. Reaction flux concerning OH in the combustion reaction system of HB5E5 (HB10E10) (HB15E15) fuels at 900 K.

In the HB10E10 system, the total OH consumption rate due to the dehydrogenation reaction of methyl oleate is 26.50%. Compared to the HB10 mixed-fuel system, the introduction of ethanol at low temperatures results in a substantial consumption of OH for reactions, leading to a higher total OH consumption rate in the HB10E10 mixed-fuel system than in the HB10 system. As illustrated in Figure 8, the total OH consumption rate for the dehydrogenation reaction of methyl oleate in the HB10 fuel reaction system is 19.08%. This increased consumption rate is attributed to the addition of ethanol, which significantly consumes OH at this temperature, thereby affecting the cracking of methyl oleate and consequently impacting OH generation.

In the 900 K environment, most of the OH in the system is converted into more stable H_2O_2. Given the low conversion rate of H_2O_2 to OH under these conditions, the enhancement in the reactivity of the reaction system is limited. Therefore, as the concentration of methyl oleate and ethanol increases in this environment, the reactivity of the reaction system gradually decreases.

3.3.2. High-Temperature OH Cross-Reaction Flux Analysis for n-Heptane–Methyl Oleate–Ethanol

Under the conditions of an initial temperature of 1400 K, initial pressure of 6 MPa, and an equivalence ratio of 1.0, the cross-reaction flux analysis concerning OH in the HB5E5 (HB10E10) (HB15E15) fuel combustion reaction system is depicted in Figure 9. Similar to the 900 K environment, stearic acid methyl ester initially undergoes a dehydrogenation reaction to form methyl oleate at 1400 K. Unlike at 900 K, during the high-temperature cracking of stearic acid methyl ester, H_2O is produced. While most water converts into more stable H_2O_2 at lower temperatures, at 1400 K, the majority of H_2O is transformed into OH through H_2O_2, providing for the reaction in the mixed-fuel system.

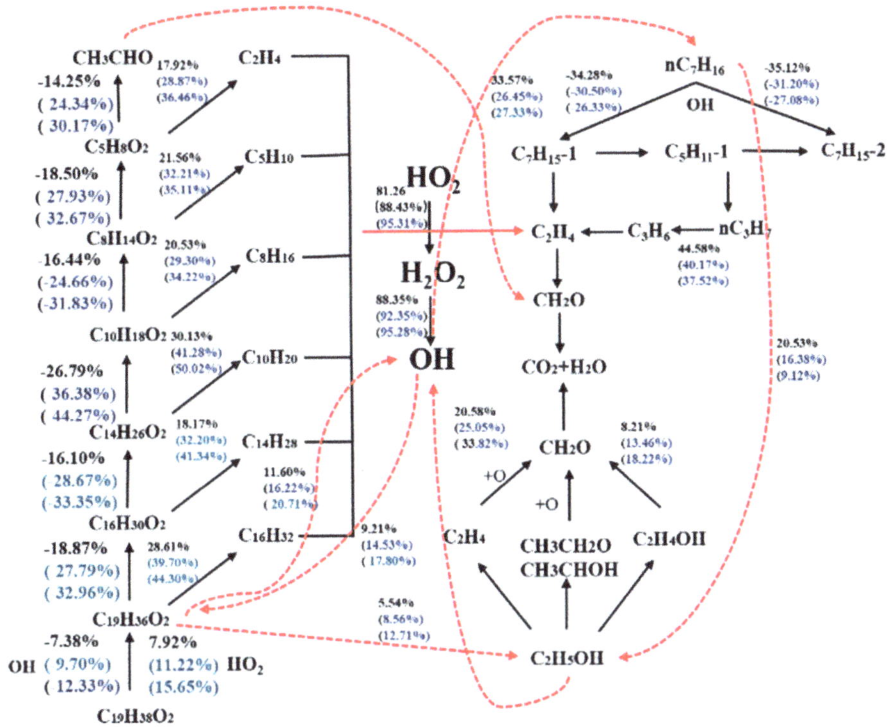

Figure 9. Reaction flux concerning OH in the combustion reaction system of HB5E5 (HB10E10) (HB15E15) fuels at 1400 K.

Methyl oleate in high-temperature conditions undergoes dehydrogenation and cracking reactions, producing ester and olefin compounds. It can be observed from the graph that the OH contribution rate produced by cracking is greater than the consumption rate by dehydrogenation, with an overall OH contribution rate of 28.97%. Contrasting with the 900 K environment where the OH generation was less than its consumption, at high temperatures, the complete cracking of methyl oleate results in OH generation exceeding consumption. Ester compounds eventually lead to the formation of acetaldehyde and formaldehyde, ultimately burning to form CO_2 and H_2O, while olefin compounds produce C_2H_4, undergoing dehydrogenation and oxidation reactions to form CH_2O and subsequently CO_2 and H_2O.

Subsequently, at high temperatures, n-heptane dehydrogenates to form C_7H_{15}-1 and C_7H_{15}-2. C_7H_{15}-1 cracks into C_2H_4 and C_5H_{11}-1, and C_5H_{11}-1 also cracks under high temperatures to form nC_3H_7 and ultimately C_2H_4. In the pathway where nC_3H_7 forms C_3H_6, the OH contribution rate is 44.58%. It is evident that although the proportion of n-heptane is decreasing, the OH contribution rate in the pathway of nC_3H_7 to C_3H_6 does not decrease significantly. This is because the higher temperatures lead to the increased production of OH in the mixed-fuel system, allowing for more thorough cracking and dehydrogenation reactions of n-heptane. This enhanced reaction progress is primarily due to the addition of ethanol and methyl oleate, which increases OH production at high temperatures, thereby promoting a more complete overall reaction.

Under high-temperature conditions, ethanol primarily undergoes cracking reactions, leading to the formation of C_2H_4OH, C_2H_4, and C_2H_5O, which further generate CH_2O. These processes produce a significant amount of OH, and as the concentration of ethanol increases, the OH provided by ethanol to the system also increases. Compared to the mixed fuel at 900 K, the cracking and dehydrogenation reactions of the three substances (ethanol, methyl oleate, and n-heptane) are different at 1400 K.

At 900 K, the OH produced by the cracking of methyl oleate and n-heptane is limited, slowing down the overall reaction. Ethanol primarily undergoes dehydrogenation reactions consuming OH, with a minor contribution to OH generation. The H_2O produced in the cracking process mostly transforms into more stable H_2O_2, with only a small portion of H_2O_2 converting to OH for the overall reaction. However, at 1400 K, most H_2O_2 transforms into OH, which is available for reactions in the system. Ethanol predominantly undergoes cracking reactions at this temperature, generating a significant amount of OH for the mixed-fuel system reaction.

As shown in Figure 9, initially, methyl oleate supplies OH for reactions with n-heptane and ethanol. As the reaction progresses, ethanol generates a large amount of OH, which eventually primarily flows towards methyl oleate and n-heptane for reactions. In the high-temperature ternary fuel system, the small-molecule reaction pathways involving active components like OH mainly include $C_2H_4 \rightarrow CH_2O \rightarrow HCO \rightarrow CO$. Additionally, intermolecular cross-reactions occur between methyl oleate and n-heptane, both producing C_2H_4 and CH_2O through dehydrogenation reactions. In low-temperature reactions, the ternary fuel system and the binary fuel system also involve the $CH_2O \rightarrow HCO \rightarrow CO$ pathway, indicating its importance in both low- and high-temperature conditions.

4. Conclusions

This study investigated the chemical kinetics mechanism representing hydrogenated biodiesel and the impact of cross-reactions on the hydrogenated biodiesel–ethanol–diesel blend. Using a constant-volume homogeneous reactor, the variations in the concentration of key components and reaction pathways in the HB5E5, HB10E10, and HB15E15 mixed-fuel systems were calculated. The findings of this study align with those of Zuolei's ternary fuel study. The main conclusions are as follows:

(1) In the MD10 (MD20) mixed-fuel system, the total consumption rate of n-heptane is 68.93% (54.31%), with OH flowing from n-heptane to methyl decanoate. In the HB10 (HB20) mixed-fuel system, the total consumption rate of n-heptane is 77.21% (60.58%), with OH flowing from the methyl oleate and stearic acid methyl ester mixture to n-heptane.

(2) At low temperatures in the HB5E5, HB10E10, and HB15E15 mixed-fuel systems, OH produced by the cracking of n-heptane is also supplied to ethanol. Due to the substantial consumption of OH by ethanol at this stage, the reactivity of the mixed-fuel system is relatively low. As the content of methyl oleate, stearic acid methyl ester, and ethanol increases, the generation of OH gradually decreases, which is unfavorable for the reaction to proceed.

(3) At high temperatures, in the HB5E5, HB10E10, and HB15E15 systems, the total generation rates are respectively 24.6%, 33.29%, and 38.65%. Ethanol predominantly undergoes cracking reactions at high temperatures, producing a substantial amount of OH. This indicates that increasing the content of hydrogenated biodiesel and ethanol within a limited range can enhance the reactivity of the system, facilitating the reaction process.

The significance of this study lies in its contribution to the fundamental understanding of biofuel combustion, particularly in the context of hydrogenated biodiesel–ethanol–diesel blends. Our findings not only illuminate the complex chemical dynamics within these blends but also pave the way for optimizing fuel formulations for enhanced combustion efficiency and reduced environmental impact. This work is a step forward in developing sustainable energy solutions, aligning with global efforts to transition towards greener fuel alternatives.

Author Contributions: Conceptualization, J.Z.; Data curation, J.Z.; Formal analysis, L.C.; Funding acquisition, J.Z.; Investigation, R.Z.; Methodology, L.C. and W.Z.; Project administration, W.Z.;

Resources, R.Z.; Software, J.Z. and R.Z.; Supervision, W.Z.; Validation, J.Z.; Writing—original draft, J.Z.; Writing—review and editing, J.Z. and W.Z. All authors have read and agreed to the published version of the manuscript.

Funding: This research was funded by the Hunan Provincial Natural Science Foundation (Grant No. 2022JJ60090), Belt and Road Project in Jiangsu Province (Grant No. BZ2022016) and Scientific research project of Department of Education of Hunan Province (Grant No. 21C1577).

Data Availability Statement: The data that support the findings of this study are available from the corresponding author upon reasonable request.

Conflicts of Interest: The authors declare no conflicts of interest.

References

1. Shi, W.; Li, J.; He, B.; Yan, F.; Cui, Z.; Wu, K.; Lin, L.; Qian, X.; Cheng, Y. Biodiesel production from waste chicken fat with low free fatty acids by an integrated catalytic process of composite membrane and sodium methoxide. *Bioresour. Technol.* **2013**, *139*, 316–322. [CrossRef] [PubMed]
2. Martos, F.J.; Soriano, J.A.; Braic, A.; Fernández-Yáñez, P.; Armas, O. A CFD Modelling Approach for the Operation Analysis of an Exhaust Backpressure Valve Used in a Euro 6 Diesel Engine. *Energies* **2023**, *16*, 4112. [CrossRef]
3. Sathyamurthy, R.; Balaji, D.; Gorjian, S.; Muthiya, S.J.; Bharathwaaj, R.; Vasanthaseelan, S.; Essa, F.A. Performance Combus-tion and Emission Characteristics of a DI-CI Diesel Engine Fueled with Corn Oil Methyl Ester Biodiesel Blends. *Sustain. Energy Technol. Assess.* **2021**, *43*, 100981.
4. Aydn, S. Comprehensive Analysis of Combustion, Performance and Emissions of Power Generator Diesel Engine Fueled with Different Source of Biodiesel Blends. *Energy* **2020**, *205*, 118074. [CrossRef]
5. Szabados, G.; Berecz Ky, Á. Experimental investigation of physicochemical properties of diesel, biodiesel and TBK-biodiesel fuels and combustion and emission analysis in CI internal combustion engine. *Renew. Energy* **2018**, *121*, 568–578. [CrossRef]
6. Mubarak, M.; Shaija, A.; Suchithra, T.V. Experimental Evaluation of Salvinia Molesta Oil Biodiesel/Diesel Blends Fuel on Combustion Performance and Emission Analysis of Diesel Engine. *Fuel* **2021**, *287*, 119526. [CrossRef]
7. Xi, S.; Xue, J.; Wang, F.; Li, X. Reduction of Large-Size Combustion Mechanisms of n-Decane and n-Dodecane with an Improved Sensitivity Analysis Method. *Combust. Flame* **2020**, *222*, 326–335. [CrossRef]
8. Nadanakumar, V.; Muthiya, J.; Prudhvi, T.; Induja, S.; Sathyamurthy, R.; Dharmaraj, V. Experimental investigation to control HC, CO & NOx emissions from diesel engines using diesel oxidation catalyst. *Mater. Today Proc.* **2021**, *43*, 434–440.
9. Sundararajan, R.; Jeyaseelan, T. Effect of Biodiesel, Biodiesel Binary Blends, Hydrogenated Biodiesel and Injection Parameters on NOx and Soot Emissions in a Turbocharged Diesel Engine. *Fuel* **2019**, *240*, 101–118.
10. Zuo, L. *Study on Combustion and Emission Characteristics of Hydrogenated Biodiesel-Ethanol-Diesel for Diesel Engine Combustion*; Jiangsu University: Zhenjiang, China, 2020.
11. Yu, Q. *Simulation Study on PCCI Diesel Combustion Process with Coupled DMC/Biodiesel Reaction Mechanism*; Jiangsu University: Zhenjiang, China, 2021.
12. Marcoberardino, G.D.; Vitali, D.; Spinelli, F.; Binotti, M.; Manzolini, G. Green Hydrogen Production from Raw Biogas: A Techno-Economic Investigation of Conventional Processes Using Pressure Swing Adsorption Unit. *Processes* **2018**, *6*, 19. [CrossRef]
13. Al-Gharibeh, E.; Kumar, K. Oxidation kinetics of methyl decanoate in a motored engine. *Fuel* **2022**, *308*, 121912. [CrossRef]
14. Herbinet, O.; Pitz, W.J.; Westbrook, C.K. Detailed chemical kinetic oxidation mechanism for a biodiesel surrogate. *Combust. Flame* **2008**, *154*, 507–528. [CrossRef]
15. Glaude, P.E.; Herbinet, O.; Bax, S.; Biet, J.; Warth, V.; Battin-Leclerc, F. Modeling of the Oxidation of Methyl Esters—Validation for Methyl Hexanoate, Methyl Heptanoate, and Methyl Decanoate in a Jet-Stirred Reactor. *Combust. Flame* **2010**, *157*, 2035–2050. [CrossRef] [PubMed]
16. Hodnett, B.K.; Verma, V. Thermodynamic vs. Kinetic Basis for Polymorph Selection. *Processes* **2019**, *7*, 272. [CrossRef]
17. Zhou, Z.; Yi, Q.; Wang, R.; Wang, G.; Ma, C. Numerical Investigation on Coal Combustion in Ultralow CO_2 Blast Furnace: Effect of Oxygen Temperature. *Processes* **2020**, *8*, 877. [CrossRef]
18. Chang, Y. *Research on the Reaction Mechanism of Fuel Skeleton for Characterization of Diesel and Biodiesel Based on Decoupling Method*; Dalian University of Technology: Dalian, China, 2016.
19. Zhai, Y. *Experimental and Model Study on the Reaction Kinetics of Biodiesel Alternative Fuel Combustion*; University of Science and Technology of China: Anhui, China, 2018.
20. Mei, D.; Zhang, Q.; Ju, Z.; Gu, M.; Yuan, Y.; Chen, C.; Tao, J. Partial hydrogenated biodiesel improves the combustion and emission of diesel engine. *China J. Highw. Transp.* **2019**, *32*, 184–190.
21. Leng, X.; Li, M.; He, Z.; Zhang, Y.; Zhong, W.; Xuan, T.; Duan, L.; Wang, P. Combustion characteristics of engine fueled by gasoline/hydrogenated biodiesel blend. *J. Harbin Eng. Univ.* **2020**, *41*, 1196–1202.
22. Sinha, A.; Thomson, M. The chemical structures of opposed flow diffusion flames of C3 oxygenated hydrocarbons (isopropanol, dimethoxy methane, and dimethyl carbonate) and their mixtures. *Combust. Flame* **2004**, *136*, 548–556. [CrossRef]

23. Adu-Mensah, D.; Mei, D.; Zuo, L.; Zhang, Q.; Wang, J. A review on partial hydrogenation of biodiesel and its influence on fuel property. *Fuel* **2019**, *251*, 660–668. [CrossRef]
24. Zuo, L.; Wang, J.; Mei, D.; Dai, S.; Adu-Mensah, D. Experimental investigation on combustion and (regulated and unregulated) emissions performance of a common-rail diesel engine using partially hydrogenated biodiesel-ethanol-diesel ternary blend. *Renew. Energy* **2022**, *185*, 1272–1283. [CrossRef]
25. Westbrook, C.K.; Naik, C.V.; Herbinet, O.; Pitz, W.J.; Mehl, M. Detailed chemical kinetic reaction mechanism for biodiesel components methyl stearate and methyl oleate. *Proc. Combust. Inst.* **2011**, *33*, 383–389.
26. Curran, H.J.; Gaffuri, P.; Pitz, W.J.; Westbrook, C.K. A Comprehensive Modeling Study of n-Heptane Oxidation. *Combust. Flame* **1998**, *114*, 149–177. [CrossRef]
27. Metcalfe, W.K.; Burke, S.M.; Ahmed, S.S.; Curran, H.J. A Hierarchical and Comparative Kinetic Modeling Study of C1 − C2 Hydrocarbon and Oxygenated Fuels. *Chem. Kinet.* **2013**, *45*, 638–675. [CrossRef]
28. Fieweger, K.; Blumenthal, R.; Adomeit, G. Self-ignition of SI engine model fuels: A shock tube investigation at high pressure. *Combust. Flame* **1997**, *109*, 599–619. [CrossRef]
29. Shen, H.-P.S.; Steinberg, J.; Vanderover, J.; Oehlschlaeger, M.A. A shock tube study of the ignition of n-heptane, n-decane, n-dodecane, and n-tetradecane at elevated pressures. *Energy Fuels* **2009**, *23*, 2482–2489. [CrossRef]
30. Herzler, J.; Jerig, L.; Roth, P. Shock tube study of the ignition of lean n-heptane/air mixtures at intermediate temperatures and high pressures. *Proc. Combust. Inst.* **2005**, *30*, 1147–1153. [CrossRef]
31. Hartmann, M.; Gushterova, I.; Fikri, M.; Schulz, C.; Schießl, R.; Maas, U. Auto-ignition of toluene-doped n-heptane and iso-octane/air mixtures: High-pressure shock-tube experiments and kinetics modeling. *Combust. Flame* **2011**, *158*, 172–178. [CrossRef]
32. Jessica, L.B.; Youngchul, R.; Rolf, D.R.; Joanna, M.; Daw, C.S. Development and Validation of a Reduced Reaction Mechanism for Biodiesel-Fueled Engine Simulations. *SAE Int. J. Fuels Lubr.* **2009**, *1*, 675–702.

processes

MDPI

Article

CFD Analysis of Counter-Rotating Impeller Performance in Mixed-Flow Pumps

Edwar L. Pérez [1], Miguel Asuaje [2] and Nicolas Ratkovich [1,*]

[1] Chemical and Food Engineering Department, Universidad de loss Andes, Bogotá 111711, Colombia; el.perez10@uniandes.edu.co
[2] Department of Energy and Automatization, Universidad Metropolitana, Caracas 1060, Venezuela; masuaje@unimet.edu.ve
* Correspondence: n.rios262@uniandes.edu.co

Abstract: This study presents a computational fluid dynamics (CFD) analysis of the performance of counter-rotating impeller systems in mixed-flow pumps. The analysis evaluates the impact of varying rotor velocity ratios and blade geometry on head rise, efficiency, and hydraulic losses. Through detailed CFD simulations, the counter-rotating system demonstrates significant improvements in head and efficiency at low flow rates, with model B achieving up to 20% higher efficiency near the best efficiency point (BEP). However, increased hydraulic losses offset efficiency gains at higher flow rates. While the findings provide valuable insights for optimizing the design of counter-rotating systems in mixed-flow pumps, experimental validation is needed to confirm the results and ensure real-world applicability. The study lays the groundwork for future work in refining counter-rotating pump designs to minimize hydraulic losses and slippage.

Keywords: counter-rotating impellers; mixed-flow pumps; computational fluid dynamics (CFD); hydraulic efficiency; head rise optimization

1. Introduction

The field of turbomachinery has a rich and significant history, evolving over the past two centuries with notable advancements occurring in the last seven decades. The concept of centrifugal impellers, first envisioned by Leonardo Da Vinci in the 16th century, laid the foundation for modern turbomachinery by utilizing centrifugal force for fluid movement. A significant breakthrough occurred in 1689 when Denis Papin developed a forced vortex system using blades within a circular or spiral housing. In 1754, Euler's theoretical work applying Newton's laws to centrifugal impellers, commonly known as Euler's law, provided a mathematical basis for further developments in fluid transport mechanisms. These innovations led to extensive research on centrifugal impellers from 1910 to 1930, enhancing theoretical and experimental understanding of flow mechanisms. [1,2].

In recent decades, the field of turbomachinery has expanded significantly, driven by the development of computational tools that enhance performance. Turbomachines play a critical role in various industries, including oil and gas (O&G), petrochemical, and renewable energy sectors, where they serve as vital components in compressors, turbines, and pumps [3,4]. Centrifugal pumps, in particular, dominate industrial applications due to their efficiency, low capital (CAPEX) and operating costs (OPEX), and versatility across a wide range of operational conditions [5]. The global centrifugal pump market is on a rapid growth trajectory, with the U.S. market projected to reach USD 8.6 billion by 2020 and China's market expected to hit USD 7.6 billion by 2027. This promising growth trend underscores the increasing demand for and potential opportunities in centrifugal pumps [6].

A 2021 report from Cognitive Market Research forecasts substantial growth in the global mixed-flow impeller pumps market, with projections showing a revenue increase of

130% in 2015 to 530% by 2027, driven mainly by demand in South America. This growth is expected to be particularly notable in the water treatment, chemical, and metallurgical industries, which are projected to grow by at least 35% by 2027 [7–9].

This study builds upon previous research on counter-rotating systems primarily focused on axial turbomachinery. However, the adaptation of such systems to centrifugal pumps presents unique challenges. Previous studies have utilized experimental and computational methods to examine counter-rotating systems in various turbomachines, including axial pumps and fans. Table 1 summarizes fundamental experimental and computational fluid dynamics (CFD)-based studies on counter-rotating systems across different turbomachines. These studies highlight the potential benefits of counter-rotating designs, including improved efficiency, increased pressure capacity, and enhanced flow dynamics [1,10–12]. The benefits of counter-rotating centrifugal pumps include improved efficiency, reduced size and weight, and the ability to explore new designs through CFD at relatively low costs [13,14].

Table 1. Experimental and CFD studies for counter-rotative systems in turbomachines.

Study	System	CFD Code/Software	Physics Modeling/Methodology	Mesh Configuration	Ref.
Kanemoto (2004)	Counter-rotative axial pump	[-]	Experimental	[-]	[15]
Momosaki et al. (2010)	Counter-rotative axial pump	ANSYS CFX-12.0	[-]	1535180 Nodes	[16]
Cao et al. (2014)	Counter-rotative axial pump	[-]	Experimental	[-]	[17]
Tosin et al. (2015)	Counter-rotative–mixed-flow pump	ANSYS CFX	RANS and Rayleigh-Plesset l multiphase model	12×10^6 Nodes	[18]
Tosin et al. (2015)	Counter-rotative–mixed-flow pump	ANSYS CFX	RANS SST turbulence model	2.1 million Nodes	[19]
Friebe et al. (2018)	Counter-rotative axial fan	CFturbo	RANS	[-]	[20]
Nguyen (2020)	Counter-rotative centrifugal compressor	STAR-CCM+	RANS	3.4–11 million cells (polyhedral)	[21]

This study explores the design and performance of counter-rotating centrifugal pumps by conducting preliminary calculations of velocity triangles under typical crude extraction conditions. The pump's geometry will be designed with blade variations and rotor-speed adjustments, and the performance of the counter-rotating system will be compared with traditional pumps through dimensional analysis and CFD simulations. This comparison will provide critical insights into the potential applications of counter-rotating systems in industries such as oil and gas, petrochemicals, and renewable energy.

While counter-rotating systems have been explored extensively in axial turbomachinery, their application in mixed-flow pumps presents distinct challenges and opportunities that must be thoroughly addressed in the existing literature. Unlike axial systems, where counter-rotating designs primarily focus on enhancing axial flow efficiency, mixed-flow pumps require a more complex interplay of radial and axial components, leading to different flow dynamics and performance considerations. The complexity of mixed-flow systems, including the interaction between centrifugal and axial forces, necessitates a tailored approach to counter-rotating designs, as traditional methods used in axial systems may not be directly transferable. This study fills a critical gap by investigating the specific impact of counter-rotating systems in mixed-flow pumps through a combination of CFD simulations and performance analysis. By optimizing the rotor speed, blade geometry, and flow dynamics, this research extends the state of the art in counter-rotating turbomachinery, offering valuable insights for industries that rely on high-performance mixed-flow pumps. This study's results can improve pump efficiency and performance in applications such as oil and gas, petrochemical, and renewable energy, thereby contributing to advancements in industrial pump design.

2. Application Experiments

2.1. Experimental Data

This study uses the BCS GN5200 centrifugal pump model, as detailed in Monteverde's research [22]. The pump operates at various rotational speeds, with the numerical model validation focusing on data at 3500 rpm. At this speed, the pump achieves a Best Efficiency Point (BEP) of 36.5 m^3/h and a head of 7.78 m per stage, with a specific speed of 1.32. Figure 1 displays the performance curves of the BCS GN5200 operating at 3500 rpm, as provided by the manufacturer.

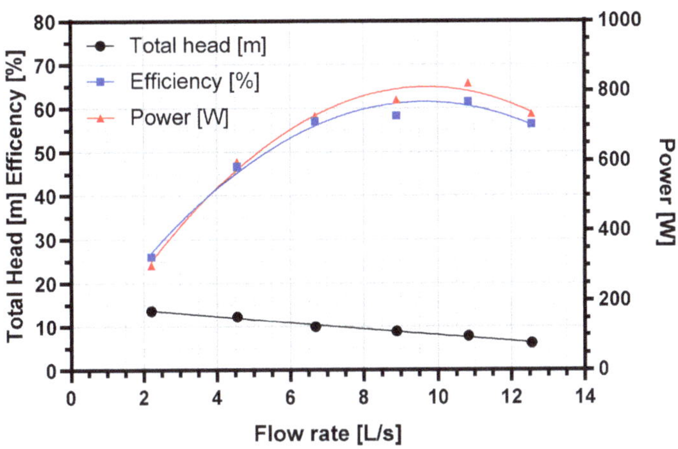

Figure 1. Performance curves of BCS GN5200 operating at 3500 rpm [22].

The impeller and stator geometrical characteristics for each stage are listed in Table 2, showing key dimensions that were used in the simulations. These parameters are critical for understanding the pump's geometry and performance characteristics:

- Number of Blades: The impeller and diffuser each have seven blades. The number of blades is critical in determining the pump's flow patterns and efficiency.
- Inlet Inner Diameter: The impeller's inlet inner diameter is 30.2 mm, while the diffuser's is larger, at 79.9 mm. This variation helps manage the fluid's transition from the impeller to the diffuser, affecting the fluid velocity and pressure distribution.
- Inlet External Diameter: The impeller's inlet external diameter is 65.3 mm, while the diffuser's is 93.3 mm. This difference reflects the change in the flow passage area, which influences the pump's ability to handle different flow rates.
- Outlet Inner Diameter: The impeller's outlet inner diameter is 79.9 mm, whereas the diffuser's is 30.2 mm. This decrease in the diffuser's outlet diameter helps to maintain pressure as the fluid moves out of the pump.
- Outlet External Diameter: The external diameter at the impeller's outlet is 93.3 mm, while the diffuser's is 65.3 mm. This difference contributes to the energy conversion as the fluid moves from the impeller to the diffuser.

Table 2. Principal dimensions of the GN5200 impeller and diffuser.

Parameter	Units	Impeller Valor	Diffuser Valor
N° of blades	[-]	7	7
Inlet inner diameter	[mm]	30.2	79.9
Inlet external diameter	[mm]	65.3	93.3
Outlet inner diameter	[mm]	79.9	30.2
Outlet external diameter	[mm]	93.3	65.3

Monteverde's study involved three different working fluids—water, a glycerin–water solution, and pure glycerin—covering a wide range of viscosities and temperatures [22,23]. For this study, five specific flow rates were chosen from the pump's operating curves: 2.21, 4.55, 6.68, 8.89, 10.89, and 12.55 L/s. Additional input data for model convergence included an inlet pressure of 1×10^8 Pa, assuming negligible temperature variations and constant density and viscosity for each fluid.

2.2. *Geometry and Numerical Modeling*

This study employed numerical simulations and geometry modeling to assess the performance of the BCS GN5200 pump. These two processes are distinct but interconnected components of the overall analysis.

2.2.1. Geometry Modeling

Three models were developed for the study: a base impeller model and two counter-rotating impeller models. The geometry was created using Autodesk Inventor 2021, which includes the impeller, stator (each with seven vanes), and pump casing.

- Base Impeller Model: This model represents a standard impeller and stator configuration, consisting of an impeller with seven vanes and a stator with an equal number of vanes. The entire assembly, including the pump casing, was modeled for analysis.
- Counter-Rotative Models: Two counter-rotating designs were introduced, each varying in the point at which the first rotor transitions to the second rotor. These designs are based on the methodology proposed by Nguyen, illustrated in Figure 2, and mathematically expressed in Equation (1).

$$LR = \frac{L_{1st}}{L_m} \tag{1}$$

where LR is the ratio of the top length of the rotor shroud (L_{1st}) to the meridional plane contour length of the base impeller (L_m). Two specific cases were analyzed:

- Model A: LR = 0.65
- Model B: LR = 0.32

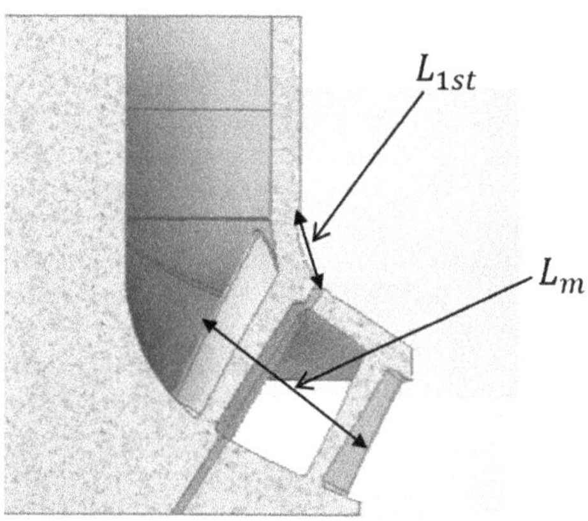

Figure 2. Geometrical representation of contour length for *LR* relation.

This formulation defines how the two rotors are geometrically split in the counter-rotating designs, with Figure 3 providing a visual representation.

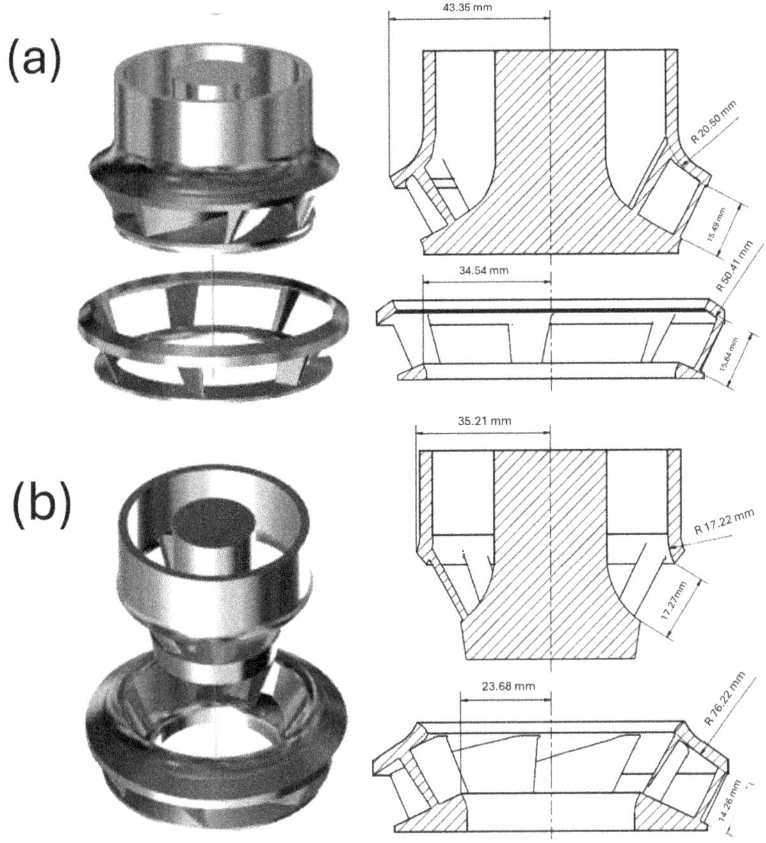

Figure 3. Principal dimensions for Model A (**a**) and B (**b**).

2.2.2. Numerical Modeling

Numerical simulations were carried out using STAR-CCM+ v19 (Siemens, Germany), which utilizes CFD techniques to evaluate the pump's performance. These simulations focused on fluid flow and mechanical interactions within the pump under various operating conditions.

Boundary conditions were applied at the pump's inlet and outlet regions, with flow rates and pressure conditions based on experimental data [22]. The numerical analysis included unsteady simulations to capture transient flow characteristics, and the Rigid Body Motion (RBM) model was used to simulate the pump's moving parts accurately.

Key performance metrics such as head, efficiency, power, and torque were calculated and analyzed, emphasizing the counter-rotating models' behavior.

The computational domain for the BCS GN5200 pump is shown in Figure 4. Three regions and boundary conditions were considered:

1. The pump inlet was modeled as a mass flow inlet, where the inlet velocity was calculated using the equation $Q = v_i \cdot a^*$, with a being the cross-sectional area at the inner rotor inlet.
2. The pump outlet was treated as a pressure outlet, with pressure values taken from Monteverde [22]. Standard atmospheric pressure (1 atm) was used as a reference.
3. The stator zone was static, and its walls were assigned a no-slip condition.

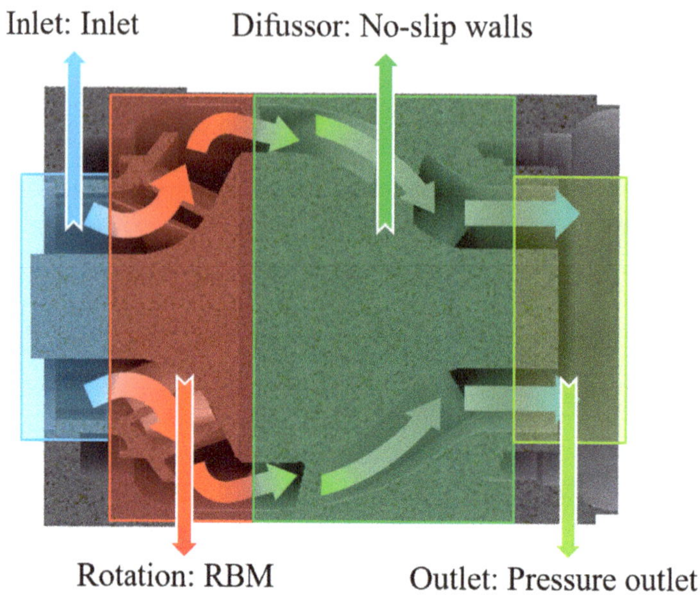

Figure 4. GN5200 Boundary conditions.

All performance data presented in the results were derived from detailed CFD simulations, with no experimental data used. Simulations were conducted at varying rotor speeds, with the "one rotor" configuration representing the internal rotor operating at 3500 RPM while the external rotor remained stationary. In counter-rotating configurations, the internal and external rotors were run at speeds between 3500 and 4375 RPM. Discrete simulation points were collected at specific flow rates, and continuous lines in the performance graphs were interpolated to highlight general trends.

The impeller was modeled using unsteady Rigid Body Motion (RBM) to accurately simulate moving components and capture transient flow effects, which are critical for studying rotating parts. The Courant–Friedrichs–Lewy (CFL) number was evaluated at a flow rate of 6.68 L/s, representing a mid-range point on the pump's performance curve. The internal and external rotors were simulated at 3500 rpm, confirming timestep independence under these conditions. Figure 5 shows the Courant number distribution, ranging from 1.32×10^{-5} and 10.

A sensitivity analysis was performed by halving and doubling the timestep size relative to the initial value to ensure timestep independence. Head rise, efficiency, and pressure distributions were compared across these simulations to verify solution convergence. The results varied by less than 1%, confirming timestep independence. While the CFL number guided timestep selection, these additional tests validated the robustness of the solution and ensured accurate capture of transient flow effects unaffected by timestep variations.

Performance metrics, including head, efficiency, power, and torque, were calculated. Torque was measured using a momentum report based on impeller rotation in the Z and -Z directions. Efficiency was determined by summing torque contributions, assuming a single shaft and mechanical power input [18,19].

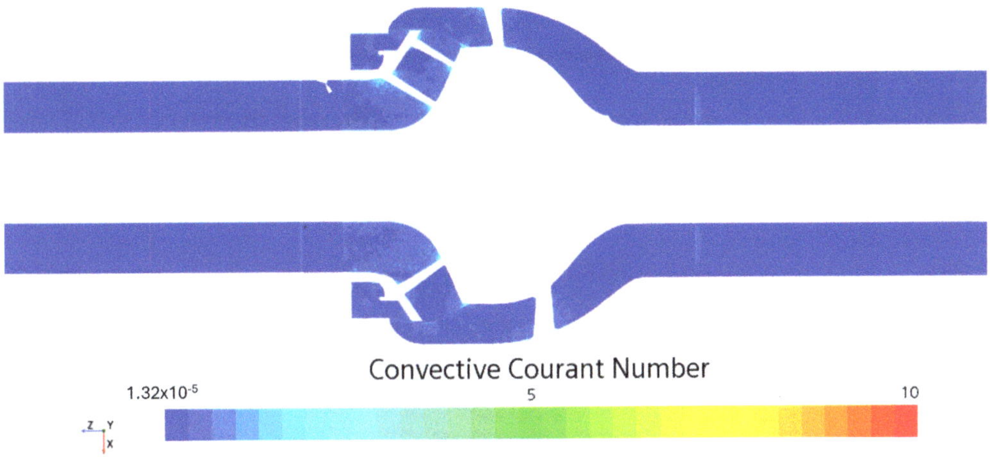

Figure 5. The courant number of GN5200 was evaluated under steady-state conditions at a flow rate of 6.68 L/s.

2.2.3. Physical Model Specification

The flow was treated as turbulent due to the high rotational velocity of the impellers. Table 3 presents the Reynolds numbers calculated using Equation (2), indicating the turbulent nature of the flow for both models.

$$Re = \frac{\rho N D^2}{\mu} \tag{2}$$

The properties and values are shown in Table 3.

Table 3. Re values for the two cases.

Impeller diameter (D) (m)	0.093	
Density (ρ) (kg/m^3)	1000	
Viscosity (μ) (Pa s)	0.003	
Rotation rate (rpm)	3500	4375
Re	1.17×10^7	1.46×10^7
Regime	Turbulent	Turbulent

Water was used as the working fluid for all simulations. A single-phase, unsteady RANS (U-RANS) approach was employed, utilizing the SST κ-ω model turbulence model to solve for eddy viscosity due to its superior performance in boundary layer predictions under adverse pressure gradients.

2.2.4. Mesh Generation

The pump geometry was discretized using a polyhedral mesh for central regions and prismatic layers for the surfaces, ensuring adequate mesh resolution. A base size of 14 mm was selected, with a total prism layer thickness of 0.1 mm. Figure 6 presents the y^+ distribution for the base impeller (a), Model A (b), and Model B (c), all evaluated with the same flow rate and Reynolds number. The variations in y^+ values are attributed to localized differences in flow behavior near the wall boundaries, particularly in flow separation and turbulence regions. Although the overall Reynolds numbers and mesh configurations were consistent, the geometry of the impellers and rotor speed differences led to distinct flow interactions, resulting in variations in y^+. The mesh was designed to ensure $y^+ < 1$, with six prism layers near the walls and a stretching ratio of 1.2 to maintain acceptable aspect ratios.

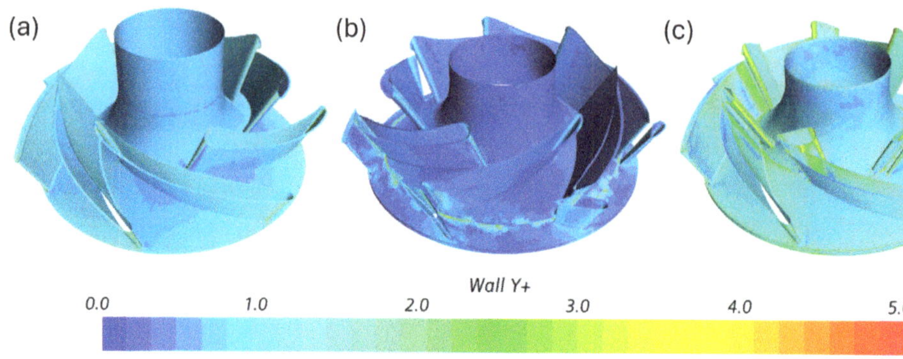

Figure 6. y^+ values for (**a**) base impeller, (**b**) Model A, and (**c**) Model B at a flow rate of 10.89 L/s and a Reynolds number of 1.32×10^7. The variations in y^+ values are due to localized differences in near-wall flow behavior, particularly in turbulence and flow separation regions. The mesh was designed with six prism layers and a stretching ratio of 1.2, ensuring $y^+ < 1$ without compromising the aspect ratio.

Mesh Independence Analysis

Mesh independence was tested by varying the cell count by ±30%. The water head's mean square error (MSE) was calculated to evaluate mesh quality using (Equation (3)).

$$\text{Total head MSE} = \sqrt{\frac{\sum_i^N \left(\frac{H_{Exp_i} - H_{CFD_i}}{H_{Exp_i}}\right)^2}{N}} \quad (3)$$

Results indicated that the base mesh provided the best compromise between accuracy and computational efficiency, as shown in Figure 7.

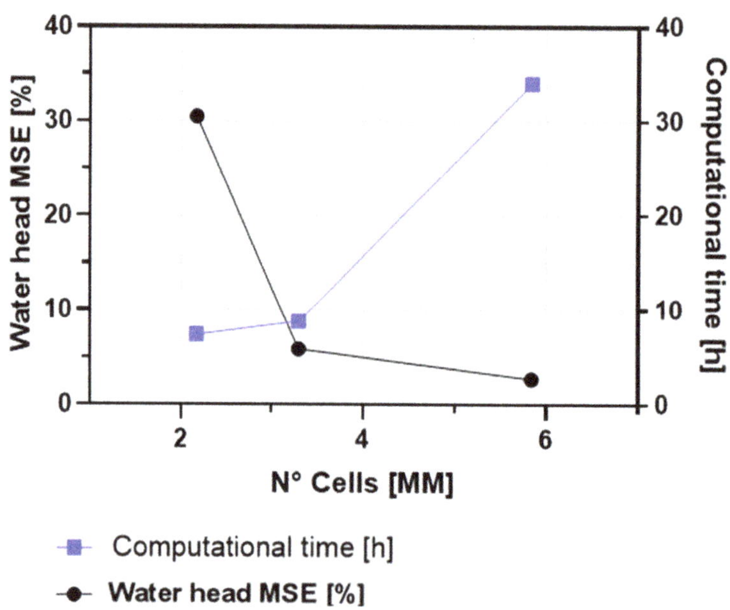

Figure 7. Mesh independence test results for water head.

A Grid Convergence Index (GCI) was computed to validate further mesh independence using refinement conditions, apparent order, and error estimations derived from Equations (4)–(8).

$$p_0 = \frac{1}{\ln(r_{21})}\left|\ln\left|\frac{\varepsilon_{32}}{\varepsilon_{21}}\right| + q(p)\right| \tag{4}$$

$$q(p_0) = \ln\left(\frac{r_{21}^{p_0} - S}{r_{32}^{p_0} - S}\right) \tag{5}$$

$$S = 1 \cdot sgn\left(\frac{\varepsilon_{32}}{\varepsilon_{21}}\right) \tag{6}$$

$$e_a^{21} = \left|\frac{\varphi_2 - \varphi_1}{\varphi_1}\right| \tag{7}$$

$$\mathrm{GCI}_{fine}^{21} = \frac{1.25 e_a^{21}}{r_{21}^{p_0} - 1} \tag{8}$$

The extrapolated curve for the water head (Equations (9) and (10)) confirmed that the base mesh was a good solution, with a maximum deviation of 3%, as shown in Figure 8.

$$\varphi_{ext}^{21} = \frac{r_{21}^{p_0}\varphi_1 - \varphi_2}{r_{21}^{p_0} - 1} \tag{9}$$

$$e_{ext}^{21} = \left|\frac{\varphi_{ext}^{21} - \varphi_1}{\varphi_{ext}^{21}}\right| \tag{10}$$

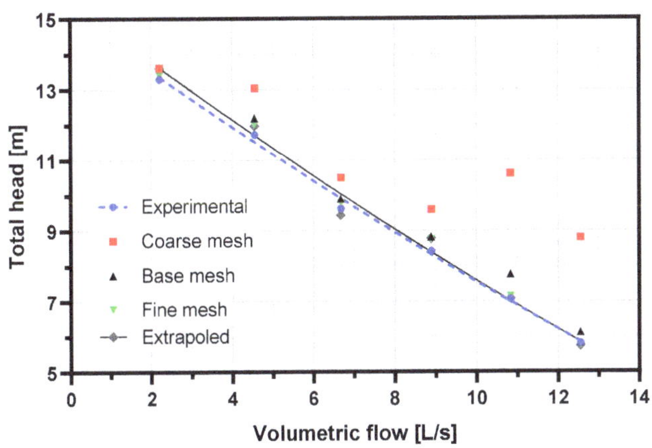

Figure 8. GCI results.

The 30% variation in cell count provided a reasonable compromise between mesh resolution and simulation time. Although a more significant variation could yield a more detailed analysis, increasing the cell count beyond 30% would significantly extend the simulation time and computational resources required. The results showed that key performance metrics—head, efficiency, and slip factor—remained consistent across different mesh sizes, with deviations of less than 1%, indicating that further refinement would minimally affect accuracy while substantially increasing computational costs

3. Results and Discussion

This section presents the performance curve results for the BCS GN5200 pump modified with a counter-rotating blade system. The analysis was conducted at six flow rates:

2.21, 4.55, 6.68, 8.89, 10.89, and 12.55 L/s. Two models (A and B) with different internal diameter ratios were examined, and three velocity ratios between the internal and external rotors were analyzed: 3500:3500 rpm, 3500:4375 rpm, and 4375:3500 rpm.

3.1. CFD Model Validation

Figure 9 compares simulated and experimental values for the base model with an unmodified impeller to validate the CFD model. The results show close agreement, with deviations in total head and efficiency within ±10%, except for one efficiency point. Experimental data from Monteverde [22] served as a benchmark, confirming the accuracy of the CFD simulations in terms of total head rise and efficiency across various flow rates. The deviations are within an acceptable range, demonstrating the robustness of the CFD model.

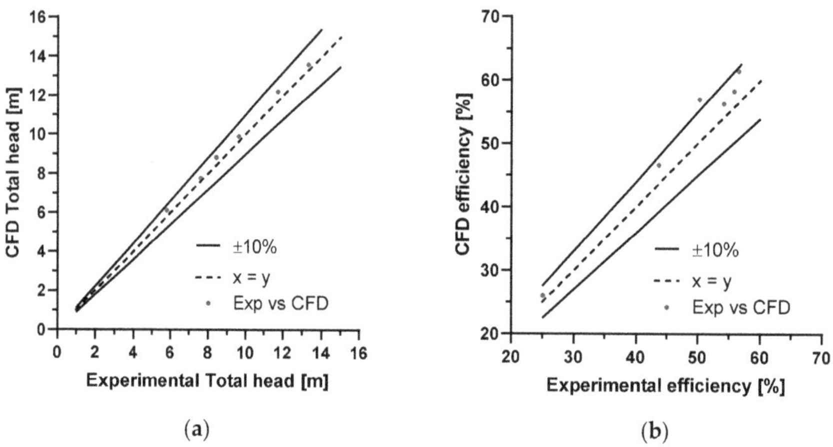

Figure 9. Comparison of CFD and experimental data for total head rise (**a**) and efficiency (**b**) across all flow rates. Simulated results are compared with experimental data from Monteverde [22].

3.2. Contra-Rotative Design Model Performance

The performance improvements of the counter-rotating impeller system were quantified through CFD simulations using standard pump performance parameters such as total head, efficiency, and slip factor. The quantification of these results involved the following steps:

1. Head Rise (H):
   The total head rise was calculated by measuring the difference between the pressure at the pump inlet and outlet, normalized by the fluid density and gravitational acceleration, as per this Equation:

$$H = \frac{(P_{outlet} - P_{inlet})}{\rho g} \quad (11)$$

   where P_{outlet} and P_{inlet} are the pressure values at the outlet and inlet, ρ is the fluid density, and g is the gravitational acceleration. The head values were then plotted as a function of flow rate and compared across different rotor velocity ratios.

2. Efficiency (η):
   Efficiency was calculated as the ratio of hydraulic power to mechanical input power. The hydraulic power was determined from the head rise and flow rate, while the mechanical power was obtained by measuring the torque applied to the rotors:

$$\eta = \frac{\rho g Q H}{T\omega} \quad (12)$$

where Q is the flow rate, T is the torque applied to the impellers, and ω is the angular velocity. The CFD simulations provided the torque and flow rate data necessary for these calculations, allowing us to track efficiency improvements as a function of rotor speed and flow rate.

3. Slip Factor:
   The slip factor quantifies the actual flow angle deviation from the impeller exit's ideal flow angle. It was calculated based on the velocity triangle formed by the relative and absolute velocities at the impeller outlet, with the slip increasing as the flow rate rises. The slip factor was measured for each rotor configuration to determine how closely the flow followed the idealized design path.

Figure 10 illustrates the total head results for the BCS GN5200 pump with counter-rotating impeller models A and B across various velocity ratios. At flow rates above 10 L/s, models A3535 and B3535 closely match the baseline model in head values but show steeper head rise slopes, highlighting the performance advantages of the counter-rotating design. Configurations with higher velocity ratios, such as A3543 and B4335, demonstrate superior performance, achieving a head increase of up to 7 m at lower flow rates and maintaining a 5 m advantage near the baseline model's BEP.

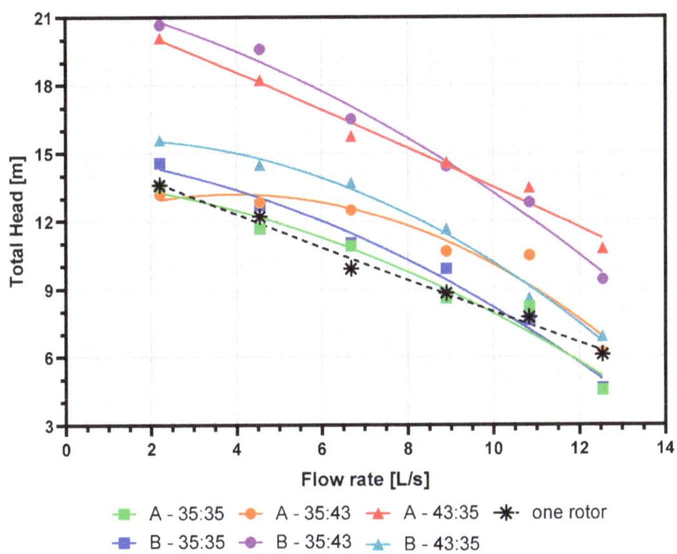

Figure 10. Total head results for models A and B at different velocity ratios compared to the base single-rotor model. Scattered data points represent discrete CFD simulation results at specific flow rates, while continuous lines represent interpolated trends. The 'one rotor' configuration corresponds to the internal rotor running at 3500 RPM.

The head increase observed in models A3543 and B3543 is due to the optimal rotor velocity ratios, where higher external rotor speeds enable more efficient energy transfer at low flow rates. In Figures 10 and 11, linear interpolation has been applied to the discrete CFD data points to illustrate general trends in total head and efficiency across various flow rates and rotor configurations. This interpolation is solely for visualization purposes and does not generate new data. The original CFD data points are also shown for reference.

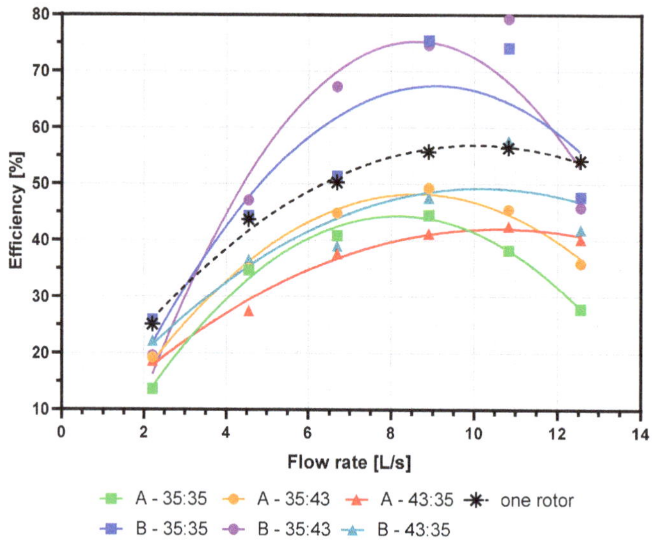

Figure 11. Overall efficiency results for models A and B at different velocity ratios compared to the base single-rotor model. Scattered data points represent discrete CFD simulation results at specific flow rates, while continuous lines represent interpolated trends. The 'one rotor' configuration corresponds to the internal rotor running at 3500 RPM.

At higher flow rates, however, these efficiency gains diminish due to increased hydraulic losses, particularly in model A, where turbulence and recirculation become more pronounced. The higher slip factor in model A4335 further contributes to the decline in efficiency at elevated flow rates.

A detailed convergence analysis was performed to ensure the accuracy of the results for the A3543 and A4335 configurations at a flow rate of 10.5 L/s. While the initial residuals for continuity, momentum, and turbulence equations fell below the convergence threshold of 10^{-4}, oscillations in the pressure and momentum equations suggested potential unsteady flow behavior. A timestep independence test revealed sensitivity in the A4335 case at this specific flow rate. By refining the timestep size, the head rise and efficiency results aligned more closely with surrounding data points. Additionally, a mesh independence test confirmed that the original mesh resolution was sufficient, with variations within 1%. The refined results for these cases are now consistent with overall trends and are reflected in the current figures.

The analysis shows that configurations with longer blades enhance performance, especially at higher rotational speeds. Model B, in particular, outperforms Model A at lower flow rates but experiences more significant head degradation as flow rates increase. The shift in head performance inversion points suggests that specific operational settings can optimize or reduce pump efficiency depending on the velocity ratio.

Comparing these results with previous studies, such as Kanemoto [15] and Cao [17], reveals similar efficiency improvements in counter-rotating axial systems. However, this study extends those findings to mixed-flow pumps, where the interaction of radial and axial forces presents additional challenges. With optimized rotor velocity ratios, mixed-flow pumps can achieve efficiency gains comparable to or greater than axial systems, though careful design is necessary to manage the complexity.

Figure 11 shows that models B3543 and B3535 exhibit up to 20% higher efficiency near the BEP than the base model. This improvement is due to model B's optimized velocity distribution and longer blades, which enhance energy transfer, especially at lower flow rates. Conversely, models with higher inlet rotor velocities (A4335 and B4335) show less efficiency gain due to increased turbulence and hydraulic losses, leading to flow instabilities and reduced energy transfer efficiency at higher flow rates.

3.3. Slip and Pump Losses

In models with higher rotational speeds, the slip factor increase was more pronounced, leading to greater energy dissipation and reduced overall efficiency. Model B, with its optimized blade length and rotor velocity ratio, exhibited a more gradual increase in slip factor, indicating better flow alignment. Hydraulic losses were highest at low flow rates due to recirculation but decreased as the flow rate approached the pump's BEP.

Figure 12 shows the theoretical slip factor as a function of flow rate for various models and velocity ratios. As expected, the slip factor increases with the flow rate in all models. Models A3535 and A4335 exhibit higher slip factors associated with lower efficiency due to poor flow alignment at higher flow rates. In contrast, model B3543 demonstrates the lowest slip factor across most flow rates, indicating better performance and reduced energy loss. This lower slip factor corresponds to the superior efficiency observed in model B3543, particularly at lower flow rates.

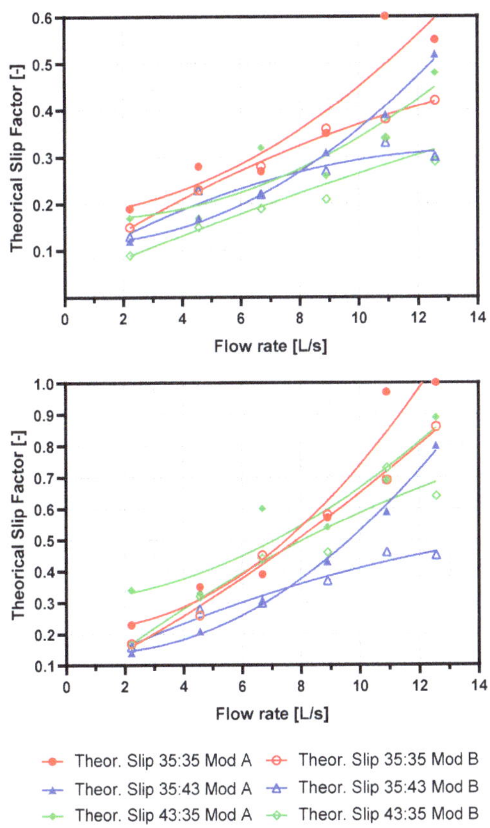

Figure 12. Total slippage for different models and numbers of rotors. Slippage is only in the inlet rotor (**up**), and total slippage is in both rotors (**down**).

Hydraulic losses were analyzed by observing the relative velocity patterns in models A and B. Figures 13 and 14 illustrate that recirculation losses are more prominent at low flow rates (e.g., 2.21 L/s), particularly in model A. In contrast, model B exhibits better flow control with fewer recirculation zones. As the flow rate increases, turbulence becomes the dominant source of hydraulic losses, especially at higher flow rates (e.g., 10.89 L/s). Model B consistently shows smoother velocity distributions, reducing turbulence and improving overall efficiency, particularly in the 3500:4375 rpm and 4375:3500 rpm velocity ratios. These results confirm that model B's blade configuration and velocity ratios are more optimized for reducing hydraulic losses, leading to better performance than model A.

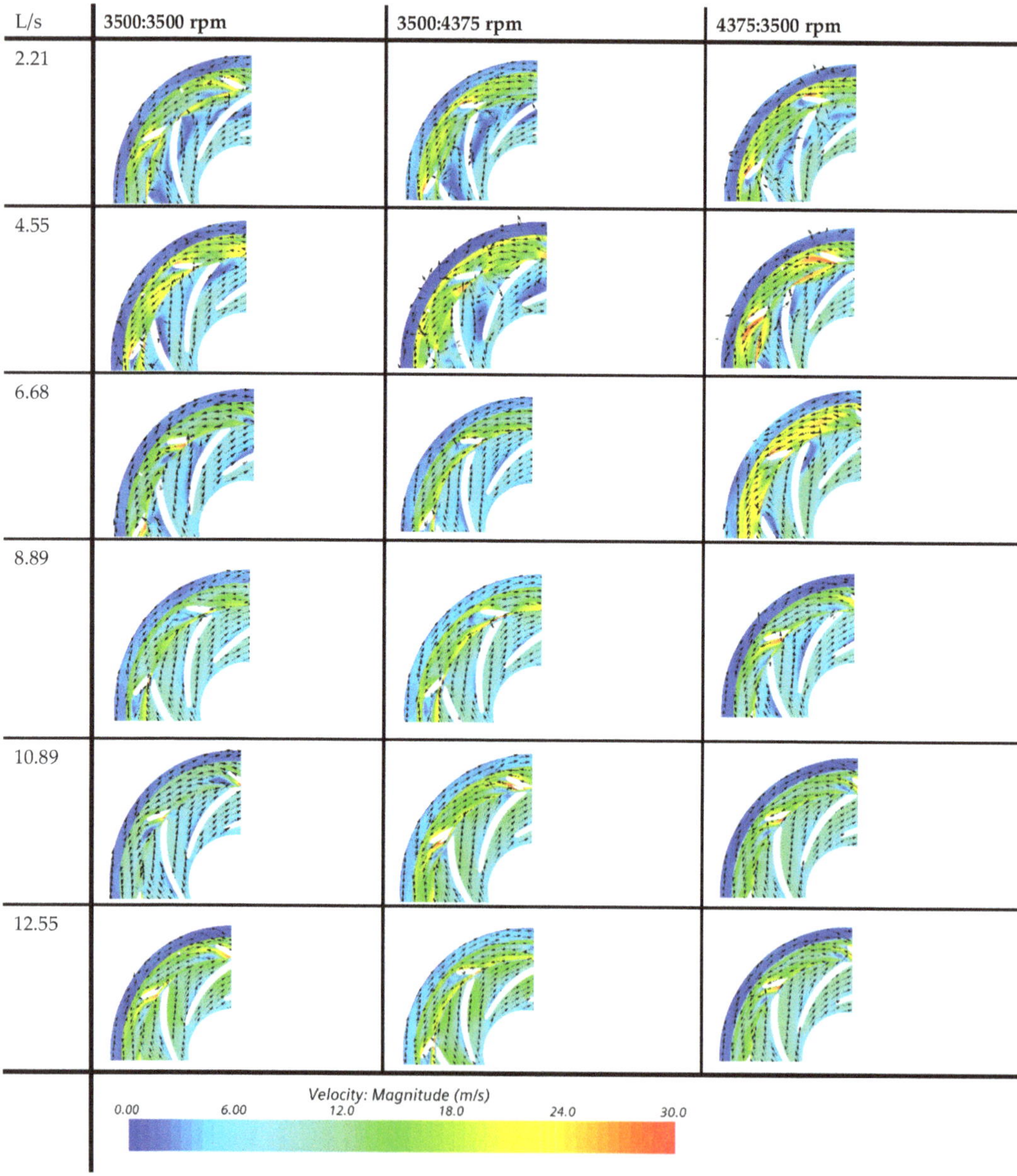

Figure 13. Relative velocity for Model A.

Figure 14. Relative velocity for Model B.

The increased slip factor in model A, especially in the A4335 configuration, results from the higher inlet rotor speed and shorter blades. Higher rotational velocity exacerbates flow instability, leading to more significant slippage. These hydraulic losses, including recirculation and turbulence, are more critical at low flow rates, contributing to overall energy inefficiency.

Momosaki observed similar slip-related inefficiencies in axial systems [16]. Our findings extend this understanding to mixed-flow pumps, where the interaction between radial and axial forces amplifies flow misalignment and slippage. Managing these forces in counter-rotating mixed-flow pumps presents unique challenges, underscoring the need for optimized designs to minimize slip and associated losses.

3.4. Head Analysis Based on Euler Theory

Figure 15 illustrates the incidence losses at the rotor outlets for different flow rates and models. The beta angle (β2), which represents the flow's exit angle relative to the rotor's tangential direction, was observed by comparing the actual flow direction to the designed flow angle. Significant β2 deviations were seen at higher velocities, particularly in models A3535 and B3535 at lower flow rates, where the misalignment led to increased slippage and reduced pump performance. Model B3535 consistently showed better flow alignment, particularly at high flow rates, with less deviation from the designed β2, aligning more closely with theoretical predictions and improving pump performance.

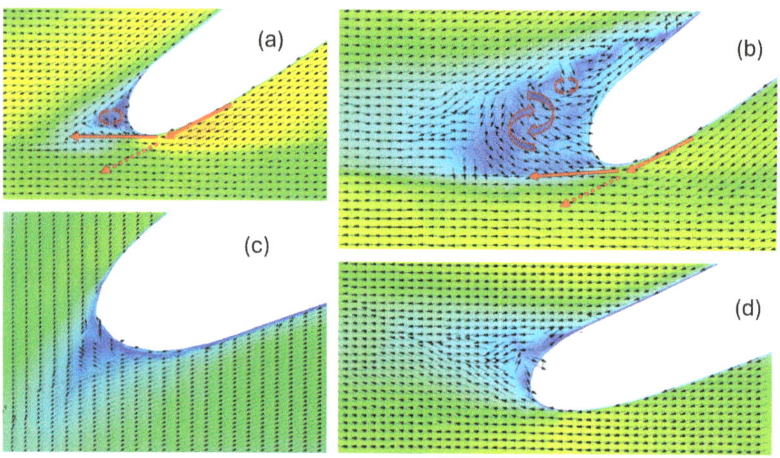

Figure 15. Incidence losses at different flow rates: (**a**) Case A3535 at 2.22 L/s, (**b**) Case B3535 at 2.22 L/s, (**c**) Case A3535 at 12.55 L/s, (**d**) Case B3535 at 12.55 L/s.

Figures 16 and 17 compare the theoretical Euler head, slip head, and CFD data for both models at different velocity ratios. Model B only achieved a higher theoretical head when operating with the internal rotor due to its more efficient design and reduced slippage. However, with both rotors in operation, Model A's performance improved, as the second rotor compensated for the initial slippage, bringing its performance closer to Model B. The CFD data closely matched the slip head predictions at lower flow rates but deviated at higher flow rates due to increased energy losses. These results suggest that optimizing rotor design and velocity ratios can significantly enhance pump performance in counter-rotating systems.

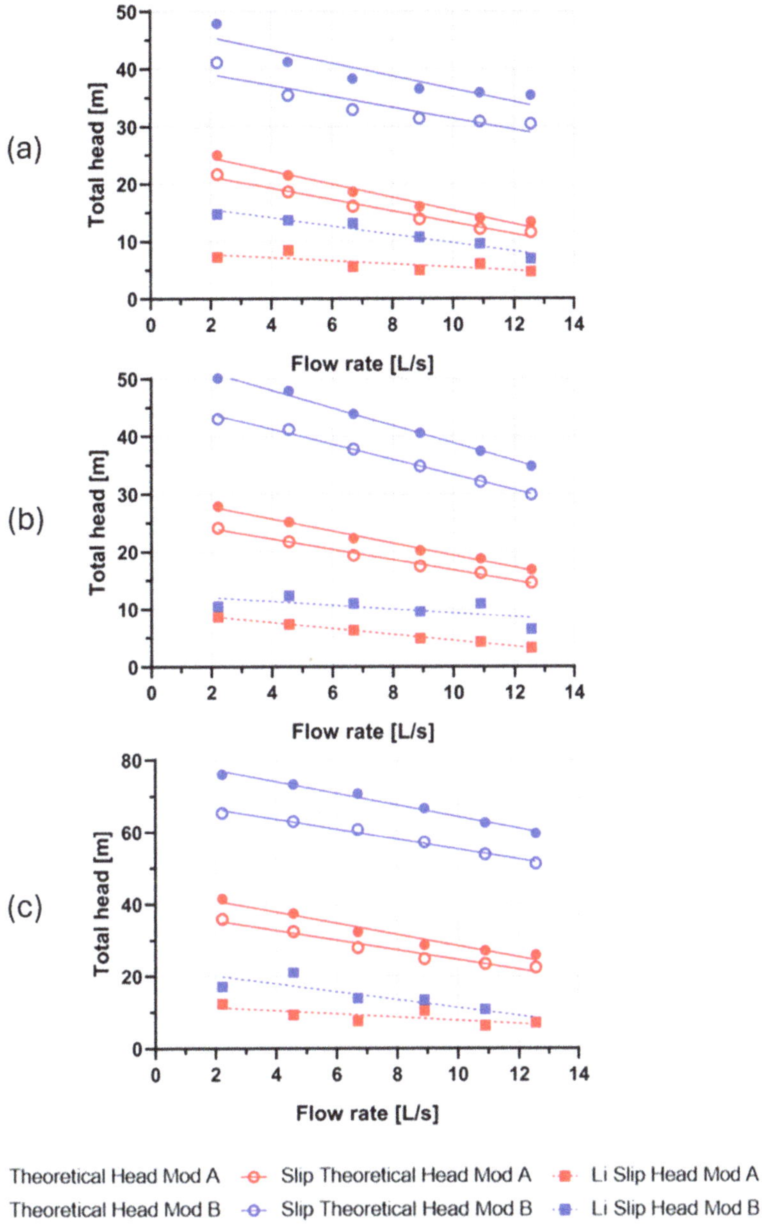

Figure 16. Theoretical Euler head and slippage head vs. CFD data for the internal rotor at different velocity ratios: (**a**) 3500:3500 rpm, (**b**) 3500:4375 rpm, (**c**) 4375:3500 rpm.

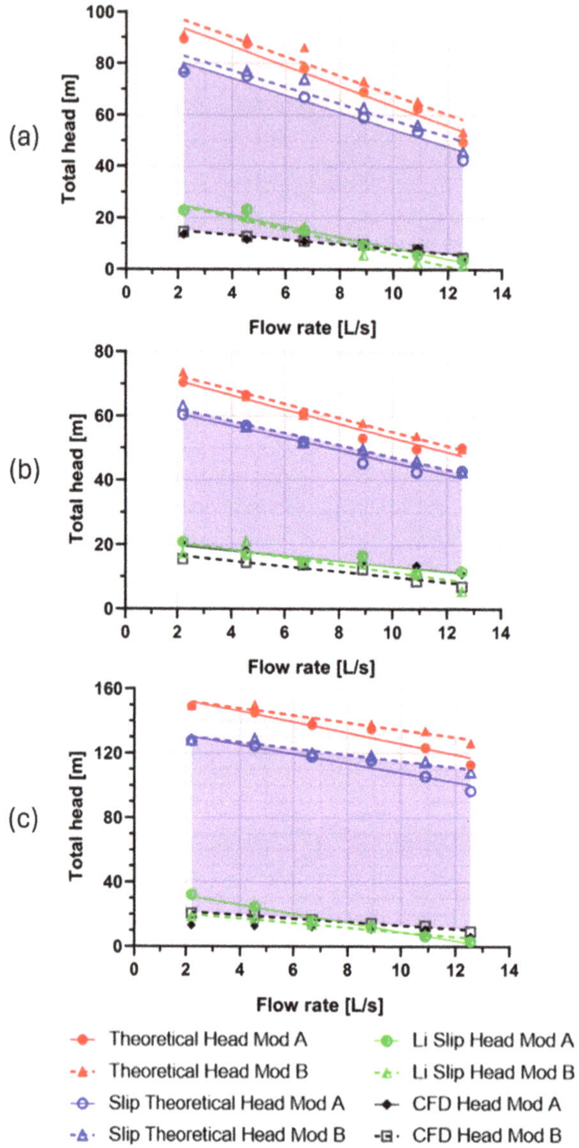

Figure 17. Theoretical Euler head and slippage head vs. CFD data for both rotors at different velocity ratios: (**a**) 3500:3500 rpm, (**b**) 3500:4375 rpm, (**c**) 4375:3500 rpm.

Significant β2 deviations were observed at higher velocities in models A4335 and B4335, leading to increased slippage and reduced efficiency, particularly at low flow rates. As flow rates increased, β2 values aligned more consistently with theoretical predictions, especially in Model B.

Nguyen reported similar β2 angle deviations [21] in counter-rotating centrifugal compressors, where misalignment was a vital issue. In mixed-flow pumps, radial forces exacerbate these deviations, emphasizing the need for optimized rotor designs to maintain flow alignment and minimize energy losses.

4. Conclusions

This study used CFD analysis to explore counter-rotating systems' performance in a mixed-flow pump. Two impeller models (A and B) with different internal diameter ratios and operating at three velocity ratios were assessed for their head rise, efficiency, slip factor, and incidence losses. The CFD model was validated with experimental data, confirming its accuracy in representing the pump's hydraulic performance.

Key findings from this study include:

1. Head and Efficiency Improvements: The introduction of counter-rotating rotors significantly enhanced the total head at low flow rates. Model B, operating at a velocity ratio of 3500:4375 rpm, demonstrated the highest head gains, particularly at low flows, achieving values up to 17 m at the BEP. However, this increase in head did not consistently translate into higher efficiency. Model B exhibited the best efficiency around the BEP, with up to 75% efficiency, whereas Model A showed diminished performance due to increased slippage at higher flow rates.
2. Slip and Hydraulic Losses: The counter-rotating configuration introduced additional slip at both rotors, particularly in Model A, which showed a steeper increase in slip factor as the flow rate increased. Model B's configuration managed slip more effectively, showing a more gradual rise in slip factor and better overall hydraulic performance. Hydraulic losses, such as recirculation and turbulence, were more significant at low flow rates but diminished at higher flows, aligning with theoretical expectations based on Euler's pump theory.
3. Incidence Losses and β Angle Deviations: Significant β angle deviations were observed in models A4335 and B4335 at higher velocities, contributing to increased slippage and reduced pump efficiency. Model B3535 exhibited the slightest deviation, with performance more closely aligned with theoretical predictions. The analysis highlighted the importance of controlling incidence losses to optimize pump performance, especially at higher operational velocities.

This study offers valuable insights into the design and operation of counter-rotating systems in mixed-flow pumps. The results demonstrate that optimizing head and efficiency through appropriate rotor speed and geometry adjustments is achievable. Future research should focus on refining rotor designs to minimize slippage and further enhance the efficiency of counter-rotating systems in centrifugal and mixed-flow pumps, expanding their industrial applications.

While this study primarily addresses the steady-state performance of counter-rotating impeller systems, time-dependent analyses could provide additional insights into the transient behavior and the time required for the flow to reach a stable state. However, conducting such analyses would require new transient simulations, which fall outside the scope of this work. Future studies will aim to incorporate transient simulations to explore the time evolution of the flow field in greater detail.

Author Contributions: Writing—original draft, E.L.P.; writing—review and editing, M.A. and N.R. All authors have read and agreed to the published version of the manuscript.

Funding: This research received no external funding.

Data Availability Statement: The original contributions presented in the study are included in the article, further inquiries can be directed to the corresponding author.

Conflicts of Interest: The authors declare no conflict of interest.

References

1. Nguyen, V.-T.; Danlos, A.; Paridaens, R.; Bakir, F. Study of the effect of the contra-rotating component on the performance of the centrifugal compressor. *Int. J. Mech. Mechatron. Eng.* **2019**, *13*, 383–392. [CrossRef]
2. Engeda, A. Early historical development of the centrifugal impeller. In Proceedings of the ASME 1998 International Gas Turbine and Aeroengine Congress and Exhibition, Stockholm, Sweden, 2–5 June 1998; Volume 5. [CrossRef]

3. Valdés, J.P.; Becerra, D.; Rozo, D.; Cediel, A.; Torres, F.; Asuaje, M.; Ratkovich, N. Comparative analysis of an electrical submersible pump's performance handling viscous Newtonian and non-Newtonian fluids through experimental and CFD approaches. *J. Petrol. Sci. Eng.* **2020**, *187*, 106749. [CrossRef]
4. Korpela, S.A. Introduction. In *Principles of Turbomachinery*, 1st ed.; John Wiley & Sons, Ltd.: Cleveland, OH, USA, 2011; pp. 1–13.
5. Stewart, M. Pump fundamentals. In *Surface Production Operations*; Elsevier: Amsterdam, The Netherlands, 2019; pp. 7–59. [CrossRef]
6. Global Centrifugal Pumps Industry. Available online: https://www.reportlinker.com/p05956265/Global-Centrifugal-Pumps-Industry.html?utm_source=GNW (accessed on 19 December 2020).
7. Global Submersible Mixed Flow Pumps Sales Market—Industry Reports. Available online: https://www.businessgrowthreports.com/global-submersible-mixed-flow-pumps-sales-market-17403859 (accessed on 29 December 2021).
8. Global Mixed Flow Impeller Pumps Market Report 2021, Market Size, Growth, CAGR, Forecast, Revenue. Available online: https://www.cognitivemarketresearch.com/machinery-&-equipment/mixed-flow-impeller-pumps-market-report#executive_summary (accessed on 29 December 2021).
9. Global Submersible Mixed-Flow Pumps Market Size and Forecast—Market Research Intellect. Available online: https://www.marketresearchintellect.com/product/global-submersible-mixed-flow-pumps-market-size-and-forecast/?utm_source=Hcnn&utm_medium=009 (accessed on 29 December 2021).
10. Wilcox, W.W. *An Analysis of the Potentialities of a Two-Stage Counterrotating Supersonic Compressor Classification Changed*; Unpublished Report; National Advisory Committee for Aeronautics: Washington, DC, USA, 1952.
11. Nouri, H.; Danlos, A.; Ravelet, F.; Bakir, F.; Sarraf, C. Experimental study of the instationary flow between two ducted counter-rotating rotors. *J. Eng. Gas Turbines Power* **2013**, *135*, 022601. [CrossRef]
12. Nuevo Sistema de Propulsión de Alta Eficiencia Energética para Buques. Result in Brief | CORDIS | European Commission. Available online: https://cordis.europa.eu/article/id/90488-new-energyefficient-ship-propulsion/es (accessed on 20 December 2020).
13. PEMEX. *El Presente Estudio Sectorial: Bombas Utilizadas por la Industria Petrolera*; Prepared for CANACINTRA and FISO; PEMEX: Mexico City, Mexico, 2018.
14. Dejour, Q.; Vo, H.D. Assessment of a novel non-axial counter-rotating compressor concept for aero-engines. In Proceedings of the ASME Turbo Expo 2018: Turbomachinery Technical Conference and Exposition, Oslo, Norway, 11–15 June 2018. [CrossRef]
15. Kanemoto, T.; Oba, S. Proposition of unique pumping system with counter-rotating mechanism. *Int. J. Rotating Mach.* **2004**, *10*, 233–240. [CrossRef]
16. Momosaki, S.; Usami, S.; Watanabe, S.; Furukawa, A. Numerical simulation of internal flow in a contra-rotating axial flow pump. *IOP Conf. Ser. Earth Environ. Sci.* **2010**, *12*, 012046. [CrossRef]
17. Cao, L.; Honda, H.; Yoshimura, H.; Watanabe, S.; Furukawa, A. Experimental investigation of blade rows interactions in contra-rotating axial flow pump. In Proceedings of the ASME 2014 4th Joint US-European Fluids Engineering Division Summer Meeting Collocated with the ASME 2014 12th International Conference on Nanochannels, Microchannels, and Minichannels, Chicago, IL, USA, 3–7 August 2014. [CrossRef]
18. Tosin, S.; Friedrichs, J.; Dreiss, A. New Design Approach for a Highly Loaded Counter-Rotating Mixed-Flow Pump in Cavitation Conditions. Available online: www.euroturbo.eu (accessed on 4 January 2022).
19. Tosin, S.; Friedrichs, J.; Dreiss, A. Experimental and numerical investigation of a counter-rotating mixed-flow single stage pump. In Proceedings of the ASME Turbo Expo 2015: Turbine Technical Conference and Exposition, Montreal, QC, Canada, 15–19 June 2015. [CrossRef]
20. Friebe, C.; Velde, O.; Krause, R.; Hackeschmidt, K. Design and Investigation of a Multistage Axial Contra-Rotating Fan. 2018. Available online: https://www.researchgate.net/publication/324687326_DESIGN_AND_INVESTIGATION_OF_A_MULTISTAGE_AXIAL_CONTRA-ROTATING_FAN (accessed on 4 January 2022).
21. Nguyen, V.T. Étude Expérimentale de L'interaction Aérodynamique Rotor-Rotor Pour un Compresseur Centrifuge Contrarotatif. Ph.D. Thesis, Laboratoire d'Ingénierie des Fluides et des Systèmes Énergétiques, Paris, France, August 2020. Available online: http://www.theses.fr/2020HESAE031 (accessed on 4 January 2022).
22. MonteVerde, W. Bombas Centrífugas Submersas: Visualização do Escomento Bifásico Gás-Líquido, Operação com Fluido Viscoso e Modelagem de Desempenho. Ph.D. Thesis, Universidade Estadual de Campinas, Campinas, Brazil, 2016.
23. Schlumberger. *REDA Electric Submersible Pump Systems Technology Catalog*; Schlumberger: Paris, France, 2017.

processes

MDPI

Article

High-Precision Numerical Investigation of a VAWT Starting Process

Ion Mălăel and Sergiu Strătilă *

Romanian Research and Development Institute for Gas Turbines COMOTI, 061126 Bucharest, Romania; ion.malael@comoti.ro
* Correspondence: sergiu.stratila@comoti.ro

Abstract: For both conventional and renewable energy conversion processes, computational fluid dynamics (CFD) has been used to address more energy-related challenges in recent decades. Using CFD to investigate vertical-axis wind turbines has become more common in recent years. The main goals of this application have been to more accurately predict the turbine's performance and to comprehend the complicated nature of the complex turbulent flow. The vertical-axis wind turbine (VAWT) simulation for energy-generating applications has several intricate components. One of them is the study of the chaotic flow that occurs during the first stages of the starting process, and which greatly influences overall effectiveness. In this article, the performance of the wind turbine was increased using a passive flow control approach. The numerical research was carried out using Large Eddy Simulation for four alternative tip speed ratios in both cases, the classic and the optimized case, equipped with a vortex trap on the extrados of the blades. The power and torque coefficient variations, as well as the velocity magnitude contours, show that the starting process may begin with a significant improvement in efficiency when flow control is used.

Keywords: vertical-axis wind turbine; large eddy simulation; tip speed ratio; vorticity magnitude; torque coefficient; vortex trap

1. Introduction

The accelerated growth in the necessity for energy in the past few decades [1,2] has raised many concerns regarding the sustainability of the present global society [3] for the future years. In order to compensate for the growth in energy needs, the preferred solution is to exploit a higher quantity of low-cost, easy-to-obtain energy sources, such as fossil fuels [4], gas, and petroleum products, resulting in a growth in combustion gases. This growth has negatively influenced the environment for the past two decades [5], being the main cause of the intensifying greenhouse effect [6] and the rise in the average Earth temperatures. The other solution to cover the growing energy demand is making use of renewable energy sources that allow for energy to be obtained from green sources, through processes that have almost null greenhouse gas emissions [7].

Renewable energy sources, however, have yet to become as efficiently harvested as environmentally unfriendly sources [8]. Solar plants are known to have reduced performance ratios during the summer [9], hydro dams are even considered to negatively impact the fauna of the region where they are constructed [10,11], and wind power plants take up big areas of land, making them unusable [10,12], as well as raising concerns regarding the impact on the local wildlife [12]. Wind turbines, which are of interest in the present paper, will also only be in use for about 85% of the year [13]. Furthermore, during the winter months, these devices are prone to ice formations, much like the airfoil of an airplane wing, affecting the aerodynamic performance [14], raising the inertia of the blade ensemble, and drastically reducing the power output of the turbine. The vibrations of these devices, especially in the case of horizontal-axis wind turbines (HAWTs), are transmitted to the ground where the turbines are installed, affecting local ground wildlife [15]. It has been shown that birds and even bats are drawn away from the region [16].

Once the aforementioned concerns with wind turbines are addressed, installing such power plants should become more appealing to meet the future energy needs of the growing population. Technologies have been developed to reduce the vibrations generated by the operation of wind turbines [17,18]. In order to remove ice formations on the blades, drones can be used to spray the airfoil with a chemical substance that defrosts the ice [19], drastically reducing the cost of removing ice formations on the blades in the case of large HAWTs [20]. By using smaller-scale vertical-axis wind turbines (VAWTs), though, this problem would not require such complex methods of defrosting since VAWTs can be installed on top of buildings and can be made easily accessible to humans [13], also providing a solution to fauna destabilization by integrating the turbines in an urban environment. If one could install smaller-scale wind turbine ensembles in large urban areas, the energy transmission losses would also be reduced by bringing the power plant directly to the consumer [21].

However, if one should consider the evolution of global energy consumption (Figure 1) and the predictions of the growth in energy demand for the next few decades for the European Union [22] and countries such as Romania [22–24], China [25], and the United Kingdom [26], simply installing more power plants would essentially lead to having no viable land space left for new plants to be mounted. The installed turbine clusters, along with the power generation module and the energy distribution system for each turbine, will have to be periodically upgraded (or even replaced) to raise the performance of the power generation ensembles to the latest trends and developments in energy efficiency. In this case, the combined solution of installing plants and developing more energy-efficient harnessing [27] and distribution systems should compensate for the growth of power requirements [2] by creating new wind power plants and, most importantly, by upgrading existing technology to reach higher efficiency.

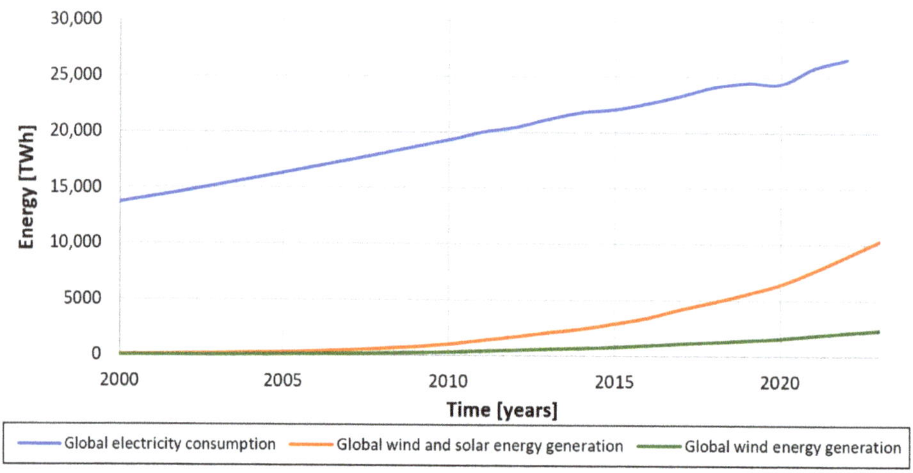

Figure 1. Global electricity consumption, global wind and solar energy generation, and global wind energy generation from 2000 to 2023 [28–30].

Many studies have been carried out regarding improving the efficiency of wind turbines, for both VAWTs and HAWTs. Due to their omnidirectional operation and ability to operate in low windspeed regimes, VAWTs are the preferred solution for WT installations in urban and suburban regions. The main advantage of the HAWTs is a higher value for the power coefficient (which will be explained further in this paper) [31]. VAWTs can be further categorized into drag-driven turbines (such as the Savonius WT) [32] and lift-driven turbines (the Darrieus turbine) [33]. The CAD model of the Darrieus VAWT (lift-based) studied in this paper can be seen in Figure 2.

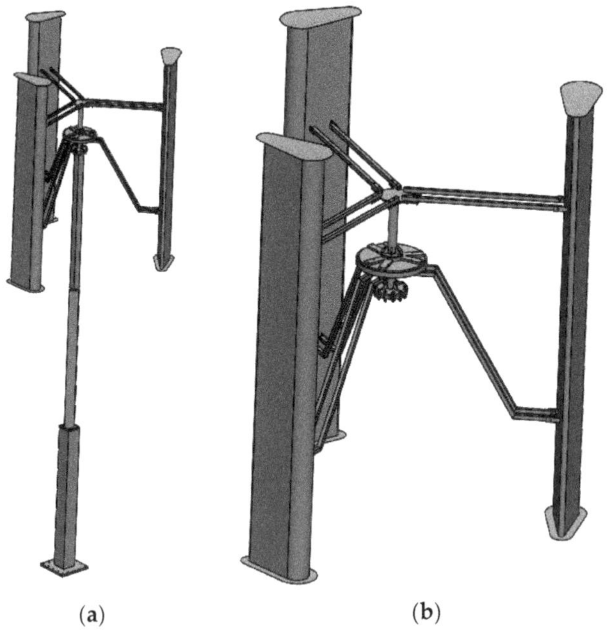

Figure 2. CAD model: (**a**) wind turbine model; (**b**) wind turbine rotor.

Furthermore, in recent studies, the performance of various blade shapes was analyzed [33–35], the impact on the power coefficient of the disposition of the blades [36] and the number of rotors was studied [37], and even the power output of multiple cluster dispositions of identical turbines was numerically simulated [38,39]. The authors of [40] reviewed the current studies in performance improvement with geometry variations regarding the Savonius VAWTs.

Wind turbines harness the kinetic energy from the wind using airfoils which generate aerodynamic forces and torques. The main reason for energy conversion loss is the formation of flow separation on the surface of the blade. This phenomenon can be diminished by the method of flow control, which can be either active, by injection of fluid at the surface of the blade, or passive, which results from varying the blade shape to take flow separation into account. Multiple solutions have been analyzed regarding active [41,42] and passive flow control [43], either by creating cavities inside the blades [44], much like the novel VAWT geometry studied in this paper, or by mounting vortex generators on the extrados [45]. The trapped vortex cavity offers the advantage of flow separation occurring inside of the cavity and reattaches towards the end of the cavity. This leads to a significant diminishing of the wake area and to the constraining of the vorticity region to the cavity, resulting in a higher lift-to-drag ratio [46,47]. To further analyze the effect of vortices and flow separation on the Cp, the authors of [48] introduced a new parameter, the Vorticity Index (VI), calculated by dividing the vorticity in the Leading Edge (LE) by the vorticity in the Trailing Edge (TE), which can help detect a possible vortex separation, either from the LE to the TE, when this index approaches a value of 1, or from the TE to the LE, when the index approaches a value around 0.2. This is known as the Imminent Vortex Separation Condition (IVSC).

The flow around the geometry of a wind turbine can be numerically studied using, much like the case of this article, the Large Eddy Simulations (LES) method, and the power coefficient (C_p) of the blade ensemble is numerically evaluated. The LES method solves the Navier–Stokes continuity equations of the fluid flow for each of the elements in the mesh, ignoring the smallest length scales which require the most computational effort in a simulation by low-pass filtering the Navier–Stokes equations [49]. The C_p represents the fraction of the wind kinetic energy that is converted to power [50]; thus, an improvement in the method of energy extraction results in the growth of the C_p value and, evidently, in the growth of the power output.

However, if the results should have the highest accuracy, the Direct Numerical Simulation (DNS) method (which has the best accuracy in terms of results) can be chosen. Greater accuracy implies a much higher computational effort, as comparisons between DNS, LES, and Unsteady Reynolds Averaged Navier–Stokes (URANS) methods have been performed [51]. The URANS method requires the least computational effort of the three methods. The time step size and the number of elements in the mesh have a great influence on the results obtained after a simulation. An analysis of this dependency was carried out by the authors of [52], showing that the numerical result of the simulation, in this case, the aerodynamic coefficients of lift and drag, will stabilize if the time step size is small enough and if the mesh is fine enough. The authors have also shown that the Transition Shear Stress Transport (SST) turbulence model is by far one of the most accurate turbulence models when compared to experimental data.

In order to improve starting process efficiency, a thorough comparison between a traditional setup and an improved vertical-axis wind turbine was conducted for this paper. It is crucial to make clear that the goal of this study was limited to the starting stages, during which tip speed ratios ranged from 0.5 to 1.25 and rotating speeds were extremely low. In order to accomplish this goal, a vertical-axis turbine design was researched numerically using the LES technique, and the outcomes of the performance parameters were compared across the configurations. The aims of this paper regarding wind turbine design are as follows:

- Generating a new passive flow control solution for an analyzed baseline geometry;
- Analyzing the numerical results of the CFD simulations for the vortex trap solution employed in this paper;
- Comparing the numerical results with the results of the baseline geometry;
- Providing conclusions and proof in favor of the modified geometry (as provided by the results) in the study of the early stages of the startup process of the VAWT studied in this paper.

2. Materials and Methods

The foundation of the research presented in this paper was a highly precise numerical investigation of the behavior of the vertical-axis wind turbine starting procedure. This turbine was designed to be utilized in residential areas with low to medium wind speeds and significant turbulence. The turbine was supposed to produce around 3 kW of power at a wind speed of 10 m/s. Peak performance is anticipated by the turbine design at a Tip Speed Ratio (TSR) of 1.9. The geometric properties of the turbine are listed in Table 1.

Table 1. VAWT geometric features.

Parameter	Value	Units
Rotor diameter	3.6	m
Rotor height	4.5	m
Nominal wind velocity	10	m/s
Blade chord	0.6	m
Number of blades	3	-
Blade airfoil	NACA0021	-
Nominal TSR	1.9	-

Three nondimensional parameters—the power coefficient (C_p), torque coefficient (C_T), and tip speed ratio (TSR)—define the performance of a vertical-axis turbine. The most significant one, C_p, which describes the efficiency of wind turbines—the fraction of the wind kinetic power that is transformed into mechanical power—is provided by Equation (1), in which P_T is the wind turbine power and P_{wind} is the maximum wind power:

$$C_{pT} = \frac{P_T}{P_{wind}} \quad (1)$$

The wind power, which represents the maximum amount of power available to be harnessed, can be expressed with the formulation in Equation (2), in which ρ is the air density, S is the reference area, and V_∞ is the wind velocity [50]:

$$P_{wind} = \frac{1}{2}\rho S V_\infty^3 \quad (2)$$

The power harnessed by the wind turbine is provided in Equation (3), in which T is the wind turbine torque, ω is the angular velocity of the turbine, L is the reference length, and C_T is the torque coefficient [50]:

$$P_T = T\omega = \frac{1}{2}\rho S V_\infty^2 L C_T \omega \quad (3)$$

Considering the definition of the tip speed ratio (TSR) λ in Equation (4):

$$\lambda = \frac{L\omega}{V_\infty} \quad (4)$$

One can use Equations (1)–(4) to reach the final equality in Equation (5) between C_T and C_p, which is no more than 0.593 according to the Betz limit [50,53,54]:

$$C_{pT} = \frac{\frac{1}{2}\rho S V_\infty^2 L C_T \omega}{\frac{1}{2}\rho S V_\infty^3} = C_T \cdot \frac{L\omega}{V_\infty} = C_T \cdot \lambda < 0.593 \cong \frac{16}{27} \quad (5)$$

Figure 3 depicts a diagram of the computational domain employed in this investigation. These dimensions were chosen to minimize the influence of boundary conditions on the final outcome. The length of the downstream domain is approximately nine times larger than the wind turbine rotor diameter, while the upstream domain length is approximately five times larger. The domain is divided into two parts: one rotating, represented by the blades and shaft, and one stationary, representing the surrounding environment. An interface was defined between these two subdomains to express the relative motion of the rotor, taking into consideration elements such as angular velocity, slip factors, and blade angles. The computational domain size was the same for both analyzed cases, as was the mesh of the stationary subdomain. The depth of the domain (the length on the z-axis perpendicular to the plane shown in Figure 3) is equal to 1 m. This length choice was made by the authors with respect to previous experience in wind turbine CFD simulations.

The grid independence investigation and domain sensitivity were included in the article [55,56]. The research was conducted for the baseline case at TSR 1, and the only change made was to the growth ratio. In other words, three meshes were examined with growth ratios of 1.05, 1.1, and 1.2, containing 269307 (Mesh 1), 208158 (Mesh 2), and 176791 (Mesh 3) components. Figure 4 depicts the torque coefficients for all three meshes. For the remaining simulated situations, we selected mesh no. 1, for which more information may be found in Table 2.

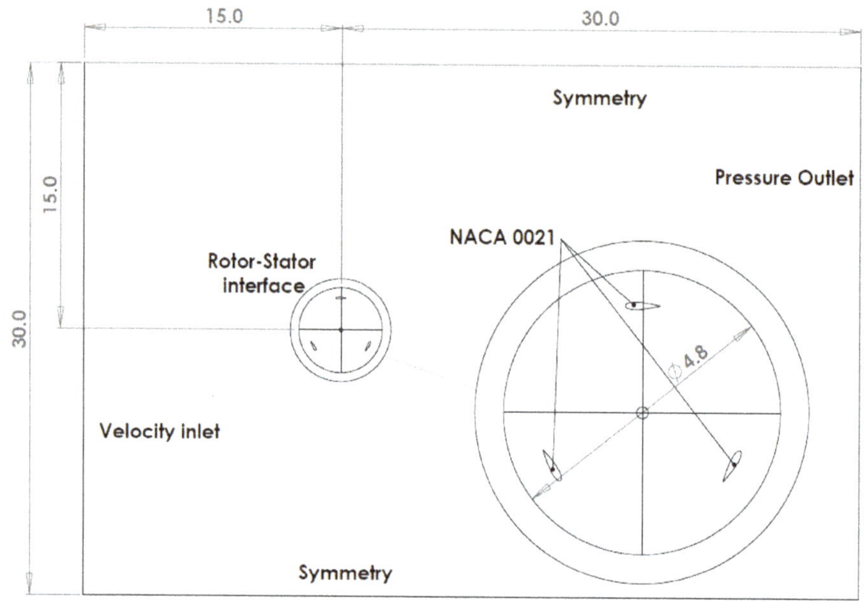

Figure 3. Computational domain dimension and the boundary conditions.

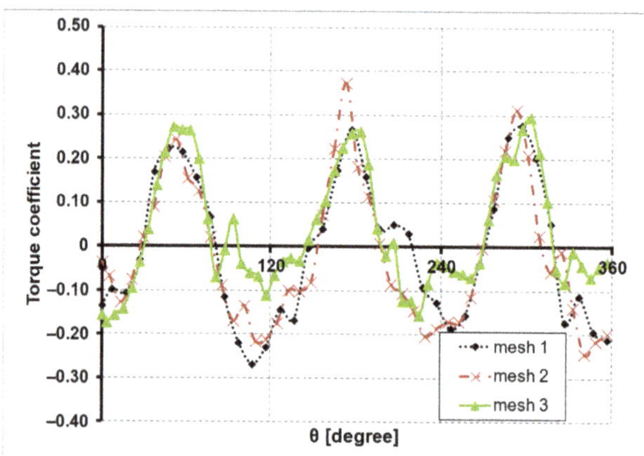

Figure 4. Mesh dependency test.

Table 2. Mesh statistics.

Case	Subdomain	No. of Nodes	No. of Elements	
			Tri	Quad
Case 1 (classic)	Rotor	237,568	387	235,071
	Stator	34,310	38	33,811
	TOTAL	271,878	425	268,882
Case 2 (optimized)	Rotor	264,226	452	261,419
	Stator	34,310	38	33,811
	TOTAL	298,536	490	295,230

To accurately resolve the turbulent boundary layer of the blades, the size of the first element of the mesh in the normal direction of the blade surface must fulfill the Y + <1 constraint, as seen in Figure 5, in which the values for Y+ can be seen to remain, for most regions, in the vicinity of 0.5, and, for some isolated cases, around the value of 1.

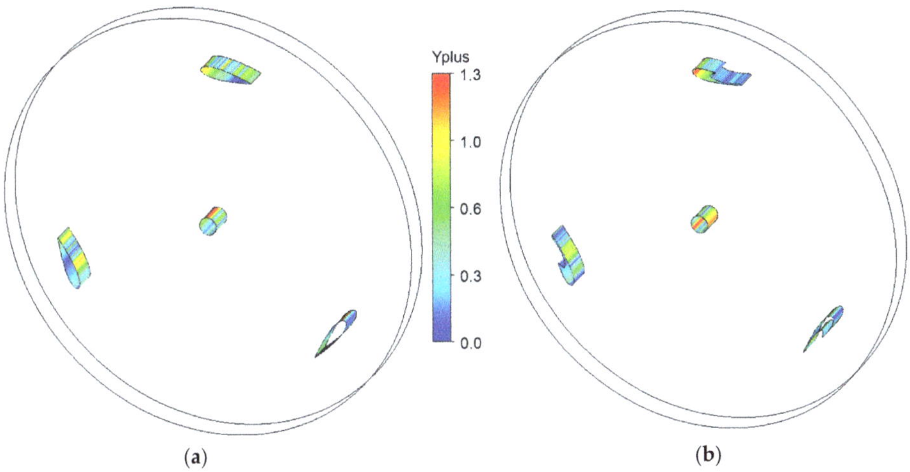

Figure 5. Values of Y+ for the first element of the mesh in the normal direction of the blade surface: (**a**) baseline geometry and (**b**) modified geometry.

The total height, growth rate, and total number of layers utilized are also important criteria in the formation of the mesh adjacent to the surface of the blades (boundary layer mesh). The growth rate is determined by the same principles that are used in every Large Eddy Simulation (LES), with values around 1.05 considered appropriate. The values for skewness are provided for the mesh cells in Figure 6.

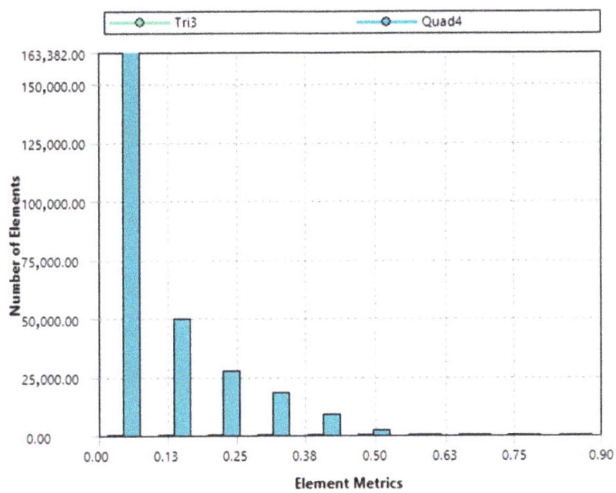

Figure 6. Skewness values for the mesh elements.

These two subdomains are linked by a numerical interface, and an inflation setting was employed throughout the meshing process. This inflation has a maximum of 15 layers on each subdomain, a smooth transition option, and a growth rate of 1.05. Figure 7 depicts a close-up of the mesh from that location.

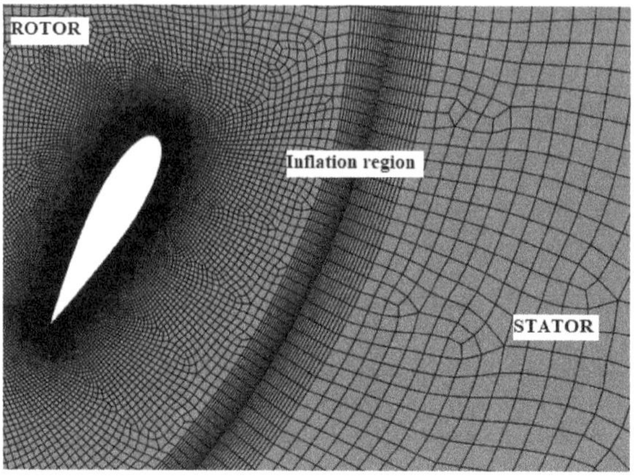

Figure 7. Mesh in the stator–rotor interface.

A cylinder containing the mesh surrounding the blades and shaft surfaces represents the rotating domain. An inflation option with 30 layers and a 1.05 growth rate was imposed to manage the mesh near each blade and the shaft. Also imposed on the blades and shaft was an edge sizing control with an element size of 1 mm. Because rotation for one degree was demanding on each iteration, an edge size with 360 divisions was employed in the interface zone on each subdomain. The ANSYS Meshing module was used to generate the mesh for each case using the same configuration. Figure 8 displays the mesh for every scenario. The new geometry was obtained by carving a NACA0021 airfoil cavity into the base geometry case, with a new chord equal to approximately half of the base geometry chord.

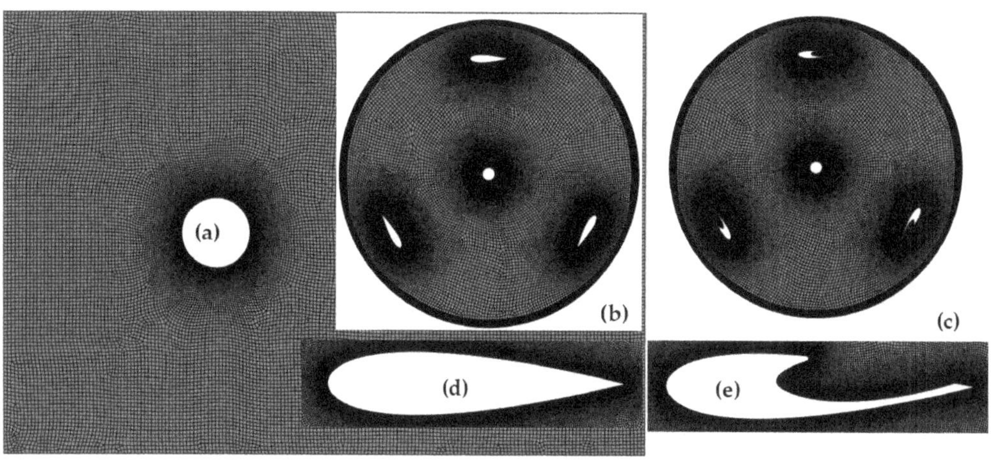

Figure 8. Mesh: (**a**) stator mesh; (**b**) classic configuration rotor (case 1); (**c**) optimized configuration rotor (case 2); (**d**) case 1 blade airfoil; (**e**) case 2 blade airfoil.

The total count of each individual element in the mesh and the overall number of mesh vertices or nodes are the basis for standard mesh statistics in computational fluid dynamics. The distribution of element types, or the division of the mesh elements into several types (triangles, quadrilaterals, tetrahedral, hexahedra, etc.), also provides significant information.

ANSYS Fluent 2022 R2, a commercial program, was used for the simulations, which assumed turbulent, transient, and incompressible flow. In the simulations, air with a constant density of 1.225 kg/m^3 was employed as the working fluid. The boundary conditions of the simulations are displayed in Table 3. A constant velocity was applied in the X direction at the inlet border. Using a fluctuating velocity method and the spectral synthesizer option, the intake turbulence was calculated with a turbulent intensity of 5% and a turbulent viscosity ratio of 10.

Table 3. ANSYS Fluent setup.

Models	Solver Viscous Model	Pressure Based	Unsteady LES	2.5 d
Materials	Air		Density, constant	
Operating Conditions		Pressure 101,325 (Pa)		
Boundary Conditions	Inlet Blades Shaft Interface Outlet Symmetry Rotor Stator		Velocity inlet: V_x = 10 m/s Wall Wall Rotor–stator interface Pressure outlet: Pressure = 0 (Pa) Symmetry Mesh motion Stationary	
Solve	Controls Initialize	Solution	SIMPLE algorithm Hybrid	
	Monitors	Residual Force	10^{-5} Torque coefficient	
	Run calculation	3600 steps	Time step size (s)—estimated at a rotation of one degree every iteration	
Report	Reference values	Inlet	Area = rotor diameter Length = rotor radius	

In the simulations, the Semi-Implicit Method for Pressure-Linked Equations (SIMPLE) was used, with transient formulation. For comparison, the viscous model utilized in both investigated situations was the Large Eddy Simulation (LES) with the Smagorinsky–Lilly subgrid-scale model due to its outstanding prediction of flow for this specific application, as previously documented in the literature. In terms of the stop criterion, 3600 time steps were chosen, equal to 10 fully completed rotations. Each time step has a maximum of 50 iterations with 5 reporting intervals and 5 profile updating intervals.

The scenarios that were simulated for various tip speed ratios (TSR) are displayed in Table 4. By raising the rotational speed of the turbine from 2.77 to 6.94 rad/s while maintaining the inflow velocity fixed at 10 m/s, the TSR of the turbine was altered. This TSR interval was chosen in accordance with the VAWT starting process step, which requires exceeding the TSR value of 1 in order to enter an auto-acceleration mode that can enable the turbine to reach its nominal operating point.

Table 4. TSR and time step size used in the numerical investigation.

TSR λ	Angular Velocity ω (rad/s)	Rotor Angular Displacement per Time Step $\Delta\theta$ (°)	Time Step Size (s)	Number of Time Steps
0.5	2.777777778	1	0.006283185	3600
0.75	4.166666667		0.004188790	
1	5.555555556		0.003141593	
1.25	6.944444444		0.002513274	

The time step size was calculated, as previously indicated, to cover exactly one degree of angular displacement of the rotor domain. Since the regimes for which the CFD simulations were employed are for low TSR values (that is, low tangential speeds of the blades caused by the rotation compared to the constant windspeed of 10 m/s), the time step should ensure that the flow of one cell of the rotor domain does not evolve past this cell between two succeeding time steps, since the number of mesh elements around the stator–rotor interface is 360. The time step values and the angular velocities according to which the steps were calculated are displayed in Table 4, and the calculation equation is provided in Equation (6):

$$\Delta t = \frac{1}{\frac{360}{2\pi}\omega[\mathrm{rad/s}]} \tag{6}$$

Figure 9 provides a schematic summary of the whole numerical investigation process.

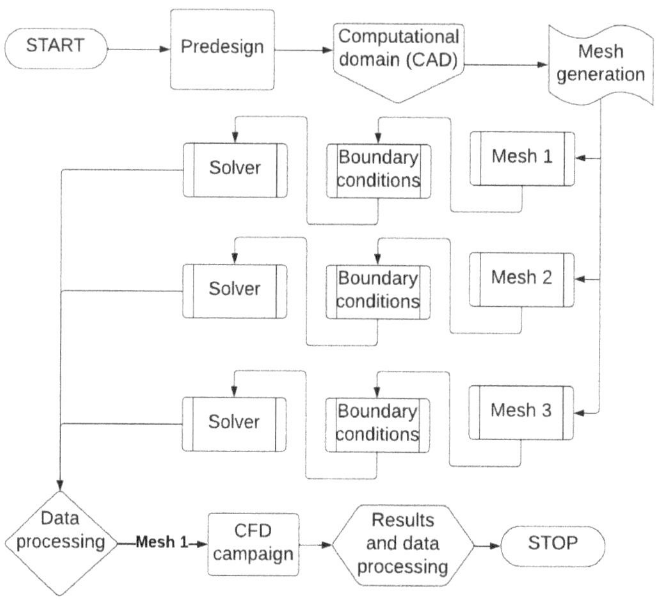

Figure 9. Block diagram.

3. Results

Using the 2.5D LES approach, the flow in the vertical-axis wind turbine, at 10 m/s wind speed, was numerically examined during its early startup phase. A vortex trap was applied to the outside of the blades as part of a passive flow control method to boost the turbine efficiency. In order to compare the results, the flow behavior was examined at several points where the vorticity magnitude was plotted.

More information on the flow development of VAWT is provided in Figures 10 and 11, where TSR = 0.5 represents the start of the process and TSR = 1 represents the point at which the turbine is about to enter the auto-acceleration mode and reach the nominal point (or peak) of its performance. Because of the wake effects that appear at this functional stage, it is anticipated that the unsteady flow field will be complex. The wake of the leading blade interacts with the blades behind it, as can be seen from both figures where the vorticity magnitude is displayed in Figures 10 and 11. However, in case 2, the separation is delayed due to the use of a vortex trap, resulting in increased efficiency.

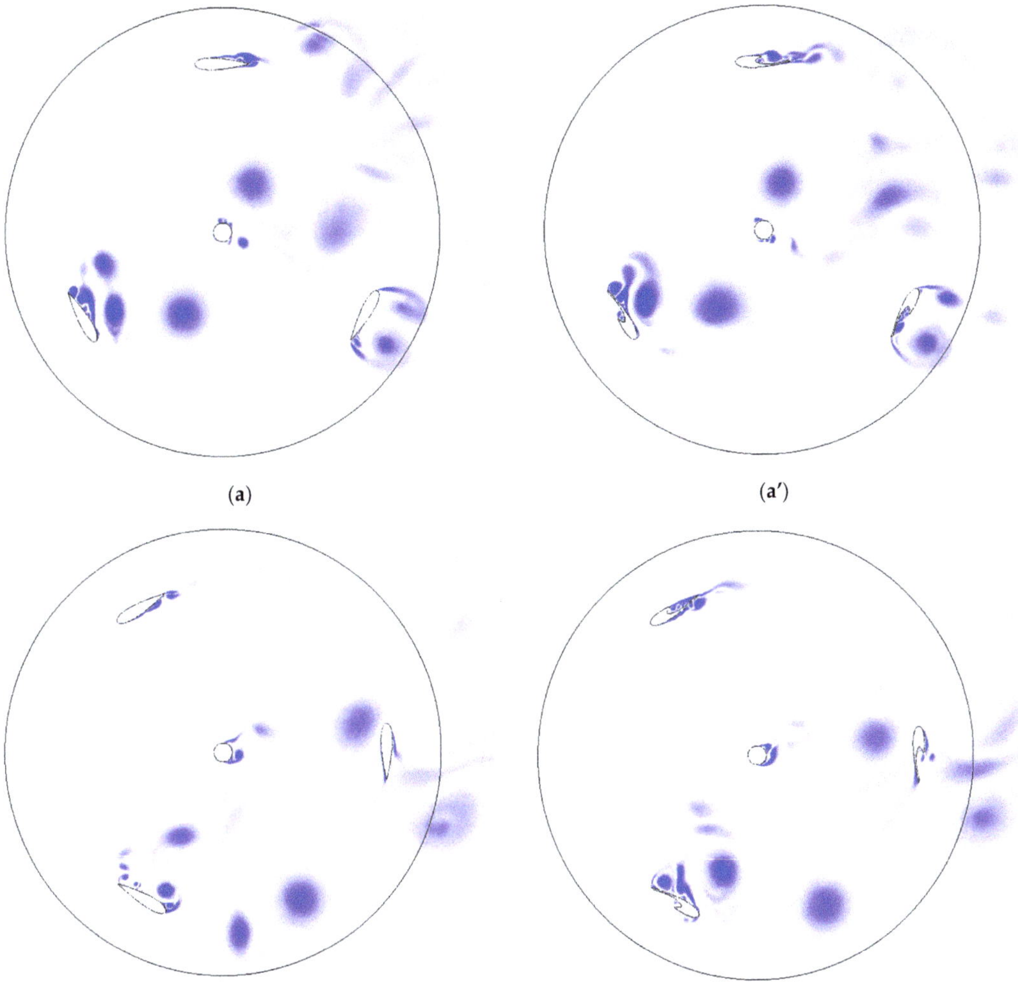

(a) (a′)

(b) (b′)

Figure 10. *Cont.*

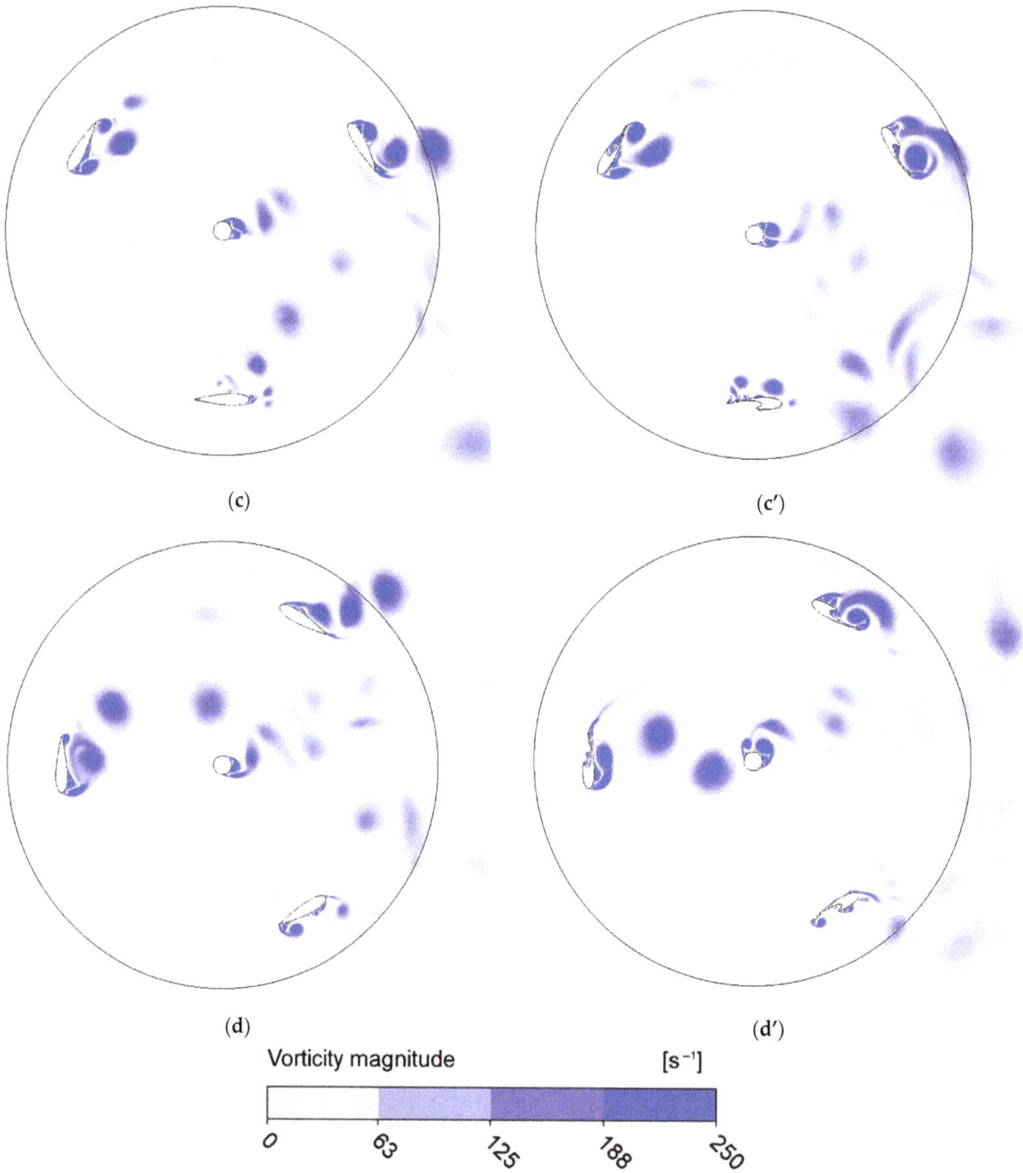

Figure 10. Vorticity magnitude TSR = 0.5: (**a**) case 1 at 0 deg.; (**b**) case 1 at 30 deg.; (**c**) case 1 at 60 deg.; (**d**) case 1 at 90 deg.; (**a′**) case 2 at 0 deg.; (**b′**) case 2 at 30 deg.; (**c′**) case 2 at 60 deg.; (**d′**) case 2 at 90 deg.

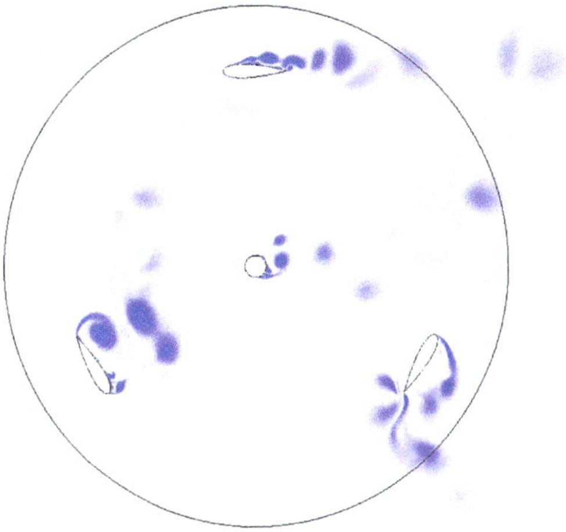

(a)

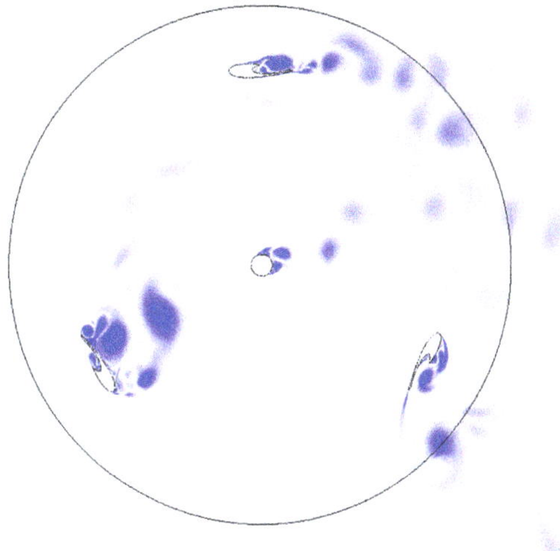

(a′)

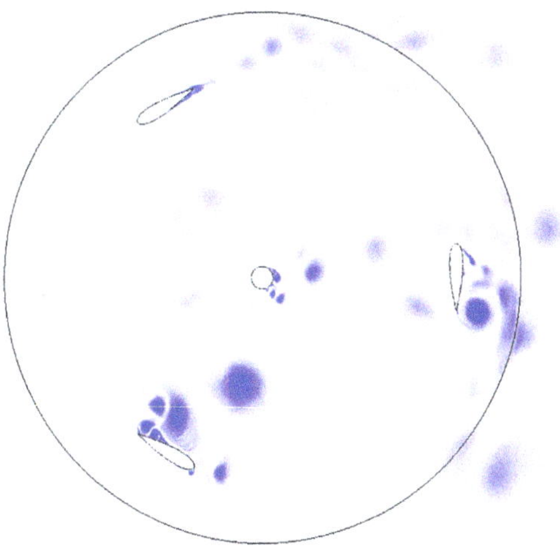

(b)

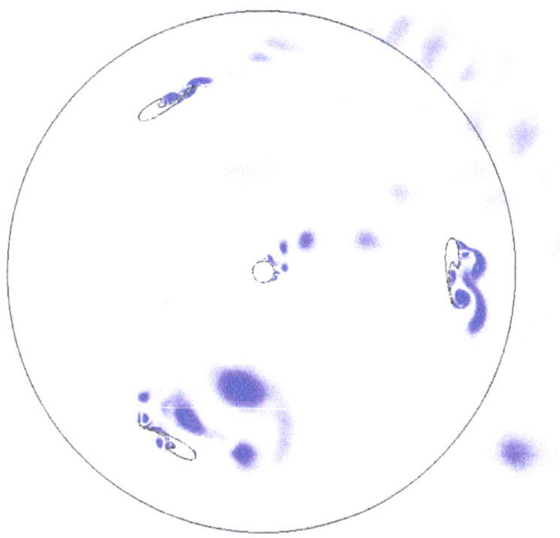

(b′)

Figure 11. *Cont.*

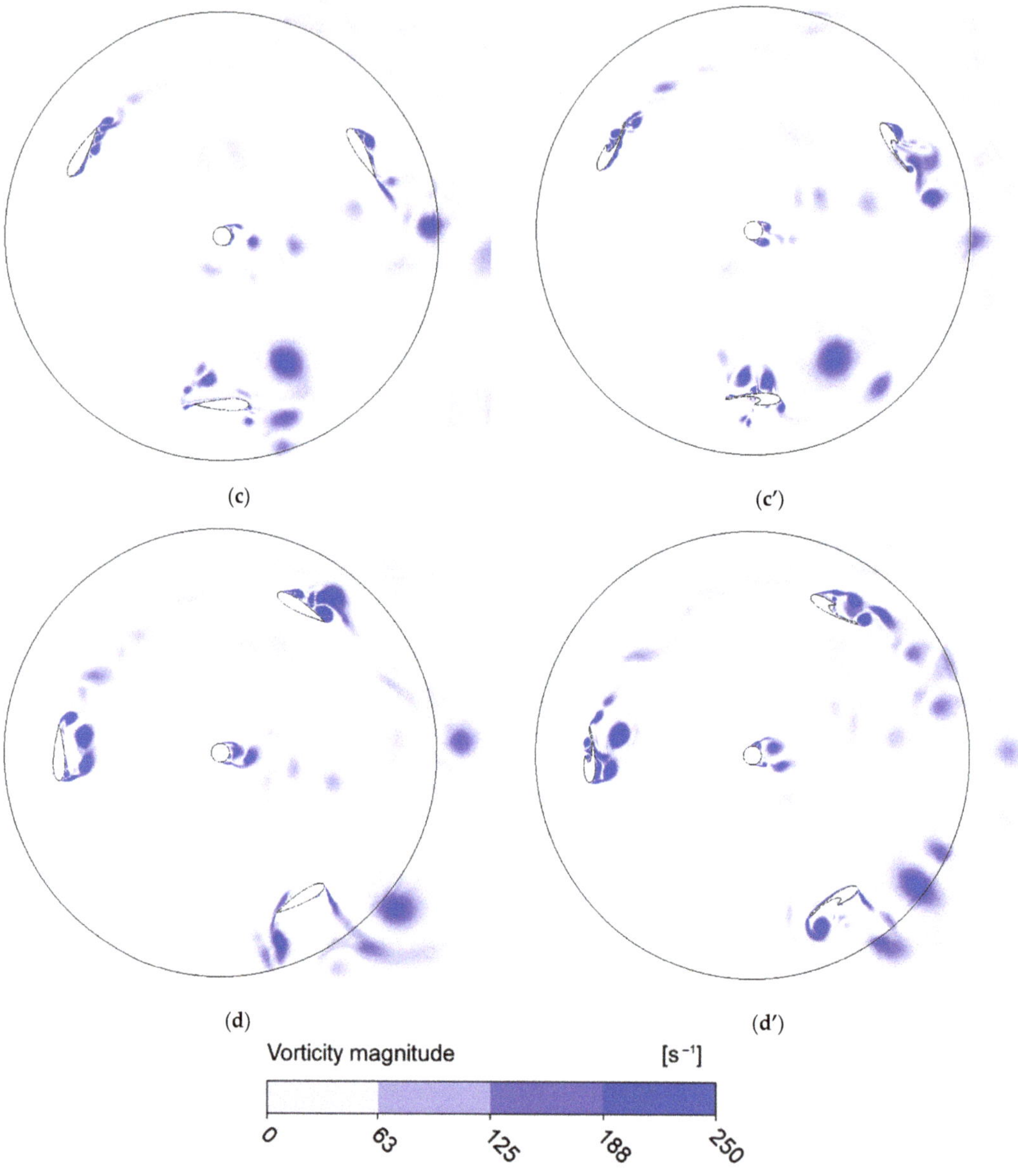

Figure 11. Vorticity magnitude TSR = 1: (**a**) case 1 at 0 deg.; (**b**) case 1 at 30 deg.; (**c**) case 1 at 60 deg.; (**d**) case 1 at 90 deg.; (**a′**) case 2 at 0 deg.; (**b′**) case 2 at 30 deg.; (**c′**) case 2 at 60 deg.; (**d′**) case 2 at 90 deg.

Figure 12 shows the flow streamlines for both cases, with the azimuth angle varying from 0 deg. To 90 deg.

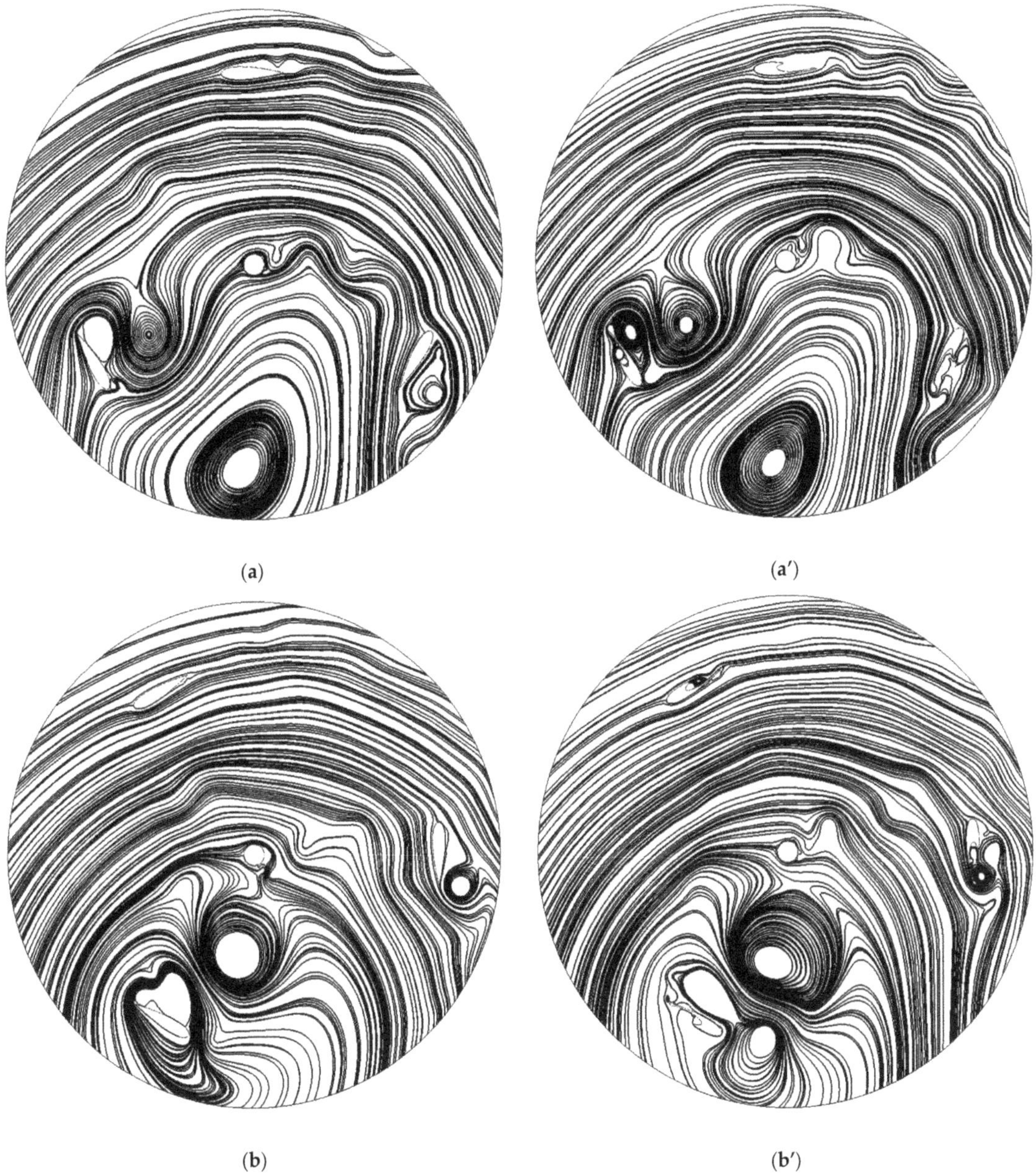

Figure 12. *Cont*.

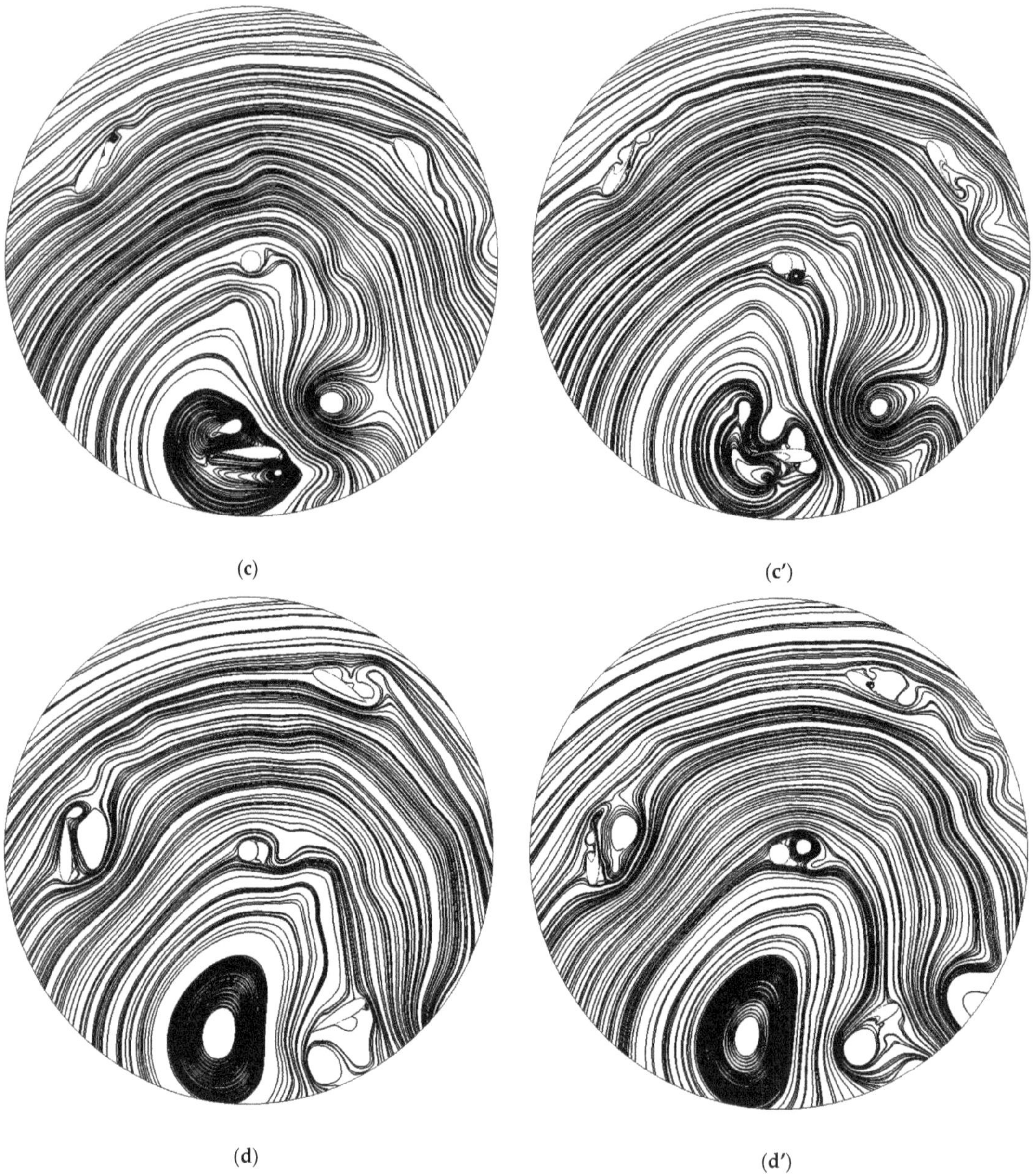

Figure 12. Streamlines at TSR = 1: (**a**) case 1 at 0 deg.; (**b**) case 1 at 30 deg.; (**c**) case 1 at 60 deg.; (**d**) case 1 at 90 deg.; (**a′**) case 2 at 0 deg.; (**b′**) case 2 at 30 deg.; (**c′**) case 2 at 60 deg.; (**d′**) case 2 at 90 deg.

Figure 13 shows the static pressure distributions on the blades of both rotors at four distinct blade locations. Considering the atmospheric pressure (101,325 Pa), according to the legend in Figure 13, the lime color that can be observed in most regions of the subfigures is associated with this pressure. The blue regions represent areas of negative (relative to atmospheric) pressure, while the orange and red areas have positive pressure values. In the case of the vortex cavity, the pressure gradient between the intrados and extrados is generally diminished. Considering the azimuthal position of 0 degrees at the top blade of Figure 13a, and the rotational direction being the same as the direction of the angular velocity turbine (counterclockwise), one can distinguish three angular position regions with different tendencies for the pressure gradient:

- For a blade situated in the angular position interval of 0 to 120 degrees, the pressure gradient is visibly diminished, but without losing the areas of negative and positive relative pressure. Inside the vortex cavity, there are two regions—one with negative pressure and one with positive (for $\theta > 60°$), which generates a pressure-based force acting on the blade in the direction of its movement.
- For the 120- to 240-degree interval, the pressure gradient of the new geometry is slightly more intense than that of the baseline geometry. The pressure gradient located in the vortex trap gradually diminishes until the position of 180°, after which the pressure gradient is directed in the opposite (relative to the blade) direction than in the previous region.
- For the last region, the pressure field around the modified blade is similar to the classic NACA blade, the exception being the intensity of the gradient—which is smaller for the modified geometry. The orientation of the gradient generates, in both cases, a force that opposes the rotational movement of the turbine and which, for the new geometry, has a smaller magnitude.

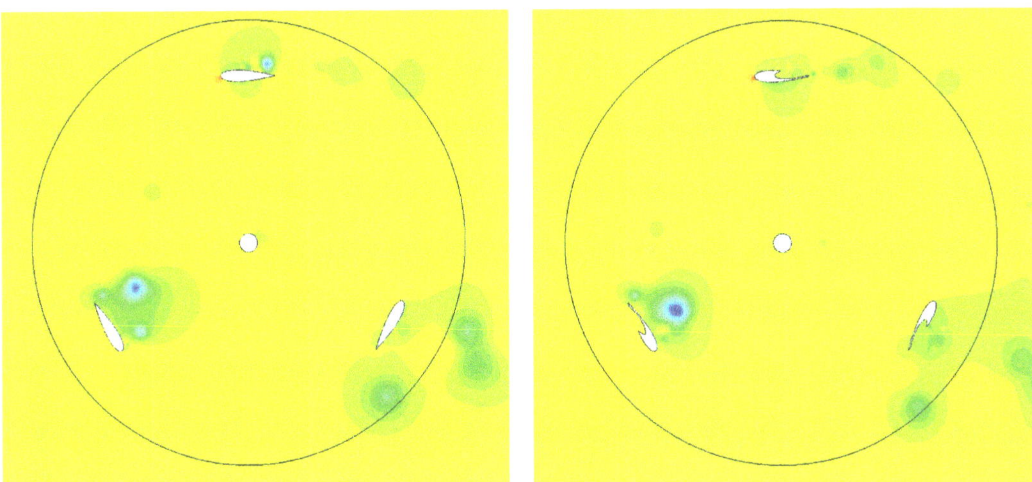

(a) (a′)

Figure 13. *Cont.*

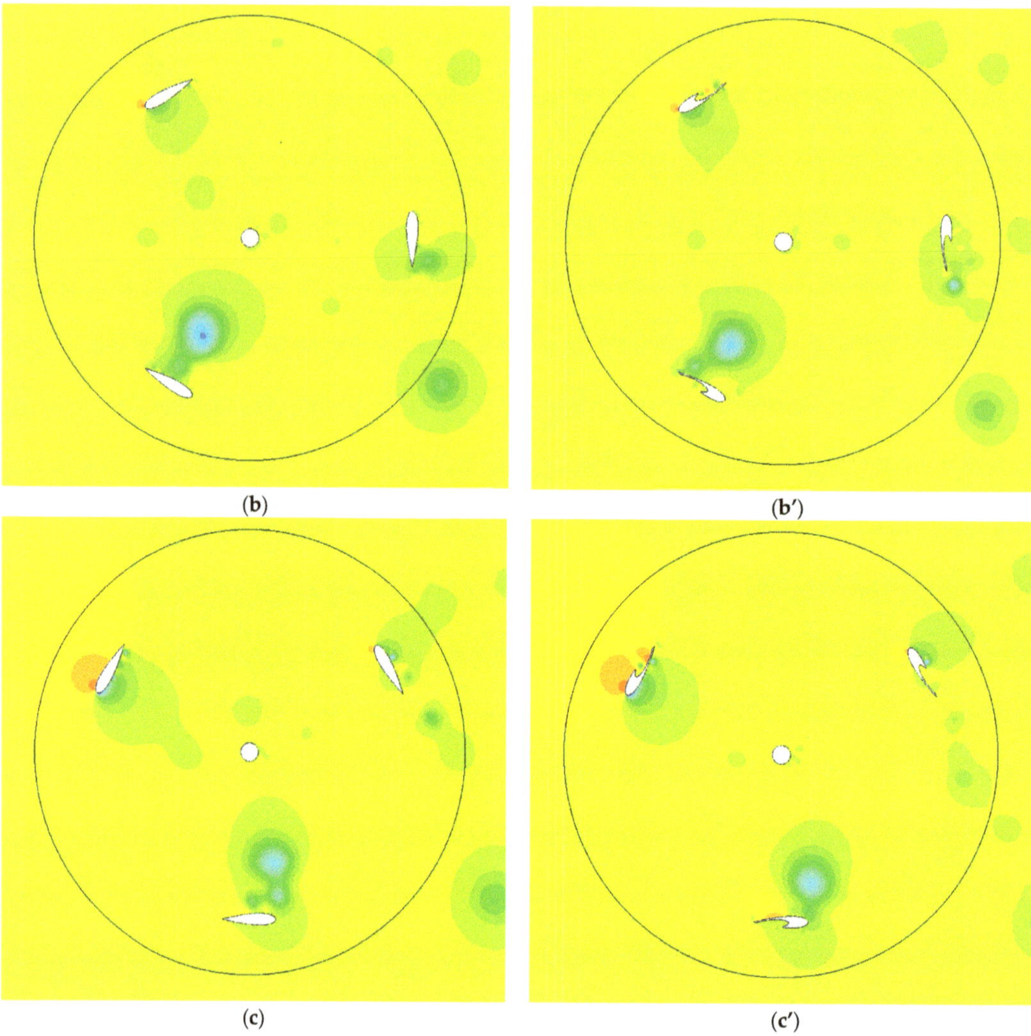

Figure 13. *Cont.*

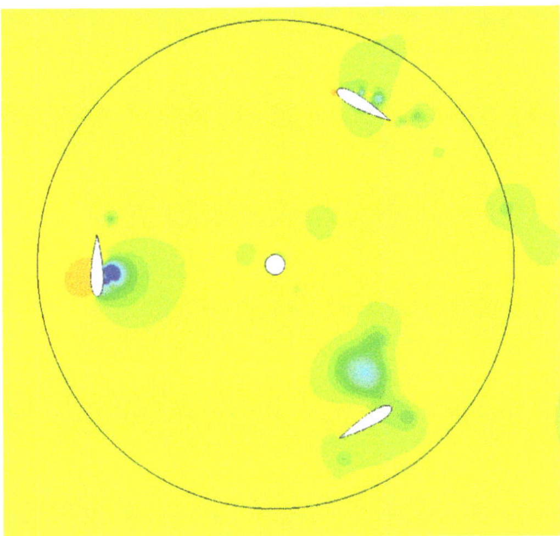

(**d**)

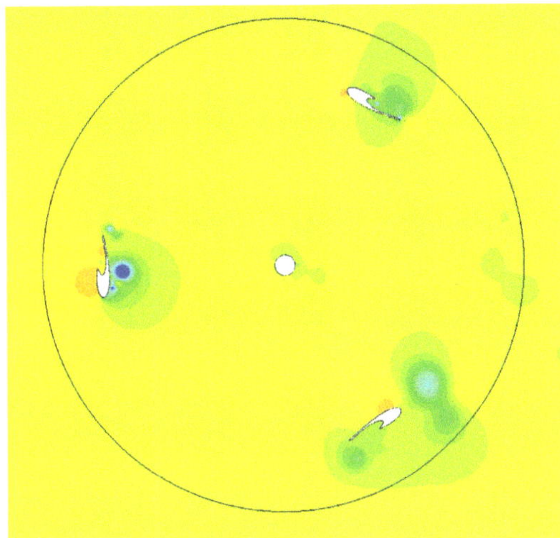

(**d′**)

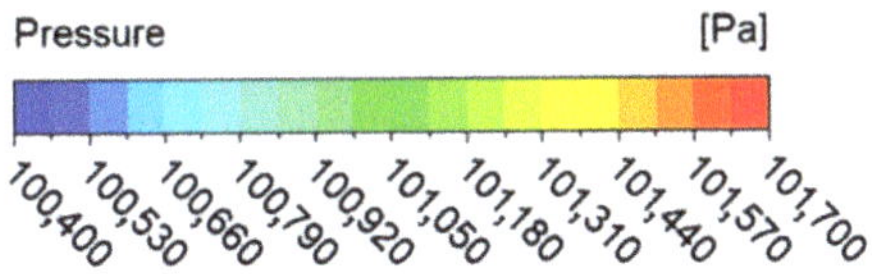

Figure 13. Pressure distribution at TSR = 1.25: (**a**) case 1 at 0 deg.; (**b**) case 1 at 30 deg.; (**c**) case 1 at 60 deg.; (**d**) case 1 at 90 deg.; (**a′**) case 2 at 0 deg.; (**b′**) case 2 at 30 deg.; (**c′**) case 2 at 60 deg.; (**d′**) case 2 at 90 deg.

During a full 360-degree rotation, the resulting aerodynamic forces acting on the blades were reduced to the overall torque coefficient. After 120 degrees, an arrangement with three blades exhibits almost periodic behavior. The torque coefficients of the numerical findings for four distinct TSRs—0.5, 0.75, 1, and 1.25 for the investigated cases—are compared in Figure 14. It is generally evident from this comparison that the second case, in which we employed the vortex trap to enhance the flow by postponing the detachments, has a greater torque coefficient.

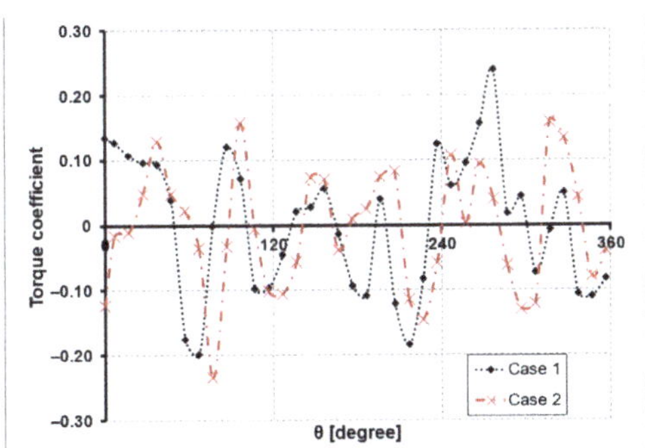

(**a**)

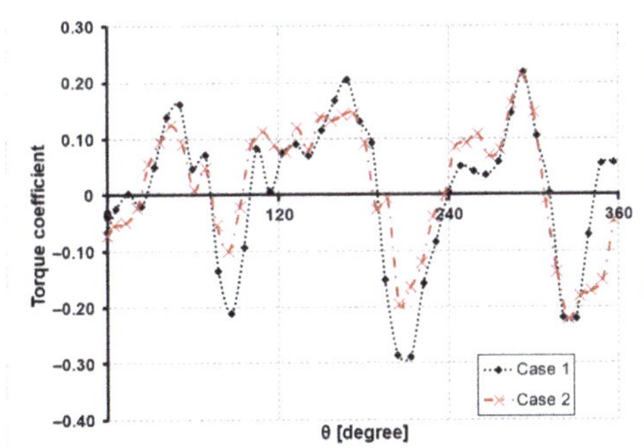

(**b**)

Figure 14. *Cont.*

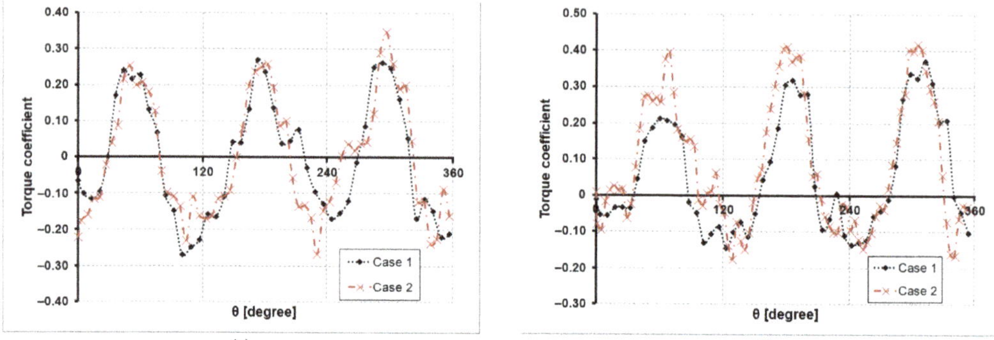

(c) (d)

Figure 14. Torque coefficient for 360 deg. revolution: (**a**) TSR 0.5; (**b**) TSR 0.75; (**c**) TSR 1; (**d**) TSR 1.25.

The evolution of the torque coefficient for the new geometry for all 10 rotations can be seen in Figure 15.

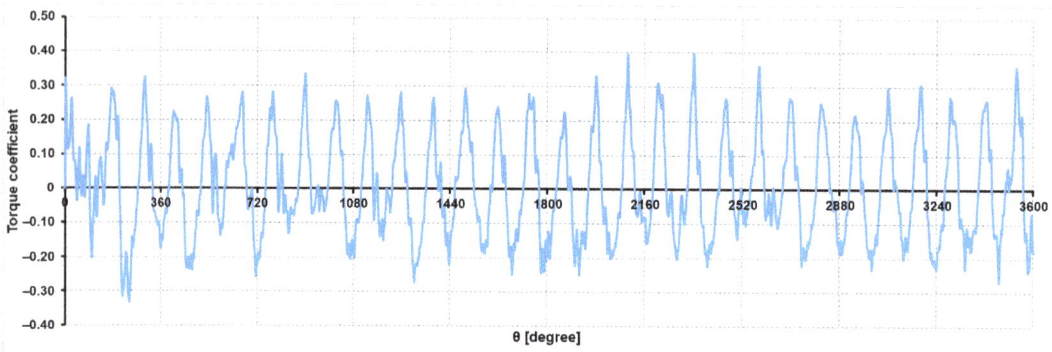

Figure 15. Torque coefficient for 10 revolutions (3600 deg) for the new geometry at TSR = 1.

The evolution of the lift coefficient for the blade ensemble can be observed in Figure 16 for a full rotation of 360 deg. of azimuthal angle, and the evolution of the lift coefficient with the incidence angle (angle of attack) can be seen in Figure 17.

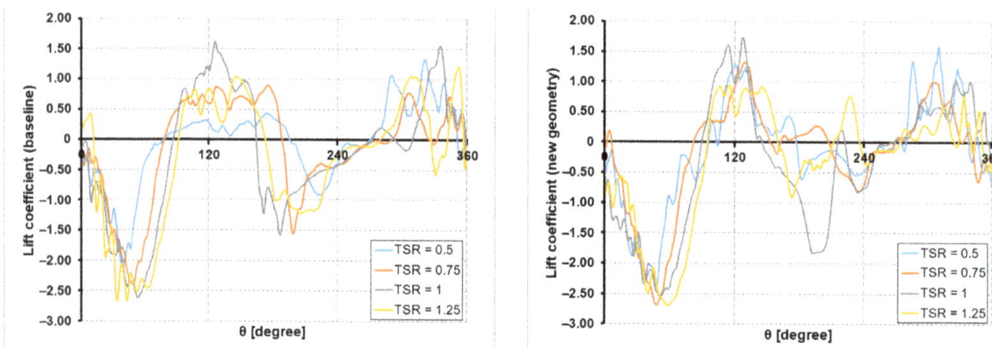

(a) (b)

Figure 16. Lift coefficient for 360 deg. revolution (azimuth angle θ): (**a**) baseline; (**b**) new geometry.

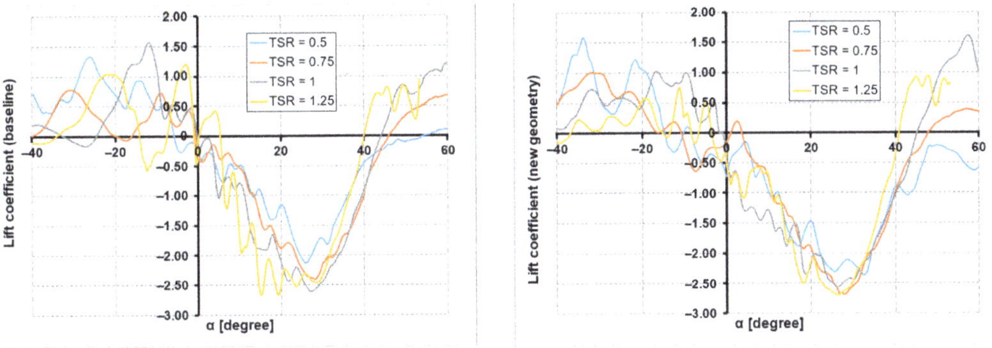

(a) (b)

Figure 17. Lift coefficient evolution with incidence angle α: (**a**) baseline; (**b**) new geometry.

The lift coefficient appears to have a similar evolution and similar values. The only exception to this is the interval of 120 to 200 deg, for which the C_L has lower values in the new case than in the base geometry.

The influence of the new geometry on the stall angle is negligible, as indicated by Figure 17.

In essence, the torque coefficient indicates how well the wind energy is being converted into mechanical torque by the VAWT. Better performance is indicated by a higher value of the torque coefficient, which suggests that the wind turbine has the capacity to attain its nominal rotational speed. In Figure 18, the torque coefficient for a single blade for each of the analyzed scenarios for TSR 0.5 and TSR 1.25 was presented in order to better illustrate the impact of the vortex trap on wind turbine efficiency.

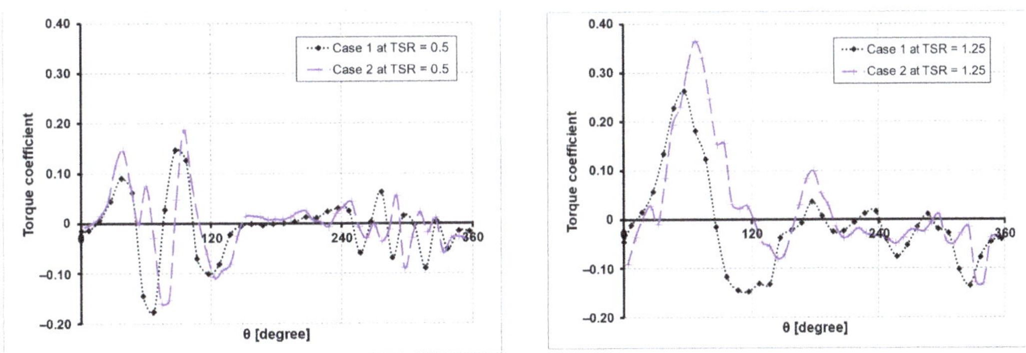

(a) (b)

Figure 18. Torque coefficient for one blade: (**a**) TSR 0.5; (**b**) TSR 1.25.

The LES technique predicts the value of C_p quite well, which is close to the trend of the curve efficiency of a vertical-axis wind turbine, as seen in many sources [57]. The average power coefficients as a function of the tip speed ratio are depicted in Figure 19. The average coefficients for the latest 360-degree revolution from the 10 completed ones studied were determined. A cohesive pattern is detected right away, indicating that utilizing the flow control approach increases the wind turbine power coefficient significantly.

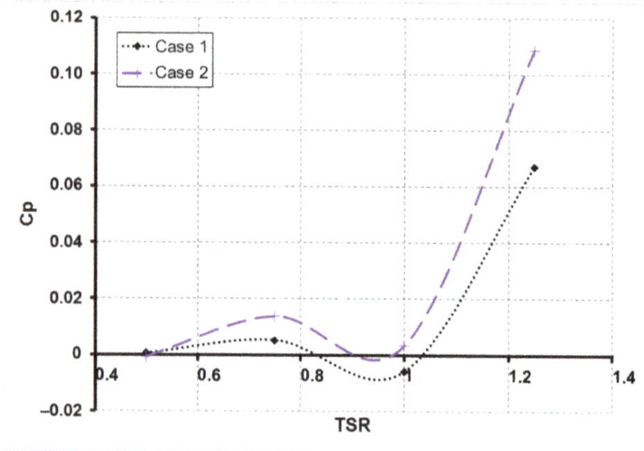

Figure 19. Power coefficient.

The values for the C_p of the turbine in case 2 were much greater than those in case 1 (Figure 19); at TSR 1.25, a maximum of 0.108 was observed, compared to just 0.06 in the conventional arrangement (case 1). This indicates that the initial wind turbine efficiency may be increased by 40% by implementing flow control. It is thought that employing a vortex trap increases performance by allowing the flow to remain attached for a longer time.

In order to validate the results, the C_P variation with TSR of the base geometry (NACA 0021) was compared to results of the same airfoil found in other research papers (Figure 20), which were scaled to the reference values used in this article, namely the reference area and wind speed.

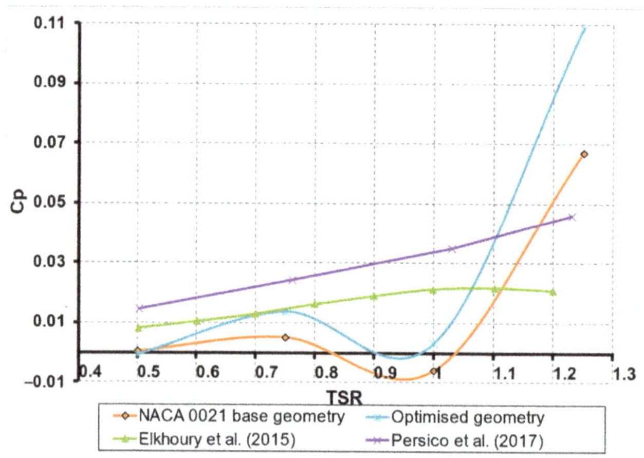

Figure 20. Power coefficient variation with TSR; comparison between the base geometry simulation results of this paper, the experimental results of Persico et al. (2017) [58], and the numerical results of Elkhoury et al. (2015) [59].

4. Discussion

Regarding what constitutes the worst drawback of these particular turbines, the wind turbine world views the self-starting mechanism of VAWTs from many points of view. If a turbine can accelerate and reach a steady-state angular velocity in response to aerodynamic forces and moments without the need for external stimulation, it is often regarded as self-starting. The duration required to attain a steady-state rotational speed commensurate with its peak performance can be utilized to ascertain its self-starting capability.

The measurable aerodynamic performance of turbines is determined by the flow physics surrounding them and the interaction between the blades. The effective angles-of-attack (AoAs) of blades are determined by their instantaneous locations, defined in Equation (7):

$$\text{AoA} = \text{atan}\frac{\sin\theta}{\cos\theta + \lambda} \quad (7)$$

Figure 21 shows how AoA fluctuates as a function of blade position (θ), indicating that the configurations suffer unusually significant effective angles-of-attack during the beginning process, and the stall angle for NACA0021.

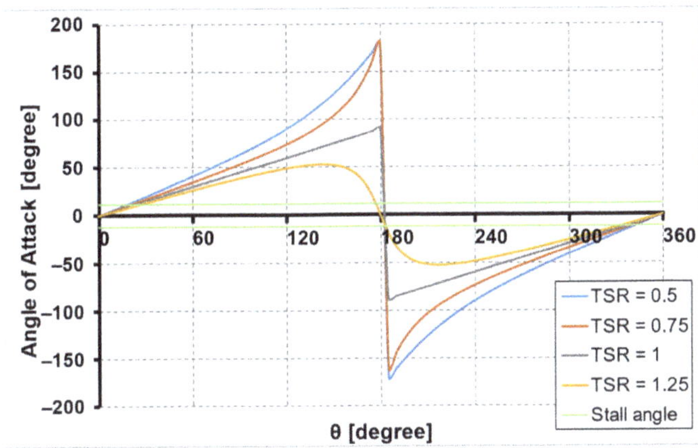

Figure 21. Angle of attack vs. blade position (θ).

When comparing these data to unstable aerodynamic forces, it can be seen that the excessive effective angles-of-attack are what create negative lift, which leads to reduced efficiency. However, the flow remains attached longer when the vortex trap is used, and even if the AoA is still quite high at low TSR, we gain an efficiency boost that is highly beneficial for the self-starting behavior.

In order to facilitate comprehension, the outcomes for each case are displayed in pairs for a single blade at various points. The vorticity contours for a single blade at a tip speed ratio of 1 are shown in Figure 22. It is possible to observe from the figure how the flow control method helps the flow stay attached to the blades for longer periods of time, thereby lowering the interactions between the blades and increasing efficiency. Additionally, the vortex trap plays a crucial role in the wind turbine's starting by developing a high torque value that can, even momentarily, convert a lift-based vertical-axis wind turbine into a drag-based vertical-axis wind turbine, which is well-known for its quick start at a low tip speed ratio.

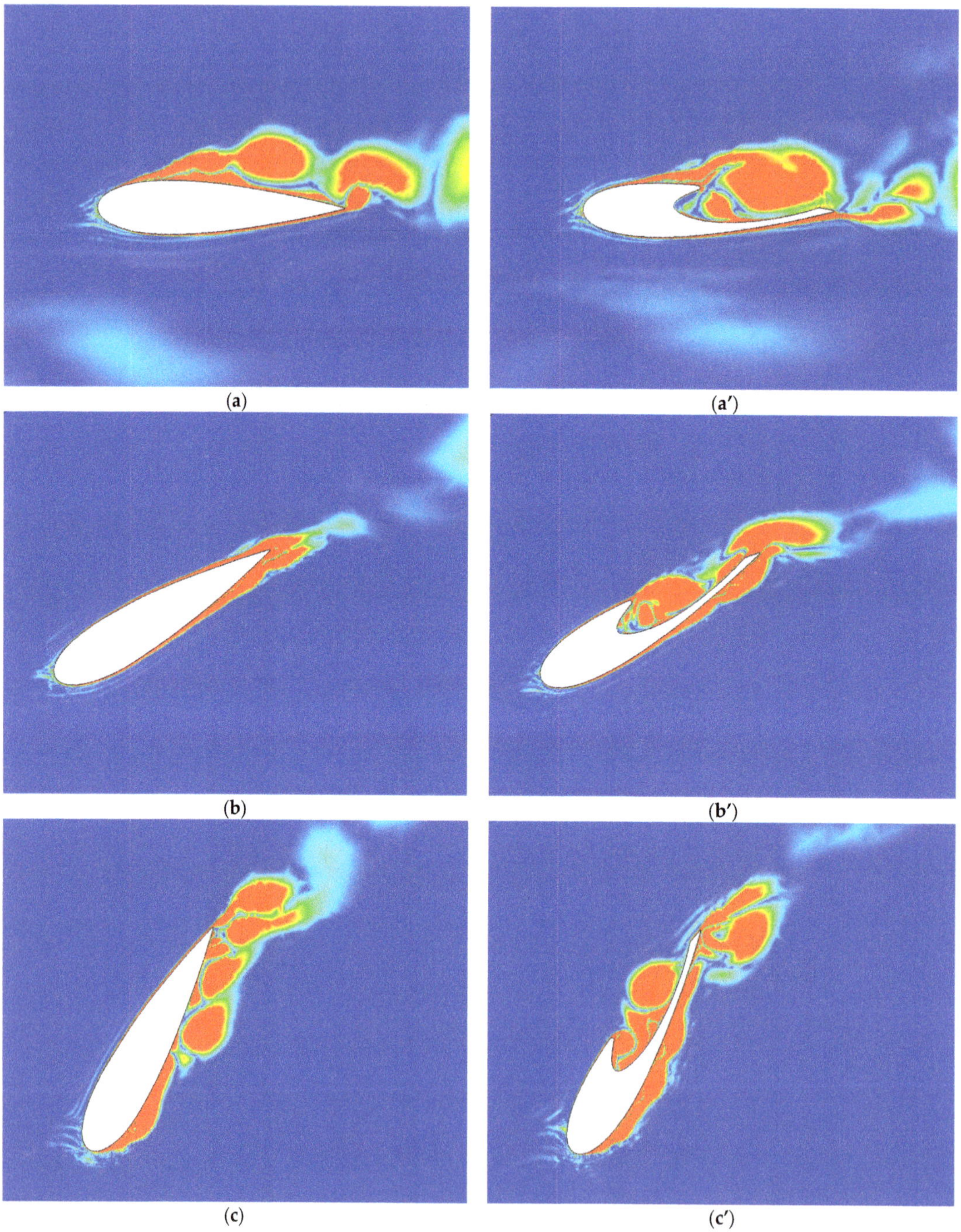

Figure 22. *Cont.*

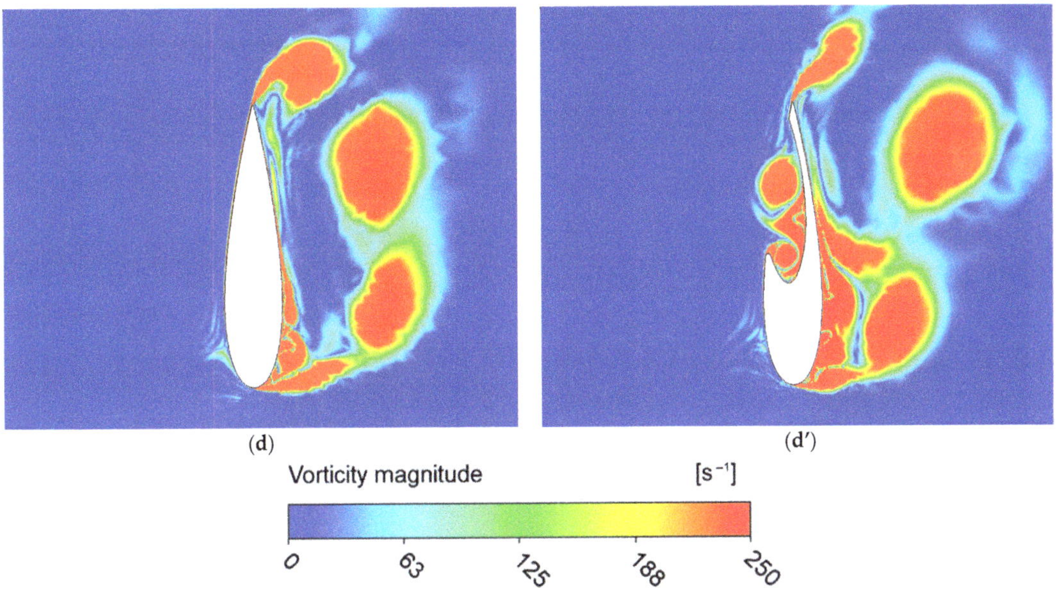

Figure 22. Vorticity magnitude for one blade at TSR = 1: (**a**) case 1_0 deg.; (**b**) case 1_30 deg.; (**c**) case 1_60 deg.; (**d**) case 1_90 deg.; (**a′**) case 2_0 deg.; (**b′**) case 2_30 deg.; (**c′**) case 2_60 deg.; (**d′**) case 2_90 deg.

To capture the flow behavior in future research, the starting procedure for various wind speed values will utilize Particle Image Velocimetry (PIV) [60,61] technology to experimentally evaluate the configurations. The experimental testing campaign will be conducted in a subsonic wind tunnel on reduced-scale models. The setup for the experimental campaign is presented in Figure 23.

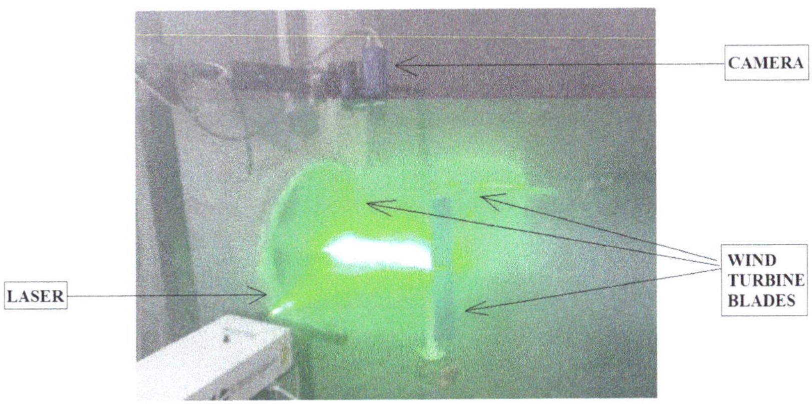

Figure 23. Experimental setup photo.

Performance measurements will be carried out using a three-phase AC Permanent Magnet Generator (position 1 in Figure 24) directly connected to the rotor axis. Electrical power will be measured after using a wind controller model (position 2 in Figure 24) connected to charge a 9AH/12VDC battery (position 3 in Figure 24). A digital multimeter (position 5 in Figure 24), placed parallel to the electric circuit, will be used to measure the electrical voltage, and a digital multimeter (position 4 in Figure 24), placed in series in the electric circuit, will be used to measure the current intensity. A diagram of the electrical circuit is presented in Figure 24.

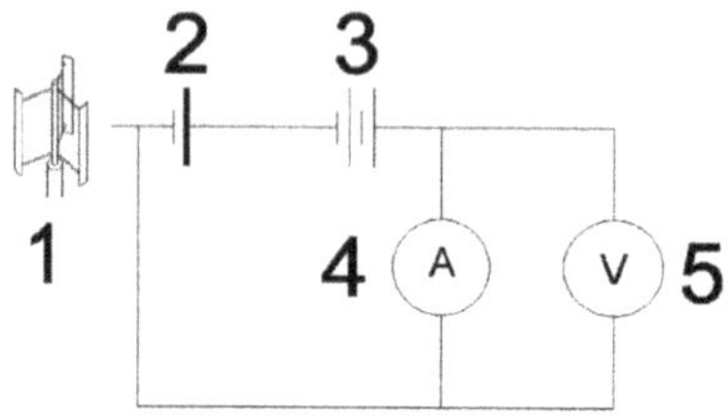

Figure 24. Electrical diagram of the circuit used for VAWT performance measurements.

5. Conclusions

In-depth research on the flow dynamics during a vertical-axis wind turbine's startup was conducted using an accurate numerical approach. The Large Eddy Simulation is highly recommended in the case of studying fully unsteady flows like those that occur in a working vertical-axis wind turbine and for the monitorization in real time for predictive maintenance in the field of rotating machines. Two computational fluid dynamics (CFD) models were developed for this work. The first model was a traditional vertical-axis wind turbine with three blades, and the second model had the same structure but was modified to include a vortex trap. Both models were numerically studied using the sliding mesh approach. In order to ensure a fair comparison, a number of factors, including the size of the computational domain, boundary conditions, LES technique, turbine characteristics, total number of elements for the stator section, and grid resolution near the blades, were maintained to be comparable in both models.

A vortex trap was mounted on the outside of the blades in order to thoroughly examine how well a flow control technique mitigated the flow separation issue in the VAWT starting process. The vortex trap, which is positioned on the outside of the blade, produces results that indicate that, even at lower TSR values, the device has little effect on turbine efficiency. However, as TSR increases, the flow begins to change, leading to a higher efficiency for the configuration in which the flow method technique was applied. Results from the vortex trap mounted outside the blade demonstrate a notable improvement in wind turbine performance. This is because the power coefficient increases to 0.106, or 40%, larger than the conventional version without a vortex trap due to the flow control managing the flow detachment, which is particularly noticeable at a high TSR of 1.25.

The flow surrounding VAWT is shown to be complex at low TSRs regardless of whether the configuration has a flow control device or not, and their performances directly affect the VAWT's capacity to achieve its nominal point of functioning.

Author Contributions: Conceptualization, I.M. and S.S.; methodology, I.M.; software, I.M.; validation, I.M. and S.S.; formal analysis, S.S.; investigation, I.M.; resources, I.M.; data curation, S.S.; writing—original draft preparation, S.S.; writing—review and editing, I.M.; visualization, S.S.; supervision, I.M.; project administration, I.M.; funding acquisition, I.M. All authors have read and agreed to the published version of the manuscript.

Funding: This work was carried out within "Nucleu" Program from The National Research, Development and Innovation Plan from 2022–2027, supported by the Romanian Ministry of Research, Innovation and Digitization, project no. PN 23.12.08.01.

Data Availability Statement: The original contributions presented in the study are included in the article, further inquiries can be directed to the corresponding author.

Conflicts of Interest: The authors declare no conflicts of interest.

References

1. Real-Time Electricity Tracker'. Available online: https://www.iea.org/data-and-statistics/data-tools/real-time-electricity-tracker (accessed on 20 August 2024).
2. Ahmad, T.; Zhang, D. A critical review of comparative global historical energy consumption and future demand: The story told so far. *Energy Rep.* **2020**, *6*, 1973–1991. [CrossRef]
3. Acheampong, A.O. Economic growth, CO_2 emissions and energy consumption: What causes what and where? *Energy Econ.* **2018**, *74*, 677–692. [CrossRef]
4. Le Quéré, C.; Andrew, R.M.; Friedlingstein, P.; Sitch, S.; Hauck, J.; Pongratz, J.; Pickers, P.A.; Korsbakken, J.I.; Peters, G.P.; Canadell, J.G.; et al. Global Carbon Budget 2018. Earth Syst. *Sci. Data* **2018**, *10*, 2141–2194. [CrossRef]
5. IEA. *CO2 Emissions in 2022*; IEA: Paris, France, 2022. Available online: https://www.iea.org/reports/co2-emissions-in-2022 (accessed on 20 August 2024).
6. IPCC. *Global Warming of 1.5 °C: IPCC Special Report on Impacts of Global Warming of 1.5 °C Above Pre-Industrial Levels in Context of Strengthening Response to Climate Change, Sustainable Development, and Efforts to Eradicate Poverty*, 1st ed.; Cambridge University Press: Cambridge, UK, 2022. [CrossRef]
7. Ntanos, S.; Ziatas, T.; Merkouri, A. Renewable energy consumption, carbon dioxide emissions and economic growth: Evidence from Europe and Greece. In Proceedings of the International Scientific Conference eRA-10, Piraeus, Greece, 23–25 September 2015.
8. Mikhaylov, A. Efficiency of renewable energy plants in Russia. *An. Acad. Bras. Ciênc.* **2022**, *94*, e20191226. [CrossRef] [PubMed]
9. Martín-Martínez, S.; Cañas-Carretón, M.; Honrubia-Escribano, A.; Gómez-Lázaro, E. Performance evaluation of large solar photovoltaic power plants in Spain. *Energy Convers. Manag.* **2019**, *183*, 515–528. [CrossRef]
10. Arutyunov, V.S.; Lisichkin, G.V. Energy resources of the 21st century: Problems and forecasts. Can renewable energy sources replace fossil fuels? *Russ. Chem. Rev.* **2017**, *86*, 777–804. [CrossRef]
11. Ledec, G.; Quintero, J.D. Good Dams and Bad Dams: Environmental Criteria for Site Selection of Hydroelectric Projects. 2003. Available online: https://api.semanticscholar.org/CorpusID:129833751 (accessed on 20 August 2024).
12. Hamed, T.A.; Alshare, A. Environmental Impact of Solar and Wind energy—A Review. *J. Sustain. Dev. Energy Water Environ. Syst.* **2022**, *10*, 1–23. [CrossRef]
13. Abohela, I.; Hamza, N.; Dudek, S. Urban Wind Turbines Integration in the Built Form and Environment. *FORUM Ejournal* **2011**, *10*, 23–39.
14. Barber, S.; Wang, Y.; Jafari, S.; Chokani, N.; Abhari, R.S. The Impact of Ice Formation on Wind Turbine Performance and Aerodynamics. *J. Sol. Energy Eng.* **2011**, *133*, 011007. [CrossRef]
15. Teff-Seker, Y.; Berger-Tal, O.; Lehnardt, Y.; Teschner, N. Noise pollution from wind turbines and its effects on wildlife: A cross-national analysis of current policies and planning regulations. *Renew. Sustain. Energy Rev.* **2022**, *168*, 112801. [CrossRef]
16. Krijgsveld, K.L.; Akershoek, K.; Schenk, F.; Dijk, F.; Dirksen, S. Collision Risk of Birds with Modern Large Wind Turbines. *Ardea* **2009**, *97*, 357–366. [CrossRef]
17. Awada, A.; Younes, R.; Ilinca, A. Review of Vibration Control Methods for Wind Turbines. *Energies* **2021**, *14*, 3058. [CrossRef]
18. Castellani, F.; Astolfi, D.; Peppoloni, M.; Natili, F.; Buttà, D.; Hirschl, A. Experimental Vibration Analysis of a Small Scale Vertical Wind Energy System for Residential Use. *Machines* **2019**, *7*, 35. [CrossRef]
19. Drones Protect Wind Turbines from Ice. Available online: https://www.fraunhofer.de/en/press/research-news/2023/december-2023/drones-protect-wind-turbines-from-ice.html (accessed on 20 August 2024).
20. Montoya, L.T.C.; Lain, S.; Ilinca, A. A Review on the Estimation of Power Loss Due to Icing in Wind Turbines. *Energies* **2022**, *15*, 1083. [CrossRef]
21. Ishugah, T.F.; Li, Y.; Wang, R.Z.; Kiplagat, J.K. Advances in wind energy resource exploitation in urban environment: A review. *Renew. Sustain. Energy Rev.* **2014**, *37*, 613–626. [CrossRef]
22. Mehedintu, A.; Soava, G.; Sterpu, M.; Grecu, E. Evolution and Forecasting of the Renewable Energy Consumption in the Frame of Sustainable Development: EU vs. *Romania. Sustain.* **2021**, *13*, 10327. [CrossRef]
23. Colesca, S.E.; Ciocoiu, C.N. An overview of the Romanian renewable energy sector. *Renew. Sustain. Energy Rev.* **2013**, *24*, 149–158. [CrossRef]
24. Bianco, V.; Manca, O.; Nardini, S.; Minea, A.A. Analysis and forecasting of nonresidential electricity consumption in Romania. *Appl. Energy* **2010**, *87*, 3584–3590. [CrossRef]
25. Yuan, X.-C.; Sun, X.; Zhao, W.; Mi, Z.; Wang, B.; Wei, Y.-M. Forecasting China's regional energy demand by 2030: A Bayesian approach. *Resour. Conserv. Recycl.* **2017**, *127*, 85–95. [CrossRef]
26. Cuce, E. An overview of domestic energy consumption in the UK: Past, present and future. *Int. J. Ambient. Energy* **2016**, *37*, 428–435. [CrossRef]
27. Boretti, A.; Castelletto, S. Trends in performance factors of large photovoltaic solar plants. *J. Energy Storage* **2020**, *30*, 101506. [CrossRef]
28. Statista. Global Electricity Consumption from 1980 to 2022. Available online: https://www.statista.com/statistics/280704/world-power-consumption (accessed on 20 August 2024).
29. Ritchie, H.; Rosado, P.; Roser, M. Electricity Generation from Wind Power. Energy. Available online: https://ourworldindata.org/grapher/wind-energy-consumption-by-region (accessed on 20 August 2024).

30. Ritchie, H.; Rosado, P.; Roser, M. Annual Change in Solar and Wind Power Consumption. Energy. Available online: https://ourworldindata.org/grapher/annual-change-in-solar-and-wind-energy-generation (accessed on 20 August 2024).
31. Al-Rawajfeh, M.A.; Gomaa, M.R. Comparison between horizontal and vertical axis wind turbine. *Int. J. Appl. Power Eng. (IJAPE)* **2023**, *12*, 13–23. [CrossRef]
32. Farajyar, S.; Ghafoorian, F.; Mehrpooya, M.; Asadbeigi, M. CFD Investigation and Optimization on the Aerodynamic Performance of a Savonius Vertical Axis Wind Turbine and Its Installation in a Hybrid Power Supply System: A Case Study in Iran. *Sustainability* **2023**, *15*, 5318. [CrossRef]
33. Stratila, S.; Glasberg, D.; Mălăel, I. Performance Analysis of a New Vertical Axis Turbine Design for Household Usage. *Eng. Technol. Appl. Sci. Res.* **2024**, *14*, 12536–12542. [CrossRef]
34. Karimian, S.; Abdolahifar, A. Performance investigation of a new Darrieus Vertical Axis Wind Turbine. *Energy* **2020**, *191*, 116551. [CrossRef]
35. Wang, Q.; Liu, B.; Hu, C.; Wang, F.; Yang, S. Aerodynamic shape optimization of H-VAWT blade airfoils considering a wide range of angles of attack. *Int. J. Low-Carbon Technol.* **2021**, *17*, 147–159. [CrossRef]
36. Malael, I.; Bucur, I.O. Numerical Evaluation of the Flow around a New Vertical Axis Wind Turbine Concept. *Sustainability* **2021**, *13*, 9012. [CrossRef]
37. Malael, I.; Dragan, V. Numerical and Experimental Efficiency Evaluation of a Counter-Rotating Vertical Axis Wind Turbine. *Eng. Technol. Appl. Sci. Res.* **2018**, *8*, 3282–3286. [CrossRef]
38. Bai, H.; Chan, C.-M. Positive interactions of two Savonius-type vertical-axis wind turbines for performance improvement. *Energy Procedia* **2019**, *158*, 625–630. [CrossRef]
39. Shaheen, M.; El-Sayed, M.; Abdallah, S. Numerical study of two-bucket Savonius wind turbine cluster. *J. Wind. Eng. Ind. Aerodyn.* **2015**, *137*, 78–89. [CrossRef]
40. Altan, B.D.; Gultekin, G.S. Investigation of Performance Enhancements of Savonius Wind Turbines through Additional Designs. *Processes* **2023**, *11*, 1473. [CrossRef]
41. Maldonado, V.; Farnsworth, J.; Gressick, W.; Amitay, M. Active control of flow separation and structural vibrations of wind turbine blades. *Wind. Energy* **2010**, *13*, 221–237. [CrossRef]
42. Shun, S.; Ahmed, N. Wind Turbine Performance Improvements using Active Flow Control Techniques. *Procedia Eng.* **2012**, *49*, 83–91. [CrossRef]
43. Aboelezz, A.; Ghali, H.; Elbayomi, G.; Madboli, M. A novel VAWT passive flow control numerical and experimental investigations: Guided Vane Airfoil Wind Turbine. *Ocean Eng.* **2022**, *257*, 111704. [CrossRef]
44. Zhu, H.; Hao, W.; Li, C.; Ding, Q.; Wu, B. A critical study on passive flow control techniques for straight-bladed vertical axis wind turbine. *Energy* **2018**, *165*, 12–25. [CrossRef]
45. Fernandez-Gamiz, U.; Zulueta, E.; Boyano, A.; Ansoategui, I.; Uriarte, I. Five Megawatt Wind Turbine Power Output Improvements by Passive Flow Control Devices. *Energies* **2017**, *10*, 742. [CrossRef]
46. De Gregorio, F.; Fraioli, G. Flow Control on a High Thickness Airfoil by Trapped Vortex Cavity. In Proceedings of the 14th International Symposium on Applications of Laser Techniques to Fluid Mechanics, Lisbon, Portugal, 7–10 July 2008.
47. Błoński, D.; Strzelecka, K.; Kudela, H. Vortex Trapping Cavity on Airfoil: High-Order Penalized Vortex Method Numerical Simulation and Water Tunnel Experimental Investigation. *Energies* **2021**, *14*, 8402. [CrossRef]
48. Acosta-López, J.G.; Blasetti, A.P.; Lopez-Zamora, S.; de Lasa, H. CFD Modeling of an H-Type Darrieus VAWT under High Winds: The Vorticity Index and the Imminent Vortex Separation Condition. *Processes* **2023**, *11*, 644. [CrossRef]
49. Smagorinsky, J. General circulation experiments with the primitive equations: I. The basic experiment. *Mon. Weather Rev.* **1963**, *91*, 99–164. [CrossRef]
50. Manwell, J.F.; McGowan, J.G.; Rogers, A.L. *Wind Energy Explained: Theory, Design and Application*, 2nd ed.; Wiley: Chichester, UK, 2009.
51. Abdelsamie, A.; Voß, S.; Berg, P.; Chi, C.; Arens, C.; Thévenin, D.; Janiga, G. Comparing LES and URANS results with a reference DNS of the transitional airflow in a patient-specific larynx geometry during exhalation. *Comput. Fluids* **2023**, *255*, 105819. [CrossRef]
52. Rogowski, K.; Królak, G.; Bangga, G. Numerical Study on the Aerodynamic Characteristics of the NACA 0018 Airfoil at Low Reynolds Number for Darrieus Wind Turbines Using the Transition SST Model. *Processes* **2021**, *9*, 477. [CrossRef]
53. Ranjbar, M.H.; Nasrazadani, S.A.; Kia, H.Z.; Gharali, K. Reaching the betz limit experimentally and numerically. *Energy Equip. Syst.* **2019**, *7*, 271–278. [CrossRef]
54. Betz, A. The Maximum of the Theoretically Possible Exploitation of Wind by Means of a Wind Motor. *Wind. Eng.* **2013**, *37*, 441–446. [CrossRef]
55. Mirmotahari, S.R.; Ghafoorian, F.; Mehrpooya, M.; Rad, S.H.; Taraghi, M.; Moghimi, M. A comprehensive investigation on Darrieus vertical axis wind turbine performance and self-starting capability improvement by implementing a novel semi-directional airfoil guide vane and rotor solidity. *Phys. Fluids* **2024**, *36*, 065151. [CrossRef]
56. Ghafoorian, F.; Mirmotahari, S.R.; Mehrpooya, M.; Akhlaghi, M. Aerodynamic performance and efficiency enhancement of a Savonius vertical axis wind turbine with Semi-Directional Curved Guide Vane, using CFD and optimization method. *J. Braz. Soc. Mech. Sci. Eng.* **2024**, *46*, 443. [CrossRef]

57. Bogateanu, R.; Dumitrache, A.; Dumitrescu, H.; Stoica, C.I. Reynolds number efects on the aerodynamic performance of small VAWTs. *UPB Sci. Bull.* **2014**, *76*, 25–36.
58. Persico, G.; Dossena, V.; Paradiso, B.; Battisti, L.; Brighenti, A.; Benini, E. Time-Resolved Experimental Characterization of the Wakes Shed by H-Shaped and Troposkien Vertical Axis Wind Turbines. *J. Energy Resour. Technol.* **2017**, *139*, 031203. [CrossRef]
59. Elkhoury, M.; Kiwata, T.; Aoun, E. Experimental and numerical investigation of a three-dimensional vertical-axis wind turbine with variable-pitch. *J. Wind. Eng. Ind. Aerodyn.* **2015**, *139*, 111–123. [CrossRef]
60. Ferreira, C.S.; van Kuik, G.; van Bussel, G.; Scarano, F. Visualization by PIV of dynamic stall on a vertical axis wind turbine. *Exp. Fluids* **2008**, *46*, 97–108. [CrossRef]
61. Arpino, F.; Cortellessa, G.; Scungio, M.; Fresilli, G.; Facci, A.; Frattolillo, A. PIV measurements over a double bladed Darrieus-type vertical axis wind turbine: A validation benchmark. *Flow Meas. Instrum.* **2021**, *82*, 102064. [CrossRef]

processes

MDPI

Article

Geometrical Evaluation of an Overtopping Wave Energy Converter Device Subject to Realistic Irregular Waves and Representative Regular Waves of the Sea State That Occurred in Rio Grande—RS

Maycon da Silveira Paiva [1], Ana Paula Giussani Mocellin [1], Phelype Haron Oleinik [1], Elizaldo Domingues dos Santos [1], Luiz Alberto Oliveira Rocha [1], Liércio André Isoldi [1,*] and Bianca Neves Machado [2]

[1] School of Engineering, Federal University of Rio Grande (FURG), Rio Grande 96203-900, RS, Brazil; mayconpaiva@furg.br (M.d.S.P.); ana.mocellin@furg.br (A.P.G.M.); elizaldosantos@furg.br (E.D.d.S.); luizrocha@furg.br (L.A.O.R.)
[2] Interdisciplinary Department, Federal University of Rio Grande do Sul (UFRGS), Tramandaí 95590-000, RS, Brazil; bianca.machado@ufrgs.br
* Correspondence: liercioisoldi@furg.br; Tel.: +55-53-99109-7356

Abstract: Among the various potential renewable energy sources, sea waves offer significant potential, which can be harnessed using wave energy converter (WEC) devices such as overtopping converters. These devices operate by directing incident waves up a ramp into a reservoir. The water then passes through a turbine coupled with an electrical generator before returning to the ocean. Thus, the present study deals with the geometrical evaluation of an overtopping WEC, where the influence of the ratio between the height and length of the device ramp (H_1/L_1) on the amount of water mass (M) that enters the reservoir was investigated. Numerical simulations were performed using ANSYS-Fluent software, 22 R1 version, to generate and propagate realistic irregular (RI) waves and representative regular (RR) waves found in the coastal region of the municipality of Rio Grande, in the state of Rio Grande do Sul, southern Brazil. Consequently, through constructal design, the optimal WEC geometry for both wave approaches were identified as the same, where $(H_1/L_1)_o = 0.30$. Thus, considering the RI waves, $M = 200{,}820.77$ kg was obtained, while, considering the RR waves, $M = 144{,}054.72$ kg was obtained.

Keywords: wave energy converter; overtopping device; geometrical evaluation; realistic irregular waves; representative regular waves; WaveMIMO methodology; computational fluid dynamics

1. Introduction

Marine renewable energy (MRE) is a viable alternative for exploring renewable energy sources, given the various forms of energy available in the oceans, including wave energy [1]. Espindola and Araújo [2] estimate that the annual energy potential of waves in deep waters off Brazil is approximately 89.97 GW, with the country's southern region having the highest average wave power. One of the possibilities for harnessing this energy is through wave energy converters (WECs) of the overtopping type. These devices operate on a partially submerged structure with a ramp that channels water into a reservoir through overtopping. The water then returns to the ocean, passing through low-head hydraulic turbines that drive coupled electrical generators [3].

Overtopping WECs have been studied in the experimental field, through real prototypes and laboratory devices, as well as in the numerical field, through computational modeling. An effective approach to investigate overtopping devices is through geometric evaluations, which can be conducted using the constructal design method. This method is based on constructal theory [4], according to which nature's flow systems follow a physical principle for the generation of their configurations and geometric patterns. This principle is called constructal law [4–6], and is being applied to the evaluation of flow systems through the constructal design. This approach allows for the analysis of the influence of geometry on system performance.

Furthermore, concerning studies carried out in the numerical field, there are different approaches for wave generation, and, therefore, several numerical investigations regarding the propagation waves have been carried out. Regarding the conventional methodology commonly applied when addressing regular waves [7,8], there are several notable studies. For instance, Lisboa et al. [9] investigated the linear and quadratic damping coefficients considered for a numerical beach tool employed to suppress wave reflections at the end of a numerical wave channel, and Zabihi et al. [10] compared two software programs, ANSYS-Fluent and Flow-3D (https://www.flow3d.com/), for generating numerical waves.

In addition, there are alternative methods for wave generation that allow the reproduction of numerical irregular waves. Some examples can be seen in Higuera et al. [11,12], who developed [11] and validated [12] a numerical methodology for realistic wave generation associated with active wave absorption in OpenFOAM software (https://www.openfoam.com/), which was based on the linear shallow-water theory; Finnegan and Goggins [13], where the generation of irregular waves on a real scale was studied; and Machado et al. [14], who developed the WaveMIMO methodology, which was verified and validated by Maciel et al. [15] through numerical simulations in the ANSYS-Fluent software and laboratory experiments; this methodology can generate irregular waves based on realistic sea state data, making it possible to reproduce the phenomenon more reliably, akin to its occurrence in nature. The WaveMIMO methodology allows the reproduction of waves generated numerically or monitored by ocean buoys.

It should be emphasized that both the experimental and numerical approaches are important and complement each other in the search for advances in the MRE field, mainly in the wave energy realm. Combined, these investigation approaches allowed several analyses and advances regarding WECs, as can be seen in Tedd and Kofoed [16], who presented a study of overtopping flow series on the Wave Dragon prototype, a low-crested overtopping device, in an inland sea in Northern Denmark; Parmeggiani et al. [17], who conducted a laboratory experimental study on a 1.5 MW Wave Dragon, identifying its response to extreme conditions typical of the DanWEC test center; and Di Lauro et al. [18], who analyzed the stability of a breakwater with a coupled overtopping device, known as the overtopping breakwater for energy conversion (OBREC), combining laboratory-scale experiments and numerical simulations. Additionally, Palma et al. [19] carried out numerical studies to optimize the hydraulic and structural performance of the OBREC device by investigating the inclusion of a berm, the shape of the sloping plate, and the crown wall shape.

In the experimental field, Contestabile et al. [20] evaluated the effectiveness of a triangular parapet placed on top of the OBREC device in reducing wave overtopping when compared to a breakwater with a traditional crown wall. Di Lauro et al. [21] proposed a vertical OBREC (OBREC-V); the hydraulic performance and stability were analyzed under hydraulic loading, comparing it with a traditional breakwater, indicating that the non-conventional geometry enhances the structure's stability. Musa et al. [22] investigated, through numerical and experimental approaches, the influence of the ramp shape parameters of an OBREC device under Malaysia's wave condition.

Moreover, Koutrouveli et al. [23] presented conceptual proof of a hybrid WEC, which addresses the operational principles of two WECs, the overtopping and the oscillating water column (OWC); the results indicated that hybridization is an effective approach. Clemente et al. [24] conducted an experimental study to analyze the performance and stability of the armor layer and the toe berm of a model of the north breakwater extension project incorporating a hybrid WEC as proposed by Koutrouveli et al. [23].

Also, Martins et al. [25] conducted a numerical study of an overtopping device, using the constructal design method to evaluate the effects of the degrees of freedom on the average dimensionless overtopping flow for devices with one and two ramps; it was found that the two-ramp configuration performed better under the incidence of irregular waves generated through the JONSWAP wave spectrum. An et al. [26] analyzed different overtopping WEC models testing the length of the substructure, i.e., the water region under the ramp and reservoir of the device, inferring that the hydraulic efficiency decreases with the substructure length but does not affect the performance of the WEC. Barros et al. [27] applied the constructal design method to evaluate the influence of a trapezoidal berm, located at the bottom of the channel and connected to the ramp of the overtopping device, showing that this configuration leads to better results than traditional overtopping WEC under the incidence of regular waves. Goulart et al. [28] performed a geometric evaluation of the overtopping device, using constructal design, through experimental and numerical studies considering the incidence of regular waves on a laboratory scale; the results showed an agreement between the approaches, validating the numerical model proposed using the ANSYS-Fluent software.

It should be highlighted that the WaveMIMO methodology [14] has also been used in investigations regarding overtopping devices, as in Hübner et al. [29], who employed it to evaluate the overtopping WEC, where they compared the fluid dynamic behavior of the device under the incidence of realistic irregular (RI) waves and representative regular (RR) waves, considering realistic sea state data of three coastal regions along the Rio Grande do Sul (RS) state in southern Brazil. However, this methodology has not yet been applied to the geometric evaluation of overtopping devices under the incidence of RI waves occurring in a specific location.

As mentioned, several studies are addressing the overtopping WEC in the literature. More examples of these are found for the experimental approach in [30–38], including the ones addressing the hybrid WEC, which combines the overtopping and OWC operating principles [35–38], while for the numerical approach there are [39–51], including more studies where the constructal design was employed to evaluate the overtopping device's geometry, such as [41,49,51]. As shown, despite the vast research on the overtopping device, few studies address realistic sea states when evaluating its geometry. This fact justifies the contribution of the present paper and is in line with what was stated by Romanowski et al. [52], who point out the lack of studies that use computational fluid dynamics (CFDs) to estimate and model irregular sea states through numerical simulations.

In this context, the present paper conducts a geometric evaluation of an overtopping device subject to the incidence of RI waves and RR waves from the sea state occurring near the Molhes da Barra breakwater, located on the coast of Rio Grande—RS, in southern Brazil. Therefore, the numerical models employed for the RI and RR waves generation were verified, comparing the numerical results obtained to realistic sea state data (RI waves) and the analytical solution (RR waves), proving that the waves were adequately reproduced. Then, the constructal design method was used to evaluate the influence of the ratio between the height and length of the ramp (H_1/L_1) on the amount of water mass (M) entering the device's reservoir and determine the geometry that maximizes converter performance.

It is important to note that the H_1/L_1 parameter has been evaluated using constructal design through both numerical [27,41,51] and experimental [28] approaches under the incidence of regular waves. Additionally, it has been evaluated using the JONSWAP wave spectrum for irregular waves [25,49]. However, in this paper, the parameter is investigated based on realistic sea state data rather than theoretical or random waves. This consideration is crucial, given that the performance of WEC devices varies according to the wave climate they are subjected to. Furthermore, a comparison was made between the results obtained, highlighting the differences and similarities in the fluid dynamic behavior of the device depending on the wave approach used. It is important to emphasize that the geometric evaluation of an overtopping WEC, employing the constructal design, under the incidence of RI waves generated using the WaveMIMO methodology is an original scientific contribution of the present paper, as is the comparison of its performance with the one that occurred due to the incidence of the RR waves from the same sea state.

2. Mathematical and Numerical Modeling

The numerical simulations carried out in the present study used the ANSYS-Fluent software [53], 22 R1 version, a CFD package that is based on the finite volume method (FVM) [54]. It should be mentioned that numerical simulations of wave channels in this software have already been validated with laboratory experiments [15,28].

The multiphase volume of fluid (VoF) model [55] was employed in the treatment of the interface between the phases considered, the water and the air. In this model, the concept of volumetric fraction (α) is used, considering that the sum of the phases contained in each control volume must always equal unity. Thus, when a computational cell contains only water or only air, is the following applies, respectively:

$$\alpha_{water} = 1, \tag{1}$$

$$\alpha_{air} = 1. \tag{2}$$

Additionally, since the VoF model is used for immiscible fluids, if a cell contains the interface between the two phases, it holds that:

$$\alpha_{water} = 1 - \alpha_{air}. \tag{3}$$

Along with using the VoF model, a set of three equations is solved, which is composed of the mass conservation equation, given as follows [54,56]:

$$\frac{\partial(\rho)}{\partial t} + \nabla\left(\rho\vec{V}\right) = 0, \tag{4}$$

where t is the time (s); $\vec{V}$ is the velocity vector (m/s); and ρ is the fluid density (kg/m^3), which is calculated for the mixture of phases (water and air) as [57]:

$$\rho = \alpha_{water}\rho_{water} + (1 - \alpha_{water})\rho_{air}. \tag{5}$$

The model also includes the conservation of the volumetric fraction, given by [54]:

$$\frac{\partial(\alpha)}{\partial t} + \nabla\left(\alpha\vec{V}\right) = 0. \tag{6}$$

Finally, the last equation that composes the VoF model is the conservation of momentum, defined as [53,56]:

$$\frac{\partial}{\partial t}\left(\rho\vec{v}\right)+\rho\left(\nabla\vec{v}\right)\vec{v}=-\nabla p+\nabla\overline{\overline{\tau}}-\rho\vec{g}+S, \tag{7}$$

where p is the static pressure (Pa); $\vec{g}$ is the gravity acceleration vector (m/s^2); and $\overline{\overline{\tau}}$ is the strain rate tensor (N/m^2). Moreover, the term S is a sink, referring to the numerical beach tool, which attributes a dissipative profile to the region of the wave channel where it is applied, avoiding the reflection effect of waves that reach the end wall of the channel, given as follows [58,59]:

$$S=-\left[C_1\rho V+\frac{1}{2}C_2\rho|V|V\right]\left(1-\frac{z-z_{fs}}{z_b-z_{fs}}\right)\left(\frac{x-x_s}{x_e-x_s}\right)^2, \tag{8}$$

where C_1 and C_2 are, respectively, the linear (s^{-1}) and quadratic (m^{-1}) damping coefficients; V is the velocity along the z direction (m/s); z_{fs} and z_b are the vertical positions of the free surface (FS) and the channel bottom (m); and x_s and x_e are the starting and ending positions of the numerical beach (m). Finally, it should be noted that the damping coefficients, C_1 and C_2, are defined, respectively, as 20 s^{-1} and 0 m^{-1}, following the indications of Lisboa et al. [9].

To perform the numerical simulations in the ANSYS-Fluent software, configurations are selected to solve the Equations (4), (6), and (7). Thus, the pressure-implicit with splitting of operators (PISO) scheme, a non-iterative transient calculation procedure based on temporal precision [60], was used to solve the pressure–velocity coupling. Furthermore, it is emphasized that when the ANSYS-Fluent software is used, the PISO scheme is the most stable option for simulation problems that use the multiphase VoF model [52]. Regarding the spatial discretization for the pressure equation, the pressure staggering option (PRESTO) scheme [53] was employed.

For the treatment of the advective terms, the discretization method adopted was the first-order upwind, as it generally leads to better convergence [53]. According to Patankar [61], with this scheme, the quantities on all faces of the volumes are determined assuming that the center of the volume for a variable field represents an average value throughout the entire volume.

Moreover, to determine the surface occupied by water, the geo-reconstruction method is used, which, in ANSYS-Fluent, presents the best accuracy [53]. In this approach, ANSYS-Fluent's standard interpolation schemes are used to obtain the flow at the faces whenever a volume is filled with one phase, water or air. However, when a volume contains a mixture of phases, the geo-reconstruction scheme is employed, which represents the interface between the fluids using a piecewise-linear approximation.

The aforementioned numerical methods, along with the remaining configuration sets, are summarized in Table 1. It is worth noting that these methods are based on methodologies used in previous studies, including numerical ones [14,25,27,41,49,51], as well as validation studies where the results were compared to laboratory experiments [15,28].

Additionally, the flow was considered under a laminar regime without addressing a turbulence model. This simplification about the real problem can be adopted without significant loss of accuracy, as indicated in Gomes et al. [8], who obtained a good agreement between the numerical model, under a laminar regime, and experimental results. When considering laminar flow, in addition to reducing processing time, the convergence of the numerical solution is facilitated, which justifies the adoption of this simplification given the total time of each simulation in the present study.

Table 1. Methods and parameters used in the present numerical simulations.

Parameters		Numerical Inputs
Solver		Pressure-Based
Pressure–Velocity Coupling		PISO
Spatial Discretization	Gradient Evaluation Pressure Momentum Volume Fraction	Green–Gauss-Cell-Based PRESTO First Order Upwind Geo-Reconstruct
Temporal Differencing Scheme		First Order Implicit
Under-Relaxation Factors	Pressure Momentum	0.3 0.7
Residual	Continuity x-velocity z-velocity	10^{-3}
Regime Flow		Laminar

2.1. Generation of Realistic Irregular Waves

To generate the RI waves addressed in the present study, the WaveMIMO methodology was used. According to Machado et al. [14], this methodology consists of imposing discrete data of the wave propagation velocity in the horizontal (u_{IR}) and vertical (w_{IR}) directions as a prescribed velocity boundary condition (BC) in a numerical wave channel. Therefore, it was necessary that realistic sea state data be processed using the Spec2Wave software, 1.2.1 version. In this software, the procedure proposed in Oleinik et al. [62] is used to transform the wave spectrum into a time series of FS elevations. This procedure approximates the spectral data through a finite sum of monochromatic waves, individually described by Airy's Linear Wave Theory [63]:

$$\eta_1 = \frac{H}{2}cos(kx - \omega t), \tag{9}$$

which allows the irregular FS elevation series to be decomposed into orbital velocity profiles u_1 and w_1 that vary with depth. Thus, the RI waves can be reproduced by applying these velocity profiles as BC in the numerical wave channel. The configurations for the computational modeling needed to employ the WaveMIMO methodology are presented in Section 3.1. More information regarding the treatment method of the spectral data can be found in Oleinik et al. [62].

The realistic data addressed in the present study comes from a database of the realistic sea state occurred in the coastal region of RS state in the year 2018. This database was generated through the TELEMAC-based operational model addressing wave action computation (TOMAWAC) spectral model. According to Awk [64], in this model, sea state data are obtained from an equation that represents the general behavior of wave propagation in an unsteady and inhomogeneous medium, given by:

$$Q(t,\ x,\ y, k_i, k_j) = \frac{\partial N}{\partial t} + \frac{\partial(\dot{x}N)}{\partial x} + \frac{\partial(\dot{y}N)}{\partial y} + \frac{\partial\left(\dot{k}_i N\right)}{\partial k_i} + \frac{\partial\left(\dot{k}_j N\right)}{\partial k_j}, \tag{10}$$

where Q is the source term (m^2/rad); N represents the directional spectrum of wave action density ($m^2/Hz/rad$); k_i the component x of the wave number vector (m^{-1}); and k_j the component z of the wave number vector (m^{-1}).

For the present study, the realistic sea state data considered refer to a point with geographic coordinates 32°11′24″ S, 52°04′45″ W, located 171.06 m away from the Molhes da Barra breakwater in the municipality of Rio Grande in southern Brazil. It is important to mention that this data is from 11 September, from 7:15 a.m. to 7:30 a.m., with more details about it presented in the next section.

2.2. Generation of Representative Regular Waves

Although a realistic sea state is composed of irregular waves, it can be represented through regular waves using statistical parameters. One of the possible metrics to address is the mean period (T_m) of the irregular waves, which essentially consists of an arithmetic average of the periods of the waves that make up the spectrum, weighted by their contribution to the spectrum's energy [64]. Another statistical parameter to be considered is the significant height (H_s), which, according to Holthuijsen [65], is commonly used to describe a sea state, as it represents the waves causing the most significant surface agitation.

Thus, the statistical parameters of H_s and T_m referring to the selected local were analyzed to establish the regular waves related to this realistic sea state. Thereby, Figure 1 presents the bivariate histogram showing the occurrence of combinations between H_s and T_m and points out the most frequent one.

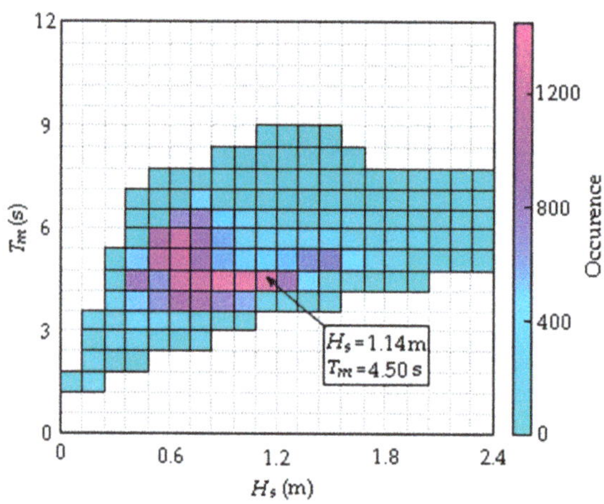

Figure 1. Bivariate histogram of H_s and T_m.

Using these characteristics and the depth (h) of the analyzed location, it is possible to establish the characteristics of the RR waves of this sea state. To do so, it is determined the wavelength, which is calculated using the dispersion relation [66]:

$$Q(t,\ x,\ y, k_i, k_j) = \frac{\partial N}{\partial t} + \frac{\partial(\dot{x}N)}{\partial x} + \frac{\partial(\dot{y}N)}{\partial y} + \frac{\partial\left(\dot{k}_i N\right)}{\partial k_i} + \frac{\partial\left(\dot{k}_j N\right)}{\partial k_j}, \tag{11}$$

where ω is the angular frequency (Hz); and k is the wave number (m^{-1}), given, respectively, by:

$$\omega = \frac{2\pi}{T}, \tag{12}$$

$$k = \frac{2\pi}{\lambda}, \tag{13}$$

where λ is the wave length (m); and T is the wave period (s).

Thus, the characteristics of the RR waves generated in the present study are shown in Table 2, where H stands for H_s and T for T_m. It should be emphasized that these characteristics are used in both the spatial and temporal discretization of the computational domain. In addition, these regular wave characteristics are similar to those observed in other parts of the world, such as China [67] and Spain [68].

Table 2. Characteristics of the RR waves.

Characteristic	Nomenclature	Magnitude
Height	H (m)	1.14
Length	λ (m)	31.50
Period	T (s)	4.50
Depth	h (m)	13.29

According to Chakrabarti [69], the regular waves of Table 2 are defined by the 2nd order Stokes theory. Therefore, the FS elevations (η) caused by the RR waves addressed in the present study are analytically described as [70]:

$$\eta_2 = \frac{H}{2}cos(kx - \omega t) + \frac{H2kcosh(kh)}{16\,sinh^3(kh)}[2 + cosh\,(2kh)]cos[2(kx - \omega t)], \tag{14}$$

while the associated velocity potential (Φ) is given by:

$$\Phi = -\frac{H}{2}\frac{g}{\omega}\frac{cosh(kh + kz)}{coshkh}sin(kx - \omega t) - \frac{3H^2\omega cosh[2k\,(h + z)]}{32\,sinh^4(kh)}\sin[2\,(kx - \omega t)]. \tag{15}$$

Furthermore, calculating the partial derivatives of Φ in the x and z directions, there are the regular wave propagation velocities, which are imposed as BC of velocity inlet in the numerical wave channel for the generation of RR waves as in [7,8,27]. Thus, these horizontal (u) and vertical (w) propagation velocities are given, respectively, by [70]:

$$u_2 = \frac{H}{2}gk\frac{cosh(kz + kh)}{\omega\,cosh(kh)}cos(kx - \omega t) + \frac{H^2}{4}\omega k\frac{cosh[2k\,(h + z)]}{sin^4(kh)}cos[2\,(kx - \omega t)], \tag{16}$$

$$w_2 = \frac{H}{2}gk\frac{sinh(kz + kh)}{\omega\,sinh(kh)}sin(kx - \omega t) + \frac{H^2}{4}\omega k\frac{sinh[2k\,(h + z)]}{cos^4(kh)}sin[2\,(kx - \omega t)]. \tag{17}$$

2.3. Model Verification

To verify the numerical models employed for the generation and propagation of the RI waves and the RR waves inside a numerical wave channel, the FS elevation was monitored using probes at: $x = 0$ m, in the wave generation zone, for the RI waves; and $x = 29.37$ m, approximately one wavelength away from the beginning of the channel, for the RR waves. Then, the results obtained, i.e., the FS elevation data from each case, were compared with data of FS elevation coming from the TOMAWAC model in the case of RI waves, or with the results of the analytical solution (Equation (14)) in the case of RR waves. As for the quantitative analysis, two metrics were considered, the mean absolute error (MAE) and root mean square error (RMSE), which are given, respectively, as follows [71]:

$$\text{MAE} = \frac{\sum_{i=1}^{n}|O_i - P_i|}{n}, \tag{18}$$

$$\text{RMSE} = \sqrt{\frac{\sum_{i=1}^{n}(O_i - P_i)^2}{n}}, \tag{19}$$

where O_i represents the numerical value (m), obtained in the ANSYS-Fluent software; P_i is the reference value (m), which comes from TOMAWAC or Equation (14); and n represents the total amount of data.

It is important to highlight that for RI waves, the verification was conducted solely through a comparison at the channel inlet (in $x = 0$ m), which may seem somewhat unusual. However, based on the previous verification and validation results for the WaveMIMO methodology presented in Machado et al. [14], where the comparison for the FS elevation was made at $x = 0$ m, as well as at the end of the wave channel through the spectral density. In the present study, only the comparison at the channel entrance was made; more details on the verification and validation of WaveMIMO can be found in Maciel et al. [15] and in Oleinik et al. [72]. Therefore, one can affirm that if the velocity profile was correctly imposed at the channel inlet, the waves are generated and propagated in an adequate way. Hence, it is possible to state that the numerical procedure was verified.

3. Problem Description

In this section, the studies conducted to achieve the objectives of this paper are described. In this sense, the first study presented is the one referring to the verification of the numerical models used for the generation of RI waves and RR waves (Section 3.1). Subsequently, the study of the geometric evaluation of the overtopping WEC device is presented (Sections 3.2 and 3.3). Figure 2 illustrates the flowchart of this study.

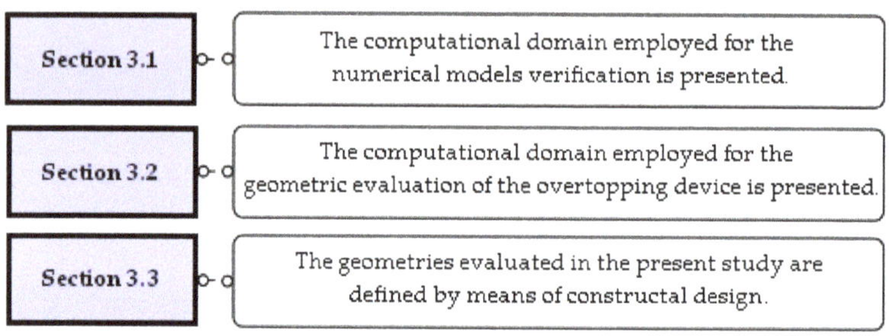

Figure 2. Illustration of the study process described.

3.1. Computational Domain for Model Verification

Two computational domains were considered for the studies carried out in the present paper. The first one, which can be observed in Figure 3, was employed to verify the accuracy of the numerical model employed. Thus, the generation of RI waves and RR waves were simulated in the wave channel without the presence of the overtopping WEC to analyze if the waves are being adequately reproduced. Moreover, Figure 3 highlights the main dimensions of the wave channel and the BC applied to the computational domain, in addition to the water level at rest (WLR).

Regarding the applied BCs, there are: in the lower portion of the left wall (red line), the velocity inlet; in the upper part of the left wall and at the top of the channel (blue dashed line), the pressure outlet as atmospheric pressure; at the bottom of the channel (black line), the non-slip and impermeability; on the right wall, also the pressure outlet, characterizing an open flow channel, where, however, a hydrostatic profile is imposed in order to prevent the wave channel from emptying. Thus, it was possible to employ the numerical beach tool (gray region), which was used to reduce the energy of the waves, avoiding their reflection when they reach the end of the channel.

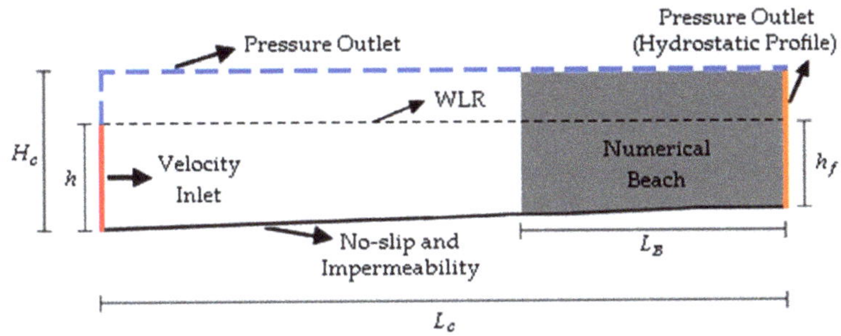

Figure 3. Computational domain employed to verify the numerical model.

Furthermore, it is worth noting that to use the WaveMIMO methodology [14], it is necessary to subdivide the region where the velocity inlet BC is imposed in smaller segments, where the discretized profiles of orbital velocities propagation of waves (u_1 and w_1) are applied. It should be mentioned that the calculation of the velocity vectors takes place in the Spec2Wave software, where their positions are defined. In accordance with the recommendations obtained in Paiva [73], the vectors were considered in the upper part of each segment of the prescribed velocity BC imposition region.

Therefore, in accordance with the recommendations of Paiva [73] for the generation of RI waves through the WaveMIMO methodology, the BC imposition region was subdivided into 15 segments. In this approach the first two segments, those closest to the WLR, have a size of $h/28$, while the remaining 13 segments, which go till the bottom of the channel, are sized $h/14$, as illustrated in Figure 4. It is worth emphasizing that in the present study this methodology is employed solely for the RI waves generation. For the RR waves generation, the BC imposition region consists of a straight segment (see Figure 3).

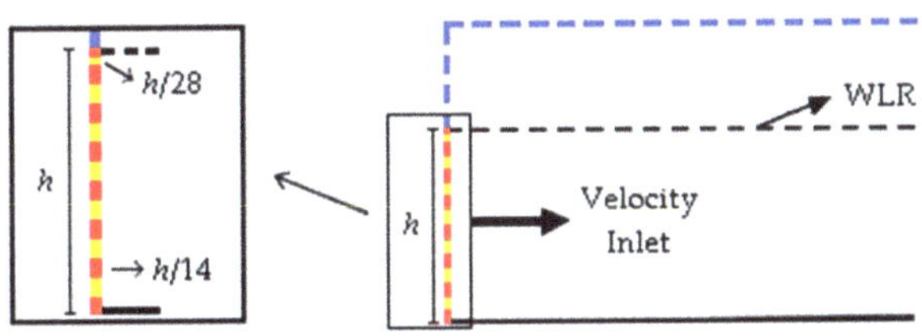

Figure 4. Illustration of the subdivision recommended by Paiva [73].

Regarding the dimensions of the wave channel, the length is $L_c = 171.06$ m, which corresponds to the distance between the location selected for analysis of the realistic sea state and the Molhes da Barra breakwater; while the numerical beach region length was $L_B = 2\lambda$, following the indication by Lisboa et al. [9]. Moreover, the height is $H_c = 21.29$ m, while the depth varies from $h = 13.29$ m to $h_f = 10.54$ m, reproducing the bathymetry found in the Rio Grande coast at the bottom of the wave channel. It should be noted that for this purpose, the bathymetric data were obtained from digitized nautical charts from the Hydrography and Navigation Directorate of the Brazilian Navy [74].

As for the spatial discretization adopted for the first computational domain (see Figure 3), a stretched mesh was used, which has been employed in Machado et al. [14] as well as in studies considering different WECs, such as Gomes et al. [8], for the OWC device, and Hübner et al. [28], for the overtopping device. Therefore, the wave channel was subdivided vertically into three regions, namely: R1, which contains only air, discretized into 20 computational cells; R2, which contains the interface between air and water, discretized into 40 cells; and R3, which contains only water, discretized into 60 cells. Also, horizontally there is the R4 region, discretized into 50 computational cells per λ. In this sense, the regions described are illustrated in Figure 5.

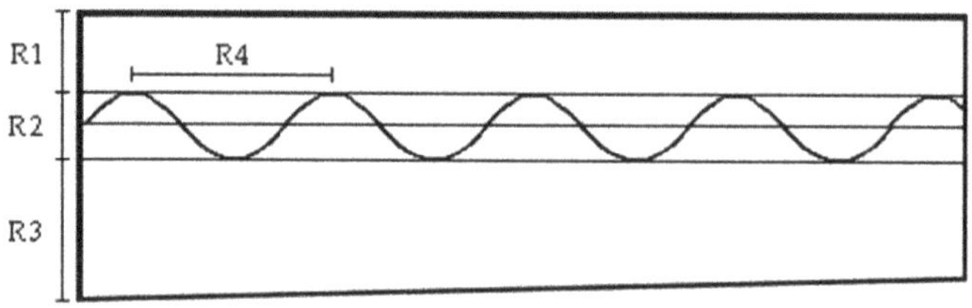

Figure 5. Illustration of the stretched mesh regions.

However, it is worth highlighting that the FS region is discretized differently in the case of RI waves (see Figures 3 and 4). Following the recommendation of Paiva [73], the R2 region (see Figure 5) is composed of 2 segments below the WLR, of size $h/28$, and 2 segments above the WLR, of the same size. In this approach, the segments closest to the WLR, those in the center of the FS region, are discretized into 20 cells, while the others are discretized into 10, totaling 60 computational cells in the region, as illustrated in Figure 6.

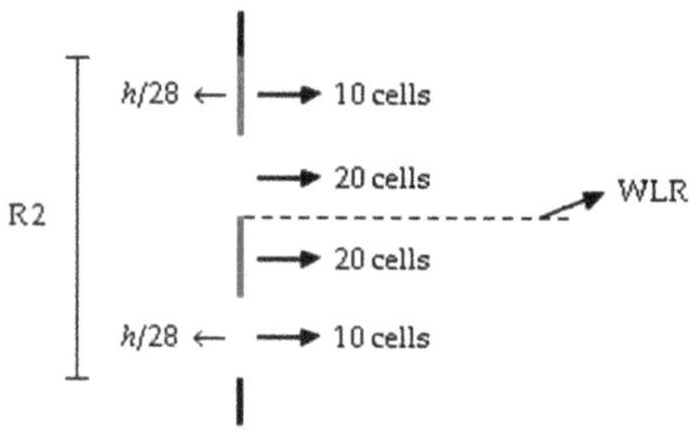

Figure 6. Illustration of the discretization applied to the FS region.

As for the temporal discretization, a time step of Δt = 0.01 s was used. This value corresponds to $\Delta t = T/450$, being in agreement with the indication of Barreiro [75], which recommends Δt between $T/300$ and $T/600$ for regular waves. Thus, the same Δt was applied for both the RI and RR waves, since it is a more refined temporal discretization than needed for the WaveMIMO methodology, which is $\Delta t = 0.05$ s according to Machado et al. [14]. Furthermore, in both studies, a total simulation time of 900 s was considered for the generation and propagation of RI waves and RR waves.

3.2. Computational Domain for Geometric Evaluation of the Overtopping Device

Regarding the second computational domain considered, Figure 7 illustrates the one employed in the geometric evaluation study, where the overtopping device is coupled to the end of the channel. In this case, the numerical beach region is absent, since the purpose of the simulations is that the waves reach the end of the numerical channel without major energy dissipations.

As before, the main dimensions and BCs applied to the wave channel are presented; the geometric characteristics of the WEC, in turn, are presented at a later moment. Moreover, the pressure outlet BC was applied at the bottom of the device's reservoir (green line) in order to keep it empty, as in Koutrouveli et al. [23], and prevent overflow, which occurred in Hübner et al. [29]. In this last study, some of the evaluated cases overflowed before the simulation ended, making it necessary to reduce the total time considered.

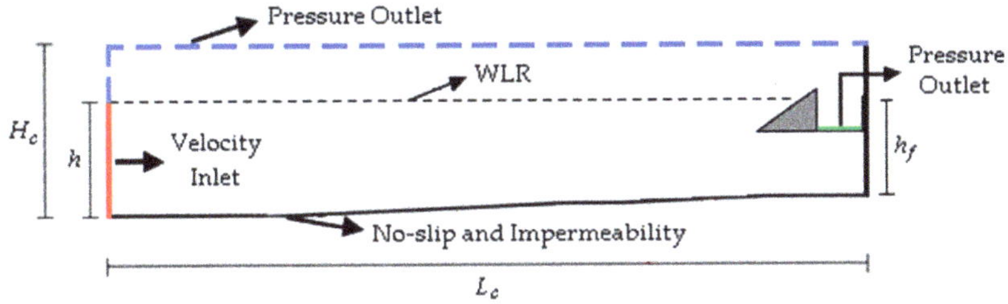

Figure 7. Computational domain employed to the geometric evaluation of the overtopping device.

Regarding the spatial discretization of the domain presented in Figure 7, the same previous configurations were used. However, the region above the ramp and the converter reservoir were discretized horizontally with mesh elements of size $\Delta x = 0.05$ m, while, inside the overtopping device reservoir, a regular mesh with quadrilaterals of size $\Delta x = 0.05$ m was used, as in Martins et al. [25]. Figure 8 presents in detail the mesh adopted for the present study in the device region. As before, regarding the temporal discretization, a time step of $\Delta t = 0.01$ s was considered for the 900 s of simulation of RI waves and RR waves.

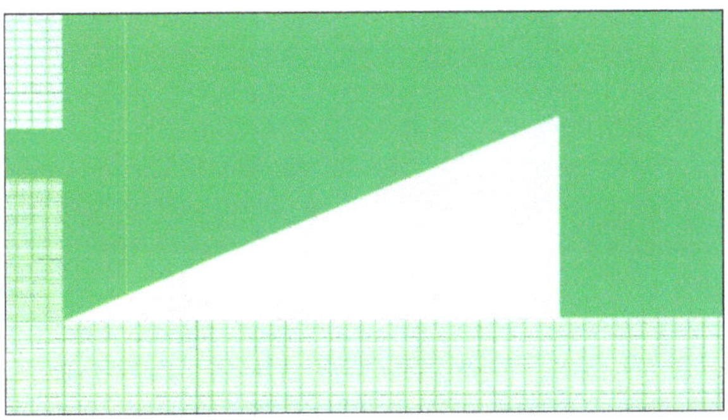

Figure 8. Spatial discretization adopted focusing on the device region.

3.3. Geometric Evaluation of the Overtopping Device Through the Constructal Design

When applying the constructal design method, the geometry is deduced by a principle of maximizing performance while being subjected to constraints, varying parameters according to the defined degrees of freedom [4–6]. Therefore, according to Dos Santos et al. [76], to employ the constructal design in the geometric study of a physical system, it is necessary to define:

- Performance indicator: a quantity to be maximized or minimized, which, in the present study, is the amount of water mass that entered in the converter reservoir;
- Geometric restrictions: the parameters that shall be kept constant, which are here the areas of the wave channel and the device ramp;
- Degrees of freedom: the geometric parameter that will be varied, in this case, is the ratio between height and length (H_1/L_1) of the overtopping device ramp.

It should be mentioned that the total water mass (M) obtained was selected as the performance indicator since it directly correlates with the potential energy generation. According to the physical operating principle of the overtopping device, the water from the incident waves that enters the reservoir subsequently returns to the ocean, thereby activating low-head turbines and generating electricity. Thus, a higher M means a greater amount of energy converted from the sea waves.

As for the overtopping device, the object of study in the present paper, Figure 9 presents its configuration, highlighting the geometric characteristics. Furthermore, the monitoring probe used to analyze the performance of the WEC is also represented (pink line), which remains aligned with the WLR in all cases evaluated, as in Barros et al. [27].

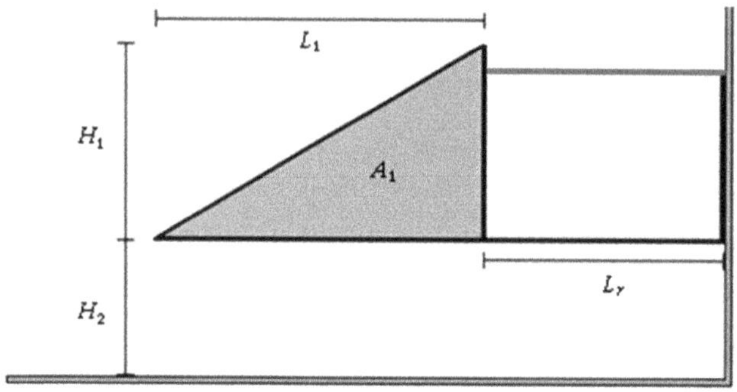

Figure 9. Geometry of the overtopping WEC device.

Regarding the dimensions adopted, the WEC studied by Barros et al. [27] was taken as a reference, which has a reservoir height $H_r = 6.50$ m and length $L_r = 20.00$ m, characteristics that are kept fixed. Concerning the submersion (H_2) of the device, it was established at $H_2 = 4.00$ m and kept fixed, being 0.50 m greater than that adopted in Barros et al. [27], due to the variation of the same magnitude in the depth of the location where the converter is considered. This geometry was used as a basis because it underwent a geometric evaluation [27], where it was subjected to regular waves with different characteristics from the RR waves employed in the present study.

As mentioned, the degree of freedom investigated was the ratio between the height and length of the device ramp, i.e., H_1/L_1. Therefore, two geometric area restrictions were considered. The channel area (A_c), defined as:

$$A_c = H_c L_c, \tag{20}$$

which, in the present study, is $A_c = 2736.96$ m^2; and the device ramp area (A_1), given by:

$$A_1 = \frac{H_1 L_1}{2}, \tag{21}$$

which, in the present study, is $A_1 = 78.4455$ m^2, as in Barros et al. [27]. Then, it was possible to define H_1/L_1 by dividing both sides of Equation (20) by ${L_1}^2$ and isolating the degree of freedom, so it is obtained:

$$\left(\frac{H_1}{L_1}\right) = \frac{2A_1}{{L_1}^2}, \tag{22}$$

allowing to determine the values of H_1 and L_1 for each ratio of H_1/L_1 considered. Also, a dimensionless area fraction can be defined by:

$$\phi = \frac{A_1}{A_c}, \tag{23}$$

being in the present study $\phi = 0.03$.

Additionally, restrictions were defined regarding the height of the device's ramp, as illustrated in Figure 10. It was established that the part of the ramp above the WLR should be lower than H_{max}, as in Figure 10a, and higher than $H_s/4$, as in Figure 10b; being

H_{max} the module of the greatest crest or trough found in the series of FS elevations from TOMAWAC, which in the present study corresponds to $H_{max} = 1.64$ m.

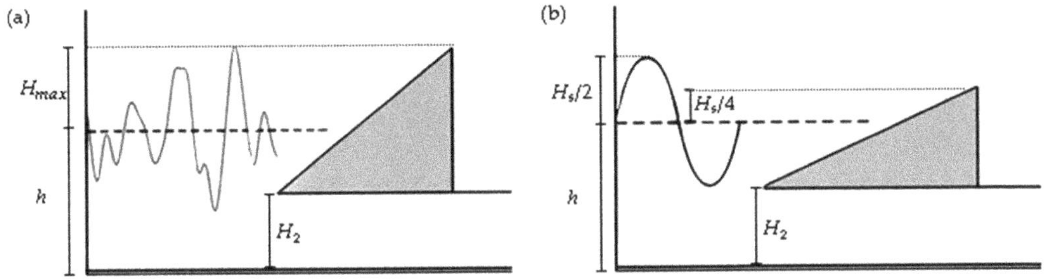

Figure 10. Illustration of the ramp height parameterization, with the height being: (**a**) maximum; (**b**) minimum.

Thus, the maximum and minimum height of the overtopping device ramp were defined as follows:

$$(H_1)_{max} < h_f + H_{max} - H_2, \tag{24}$$

$$(H_1)_{min} > h_f + \frac{H_s}{4} - H_2. \tag{25}$$

Once the upper and lower limits of the height of the overtopping device ramp were defined, 13 ratios were established for the degree of freedom H_1/L_1, ranging from 0.30 to 0.42, which are presented in Table 3, along with the dimensions adopted for H_1 and L_1 in each geometry. It is important to emphasize that a ratio higher than 0.42 or lower than 0.30 would infringe the maximum and minimum ramp height restrictions. Since the submersion of the device was kept fixed, lower values for H_1/L_1 would leave the device completely submerged, while higher values would turn the device into a breakwater. In both cases, the device would cease to function as a WEC.

Table 3. Values considered for the degree of freedom H_1/L_1 investigated.

H_1/L_1	H_1 (m)	L_1 (m)
0.30	6.8606	22.8685
0.31	6.9740	22.4967
0.32	7.0856	22.1424
0.33	7.1954	21.8043
0.34	7.3036	21.4813
0.35	7.4103	21.1726
0.36	7.5154	20.8760
0.37	7.6190	20.5920
0.38	7.7213	20.3192
0.39	7.8222	20.0570
0.40	7.9219	19.8047
0.41	8.0203	19.5617
0.42	8.1175	19.3274

It should be mentioned that the importance of evaluating the degree of freedom H_1/L_1 lies in the fact that how waves break on a structure, such as an overtopping device, depends on the wave conditions and the slope of the ramp. Thus, Figure 11 illustrates the effect of varying the degree of freedom investigated on the design of the device geometry, where it is possible to observe changes in the slope of the WEC ramp when considering both extremes and an intermediate geometry.

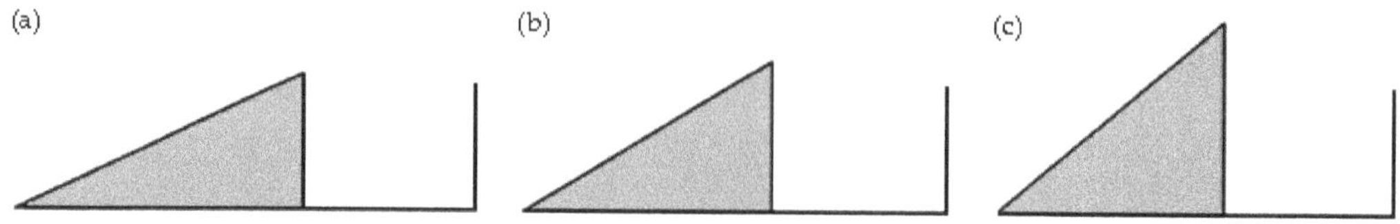

Figure 11. Geometric configurations of the overtopping WEC considering: (**a**) $H_1/L_1 = 0.30$; (**b**) $H_1/L_1 = 0.36$; and (**c**) $H_1/L_1 = 0.42$.

Thereby, with variations in the degree of freedom H_1/L_1, it was possible to analyze the influence of the device geometry over the performance of the converter subject to RI and RR waves occurring in the municipality of Rio Grande, in addition to determining the best geometric configuration. As mentioned, the performance of the overtopping WEC geometries is evaluated by the amount of water mass that entered the device's reservoir. Therefore, a mass flow monitoring probe was used at the reservoir inlet (see the pink line in Figure 9), which made it possible to calculate the total water mass, given as follows:

$$M = \frac{1}{t_f} \int_0^{t_f} \dot{M} dt, \tag{26}$$

where, M is the total mass of water that entered in the reservoir (kg); $\dot{M}$ is the instantaneous mass flow rate (kg/s); and t_f is the time interval (s) between the first overtopping registered and the end of the simulation.

4. Results and Discussions

In this section, the results of the conducted studies are presented and discussed. To do so, it is divided into four sections. The first results presented refer to the verification of the numerical models employed (Section 4.1), followed by the geometric evaluation of the overtopping device (Section 4.2). Then, there is a more detailed analysis of the results monitored for selected geometries (Section 4.3). Finally, a graphical representation of the physical phenomenon is shown (Section 4.4).

4.1. Wave Generation Models Verification

As mentioned, the initial results presented are from studies verifying the numerical models used for generating and propagating RI waves and RR waves, which were simulated in channels without the presence of the overtopping WEC device. In this context, Figure 12a shows the qualitative comparison between the irregular FS elevation monitored in ANSYS-Fluent and the irregular FS elevation data from the TOMAWAC spectral model over the 900 s simulated, while Figure 12b focuses on the first 300 s of the simulation in order to provide a clearer visualization of the results.

As might be seen in Figure 12, the RI waves generated through the WaveMIMO methodology adequately reproduced the sea state found off Rio Grande. However, one can note that the waves generated in the ANSYS-Fluent software do not reach some of the crests and troughs present in the FS elevation from the TOMAWAC model, a phenomenon also observed in [14,73]. Furthermore, the effects of the initial inertia of the fluids were not observed in the generation of RI waves due to the fact that the position of the monitoring probe is in the wave generation zone, i.e., at $x = 0$ m. As for the quantitative evaluation of the results, metrics MAE = 0.101 m and RMSE = 0.131 m were obtained, indicating good results when compared to those present in the literature.

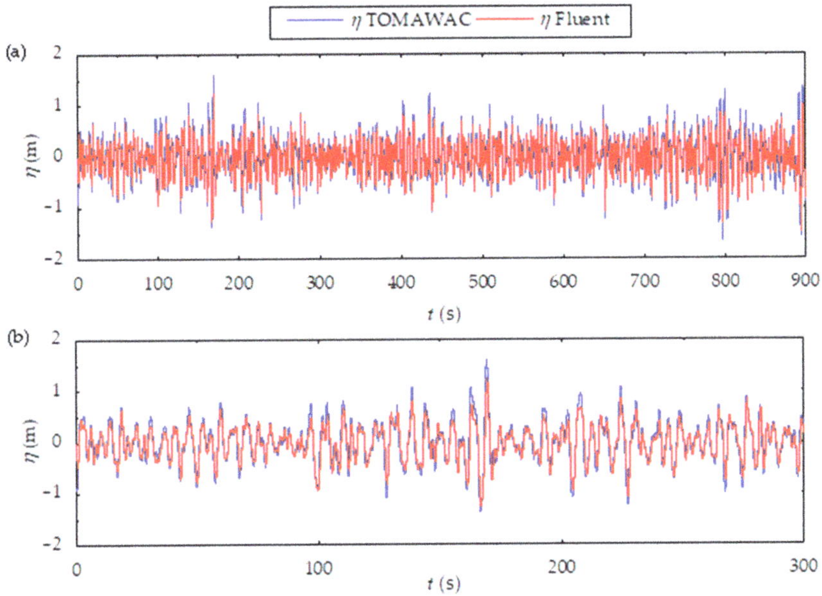

Figure 12. Qualitative comparison of the results obtained for the RI waves considering: (**a**) the 900 s of simulation; (**b**) the first 300 s of simulation.

Continuing the verification of the numerical models, regarding the RR waves, Figure 13a presents the qualitative comparison between the FS elevation monitored in ANSYS-Fluent and the results obtained analytically (Equation (14)) over the 900 s of simulation; as done before, the first 300 s of the simulation are highlighted in Figure 13b to provide a better visualization of the results found.

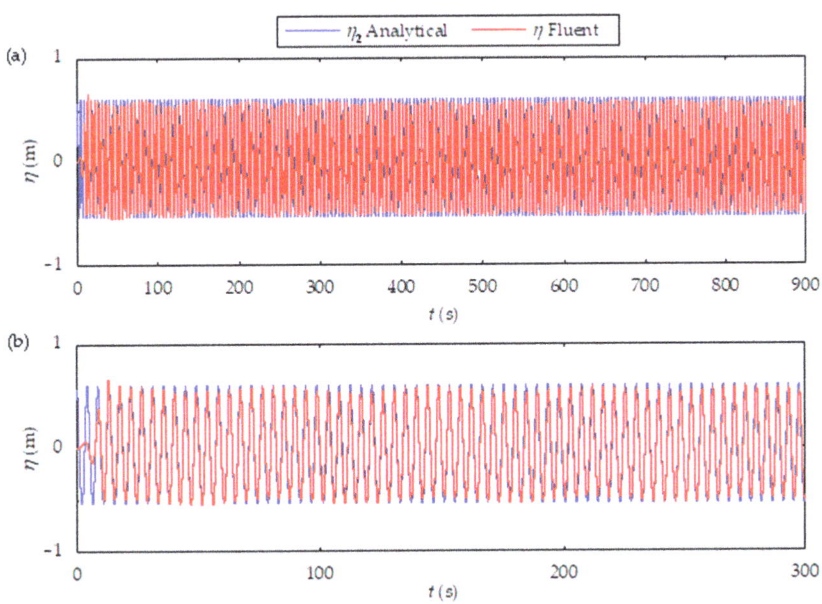

Figure 13. Qualitative comparison of the results obtained for the RR waves considering: (**a**) the 900 s of simulation; (**b**) the first 300 s of simulation.

Analogously to the previous case, it is possible to observe in Figure 13 that the generation of RR waves of the sea state that occurred in Rio Grande accurately reproduces the analytical waves; however, a difference is noted in the first 10 s of simulation. This is due to the fact that the FS elevation was monitored at $x = 29.37$ m and, due to the initial condition of fluid inertia, the flow starts from rest, which causes damping in the first waves generated in the channel. Thus, in order to verify this numerical model, the results monitored between $10 \leq t \leq 900$ s were considered, achieving metrics of MAE $= 0.062$ m and RMSE $= 0.085$ m, indicating, once again, good quantitative results when compared to those present in the literature.

After that, for the visualization of the physical phenomenon reproduced numerically, in Figure 14 the volumetric fraction field is presented, where the water phase is represented in blue, while the air phase is in red. In this sense, Figure 14a shows the initial instant at $t = 0$ s where the fluids are at rest, i.e., there are no FS elevations. This fluid-dynamic behavior occurs due to the initial condition of inertia applied for both wave approaches. Furthermore, the final moments of generation and propagation of the waves, at $t = 900$ s, are depicted in Figures 14b and 14c, respectively, for the RI and RR waves.

Figure 14. Representation of the physical phenomenon at: (**a**) $t = 0$ s, for both cases; (**b**) $t = 900$ s, for RI waves; (**c**) $t = 900$ s, for RR waves.

Therefore, it is possible to observe in Figure 14 that at the end of each simulation, at $t = 900$ s, the difference between the wave approaches addressed is evident. In Figure 14b there is irregular behavior in the FS elevations, reproducing the realistic sea state that occurred in Rio Grande in 2018. However, in Figure 14c a regular oscillatory behavior is observed, which, in turn, represents the same sea state through the combination of the most frequent H_s and T_m characteristics. Moreover, it can be seen in Figure 14b,c the operation of the numerical beach tool at the end of the numerical channel, where there are no FS elevations, since the waves reaching this region are damped.

Thereby, based on the results of MAE and RMSE, as well as from Figures 12–14, one can infer that the computational models for the generation of RI and RR waves were properly verified. Regarding the computational modeling of the overtopping device operation, it is important to emphasize that the numerical model used in the present study is based on the model validated by Goulart et al. [28], where a geometric evaluation was carried out through experimental and numerical studies. To do so, a numerical wave channel, with the device inserted in it, was simulated considering regular waves with the same parameters as laboratory experiments. The numerical model proposed was validated regarding the wave propagation, instantaneous mass flow rates, and the effect of H_1/L_1 on the device performance.

4.2. Geometric Evaluation of the Overtopping Wave Energy Converter Device

Once the numerical models employed were verified, it was possible to carry out the geometric investigation of the overtopping WEC. First of all, it is worth highlighting that the constructal design method is a method for geometric evaluation. In this study, it allowed to analyze the influence of an overtopping device's geometry on its performance. Thus, a qualitative analysis of the results was performed to evaluate their tendencies.

In this context, Figure 15 shows the influence of the degree of freedom H_1/L_1 on the total mass of water (M) entering the device's reservoir when subjected to RI waves and RR waves from the sea state occurring near the Molhes da Barra breakwater in Rio Grande—RS. Highlighting that as this is a two-dimensional domain, the third dimension is unitary, i.e., it was considered 1 m in the y direction.

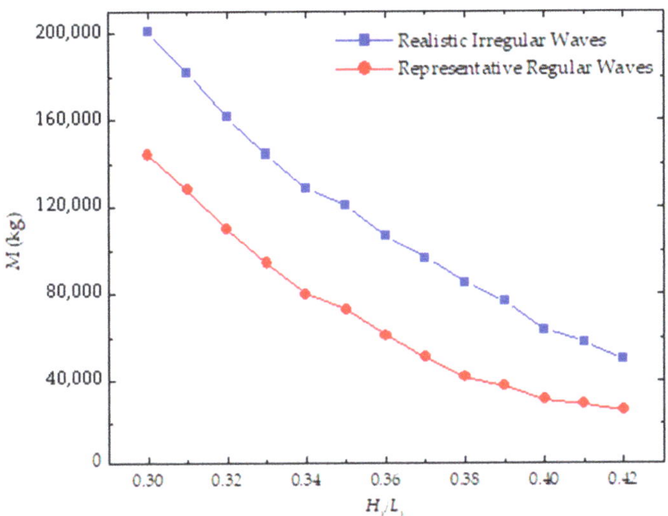

Figure 15. Qualitative comparison of the H_1/L_1 influence on the total water mass (M) that entered in the device's reservoir.

As can be seen in Figure 15, regarding the geometric evaluation of the overtopping WEC device, there is an inversely proportional behavior between the total water mass and the values considered for the degree of freedom. This trend is explained since a greater ratio for H_1/L_1 means that a greater portion of the ramp is above the WLR, making it more difficult for the waves to overtop it. This kind of behavior has been previously seen in numerical studies, such as [25,27,41] as well as in [28], which is a numerical and experimental study. However, it remains valid to investigate the ratio H_1/L_1 when considering a different wave climate, as the amount of water entering the device's reservoir is associated with the way the wave breaks on the WEC ramp. This, in turn, depends on both the wave characteristics and the structure's geometry. For instance, in da Silva [51], two geometric evaluations of the overtopping device were carried out through the constructal design. In each evaluation, a different wave climate and device geometries were considered; in both cases, the presence of local minima was found for the lowest values of H_1/L_1.

In other words, the curve that illustrates the influence of the degree of freedom H_1/L_1 on M has the tendency of a descending function. Thus, for both the RI waves and RR waves, the best geometry is that with $(H_1/L_1)_o = 0.30$, while the worst one has $H_1/L_1 = 0.42$. Moreover, when observing the curves illustrated in Figure 15, it is noted that they are, essentially, parallel. Other than showing a decreasing trend, for instance, it is possible to observe a change in their inclinations recorded for the same geometry, being the one with $H_1/L_1 = 0.35$. The parallel behavior, however, does not occur for the higher values of H_1/L_1 (from 0.39), when the device performance shows more stability under the incidence of RR waves than for RI waves.

Additionally, as noted in Figure 15, all geometries evaluated obtained higher M when subjected to RI waves than to RR waves. However, it is important to mention that this behavior is the opposite of what was found in Hübner et al. [29], where a single geometry of the overtopping device was subjected to different wave climates found in three coastal cities of the RS state. To better understand this phenomenon, the instantaneous mass flow rates of the water entering the device's reservoir are analyzed later, for selected geometries. Despite the differences in the magnitude of M obtained for each wave approach, the curves approximate for geometries with the higher values investigated, i.e., $0.40 \leq H_1/L_1 \leq 0.42$. As previously mentioned, the higher the H_1/L_1 ratio is, the greater the portion of the ramp above the WLR, making wave overtopping more difficult, which explains the tendency of proximity between the results obtained with the RI and RR waves approaches.

To conduct a quantitative evaluation of these results, Table 4 presents the M values achieved for each geometry of the overtopping device over the 900 s of generation and propagation of RI waves and RR waves that occurred in Rio Grande—RS. In this context, Table 4 also presents relative differences (D_R) found between the results obtained, which indicate the variation of M between the geometries evaluated. Therefore, the performance of each geometry is always compared with that located on the line immediately above in the table. This quantification elucidates the effect of the investigated degree of freedom on the converter's performance. Furthermore, the difference regarding the wave approaches (D_W) is also found in Table 4, where the values of M obtained in cases with the same H_1/L_1 are compared, taking the RI waves approach as the reference.

Table 4. Quantitative comparison of the influence of H_1/L_1 on the M that entered in the overtopping device's reservoir.

H_1/L_1	M—RI Waves (kg)	D_R (%)	M—RR Waves (kg)	D_R (%)	D_W (%)
0.30	200,820.77	-	144,054.72	-	−28.26
0.31	182,093.85	−9.32	128,128.68	−11.05	−29.63
0.32	161,409.70	−11.35	109,880.09	−14.24	−31.92
0.33	144,289.64	−10.60	94,068.96	−14.38	−34.80
0.34	128,352.21	−11.04	79,899.03	−15.06	−37.75
0.35	120,552.95	−6.07	72,727.34	−8.97	−39.67
0.36	106,433.13	−11.71	60,828.26	−16.36	−42.84
0.37	96,307.13	−9.51	50,905.02	−16.31	−47.14
0.38	85,269.83	−11.46	41,566.58	−18.34	−51.25
0.39	76,661.55	−10.09	37,428.83	−9.95	−51.18
0.40	63,339.88	−17.38	30,986.83	−17.21	−51.07
0.41	57,760.40	−8.80	29,009.56	−6.38	−49.77
0.42	49,925.58	−13.56	26,215.31	−9.63	−47.49

Therefore, observing the variations of D_R in Table 4, it is possible to identify the tendencies previously observed for the curves that describe the influence of the device geometry on M (see Figure 15). It is noted that, when considering the RI waves, the variations in the total water mass that entered the device reservoir remain around 11% for most of the geometries evaluated. An exception occurs for $H_1/L_1 = 0.35$, where there is a smaller drop, corresponding to the change observed in the slope of the blue curve in Figure 15. Other examples occur for the geometries with $H_1/L_1 = 0.40$ and $H_1/L_1 = 0.42$, where there are sharper drops in the values obtained for M.

Regarding the incidence of RR waves, the variations are more unstable and generally higher than those obtained for the RI waves; however, it is possible to say that they were around 15% for most geometries. In this case, the exceptions refer to lower variations than the rest, such as for the geometry with $H_1/L_1 = 0.35$, where a change in the slope of the red curve illustrated in Figure 15 was previously highlighted. Additionally, low variations were recorded for some geometries with higher H_1/L_1 ratios, where the device showed stability in its performance, such as for $H_1/L_1 = 0.39$, $H_1/L_1 = 0.41$, and $H_1/L_1 = 0.42$.

In turn, observing the variations in D_W in Table 4, one can note that the differences caused by the wave approach are accentuated as the height of the ramp and its portion above the WLR increase. This occurs until it becomes more difficult for the waves to overtop the ramp and enter the device reservoir. Thus, D_W varies between approximately 28% for $H_1/L_1 = 0.30$ and 51% for $H_1/L_1 = 0.38$. Then, from $H_1/L_1 = 0.39$, it reverses the growth trend and gradually decreases to 47% at $H_1/L_1 = 0.42$, which is when the curves approximate in Figure 15. Despite the proximity illustrated in Figure 15 for the higher values of the degree of freedom ($0.40 \leq H_1/L_1 \leq 0.42$), the quantitative evaluation shows that the difference in the overtopping device performance caused by waves approached is relatively smaller considering the lowest ratios ($0.30 \leq H_1/L_1 \leq 0.32$). This highlights the importance of performing both qualitative and quantitative analyses when applying the constructal design method.

Therefore, the constructal design method allowed finding the best geometry for the device, that is, the one that maximizes the M that entered in the overtopping WEC reservoir. Accordingly, it is highlighted that, for both wave approaches, the best geometry was the one with $(H_1/L_1)_o = 0.30$ ($H_1 = 6.8606$ m and $L_1 = 22.8685$ m); while, the worst geometry was the one with $H_1/L_1 = 0.42$. Thus, considering the incidence of RI waves, $M = 200,820.77$ kg entered the converter's reservoir. This amount of water is 4.02 times superior to the total obtained for the worst geometry evaluated. On the other hand, when RR waves are considered, $M = 144,054.72$ kg was obtained in the device's reservoir, showing an even greater difference, corresponding to an amount 5.49 times of what was obtained by the worst geometry evaluated.

4.3. Detailed Analysis of the Monitored Results

In the present section, a detailed analysis of the results monitored by the probe at the entrance of the device's reservoir is carried out. Thus, the overtopping device's performance was analyzed considering the best geometry, $(H_1/L_1)_o = 0.30$; a geometry with intermediate performance, $H_1/L_1 = 0.35$; and the worst-performing geometry, $H_1/L_1 = 0.42$ (see Figure 11). In this context, Figure 16 shows the instantaneous mass flow rates ($\dot{M}$) entering the device's reservoir, considering (a) RI waves and (b) RR waves.

The first noticeable aspect in Figure 16 is the difference in the scale of the $\dot{M}$ peaks, which are higher considering the incidence of RI waves. Additionally, the frequency of the flows also differs; in this case, it is higher with the incidence of RR waves. Thus, it can be observed that the hydrodynamic behavior of the overtopping device varies depending on the wave approach. There are flows of variable magnitude, with higher peaks but less frequent occurrences for RI waves. In contrast, there are lower magnitude flows, practically constant, occurring more frequently with RR waves. It can be partially explained by observing Figures 12 and 13, in Section 4.2. As seen in Figure 12, there are moments where the FS elevation reached for the RI waves exceeds what is achieved for the RR waves, shown in Figure 13; these waves cause the highest $\dot{M}$ peaks. Furthermore, there are also times when the elevations shown in Figure 12 (RI waves) are lower than those in Figure 13 (RR waves); these waves are not able to overtop the ramp, thus causing the difference in flow frequency.

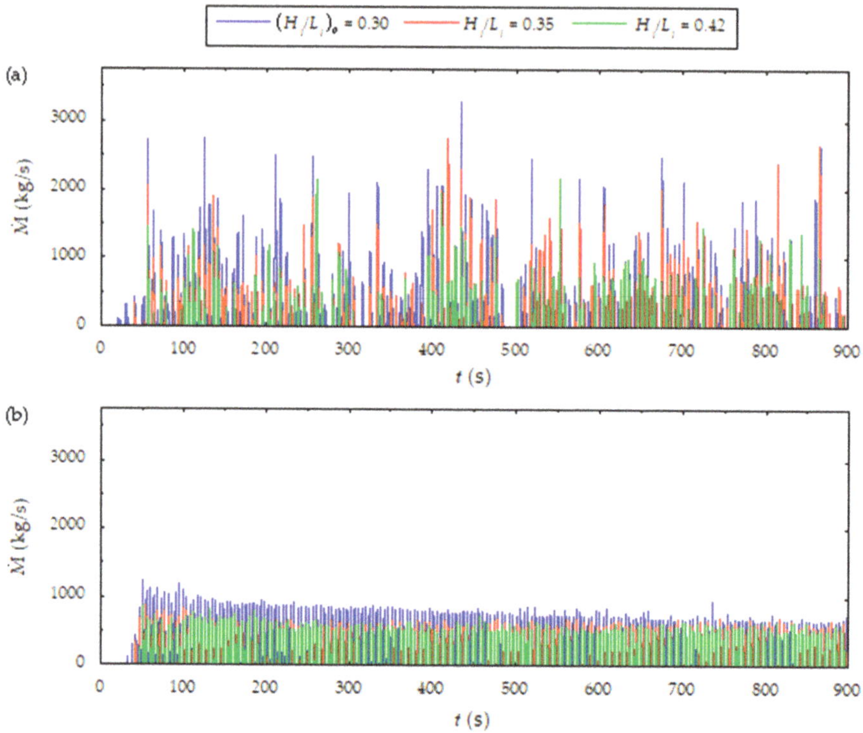

Figure 16. Instantaneous mass flow rates ($\dot{M}$) monitored throughout the 900 s of simulation of: (**a**) RI waves; and (**b**) RR waves.

Another point that should be highlighted is that when considering RR waves, the peaks of mass flow rates decreased over time for geometries with lower H_1/L_1 ratios, as seen for $(H_1/L_1)_o = 0.30$ in Figure 16b; while they maintained a more constant behavior for geometries with higher H_1/L_1 ratios, as seen for $H_1/L_1 = 0.42$ in Figure 16b. Thus, one can infer that those geometries with lower ramps suffered more from the effect of wave reflection. Since a greater amount of water mass was frequently able to overtop the ramp, the subsequent waves faced reflection. On the other hand, this behavior does not occur when considering the incidence of RI waves, even for geometries with lower H_1/L_1 ratios, as observed in Figure 16a. Since water mass overtopping is less frequent due to waves that are unable to overtop the ramp, the subsequent waves do not face the same effects of reflection. Thus, large peaks of water mass entering the device's reservoir can be observed throughout the simulation.

In a quantitative analysis of the results, considering the RI waves (Figure 16a), the first overtopping, or water mass flow rates, occurred close to: $t = 20$ s for $(H_1/L_1)_o = 0.30$; $t = 40$ s for $H_1/L_1 = 0.35$; and $t = 55$ s for $H_1/L_1 = 0.42$. Besides that, the largest mass flow peaks occurred in different instants for the three geometries considered, which correspond to: $\dot{M} = 3295.17$ kg/s at $t = 434.34$ s, for $(H_1/L_1)_o = 0.30$; $\dot{M} = 2748.39$ kg/s at $t = 418.80$ s, for $H_1/L_1 = 0.35$; and $\dot{M} = 2174.47$ kg/s at $t = 261.14$ s, for $H_1/L_1 = 0.42$. Thus, it was noted that for $(H_1/L_1)_o = 0.30$ and $H_1/L_1 = 0.35$ these peaks occurred close to 50% of the total simulation time, while for $H_1/L_1 = 0.42$ it was close to 30%. Furthermore, the difference in the magnitude of the peaks was practically the same when comparing one geometry with the other, being around $\dot{M} = 550$ kg/s.

Regarding the incidence of RR waves (Figure 16b), the first overtopping also occurs first for the best geometry, $(H_1/L_1)_o = 0.30$. However, the range for the other geometries is smaller, it started close to: $t = 36$ s for $(H_1/L_1)_o = 0.30$; $t = 41$ s for $H_1/L_1 = 0.35$; and $t = 50$ s for $H_1/L_1 = 0.42$. Moreover, the largest mass flow peaks were monitored close to $t = 50$ s for the three geometries covered in this analysis. These peaks correspond to: $\dot{M} = 1258.65$ kg/s at $t = 50.02$ s, for $(H_1/L_1)_o = 0.30$; $\dot{M} = 934.81$ kg/s at $t = 54.25$ s, for $H_1/L_1 = 0.35$; and $\dot{M} = 948.83$ kg/s at $t = 50.01$ s, for $H_1/L_1 = 0.42$. Thus, it is important to mention that the worst geometry, $H_1/L_1 = 0.42$, recorded a higher $\dot{M}$ peak than the intermediate geometry, $H_1/L_1 = 0.35$.

Continuing with the analysis of the overtopping WEC performance, Figure 17 depicts the evolution of the amount of water mass (M) that entered in the overtopping device's reservoir over the 900 s simulation for both wave approaches considered. The same geometries were considered for this analysis. Moreover, the fitted curves for the growth of M (dotted black lines) are also included in Figure 17.

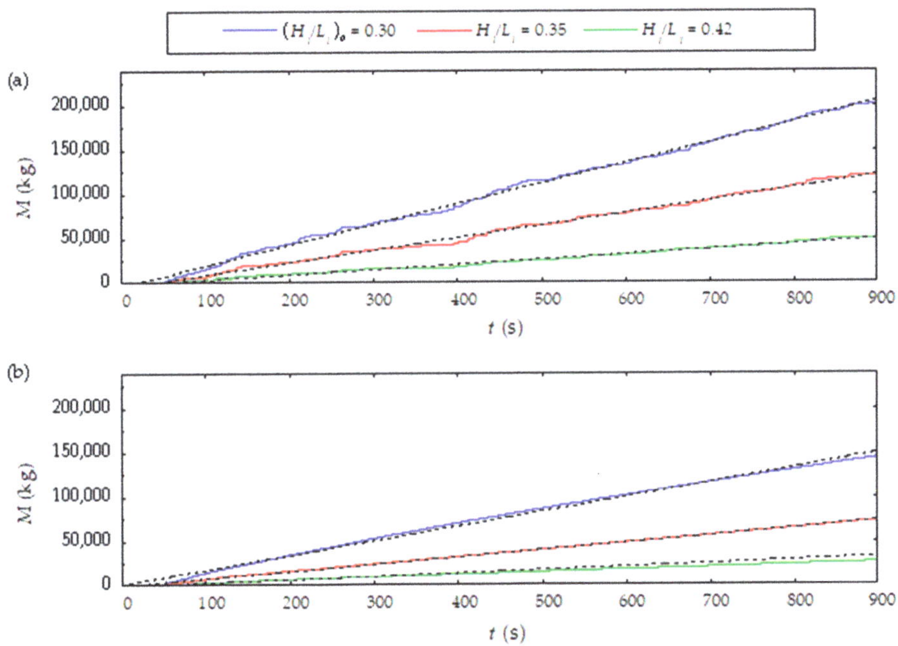

Figure 17. Evolution of the amount of water mass (M) monitored throughout the 900 s of simulation of: (**a**) RI waves; and (**b**) RR waves.

As expected, the curves illustrated in Figure 17 differ depending on the wave approach used. For the RI waves (Figure 17a), the curves resemble steps, with plateau intervals where M remains constant between growth intervals. For instance, some examples of intervals where it occurs are: 485.89 s $\leq t \leq$ 505.29 s, for $(H_1/L_1)_o = 0.30$; 377.90 s $\leq t \leq$ 387.31 s, for $H_1/L_1 = 0.35$; and 162.53 s $\leq t \leq$ 186.62 s, for $H_1/L_1 = 0.42$. On the other hand, the amount of M increases steadily over time when considering the RR waves (Figure 17b). This is a direct consequence of what was observed in Figure 16,b, where it was noted that the frequency of wave overtopping depends on the wave approach. For the same reason, this behavior was observed in Hübner et al. [30]. Additionally, it is possible to observe that RI waves led to better results, as previously presented (see Figure 15 and Table 4 for full results). Thus, it is safe to state that the higher frequency of overtopping considering the incidence of RR waves was not enough to compensate for the higher mass flow rates $(\dot{M})$ obtained when the overtopping device was subjected to RI waves.

To draw the fitted curves shown in Figure 17a,b, a first-degree polynomial was defined, relating the amount of water entering the reservoir with the time elapsed, which characterizes a linear correlation between the variables [77]. Thus, it was determined the correlation coefficient (R^2) for each polynomial. In this context, Table 5 presents the equations for the polynomial approximation of the water mass (M_p), as well as the correlation coefficient for the highlighted geometries.

Table 5. Statistical parameters of fitting curves.

Geometry	Wave Approach	First-Degree Polynomial	R^2
$(H_1/L_1)_o = 0.30$	RI	$M_p = 233.27\,t - 4161.20$	0.9979
	RR	$M_p = 165.44\,t + 708.64$	0.9952
$H_1/L_1 = 0.35$	RI	$M_p = 141.17t - 5629.30$	0.9964
	RR	$M_p = 82.17t - 951.74$	0.9992
$H_1/L_1 = 0.42$	RI	$M_p = 57.96\,t - 2639.10$	0.9925
	RR	$M_p = 29.79\,t - 23.856$	0.9956

The results from Table 5 indicates that the doted black curves presented in Figure 17 are an adequate approximation to the numerical results found in the geometric evaluation. Since $R^2 \geq 0.99$, one can infer that all the fitted curves represent a strong linear relationship between M_p and t [77]. Thus, it allows the use of the polynomial approximation for the water mass that enters the reservoir in future studies considering longer activity times of the overtopping device.

4.4. Visualization of Physical Phenomenon

Finally, Figure 18 shows the behavior of wave flow over the best geometry evaluated for the overtopping device, for both wave approaches, at the start and end of the simulations. As before, the water phase is represented in blue while the air phase is represented in red. Additionally, the overtopping WEC ramp and the bottom of the reservoir are represented in gray.

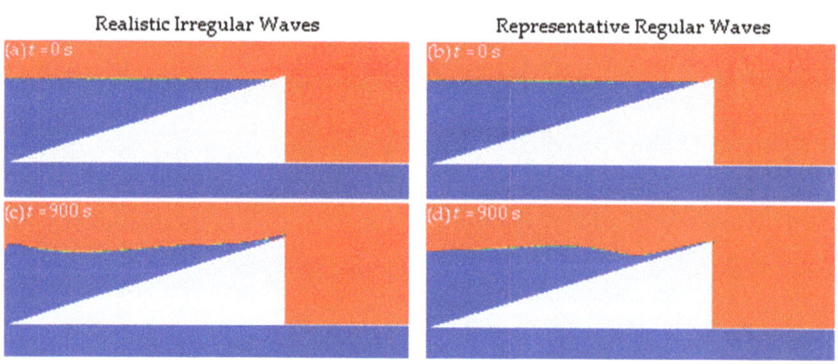

Figure 18. Wave flow behavior over the overtopping device at: (**a**) $t = 0$ s, RI waves; (**b**) $t = 0$ s, RR waves; (**c**) $t = 900$ s, RI waves; (**d**) $t = 900$ s, RR waves.

Thus, there are the volumetric fraction field for $(H_1/L_1)_o = 0.30$ subjected to RI waves and RR waves, respectively, at: $t = 0$ s, in Figure 18a,b; and $t = 900$ s, in Figure 18c,d. It is possible to observe in Figure 18a,b the initial instant of the simulations, where no waves were impinging on the device due to the initial condition of fluids at rest in the channel. On the other hand, Figure 18c shows a wave reaching the overtopping WEC device at the final instant of the simulations, while Figure 18d shows a wave returning down the ramp after its overtopping. As seen in Figure 16b, in Section 4.3, there was a mass flow entering the device's reservoir in the final moment of the simulation of the RR waves.

Additionally, it should be noted that due to the pressure outlet BC imposed at the bottom of the reservoir, there is no accumulation of water. Thus, it is recommended that in future studies, the overtopping device reservoir should be reduced in order to decrease the total processing time of the simulations. Since it is a very refined region of the computational domain, discretized with quadrilateral computational cells of size $\Delta x = 0.05$ m (see Figure 8), reducing this area will result in a significant reduction in the total number of computational cells employed.

5. Conclusions

The present study aimed to investigate the influence of the geometry of the overtopping WEC device on its performance when subjected to RI waves and RR waves from the sea state occurring near the Molhes da Barra breakwater in Rio Grande—RS, southern Brazil, in 2018. Thus, in both studies, numerical simulations were conducted to generate and propagate waves impinging on the overtopping WEC in a channel. To do so, the ANSYS-Fluent software was employed, which is based on the FVM and applies the multiphase VoF model to treat the interface between the air and water.

The numerical models used were verified quantitatively through the MAE and RMSE metrics, and qualitatively through a visual comparison. Both the RI waves and the RR waves generated numerically provided an accurate representation of the physical phenomenon addressed. It was found that the geometries that maximize the overtopping device's performance are the same for both wave approaches. Additionally, it was noted that using RR waves underestimates the device's performance.

Regarding geometric evaluation, constructal design was applied in order to investigate the influence of the degree of freedom H_1/L_1 on M, that is, the influence of the ratio between the height and length of the ramp on the total mass of water that entered the overtopping device reservoir. Thus, it was possible to infer that the best geometry evaluated was $(H_1/L_1)_o = 0.30$, which has $H_1 = 6.8606$ m and $L_1 = 22.8685$ m; while the worst was the one with $H_1/L_1 = 0.42$. That is, it was observed an inversely proportional behavior between the total water mass and the values considered for the degree of freedom, as seen in [25,27,28,41]. As for the total mass of water that entered the reservoir throughout the 900 s of simulation, $M = 200{,}820.77$ kg was obtained for the RI waves, an amount 4.02 times higher than the worst geometry analyzed. $M = 144{,}054.72$ kg was obtained for the RR waves, an amount 5.49 times greater than the worst geometry analyzed.

As noted, all the geometries analyzed had their performance underestimated when subjected to RR waves, that is, all obtained higher M when subjected to RI waves. For instance, considering the optimized geometric configuration of the overtopping WEC, the RI waves achieved a total mass of water in the reservoir approximately 40% higher than the water amount reached by the RR approach. However, as mentioned, this behavior is the opposite of what was found in Hübner et al. [29]. This highlights the importance of geometric evaluation studies for WECs being carried out considering the realistic sea state corresponding to the location where they will be installed.

Finally, regarding future studies, it is suggested to investigate the influence of the device's submersion (H_2) preserving the maximum and minimum height restrictions (H_1) of the ramp that were established based on the characteristics of the realistic sea state addressed. Thus, different values for the area of the converter ramp (A_1) can be evaluated, leading to different values for the area fraction (ϕ). Additionally, it is suggested to carry out analyses considering the sea state in other regions of Rio Grande do Sul, with the aim of identifying whether there is an ideal geometry to be reproduced along its coast.

Author Contributions: Conceptualization, M.d.S.P., A.P.G.M., P.H.O., L.A.I. and B.N.M.; methodology, M.d.S.P., A.P.G.M., P.H.O., L.A.O.R., E.D.d.S., L.A.I. and B.N.M.; software, L.A.O.R., E.D.d.S., L.A.I. and B.N.M.; validation, M.d.S.P., A.P.G.M., P.H.O., L.A.I. and B.N.M.; formal analysis, M.d.S.P., A.P.G.M., L.A.I. and B.N.M.; investigation, M.d.S.P., A.P.G.M., L.A.I. and B.N.M.; resources, L.A.O.R., E.D.d.S., L.A.I. and B.N.M.; data curation, M.d.S.P., A.P.G.M. and P.H.O.; writing—original draft preparation, M.d.S.P. and A.P.G.M.; writing—review and editing, M.d.S.P., A.P.G.M., L.A.I. and B.N.M.; visualization, L.A.O.R. and E.D.d.S.; supervision, L.A.O.R., E.D.d.S., L.A.I. and B.N.M.; project administration, L.A.O.R., E.D.d.S., L.A.I. and B.N.M.; funding acquisition, L.A.O.R., E.D.d.S., L.A.I. and B.N.M. All authors have read and agreed to the published version of the manuscript.

Funding: This research was funded by the Brazilian Coordination for the Improvement of Higher Education Personnel—CAPES (Finance Code 001), Research Support Foundation of the State of Rio Grande do Sul—FAPERGS (Public Call FAPERGS 07/2021—Programa Pesquisador Gaúcho—PqG—21/2551-0002231-0), Brazilian National Council for Scientific and Technological Development—CNPq (Processes: 307791/2019-0, 308396/2021-9, 309648/2021-1, and 403408/2023-7), and by the institutional research assistance program for newly hired professors at the Federal University of Rio Grande do Sul—UFRGS.

Data Availability Statement: The data presented in this study are available on request from the corresponding author. The data are not publicly available due to privacy reasons.

Acknowledgments: M.d.S.P. and A.P.G.M. thank CAPES and CNPq for their master's and doctorate scholarships. L.A.O.R., E.D.d.S. and L.A.I. thank CNPq for research grants. All authors thank FAPERGS and CNPq.

Conflicts of Interest: The authors declare no conflicts of interest. The funders had no role in the design of the study, in the collection, analyses, or interpretation of data, or in the writing of the manuscript, as well as in the decision to publish the results.

Nomenclature

A_c	channel area [m^2]
A_1	overtopping device ramp area [m^2]
C_1	linear damping coefficient [s^{-1}]
C_2	quadratic damping coefficient [m^{-1}]
D_R	relative differences [%]
D_W	difference relative to the waves considered [%]
$\vec{g}$	gravity acceleration vector [m/s^2]
h	depth [m]
H_1	height of the ramp [m]
H_2	device submersion [m]
H_c	wave channel height [m]
H_{max}	maximum height of RI waves [m]
H_r	reservoir height [m]
H_S	significant height of RI waves [m]
h_f	final depth of the wave channel [m]
k	wave number [m^{-1}]
k_x	component x of the wave number vector [m^{-1}]
k_z	component z of the wave number vector [m^{-1}]
L_1	length of the ramp [m]
L_B	numerical beach length [m]

L_c	wave channel length [m]
L_r	reservoir length [m]
M	water mass [kg]
$\dot{M}$	mass flow rate [kg]
M_p	approximate value of M [kg]
n	total number of data [-]
N	directional spectrum of wave action density [m^2/hz/rad]
O_i	numerical value [m]
p	static pressure [Pa]
P_i	reference value [m]
Q	source term [m^2/rad]
R^2	correlation coefficient [-]
S	numerical beach sink term [N/m^2]
T_m	mean period of the RI waves [s]
T	wave period [s]
t	time [s]
t_f	time interval considered [s]
u_1	horizontal velocity profile [m/s]
u_2	horizontal velocity component of 2nd Order Stokes Wave [m/s]
V	velocity along the z direction [m/s]
$\vec{V}$	velocity vector [m/s]
Δ_x	mesh elements [m]
Δ_t	time step [s]
x	horizontal coordinate axis [m]
x_s	starting position of the numerical beach [m]
x_e	ending position of the numerical beach [m]
w_1	vertical velocity profile [m/s]
w_2	vertical velocity component of 2nd Order Stokes Wave [m/s]
z	vertical coordinate axis [m]
z_b	vertical position of the channel bottom [m]
z_{fs}	vertical positions of the FS [m]
α	volume fraction [-]
η_1	FS elevation caused by Airy's Linear Wave [m]
η_2	FS elevation caused by 2nd Order Stokes Wave [m]
λ	wave length [m]
ρ	fluid density [kg/m^3]
$\overline{\overline{\tau}}$	strain rate tensor [n/m^2]
ϕ	dimensionless area fraction [-]
Φ	velocity potential [m^2/s]
ω	angular frequency [Hz]

References

1. Narula, K. *The Mairtime Dimension of Sustainable Energy Security*; Springer Nature: Singapore, 2018.
2. Espindola, R.L.; Araújo, A.M. Wave energy resource of Brazil: An analysis from 35 years of ERA—Interim reanalysis data. *PLoS ONE* **2017**, *12*, e0183501. [CrossRef] [PubMed]
3. Pecher, A.; Kofoed, J.P. *Handbook of Ocean Wave Energy*; Springer Open: Cham, Switzerland, 2017.
4. Bejan, A. *Shape and Structure, from Engineering to Nature*; Cambridge University Press: Cambridge, UK, 2000.
5. Bejan, A.; Lorente, S. Constructal Theory of generation of configuration in nature and engineering. *J. Appl. Phys.* **2006**, *100*, 5. [CrossRef]
6. Bejan, A.; Lorente, S. *Design with Constructal Theory*; Wiley: Hoboken, NJ, USA, 2008.
7. Horko, M. CFD Optimisation of an Oscillating Water Column Energy converter. Master's Thesis, Ocean Engineering, School of Mechanical Engineering, The University of Western Australia, Perth, Australia, 2007.

8. Gomes, M.; Das, N.; Lorenzini, G.; Rocha, L.A.O.; Dos Santos, E.D.; Isoldi, L.A. Constructal Design Applied to the Geometric Evaluation of an Oscillating Water Column Wave Energy Converter Considering Different Real Scale Wave Periods. *J. Eng. Thermophys.* **2018**, *27*, 173–190. [CrossRef]
9. Lisboa, R.C.; Teixeira, P.R.; Didier, E. Regular and Irregular wave propagation analysis in a flume with numerical beach using a Navier-stokes based model. *Defect Diffus. Forum* **2017**, *372*, 81–90. [CrossRef]
10. Zabihi, M.; Mazaheri, S.; Rezaee Mazyak, A. Wave generation in a numerical wave tank. *Int. J. Coast. Offshore Environ. Eng.* **2017**, *2*, 33–43. [CrossRef]
11. Higuera, P.; Lara, J.L.; Losada, I.J. Realistic wave generation and active wave absorption for Navier–Stokes models. *Coastal Eng.* **2013**, *71*, 102–118. [CrossRef]
12. Higuera, P.; Lara, J.L.; Losada, I.J. Simulating coastal engineering processes with OpenFOAM®. *Coastal Eng.* **2013**, *71*, 119–134. [CrossRef]
13. Finnegan, W.; Goggins, J. Linear irregular wave generation in a numerical wave tank. *Appl. Ocean Res.* **2015**, *52*, 188–200. [CrossRef]
14. Machado, B.N.; Oleinik, P.H.; Kirinus, E.P.; Dos Santos, E.D.; Rocha, L.A.O.; Gomes, M.; Das, N.; Conde, J.M.P.; Isoldi, L.A. WaveMIMO Methodology: Numerical Wave Generation of a Realistic Sea State. *J. Appl. Comput. Mech.* **2021**, *1*, 1–20. [CrossRef]
15. Maciel, R.P.; Fragassa, C.; Machado, B.N.; Rocha, L.A.O.; Dos Santos, E.D.; Gomes, M.; Das, N.; Isoldi, L.A. Verification and Validation of a Methodology to Numerically Generate Waves Using Transient Discrete Data as Prescribed Velocity Boundary Condition. *J. Mar. Sci. Eng.* **2021**, *9*, 896. [CrossRef]
16. Tedd, J.; Kofoed, J.P. Measurements of overtopping flow time series on the Wave Dragon, wave energy converter. *Renew. Energy* **2009**, *34*, 711–717. [CrossRef]
17. Parmeggiani, S.; Kofoed, J.P.; Friis-Madsen, E. Experimental Study Related to the Mooring Design for the 1.5 MW Wave Dragon WEC Demonstrator at DanWEC. *Energies* **2013**, *6*, 1863–1886. [CrossRef]
18. Di Lauro, E.; Lara, J.L.; Maza, M.; Losada, I.J.; Contestabile, P.; Vicinanza, D. Stability analysis of a non-conventional breakwater for wave energy conversion. *Coast. Eng.* **2019**, *145*, 36–52. [CrossRef]
19. Palma, G.; Formentin, S.M.; Zanuttigh, B.; Contestabile, P.; Vicinanza, D. Numerical Simulations of the Hydraulic Performance of a Breakwater-Integrated Overtopping Wave Energy Converter. *J. Mar. Sci. Eng.* **2019**, *7*, 38. [CrossRef]
20. Contestabile, P.; Crispino, G.; Russo, S.; Gisonni, C.; Cascetta, F.; Vicinanza, D. Crown Wall Modifications as Response to Wave Overtopping under a Future Sea Level Scenario: An Experimental Parametric Study for an Innovative Composite Seawall. *Appl. Sci.* **2020**, *10*, 2227. [CrossRef]
21. Di Lauro, E.; Maza, M.; Lara, J.L.; Losada, I.J.; Contestabile, P.; Vicinanza, D. Advantages of an innovative vertical breakwater with an overtopping wave energy converter. *Coast. Eng.* **2020**, *159*, 103713. [CrossRef]
22. Musa, M.A.; Roslan, M.F.; Ahmad, M.F.; Muzathik, A.M.; Mustapa, M.A.; Fitriadhy, A.; Mohd, M.H.; Rahman, M.A.A. The Influence of Ramp Shape Parameters on Performance of Overtopping Breakwater for Energy Conversion. *J. Mar. Sci. Eng.* **2020**, *8*, 875. [CrossRef]
23. Koutrouveli, T.I.; Di Lauro, E.; das Neves, L.; Calheiros-Cabral, T.; Rosa-Santos, P.; Taveira-Pinto, F. Proof of concept of a breakwater-integrated hybrid wave energy converter using a composite modelling approach. *J. Mar. Sci. Eng.* **2021**, *9*, 226. [CrossRef]
24. Clemente, D.; Calheiros-Cabral, T.; Rosa-Santos, P.; Taveira-Pinto, F. Hydraulic and Structural Assessment of a Rubble-Mound Breakwater with a Hybrid Wave Energy Converter. *J. Mar. Sci. Eng.* **2021**, *9*, 922. [CrossRef]
25. Martins, J.C.; Fragassa, C.; Goulart, M.M.; dos Santos, E.D.; Isoldi, L.A.; das Neves Gomes, M.; Rocha, L.A.O. Constructal Design of an Overtopping Wave Energy Converter Incorporated in a Breakwater. *J. Mar. Sci. Eng.* **2022**, *10*, 471. [CrossRef]
26. An, S.H.; Kim, G.G.; Lee, J.H. Optimal Design of the Overtopping Wave Energy Converter Based on Fluid–Structure Interaction Simulation. *J. Coast. Res.* **2023**, *116*, 578–582. [CrossRef]
27. Barros, A.S.; Fragassa, C.; Paiva, M.D.S.; Rocha, L.A.O.; Machado, B.N.; Isoldi, L.A.; Gomes, M.D.N.; dos Santos, E.D. Numerical Study and Geometrical Investigation of an Onshore Overtopping Device Wave Energy Converter with a Seabed Coupled Structure. *J. Mar. Sci. Eng.* **2023**, *11*, 412. [CrossRef]
28. Goulart, M.M.; Martins, J.C.; Gomes, A.P.; Puhl, E.; Rocha, L.A.O.; Isoldi, L.A.; Gomes, M.d.N.; dos Santos, E.D. Experimental and numerical analysis of the geometry of a laboratory-scale overtopping wave energy converter using constructal design. *Renew. Energy* **2024**, *236*, 121497. [CrossRef]
29. Hübner, R.G.; Fragassa, C.; Paiva, M.S.; Oleinik, P.H.; Gomes, M.N.; Rocha, L.A.O.; Santos, E.D.; Machado, B.N.; Isoldi, L.A. Numerical Analysis of an Overtopping Wave Energy Converter Subjected to the Incidence of Irregular and Regular Waves from Realistic Sea States. *J. Mar. Sci. Eng.* **2022**, *10*, 1084. [CrossRef]
30. Liu, Z.; Shi, H.; Cui, Y.; Kim, K. Experimental study on overtopping performance of a circular ramp wave energy converter. *Renew. Energy* **2017**, *104*, 163–176. [CrossRef]

31. Maliki, A.Y.; Musa, M.A.; Ahmad, M.F.; Zamri, I.; Omar, Y. Comparison of numerical and experimental results for overtopping discharge of the OBREC wave energy converter. *J. Eng. Sci. Technol.* **2017**, *12*, 1337–1353.
32. Liu, Z.; Han, Z.; Shi, H.; Yang, W. Experimental study on multi-level overtopping wave energy convertor under regular wave conditions. *Int. J. Nav. Archit. Ocean Eng.* **2018**, *10*, 651–659. [CrossRef]
33. Hur, D.S.; Jeong, Y.M.; Lee, J.L.; Kim, I.H.; Lee, W.D. Energy generation efficiency due to wave overtopping on floating-overflow-type wave energy converter. *J. Coast. Res.* **2018**, *85*, 1341–1345. [CrossRef]
34. Iuppa, C.; Cavallaro, L.; Musumeci, R.E.; Vicinanza, D.; Foti, E. Empirical overtopping volume statistics at an OBREC. *Coast. Eng.* **2019**, *152*, 103524. [CrossRef]
35. Rosa-Santos, P.; Taveira-Pinto, F.; Clemente, D.; Cabral, T.; Fiorentin, F.; Belga, F.; Morais, T. Experimental Study of a Hybrid Wave Energy Converter Integrated in a Harbor Breakwater. *J. Mar. Sci. Eng.* **2019**, *7*, 33. [CrossRef]
36. Cabral, T.; Clemente, D.; Rosa-Santos, P.; Taveira-Pinto, F.; Morais, T.; Belga, F.; Cestaro, H. Performance Assessment of a Hybrid Wave Energy Converter Integrated into a Harbor Breakwater. *Energies* **2020**, *13*, 236. [CrossRef]
37. Calheiros-Cabral, T.; Clemente, D.; Rosa-Santos, P.; Taveira-Pinto, F.; Ramos, V.; Morais, T.; Cestaro, H. Evaluation of the annual electricity production of a hybrid breakwater-integrated wave energy converter. *Energy* **2020**, *213*, 118845. [CrossRef]
38. Simonetti, I.; Esposito, A.; Cappietti, L. Experimental Proof-of-Concept of a Hybrid Wave Energy Converter Based on Oscillating Water Column and Overtopping Mechanisms. *Energies* **2022**, *15*, 8065. [CrossRef]
39. Jungrungruengtaworn, S.; Hyun, B.S. Influence of slot width on the performance of multi-stage overtopping wave energy converters. *Int. J. Nav. Archit. Ocean Eng.* **2017**, *9*, 668–676. [CrossRef]
40. Han, Z.; Liu, Z.; Shi, H. Numerical study on overtopping performance of a multi-level breakwater for wave energy conversion. *Ocean Eng.* **2018**, *150*, 94–101. [CrossRef]
41. Martins, J.C.; Goulart, M.M.; Gomes, M.D.N.; Souza, J.A.; Rocha, L.A.O.; Isoldi, L.A.; Dos Santos, E.D. Geometric evaluation of the main operational principle of an overtopping wave energy converter by means of Constructal Design. *Renew. Energy* **2018**, *118*, 727–741. [CrossRef]
42. Kralli, V.E.; Theodossiou, N.; Karambas, T. Optimal design of Overtopping Breakwater for Energy Conversion (OBREC) Systems using the harmony search algorithm. *Front. Energy Res.* **2019**, *7*, 80. [CrossRef]
43. Mustapa, M.A.; Yaakob, O.B.; Ahmed, Y.M. Numerical simulation of the overtopping-ramp design of a multistage overtopping wave energy breakwater hybrid device. *J. Inn. Tech. Expl. Eng.* **2019**, *9*, 4902. [CrossRef]
44. Cavallaro, L.; Iuppa, C.; Castiglione, F.; Musumeci, R.E.; Foti, E. A Simple Model to Assess the Performance of an Overtopping Wave Energy Converter Embedded in a Port Breakwater. *J. Mar. Sci. Eng.* **2020**, *8*, 858. [CrossRef]
45. Vidura, A.; Nurjaya, I.W.; Iqbal, M.; Jaya, I. Ocean wave measurement and wave energy calculation using overtopping power plant scheme. *IOP Conf. Ser. Earth Environ. Sci.* **2020**, *429*, 012047. [CrossRef]
46. Li, Q.; Mi, J.; Li, X.; Chen, S.; Jiang, B.; Zuo, L. A self-floating oscillating surge wave energy converter. *Energy* **2021**, *230*, 120668. [CrossRef]
47. Jungrungruengtaworn, S.; Thaweewat, N.; Hyun, B.S. Three-dimensional effects on the performance of multi-level overtopping wave energy converter. *IOP Conf. Ser. Mater. Sci. Eng.* **2021**, *113*, 012016. [CrossRef]
48. Izzul, M.; Fadhli, M.; Azlan, M.; Noor, M. Parametric study on wave overtopping due to wedge angle and freeboard of wavecat wave energy converter. *UMT J. Undergrad. Res.* **2022**, *4*, 37–50. [CrossRef]
49. Martins, J.C.; Goulart, M.M.; dos Santos, E.D.; Isoldi, L.A.; Gomes, M.D.N.; Rocha, L.A.O. Constructal Design of a Two Ramps Overtopping Wave Energy Converter Integrated into a Breakwater: Effect of the Vertical Distance between the Ramps over its Performance. *Def. Dif. Forum* **2022**, *420*, 242–258. [CrossRef]
50. Liu, Z.; Zhang, G. Overtopping performance of a multi-level CROWN wave energy convertor: A numerical study. *Energy* **2024**, *294*, 130795. [CrossRef]
51. da Silva, S.A.; Martins, J.C.; dos Santos, E.D.; Rocha, L.A.O.; Machado, B.N.; Isoldi, L.A.; das Neves Gomes, M. Constructal Design Applied to an Overtopping Wave Energy Converter Locate on Paraná Coast in Brazil. *Sustain. Mar. Struct.* **2024**, *6*, 1–14. [CrossRef]
52. Romanowski, A.; Tezdogan, T.; Turan, O. Development of a CFD methodology for the numerical simulation of irregular sea-states. *Ocean Eng.* **2019**, *192*, 106530. [CrossRef]
53. ANSYS. *ANSYS Fluent Theory Guide, Release 20.2*; ANSYS Inc.: Canonsburg, PA, USA, 2017.
54. Versteeg, H.K.; Malalasekera, W. *An Introduction to Computational Fluid Dynamics—The Finite Volume Method*; Pearson Education Ltd.: London, UK, 2007.
55. Hirt, C.W.; Nichols, B.D. Volume of fluid (VoF) method for the dynamics of free boundaries. *J. Comput. Phys.* **1981**, *39*, 201–225. [CrossRef]
56. Schlichting, H. *Boundary Layer Theory*, 7th ed.; McGraw-Hill: New York, NY, USA, 1979.
57. Srinivasan, V.; Salazar, A.J.; Saito, K. Modeling the disintegration of modulated liquid jets using volume-of-fluid (VoF) methodology. *Appl. Math. Model.* **2011**, *35*, 3710–3730. [CrossRef]

58. Park, J.C.; Kim, M.H.; Miyata, H.; Chun, H.H. Fully nonlinear numerical wave tank (NWT) simulations and wave run-up prediction around 3-D structures. *Ocean Eng.* **2003**, *30*, 1969–1996. [CrossRef]
59. Foyhirun, C.; Kositgittiwong, D.; Ekkawatpanit, C. Wave Energy Potential and Simulation on the Andaman Sea Coast of Thailand. *Sustainability* **2020**, *12*, 3657. [CrossRef]
60. Issa, R.I. Solution of the implicitly discretized fluid flow equations by operator-splitting. *J. Comput. Phys.* **1986**, *62*, 40–65. [CrossRef]
61. Patankar, S.V. *Numerical Heat Transfer and Fluid Flow*; McGraw Hill: New York, NY, USA, 1980.
62. Oleinik, P.H.; Tavares, G.P.; Machado, B.N.; Isoldi, L.A. Transformation of water wave spectra into time series of surface elevation. *Earth* **2021**, *2*, 997–1005. [CrossRef]
63. Airy, G.B. *Tides and Waves*; Encyclopædia Metropolitana: London, UK, 1845.
64. Awk, T. *TOMAWAC User Manual Version 7.2.*, 7.2.3 ed.; The Telemac-Mascaret Consortium; Telemac: Chambéry, France, 2017.
65. Holthuijsen, L.H. *Waves in Oceanic and Coastal Waters*; Cambridge University Press: Cambridge, UK, 2007.
66. McCormick, M.E. *Ocean Engineering Mechanics: With Applications*; Cambridge University Press: New York, NY, USA, 2010.
67. Ye, Z.; Ma, X.; Yang, N.; Cui, L. Assessment of wave energy resources in the Pearl River estuary of China. *Desalin. Water Treat.* **2023**, *298*, 222–232. [CrossRef]
68. Arguilé-Pérez, B.; Ribeiro, A.S.; Costoya, X.; deCastro, M.; Gómez-Gesteira, M. Suitability of wave energy converters in northwestern Spain under the near future winter wave climate. *Energy* **2023**, *278*, 127957. [CrossRef]
69. Chakrabarti, S.K. *Handbook of Offshore Engineering*; Elsevier: Amsterdam, The Netherlands, 2005.
70. Dean, R.G.; Dalrymple, R.A. *Water Wave Mechanics for Engineers and Scientists*; World Scientific Publishing Company: Bukit Batok, Singapore, 1991.
71. Chai, T.; Draxler, R.R. Root mean square error (RMSE) or mean absolute error (MAE)? Arguments against avoiding RMSE in the literature. *Geosci. Model Dev.* **2014**, *7*, 1247–1250. [CrossRef]
72. Oleinik, P.H.; Maciel, R.P.; dos Santos, E.D.; Rocha, L.A.O.; Machado, B.N.; Isoldi, L.A. Numerical method for the characterization of sea states using realistic irregular waves on computational fluid dynamics simulations for application on wave energy converters. *Sustain. Energy Tech. Assess.* **2025**, *73*, 104093. [CrossRef]
73. Paiva, M.S. Simulação Numérica de Um Conversor de Energia das Ondas do Mar em Energia Elétrica do Tipo Galgamento Considerando Dados Realísticos de Estado de Mar Encontrados na Costa do Município de Rio Grande. Master's Thesis, Matemática Aplicada, Universidade Federal do Rio Grande do Sul, Porto Alegre, Brazil, 2023. (In Portuguese)
74. Cardoso, S.D.; Marques, W.C.; Kirinus, E.D.P.; Stringari, C.E. Levantamento Batimétrico Usando Cartas Náuticas. In Proceedings of the 13ª Mostra da Produção Universitária, Rio Grande, Brazil, 14–17 October 2014. (In Portuguese)
75. Barreiro, T.G. Estudo da Interação de Uma Onda Monocromática Com um Conversor de Energia. Master's Thesis, Engenharia Mecânica, Universidade de Nova Lisboa, Lisbon, Portugal, 2009. (In Portuguese)
76. Dos Santos, E.D.; Isoldi, L.A.; Gomes, M.D.N.; Rocha, L.A. The constructal design applied to renewable energy systems. In *Sustainable Energy Technologies*; CRC Press: Boca Raton, FL, USA, 2017.
77. Schiefer, H.; Schiefer, F. *Statistics for Engineers—An Introduction with Examples from Practice*; Springer: Wiesbaden, Germany, 2021.

processes MDPI

Article

Research on the Cavitation Characteristics of Pump Turbines Based on Mode Decomposition

Jiaxing Lu [1,2,3], Jiarui Li [1,2,3,*], Chuan Zhang [1,2,3], Yuzhuo Zhou [1,2,3] and Yanjun He [1,2,3]

1 Key Laboratory of Fluid and Power Machinery, Ministry of Education, Xihua University, Chengdu 610039, China
2 School of Energy and Power Engineering, Xihua University, Chengdu 610039, China
3 Key Laboratory of Fluid Machinery and Engineering, Xihua University, Chengdu 610039, China
* Correspondence: ljr17381574805@163.com

Abstract: The cavitation phenomenon significantly impacts the performance of pump turbines, necessitating in-depth research on their cavitation characteristics. This study investigates the performance characteristics of a pump turbine through experimental and numerical simulation methods, with consistent results verifying the accuracy of the numerical simulations. The cavitation flow field is numerically analyzed to compare the cavitation distribution and velocity streamlines at different stages of cavitation development. The Q criterion and entropy production method are employed to identify vortex structures and energy loss regions, respectively, exploring the correlation between vortices and energy losses in the cavitation flow field under low-flow pump conditions. The results demonstrate that intensified cavitation generates more multi-scale vortices in the flow field, leading to increased entropy production and reduced energy efficiency. Proper orthogonal decomposition (POD) and dynamic mode decomposition (DMD) methods were subsequently applied to extract vorticity characteristics from transient cavitation flow fields, revealing primary energy loss regions and elucidating the evolution and distribution patterns of vortices. The POD analysis shows that low-order modes represent dominant vortex structures, while intensified cavitation increases both the quantity of vortices and their complexity in scale, distribution, and evolutionary frequency. The DMD results further indicate distinct evolutionary patterns for vortices of different scales. This research provides insights into the instability characteristics of cavitation flow fields in pump turbines under low-flow pump conditions and offers theoretical support for optimizing the design of pump turbines to expand their high-efficiency operational range.

Keywords: pump turbine cavitation; vortex; entropy production method; mode decomposition

1. Introduction

To mitigate the global reliance on fossil fuels, reduce carbon emissions, and improve ecological sustainability, renewable energy sources, such as wind and solar power, have garnered significant attention. These alternatives are expected to gradually replace thermal power generation in electrical grids, emerging as primary energy suppliers. However, wind and solar energy exhibit inherent intermittency and stochastic fluctuations due to seasonal variations and diurnal cycles, potentially compromising the operational stability of power systems [1–3]. Consequently, the increasing integration of renewable energy necessitates enhanced grid flexibility in terms of peak shaving, frequency regulation, and energy storage capabilities. Compared with other energy storage solutions, pumped storage hydropower plants demonstrate superior technological maturity, operational stability, and scalability for large-scale deployment. This positions pumped storage hydropower stations as the most viable regulating power source for novel power systems, capable of optimizing energy portfolios while ensuring grid resilience [4–6].

The efficient and stable operation of the pump turbine, as a critical component of pumped storage power stations, is essential for ensuring the safe and effective functioning of the facility. With advancements in the hydropower sector, pumped storage power stations are increasingly evolving towards high-head and high-speed operations, resulting in frequent occurrences of cavitation phenomena affecting the units. Numerous studies have indicated that cavitation has a more significant impact on the operational conditions of pumps than on those of turbines. The cavitation performance of pump turbines under pump conditions is deemed more critical than under turbine conditions [7–9]; consequently, it is essential to conduct in-depth research on the cavitation flow field specific to pump turbines operating in pump mode. Li et al. [10] employed numerical simulation techniques to investigate the cavitation characteristics of pump turbine runners in high-head regions under varying flow rates and heads for both pump and turbine conditions. Zhang et al. [9] conducted a study on the cavitation flow field and performance of pump turbine operating conditions, determining the critical cavitation bubble volume fraction; their findings indicated that varying flow rates influence the location of bubble formation at the inlet. Tao et al. [8] conducted a study on the cavitation behavior of pump turbines under various operating conditions; the research compared the inception and critical cavitation points across different scenarios, concluding that utilizing an initial cavitation standard allows for a more timely detection of cavitation bubbles, thereby facilitating the early mitigation of cavitation phenomena.

Cavitation can also lead to adverse phenomena such as vibrations and noise within the machinery, which may result in a reduction in mechanical lifespan, a decrease in maintenance intervals, and an increase in the frequency of unit startups and shutdowns, ultimately affecting the stability of the power station. Hao et al. [11] conducted a study on the impact of different impeller tip clearances in the cavitating flow field of mixed-flow pumps operating under pump conditions, specifically focusing on the radial forces acting on the shaft in pumped storage units. The findings indicate that when the apex gap is asymmetrical, the cavitation performance is at its lowest. In this scenario, the amplitude of the radial force increases with the intensification of cavitation and, due to the asymmetric distribution of the cavitation region, the direction of the radial force also varies as cavitation worsens. Liu et al. [7] examined the impact of cavitation on the radial forces experienced by water pump turbines under pumping conditions. Their study revealed that the number of cavitation bubbles on the blade surfaces increased as cavitation intensified, resulting in a reduction in blade load and a decrease in the regularity of pressure fluctuations. This led to a significant alternation in radial forces, contributing to the emergence of fatigue cracks in high-stress areas of the blades. However, due to the periodic symmetrical structure of the guide vanes and the runner, the distribution of radial forces remained relatively uniform.

The unsteady flow structures induced by internal cavitation are one of the critical factors leading to adverse phenomena such as structural vibrations and noise in the unit, as well as significant performance degradation. Meng et al. [12] investigated the dynamic characteristics of cavitation-induced vortices in centrifugal pumps based on the vorticity transport method, revealing that vortex structures form at the trailing edge of cavitation regions in flow channels during severe cavitation stages. Wu et al. [13] employed numerical simulations to analyze two-dimensional vortex structures on cross-sections through velocity component analysis and extracted vorticity on pump blade surfaces using the new omega vortex identification method proposed by Zhang et al. [14]. The results demonstrated that as the net positive suction head decreases, the vapor volume expands, leading to flow field disorder and localized vortex formation. These vortices detach from the blade surfaces as cavitation cavities roll up, with the vorticity transport method effectively capturing large-scale high-vorticity regions. Ruan et al. [15] investigated the cavitation characteristics of pump turbines under various flow conditions in pump mode. The findings indicated that under low-flow conditions, intense cavitation on the suction side of the blade trailing edge induces flow separation, disrupting the flow field structure and generating large-scale flow separation and vortex structures, resulting in a significant reduction in pump turbine efficiency.

The decline in unit performance and efficiency is attributed to increased internal energy dissipation. To identify the distribution and magnitude of energy dissipation regions, Gong et al. [16] pioneered the application of entropy production theory to flow field analysis. Li et al. [17] utilized this theory to investigate energy loss characteristics in pump turbines operating within the hump region under pump mode. Yu et al. [18] further explored energy dissipation in cavitating flows within micro-pumps using the entropy production method, establishing correlations among cavitation, vorticity, and entropy production. Their results revealed that during cavitation development, mass exchange between gas and liquid phases and momentum transfer between fluids contribute significantly to hydraulic losses.

Traditional methods have difficulty in capturing the dominant features of unsteady flow fields. Therefore, to effectively process complex flow field data, many reduced-order models have been proposed by researchers in recent years. The most commonly used methods are proper orthogonal decomposition (POD) [19] and dynamic mode decomposition (DMD) [20].

The POD (proper orthogonal decomposition) method decomposes complex flow fields into a series of spatial orthogonal modes, ranking them based on their energy content (i.e., their eigenvalues) to emphasize the dominant flow structures. Due to its ability to accurately capture coherent structures within the flow field and significantly simplify the analysis, POD has been widely applied in flow field studies. For instance, Lu et al. [21] utilized the POD method to decompose and reconstruct the flow field in the tongue region of a centrifugal pump, establishing the relationship between pressure fluctuations and the dominant flow structures. Yang et al. [22,23] employed the POD method to investigate the evolution frequency of the stall region in the flow field of the guide vane when the pump operates in the hump region.

In contrast to the POD method, the DMD technique allows for the decomposition of the flow field into a series of distinct frequency modes, thereby facilitating the observation of flow structures evolving at various frequencies. Xie et al. [24] employed the DMD method to analyze the velocity field of the cavitating flow around a hydrofoil, finding that the frequency of the second mode matched the bubble shedding frequency, while the frequencies of the third and fourth modes were harmonics of the second mode. For rotating fluid machinery, Lu et al. [25] utilized a wavelet transform to analyze pressure fluctuation data at the inlet and outlet of a centrifugal pump and combined this with numerical simulations to decompose the internal flow field data using DMD. The results indicated the presence of complex pressure fluctuation characteristics within the centrifugal pump, dominated by different frequencies.

To date, few studies have combined two mode decomposition methods and analyzed the unsteady cavitation characteristics of the flow field in the runner and guide vanes of a pump turbine under low-flow conditions. In this study, the POD method and the DMD method are combined to study the unsteady characteristics of vortices under different cavitation conditions in the low-flow pump operation of a pump turbine. The results of the two mode decomposition methods are compared and verified to gain a deeper understanding of the unsteady characteristics of the cavitation flow field.

2. Research Object

2.1. Pump Turbine Parameters

This study focuses on a model of a pump turbine, as illustrated in Figure 1a. The unit primarily consists of several flow components, including the volute, fixed guide vanes, movable guide vanes, a runner, and a draft tube. A schematic representation of the meridional flow surface is provided in Figure 1b. Specific parameters of the unit, such as the runner speed, design flow rate, and design head, are detailed in Table 1.

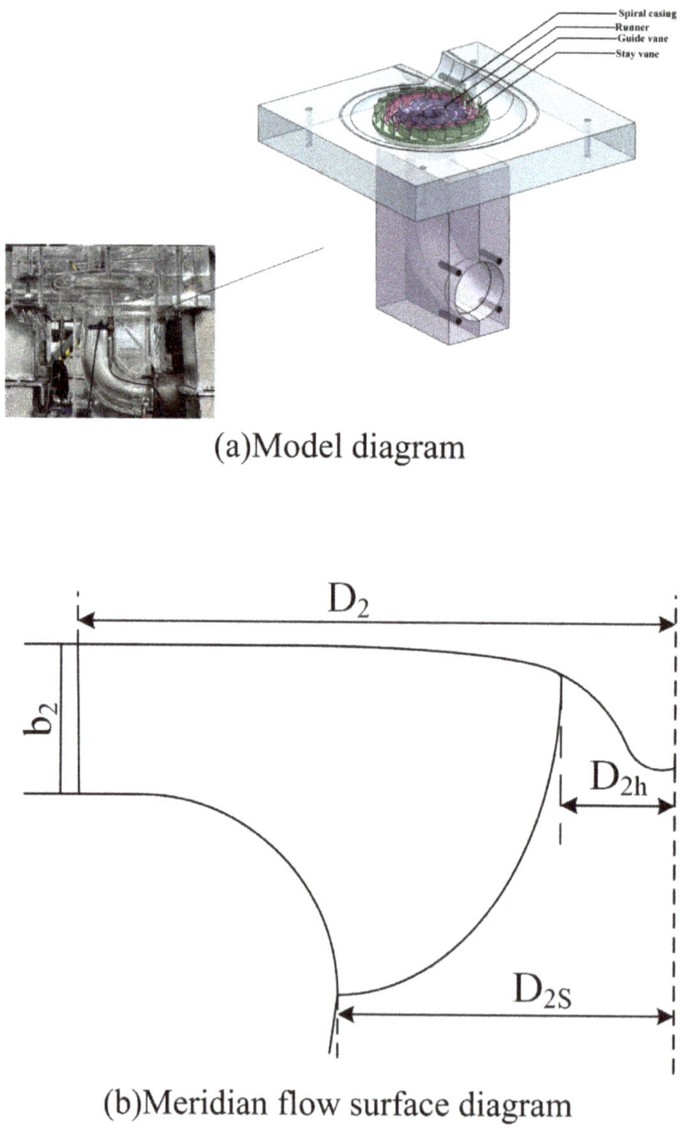

Figure 1. Model pump turbine structure diagram.

Table 1. Model parameter.

Parameters	Value	Unit
Draft tube inlet diameter D1	164	mm
Runner diameter D2	200	mm
Hub diameter D2h	22.8	mm
Shroud diameter D2S	57.41	mm
Runner outlet width b2	23.88	mm
Runner blades ZR	7	pcs
Stay vane ZS	20	pcs
Guide vane ZG	19	pcs
Rotational speed n	1000	r/min
Design flow Qd	15.25	L/s
Design head Hr	4.5	m

2.2. Experimental Method

The experiment was conducted on a closed-loop water pump turbine test rig, as illustrated in Figure 2. The experimental setup included force sensors, data acquisition instruments, and solenoid valves. Pressure gauges were installed at the inlet and outlet of the water pump turbine unit to measure pressure, allowing for the calculation of the pressure differential across the pump turbine. Additionally, a flow meter was utilized to determine the flow rate within the system.

(a) Experimental bench

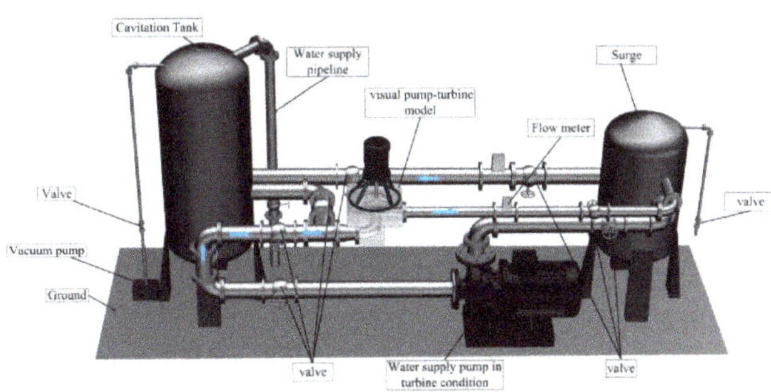

(b) Technical route

Figure 2. Experimental technology roadmap.

The specific experimental procedure is described as follows: (1) close the valves on the inlet and outlet pipelines of the water supply pump while simultaneously opening the valves on the inlet and outlet pipelines under pump operation, as well as the return water valve. (2) Verify the opening and closing status of the relevant valves during the experiment and ensure that the system is sealed. (3) Fill the left-side cavitation tank with water to a level that submerges the turbine runner; then, activate the motor. Next, fill the right-side pressure tank with water. Once the right-side pressure tank is full, close its valve. (4) Release any air trapped in the system through the valve of the left-side cavitation tank and adjust the valve on the return water line to its maximum opening. (5) Collect data from the flow meter and pressure gauge. (6) Gradually reduce the valve opening. (7) After the flow meter data stabilizes, repeat steps (5) and (6) until all data have been collected.

3. Numerical Simulation Methods

3.1. Numerical Simulation Setup

Numerical simulation of the flow field was conducted using the ANSYS Fluent 2020 R2 computational software. Figure 3 illustrates the computational domain and the grid schematic, which consists of five flow components, namely the volute, fixed guide vanes, movable guide vanes, a runner, and a tailwater pipe. The computational domain was pre-processed for grid generation using commercial software. Hexahedral mesh elements were employed for the fixed guide vanes, movable guide vanes, and tailwater pipe, while tetrahedral mesh elements were utilized for the volute and runner, as depicted in Figure 3a. To minimize computational resource expenditure while ensuring the accuracy of the results, the independence of the grid was verified by examining the head rise. As shown in Figure 3b, the increase in the head rise of the pump turbine diminishes progressively with an increase in the number of grid elements.

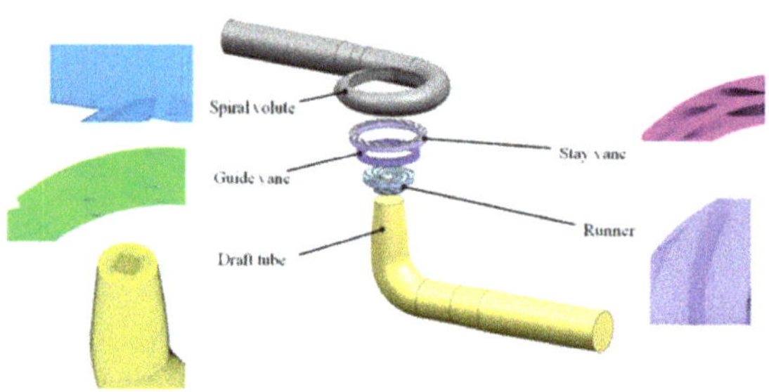

(a) Calculation domain and its local grid diagram

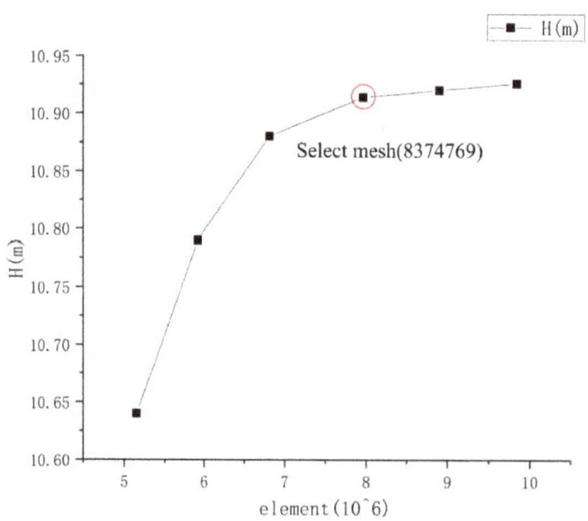

(b) Grid independence verification

Figure 3. Calculation model and grid diagram.

Considering the costs associated with simulation and the required computational accuracy, this study selects a grid count of 8,374,769 for the calculations. Table 2 presents the types and quantities of grid cells for each flow region.

Table 2. Number and type of grid.

Part	Grid Cell	Grid Type
Spiral volute	1,261,554	Tetrahedral
Stay vane	1,196,164	Hexahedral
Guide vane	1,937,354	Hexahedral
Runner	1,505,854	Tetrahedral
Draft tube	2,473,843	Hexahedral
Design head Hr	8,374,769	/

In the ANSYS Fluent 2020 R2, the rotor region is designated as a rotating area based on multiple reference frames (MRFs), while the remaining regions are classified as stationary. The inlet of the tailwater pipe is configured as a total pressure inlet, the volute outlet is set as a flow outlet, and the wall surfaces are defined as non-slip boundaries. The governing equations consist of the continuity equation, the energy conservation equation, and the momentum conservation equation. The calculations employ the RNG k-ε turbulence model, the mixture multiphase flow model, and the Schnerr–Sauer cavitation model, which is based on the Rayleigh–Plesset bubble dynamics equation. The fluid medium is specified as water and water vapor, with the saturation vapor pressure of water set at Pv = 3540 Pa and the ambient pressure at 0 Pa. A convergence criterion of a residual less than 10^{-5} is established. The mid-plane flow surface between the rotor and the movable guide vanes is captured as Plane A, as illustrated in Figure 4. The time step is set to one step for every 3° of rotation, totaling 120 steps for one complete rotation with a time increment of 0.0005 s. Starting from the last step of the transient calculation at the 10th rotation of the rotor, a sequence of 121 data sets, spaced by two time steps, is utilized for subsequent flow field analysis.

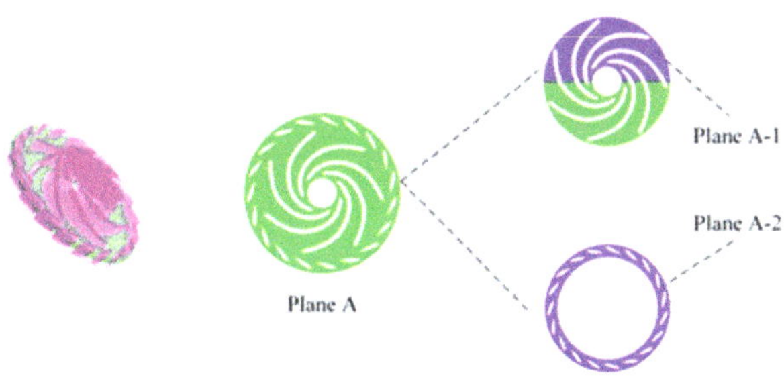

Figure 4. Calculation plane.

3.2. *Snapshot POD Method*

When the number of grid nodes and elements significantly exceeds the number of sampling instances, the classic POD method can become excessively resource-intensive and time-consuming. Consequently, this study employs the snapshot POD approach proposed by Sirovich [26]. The underlying principle is as follows: a set of snapshots is formed by selecting flow field data from M grid elements at N instantaneous moments within a specified flow duration, and the time-averaged data for each grid element during this period are extracted.

$$V(x_i, t_j) = \begin{bmatrix} v(x_1, t_1) & v(x_1, t_2) & \cdots & v(x_1, t_N) \\ v(x_2, t_1) & v(x_2, t_2) & \cdots & v(x_2, t_N) \\ \vdots & \vdots & \vdots & \vdots \\ v(x_M, t_1) & v(x_M, t_2) & \cdots & v(x_M, t_N) \end{bmatrix} \tag{1}$$

$$\overline{V}(x_i, t_j) = \frac{1}{N} V(x_i, t_j) \tag{2}$$

By applying Equation (3) to remove the time-averaged data corresponding to each grid cell from the original flow field at each instant, the resulting data reflect the fluctuations within the flow field:

$$V(x_i, y_j) = V(x_i, y_j) - \overline{V}(x_i) \tag{3}$$

The covariance matrix is derived from the matrix formed by the fluctuating data of the flow field, followed by the calculation of the eigenvalues and corresponding eigenvectors of the covariance matrix:

$$S = \frac{1}{N} V^T V \tag{4}$$

$$SA_j = \lambda_j A_j \tag{5}$$

The corresponding spatial basis modes and temporal mode coefficients of the POD can be derived from Equations (6) and (7):

$$\varphi_j(x) = \frac{1}{\sqrt{\lambda_i}} VA_j \tag{6}$$

$$a_j(t) = V(x_i, t_j) \times \varphi_j(x) \tag{7}$$

The temporal average mode, in conjunction with a set of base modes, can reconstruct the flow field at any given moment using different order modes, as expressed in Equation (8):

$$V(x, t) = \overline{V}(x) + \sum_{j=1}^{N} \varphi_j(x) a_j(t) \tag{8}$$

3.3. DMD Method

The principle of DMD, as proposed by Schmid [20], is outlined as follows: the computational results are organized in chronological order, forming a set of snapshot matrices, as represented in Equation (9):

$$X_i^N = \{x_1, x_2, x_3, \ldots, x_N\}, \tag{9}$$

Let x_i represent the i-th snapshot and N denote the total number of snapshots. It is assumed that there exists a linear mapping from the snapshot matrix X_i to the matrix X_{i+1}, expressed as follows:

$$X_{i+1} = AX_i \tag{10}$$

Consequently, all subsequent snapshots can be represented using the matrix A and the vector x_1, as indicated in Equation (11).

$$X_i^N = \left\{x_{1,} Ax_{1,} A^2 x_{1,} \ldots, A^{N-1} x_{1,}\right\} \tag{11}$$

Matrix A can be defined as follows:

$$A = XX^+ \tag{12}$$

The matrix X^+ represents the pseudoinverse of the matrix X, while the eigenvalues and eigenvectors of matrix A serve as the eigenvalues and modes in the context of DMD.

The singular value decomposition of matrix X yields matrices U, Σ, and V^T, as indicated in Equation (13). By retaining the first r rows and r columns of these matrices, new matrices U_r, Σ_r, and V_r are obtained. Subsequently, $\widetilde{A}$ is computed using Equation (14).

$$X = U\Sigma V^T \tag{13}$$

$$\widetilde{A} = U_r^T A U_r = U_r^T X V_r \Sigma_r^{-1} \epsilon R^{r\times r} \tag{14}$$

The eigenvalues and eigenvectors of matrix $\widetilde{A}$ are denoted as μ_j and $\widetilde{v}_j$, respectively, and the corresponding DMD modes can be derived using Equation (15):

$$v_i = \mu^{-1} X V_r \Sigma_r^{-1} \widetilde{v}_i \tag{15}$$

The logarithm of the eigenvalues, when considered in relation to the time step size, has its imaginary and real components corresponding to the magnitudes of the frequencies and the growth/decay rates of the various modes in DMD.

$$\lambda_j = \frac{1}{\Delta t} \times \log(u_j) \tag{16}$$

4. Results and Discussion

4.1. Analysis of Cavitation Characteristics

In this study, head values under different flow conditions were obtained experimentally, and corresponding numerical simulations were conducted for each test case to derive the flow–head (Q-H) curve. Due to system leakage and limitations in instrument precision, the numerical simulations could not fully replicate experimental conditions, resulting in discrepancies between the two datasets. However, as shown in Figure 5, the relative error remained within 5%, which is acceptable and confirms the accuracy of the numerical simulations, thereby validating their use for subsequent analyses. The Q-H curve reveals that as the flow rate decreases, the head gradually increases until reaching approximately 60% of the design flow rate (0.6 Q_d), where the curve exhibits a positive slope region—the hump region. Giovanna et al. [27] identified a strong correlation between hump characteristics and rotor–stator interactions in this region, prompting this study to focus on the cavitation flow field under the 0.6 Q_d condition for in-depth investigation.

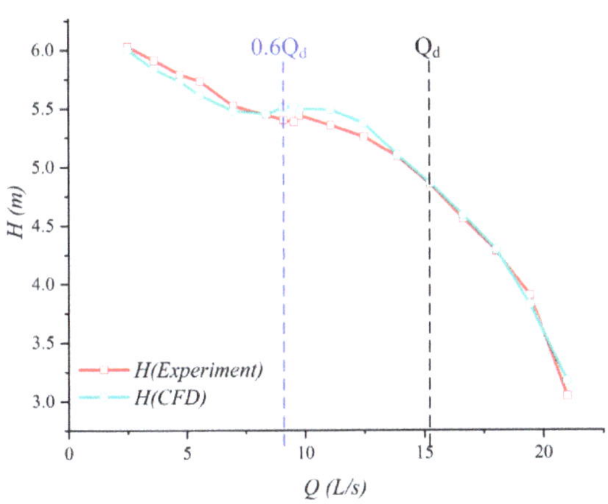

Figure 5. Q-H.

Numerical simulations of cavitation characteristics were conducted for the conditions of 0.6 times Q_d and Q_d. The resulting cavitation characteristic curves are illustrated in Figure 6, where the horizontal axis represents the cavitation coefficient defined in Equation (17), and the vertical axis indicates the head corresponding to the respective cavitation number.

$$\sigma = \frac{(P_{in} - P_b)}{0.5\rho v^2} \tag{17}$$

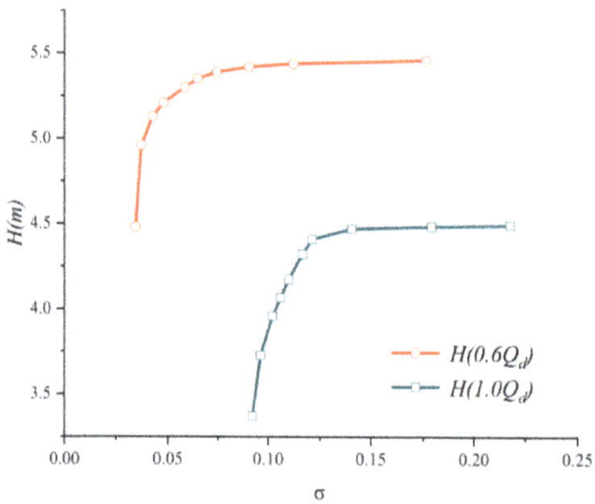

Figure 6. σ-H.

It is evident that as the cavitation number decreases, there is little variation in the head during the initial phase until a critical point is reached (referred to as critical cavitation), at which point a noticeable decline in head begins. Subsequently, the slope of the curve increases rapidly, leading to a sharp decrease in head, significantly impacting the performance of the unit.

Figure 7 illustrates the total volume of cavitation bubbles in the impeller region under the conditions of 0.6 Q_d and Q_d at varying levels of cavitation. Figure 8 depicts the iso-surface formed by the cavitation volume fraction equal to 10% under the corresponding cavitation conditions. A comparison of these two figures reveals that when the cavitation number is relatively high and bubbles are just beginning to form, the volume of the bubbles is minimal, occupying a small portion of the flow channel and causing negligible disturbance to the flow field. As the cavitation number decreases to a critical point, cavitation develops to a certain extent, resulting in bubbles occupying part of the flow channel and affecting the flow field, which leads to a reduction in head. Conversely, when cavitation is fully developed, the bubbles occupy a significant portion of the space, severely impeding the normal flow of water.

In the Q_d operating condition, cavitation occurs predominantly on the pressure side, whereas in the 0.6 Q_d condition, significant cavitation is observed on both the suction and pressure sides. Notably, the cavitation on the suction side is primarily concentrated near the upper cover plate, while the pressure-side cavitation is mainly found close to the hub. Additionally, due to the dynamic interference between the rotating blades and the stationary guide vanes, cavitation begins to manifest at the outlet of the rotating blades and in the no-blade region. As cavitation progresses, the volume of the vapor bubbles in the no-blade area increases, adversely affecting the flow in that region. This phenomenon leads to a sharp decline in head and a reduction in efficiency.

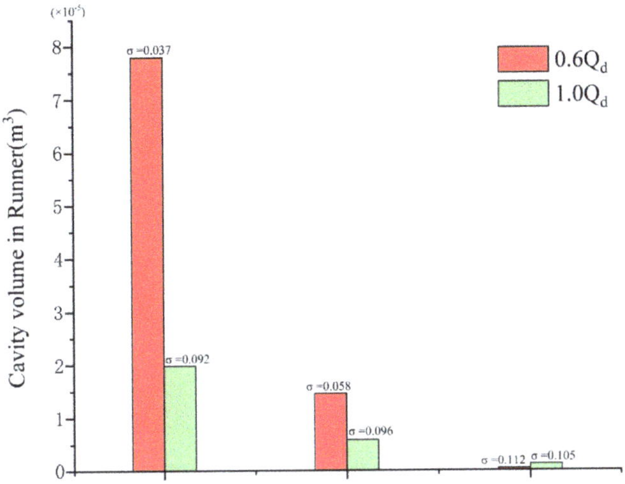

Figure 7. The total volume of the vapor.

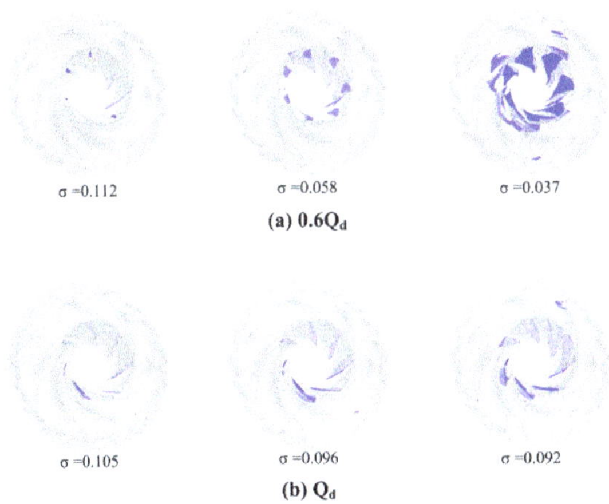

Figure 8. The vapor iso-surface of 10% vapor fraction.

Figures 9 and 10 illustrate the velocity contours and streamlines on Plane A for different cavitation levels at 0.6 Q_d and Q_d operating conditions. The flow separation on the suction side of the blade exit and the recirculation in the variable guide vane region result in the formation of large-scale vortex structures in the blade-free zone due to internal viscous forces. Compared to the Q_d flow rate condition, the 0.6 Q_d flow rate condition exhibits more severe backflow phenomena in the guide vane region, with vortex structures of significantly larger scales. As cavitation intensifies, the vortex scale in the bladeless region expands, the flow separation on the suction side of the blade increases, and water flows backward into the runner, forming a vortex near the suction side of the blade and runner. The vortices present in both the blade-free and runner areas significantly impede the flow within the field, ultimately resulting in a decline in the performance of the unit.

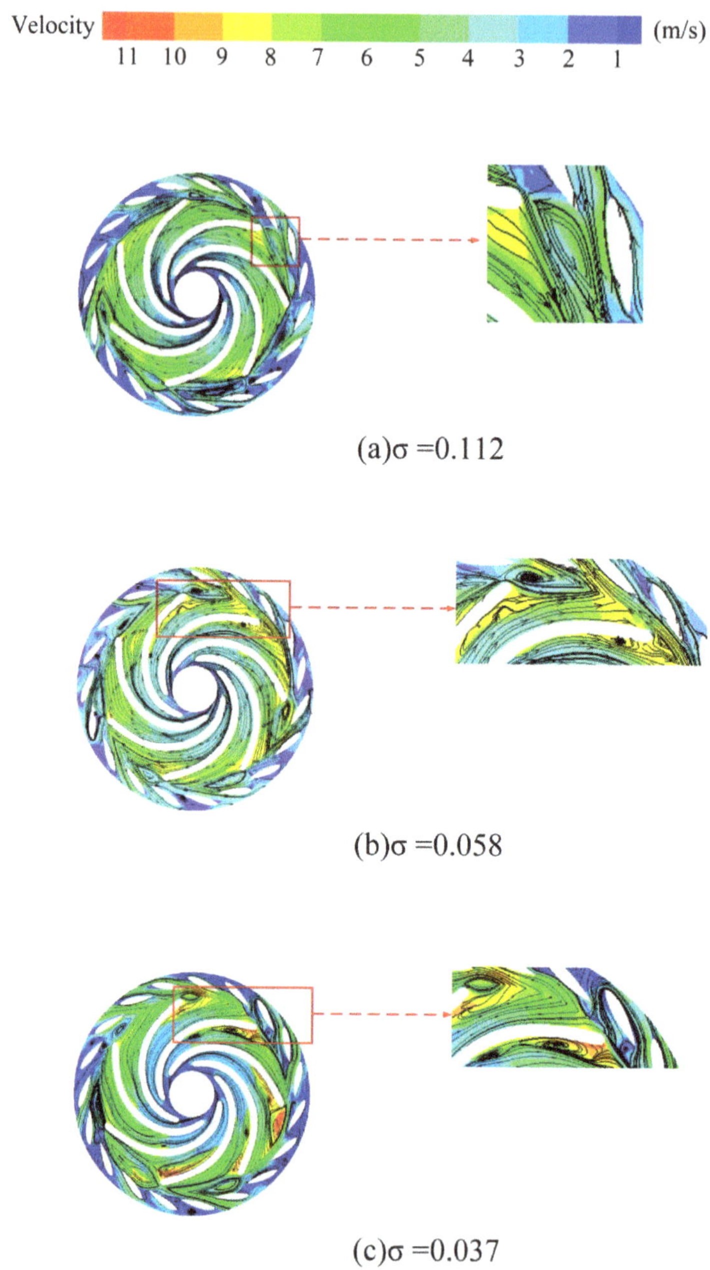

Figure 9. Velocity streamline of cavitation flow field at 0.6 Q_d.

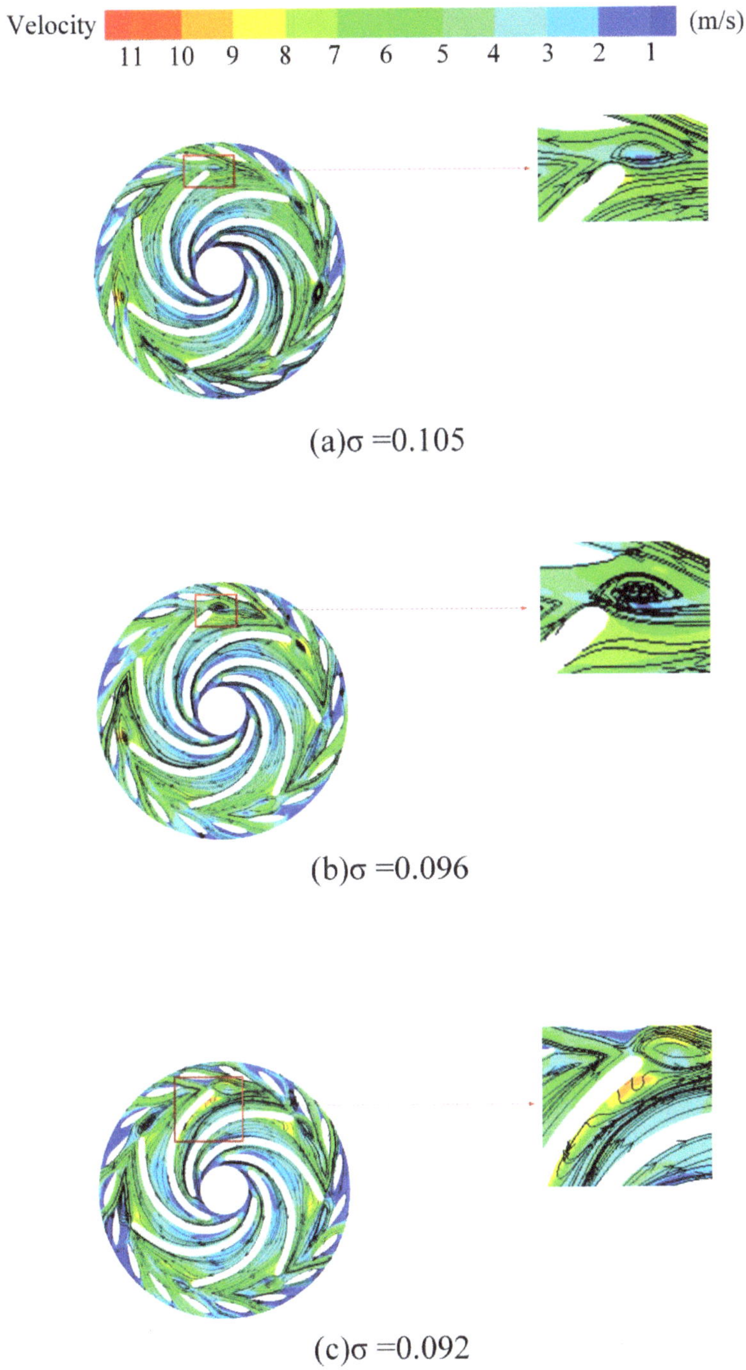

Figure 10. Velocity streamline of cavitation flow field at Q_d.

4.2. Analysis of Energy Dissipation Characteristics in Cavitating Flow Fields

To facilitate a more intuitive observation of the vortices and the associated energy loss distribution in low-flow conditions, the Q criterion vortex identification method was employed to capture the vortices in the flow field of Plane A. Figure 11 illustrates the vorticity and its corresponding entropy generation distribution, along with the average entropy generation in Plane A's flow field. As indicated by the comparison of the average entropy generation in Plane A's flow field shown in Figure 11b, it is evident that with the progression of cavitation, the entropy generation increases, leading to greater energy loss. Figure 11a presents the vortex map identified using the Q criterion, revealing that during the initial stages of cavitation, high vorticity primarily occurs in the blade exit region and the blade-free zone between the moving guide vanes. As cavitation intensifies, the scale of the high-intensity vortices in the blade-free zone increases, and the number of vortices in the impeller flow field also rises, resembling the vortex structure distribution depicted in Figure 9, thereby accurately identifying the vortex structures. Furthermore, Figure 11a also displays the entropy generation distribution, which upon comparison is predominantly concentrated in the vortex regions. Consequently, the cavitation vortices in the low-flow operating conditions of the pump turbine are identified as the primary cause of energy loss.

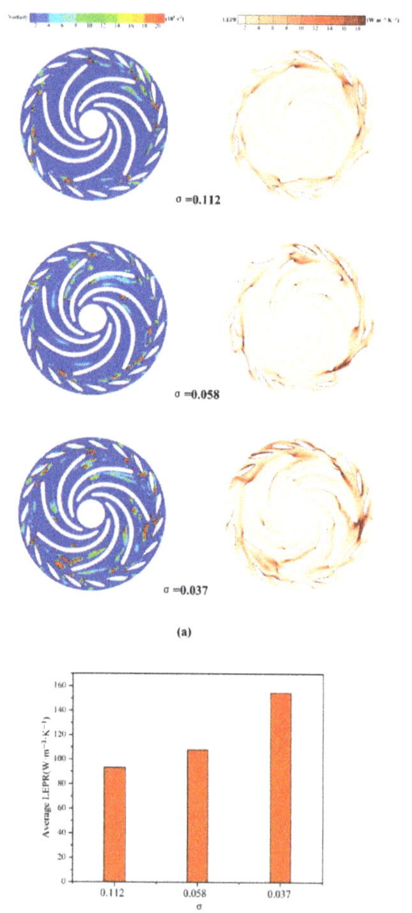

Figure 11. Distribution of vorticity and entropy production.

4.3. POD Analysis of Vorticity Field

To investigate the evolution of vortices, Figure 12 illustrates the vortex evolution of the flow field at Plane A-1 over one rotation period under various cavitation conditions. When σ is 0.112, the changes in the blade tip vortex during one cycle are minimal, indicating that this cavitation condition primarily governs the vortex distribution in the impeller region. At the impeller outlet, vortices resulting from the shedding of the blade tip are present, with an evolution period approximately equal to one rotation period T. As the cavitation coefficient decreases to $\sigma = 0.058$, Plane A-1 generates vortices with consistent evolution periods at the same location. However, due to the development of cavitation, the intensity of these vortices is significantly higher than during the initial stages of cavitation. Additionally, other flow passages exhibit vortices of varying sizes that lack a clear evolutionary pattern. With the further reduction in the cavitation coefficient to $\sigma = 0.058$, cavitation becomes fully developed, resulting in a substantial presence of vortices with unstable evolution periods throughout the entire rotation cycle at Plane A-1.

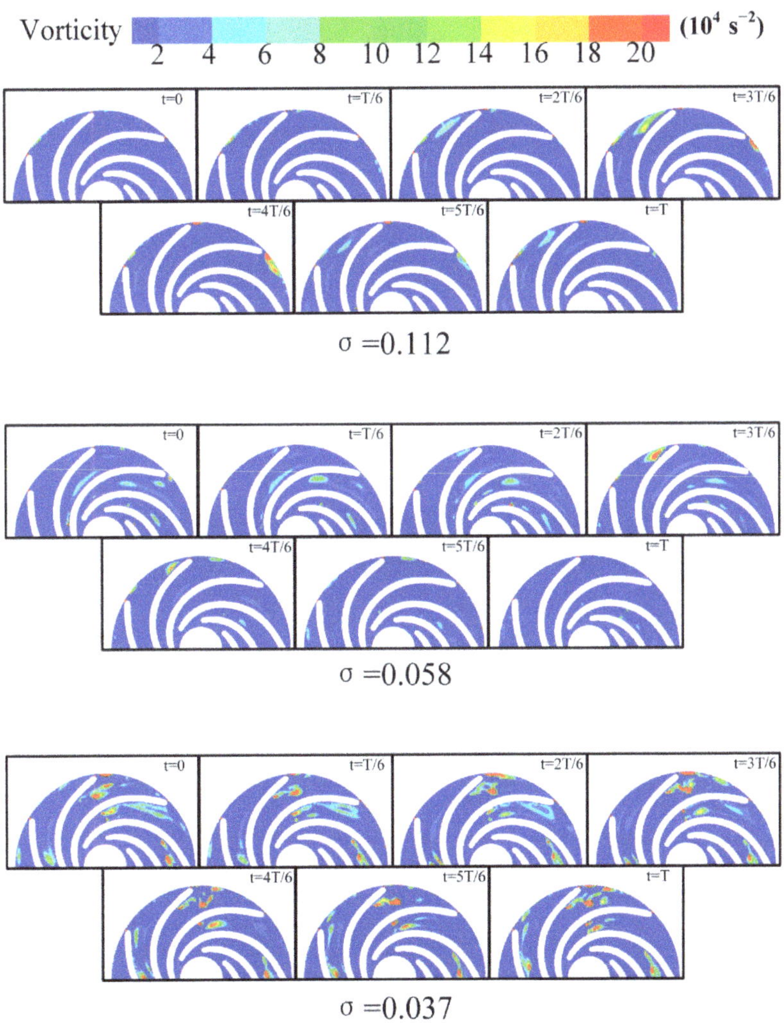

Figure 12. Evolution of vortices in Plane A-1.

Figure 13 illustrates the evolution of vortices in the flow field of Plane A-2 over a single rotation period under various cavitation conditions. It is observed that large-scale vortices are predominantly located in the blade-free region and the trailing edge of the movable guide vanes. At the onset of cavitation, large-scale vortices are present in the blade-free area. As cavitation progresses, its influence on the flow field intensifies, leading to an expansion in vortex size in the blade-free region of Plane A-2, an increase in the number of vortex structures, and the emergence of small-scale, low-intensity vortices with varying periodicity.

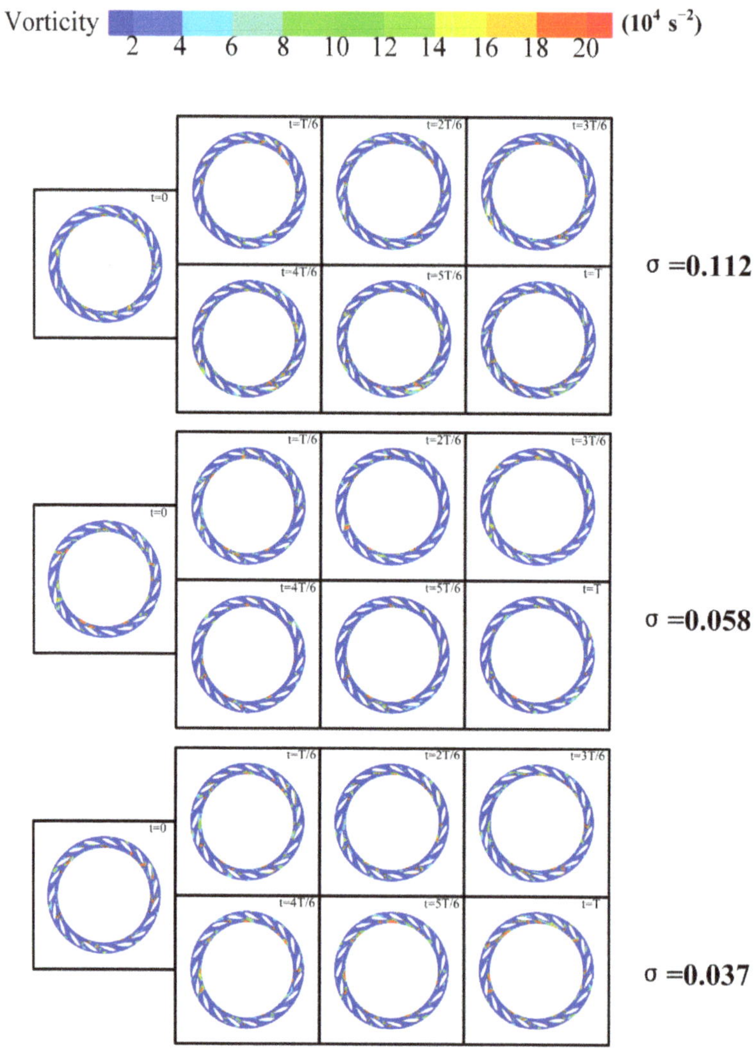

Figure 13. Evolution of vortices in Plane A-2.

The instability of cavitation vortex structures leads to variability in the duration of vortex evolution cycles, making it challenging to observe the periodicity of these cycles. To address this, POD was employed to perform dimensionality reduction on snapshots of the vorticity field in a low-flow condition over two rotation periods. This approach allowed for the extraction of the primary coherent vortex structures on Plane A based on the energy contributions of the various mode orders.

Figure 14 illustrates the proportion of energy contained in various-order POD modes, represented as the ratio of the mode characteristic values to the sum of the eigenvalues. It is evident that the first-order mode possesses the highest energy proportion, significantly surpassing that of the second-order and higher modes, indicating that it represents the predominant coherent structures within the vorticity field. Furthermore, for modes of order greater than eight, the energy proportion falls below 1%, and this proportion decreases gradually with increasing mode order. This suggests that the corresponding mode vorticity fields are influenced by the unsteady characteristics of the flow, rendering their contributions negligible. Additionally, across different levels of cavitation, the distribution of energy proportions among the various modes remains largely consistent. Notably, the energy proportion of the first-order mode diminishes as the degree of cavitation intensifies, indicating that the progression of cavitation enhances the unsteady characteristics of the vorticity field.

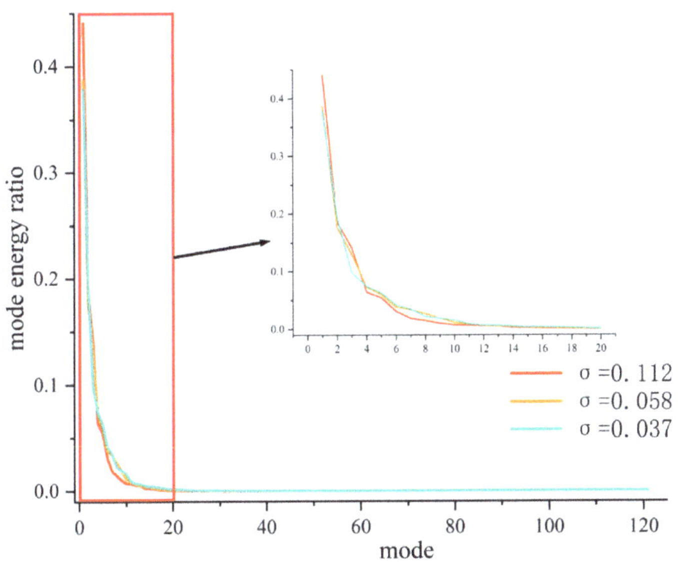

Figure 14. POD mode energy contribution diagram.

The time mode coefficients for the first, fourth, and eighth mode were subjected to a fast Fourier transform (FFT), resulting in the time evolution graphs and corresponding frequency spectra illustrated in Figure 15a–c. In Figure 15a, the evolution of the first-order POD mode's time mode coefficients reveals a predominantly low frequency. This is attributed to a high energy proportion, indicating that the mode encompasses a significant amount of instability. As the instability of the cavitating flow field increases, the mode incorporates more unstable elements, leading to a chaotic evolution of the vortices. Conversely, as the mode order increases, the primary frequency in the time mode coefficients rises, while the energy proportion decreases, resulting in a reduction in the flow field information contained within the mode, thus rendering the frequency more distinct and clear.

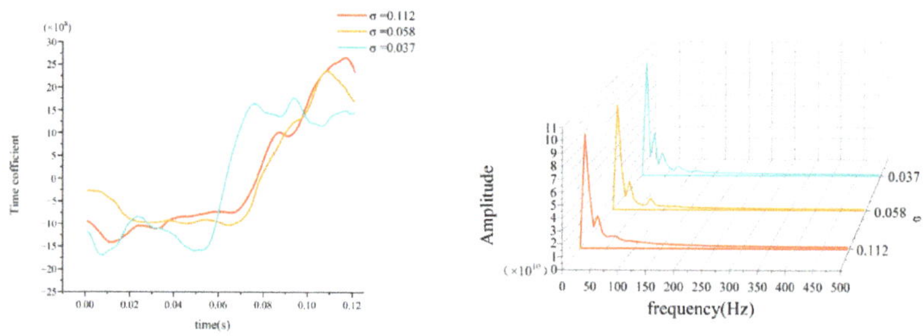

(a) mode1

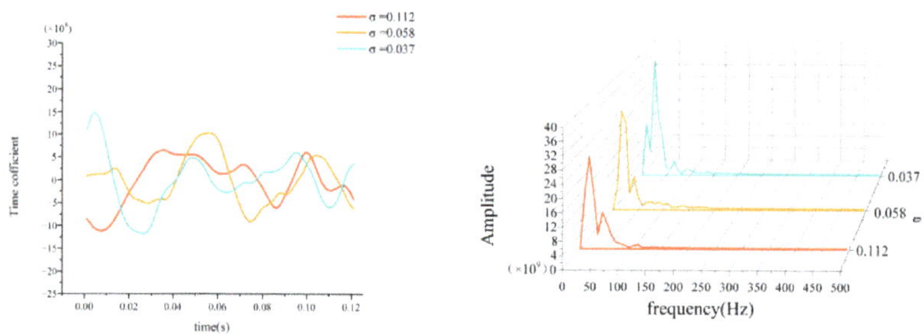

(b) mode4

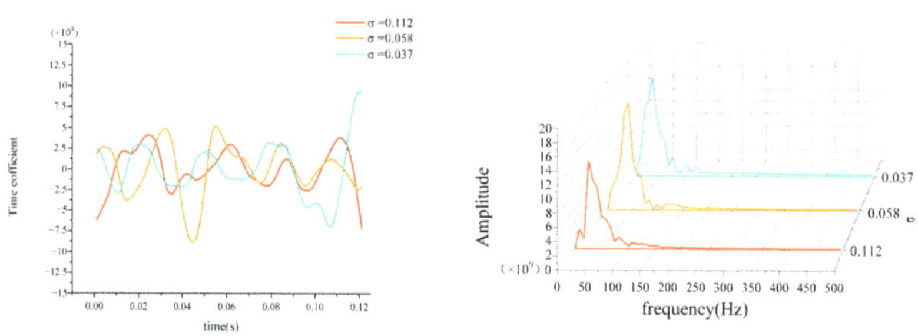

(c) mode8

Figure 15. Time mode coefficient time and frequency diagram.

To provide a more intuitive and concrete observation of the coherent vortex structures at various cavitation levels, Figure 16 illustrates the spatial distribution of the first eight mode shapes. When the cavitation number σ is set at 0.112, mode 1 represents the coherent vortex attached to the outlet of the impeller blades. Modes 2–6 depict the diffusion of vorticity from the blade outlet into the non-bladed region. As the mode order increases to mode 7 and mode 8, the wake vortices at the blade outlet expand, leading to dynamic interactions with the movable guide vanes. Concurrently, numerous small-scale vortices emerge in the surrounding non-bladed area, and vorticity is also generated on both the pressure and suction sides of the blades. Therefore, at a cavitation number of σ = 0.112, the wake vortices at the blade outlet are distributed across all mode orders, representing the inherent modes of the flow field. The wake vortices from the blades are identified as the primary source of energy dissipation, while the small-scale vortices in the non-bladed region predominantly appear in higher mode orders, a phenomenon attributed to the characteristics of the unsteady flow field. Although the contribution of small-scale vortices to the main flow structure is relatively minor, their high-frequency evolution significantly enhances the energy dissipation within the flow field.

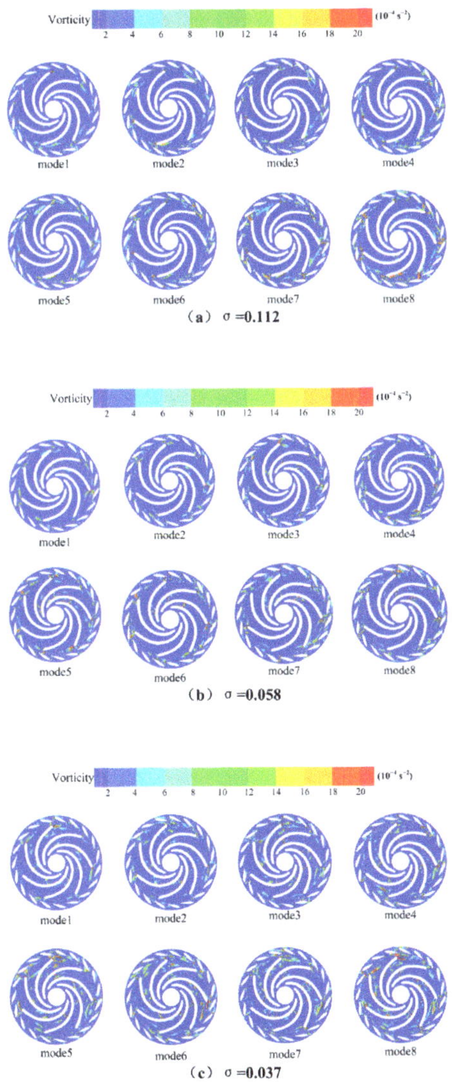

Figure 16. The distribution of the first eight POD modes.

As cavitation progresses, the scale of the vortices represented by mode 1 increases, leading to a greater number of vortices adhering to the blade surfaces within the rotor passage. When severe cavitation occurs, vortices that detach from the suction side of the blades are distributed across all mode orders, particularly near the rotor exit. Concurrently, smaller-scale vortices begin to transition towards lower-order modes, which possess a higher energy contribution, thereby enhancing their impact on the flow field. Consequently, with the advancement of cavitation, the vortices associated with the blade wake, as represented by the lower-order modes, increase in size and number, resulting in heightened energy losses. This phenomenon is accompanied by a reduction in the energy contribution of mode 1, while the contributions from mode 2 and higher modes to the flow field become more significant, intensifying the unsteady characteristics of the flow and accelerating energy dissipation.

4.4. DMD Analysis of Vorticity Field

The POD method primarily identifies dominant coherent structures within a flow field by arranging modes based on their energetic contributions. The modes exhibit complex frequency components, and the DMD technique is employed for dimensionality reduction and decomposition of the flow field, allowing for the extraction of coherent structures at specific characteristic frequencies. This facilitates a comparative analysis of the dynamic characteristics of evolving vortex structures across different scales. Figure 17 illustrates the distribution of the DMD mode eigenvalues in the complex plane, where the horizontal axis represents the real part of the eigenvalues and the vertical axis represents the imaginary part. A majority of the eigenvalues cluster near the unit circle, indicating that the corresponding modes are relatively stable and represent the primary coherent structures of the vorticity field. Conversely, modes located within the unit circle are deemed unstable and do not correspond to the main vortex structures of the vorticity field. The parameters that describe the flow field information of each mode are the real parts of the spatial basis modes, with each set of conjugate eigenvalues corresponding to conjugate modes.

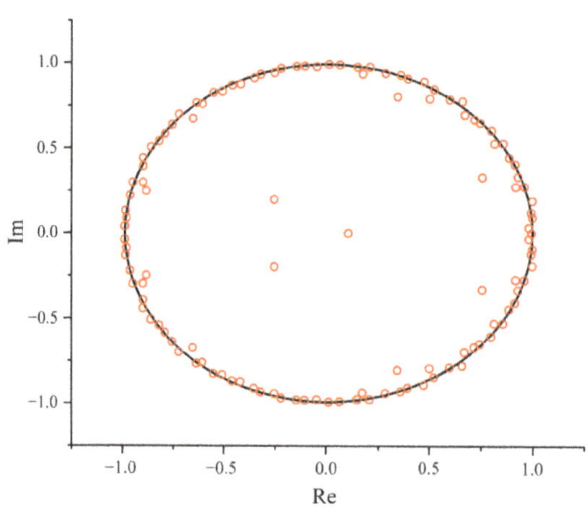

Figure 17. Eigenvalue distribution.

The mode frequencies and their corresponding correlation coefficients are illustrated in Figure 18. The analysis was conducted by extracting mode characteristics at three specific frequencies as follows: 0 Hz, the shaft passing frequency at f_0 = 16.66 Hz, and the blade passing frequency at f_d = 116.66 Hz, based on varying wheel speeds and the number of blades.

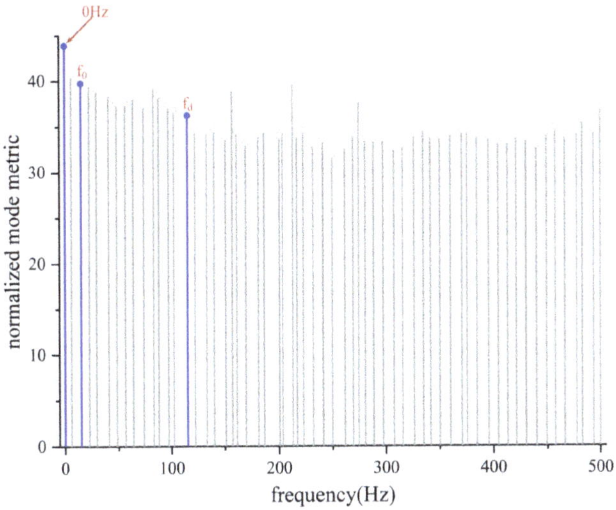

Figure 18. DMD frequency spectral diagram.

The spatial distribution and comparison of three DMD modes under various cavitation conditions are illustrated in Figure 19. The spatial distribution of the 0 Hz frequency mode represents the dominant time-averaged flow field within the flow, highlighting the primary regions of vortex distribution. The coherent vortex locations in the f_0 and f_d frequency modes correspond to those observed in the 0 Hz frequency mode. The f_0 frequency mode predominantly showcases large-scale coherent vortices evolving at the rotational frequency of the main axis. In contrast, the f_d frequency mode depicts coherent vortices evolving at the blade passage frequency, characterized by smaller scales compared to the previous two modes. Consequently, the vortices in the cavitating flow field evolve primarily at the main axis rotational frequency, while the instability induced by cavitation also gives rise to coherent vortices evolving at higher frequencies, such as the blade passage frequency f_d. As the degree of cavitation intensifies, there is an increase in the number of small-scale vortical structures.

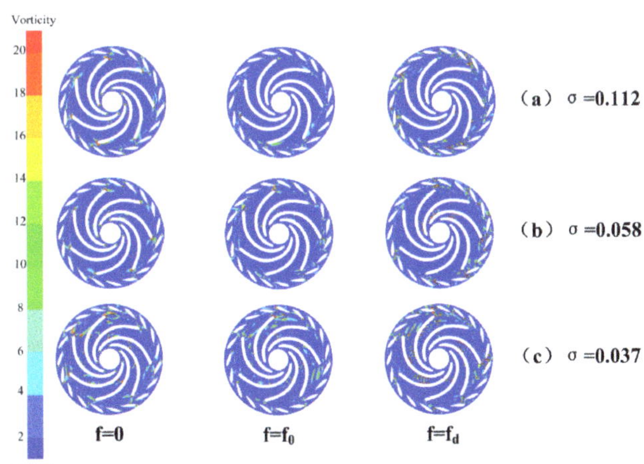

Figure 19. DMD mode distribution.

5. Conclusions

This study investigated the impact of cavitation development on flow field structures through experimental and numerical methods, explored the correlation between energy loss distribution and vorticity distribution, and established the influence of cavitation evolution on vortex dynamics using mode decomposition methods. The key conclusions are as follows:

1. The experimental and numerical performance curves align closely, with errors within acceptable limits, validating the accuracy of the numerical simulations. Using these simulations, cavitation characteristics under both the design flow rate (Q_d) and 60% Q_d conditions were investigated. Compared to the Q_d condition, the 60% Q_d condition exhibited more severe cavitation phenomena and larger-scale vortex structures.
2. The Q criterion and entropy production results demonstrate that high-entropy-production regions closely coincide with high-intensity vortex regions under identical cavitation conditions; consequently, vortices in cavitation flow fields are identified as the primary cause of energy loss in hydraulic machinery.
3. As the mode order increases, the energy content percentage of each mode decreases. The cumulative energy contribution of the first eight POD modes exceeds 99%, encompassing the dominant flow field structures. Specifically, low-order POD modes represent large-scale vortices with low-frequency characteristics and complex frequency characteristics, while high-order modes correspond to small-scale vortices characterized by high-frequency components and simpler frequency components. Intensified cavitation enhances flow field instability, leading to a reduction in energy contribution from low-order modes and an increase in high-order mode energy, accompanied by a rise in the vortex evolutionary frequency mode.
4. The DMD method is employed to decompose the flow field from a frequency perspective, allowing for the extraction of coherent vortices. Cavitation vortices evolve at the rotational frequency of the main shaft f_0, while the instability characteristics of the flow field induced by cavitation are accompanied by coherent vortices that evolve at higher frequencies, such as the blade passage frequency f_d. Furthermore, as the degree of cavitation intensifies, there is an increase in the number of small-scale vortex structures.

The mode decomposition method proposed in this paper can be extended to other turbine flow field analyses, and the conclusions obtained can provide some reference for the design of pump turbine units. As this study did not use a certain linear combination of modes to perform flow field reconstruction, it is possible to further develop modes to combine the advantages of POD and DMD to form a new mode decomposition method and perform fast flow field reconstruction. Through the use of artificial intelligence to analyze and reconstruct the flow field, a function to predict the cavitation flow field quickly and accurately can be formed, and the cavitation performance of the machinery can be evaluated according to the corresponding indicators such that the fluid machinery can be optimized to the greatest extent in a very short time and the time cost can be greatly reduced.

Author Contributions: Analyzed the data and wrote the paper, J.L. (Jiaxing Lu) and J.L. (Jiarui Li); C.Z., Y.Z. and Y.H. designed the experiment and performed the simulation. All authors have read and agreed to the published version of the manuscript.

Funding: This work was funded by the Sichuan Province natural science Foundation project and the Joint Fund of National Natural Science Foundation (no. 2024NSFSC0214 and no. U23A20669).

Data Availability Statement: The data presented in this study are available in the main text of the article.

Conflicts of Interest: The authors declare no conflict of interest.

References

1. Ambec, S.; Crampes, C. Electricity Provision with Intermittent Sources of Energy. *Resour. Energy Econ.* **2012**, *34*, 319–336. [CrossRef]
2. Guezgouz, M.; Jurasz, J.; Bekkouche, B.; Ma, T.; Javed, M.S.; Kies, A. Optimal Hybrid Pumped Hydro-Battery Storage Scheme for off-Grid Renewable Energy Systems. *Energy Convers. Manag.* **2019**, *199*, 112046. [CrossRef]
3. Pommeret, A.; Schubert, K. Optimal Energy Transition with Variable and Intermittent Renewable Electricity Generation. *J. Econ. Dyn. Control* **2022**, *134*, 104273. [CrossRef]
4. Rehman, S.; Al-Hadhrami, L.M.; Alam, M.M. Pumped Hydro Energy Storage System: A Technological Review. *Renew. Sustain. Energy Rev.* **2015**, *44*, 586–598. [CrossRef]
5. Widén, J.; Carpman, N.; Castellucci, V.; Lingfors, D.; Olauson, J.; Remouit, F.; Bergkvist, M.; Grabbe, M.; Waters, R. Variability Assessment and Forecasting of Renewables: A Review for Solar, Wind, Wave and Tidal Resources. *Renew. Sustain. Energy Rev.* **2015**, *44*, 356–375. [CrossRef]
6. Xu, L.; Kan, K.; Zheng, Y.; Liu, D.; Binama, M.; Xu, Z.; Yan, X.; Guo, M.; Chen, H. Rotating Stall Mechanism of Pump-Turbine in Hump Region: An Insight into Vortex Evolution. *Energy* **2024**, *292*, 130579. [CrossRef]
7. Liu, Y.; Gong, J.; An, K.; Wang, L. Cavitation Characteristics and Hydrodynamic Radial Forces of a Reversible Pump–Turbine at Pump Mode. *J. Energy Eng.* **2020**, *146*, 04020066. [CrossRef]
8. Tao, R.; Xiao, R.; Wang, F.; Liu, W. Cavitation Behavior Study in the Pump Mode of a Reversible Pump-Turbine. *Renew. Energy* **2018**, *125*, 655–667. [CrossRef]
9. Zhang, L.; Jing, X.; Wang, Z.; Chang, J.; Peng, G. Analysis of Francis Pump-Turbine Runner Cavitation Flows in Pump Mode. In Proceedings of the ASME 2009 Fluids Engineering Division Summer Meeting, Vail, CO, USA, 2–6 August 2009; pp. 131–134.
10. Li, G.; Hou, W.; Wang, H.; Qin, H.; Zhu, G. Study on Cavitation Performance of High Water-Head Pump Turbine Based on CFD. *Power Syst. Clean Energy* **2017**, *33*, 131–136. [CrossRef]
11. Hao, Y.; Tan, L. Symmetrical and Unsymmetrical Tip Clearances on Cavitation Performance and Radial Force of a Mixed Flow Pump as Turbine at Pump Mode. *Renew. Energy* **2018**, *127*, 368–376. [CrossRef]
12. Meng, Q.; Shen, X.; Zhao, X.; Yang, G.; Zhang, D. Numerical Investigation on Cavitation Vortex Dynamics of a Centrifugal Pump Based on Vorticity Transport Method. *JMSE* **2023**, *11*, 1424. [CrossRef]
13. Wu, J.; Qiu, N.; Zhu, H.; Xu, P.; Si, Q. Numerical Analysis of Vortex Structure in Centrifugal Pump Based on Unsteady Cavitation Flow. *J. Xihua Univ.* **2023**, *42*, 90–99.
14. Zhang, Y.; Liu, K.; Li, J.; Xian, H.; Du, X. Analysis of the Vortices in the Inner Flow of Reversible Pump Turbine with the New Omega Vortex Identification Method. *J. Hydrodyn* **2018**, *30*, 463–469. [CrossRef]
15. Ruan, H.; Guo, P.; Yu, L.; Zhou, C.; Chao, W.; Li, X. Cavitation induced flow instability mechanism of pumptubine under pump conditions. *J. Drain. Irrig. Mach. Eng.* **2021**, *41*, 779–786.
16. Gong, R.; Wang, H.; Chen, L.; Li, D.; Zhang, H.; Wei, X. Application of Entropy Production Theory to Hydro-Turbine Hydraulic Analysis. *Sci. China Technol. Sci.* **2013**, *56*, 1636–1643. [CrossRef]
17. Li, D.; Gong, R.; Wang, H.; Xiang, G.; Wei, X.; Qin, D. Entropy Production Analysis for Hump Characteristics of a Pump Turbine Model. *Chin. J. Mech. Eng.* **2016**, *29*, 803–812. [CrossRef]
18. Yu, A.; Wang, Y.; Lv, S.; Tang, Q. Numerical Analysis of the Cavity Vorticity Transport and Entropy Production in a Micropump. *Int. Commun. Heat Mass Transf.* **2024**, *159*, 108144. [CrossRef]
19. Lumley, J. The structure of inhomogeneous turbulent flows. *Atmos. Turbul. Radio Wave Propag.* **1967**, 166–178.
20. Schmid, P.; Sesterhenn, J. Dynamic Mode Decomposition of Numerical and Experimental Data. *J. Fluid Mech.* **2008**, *656*, 5–28. [CrossRef]
21. Lu, J.; Wu, F.; Liu, X.; Zhu, B.; Yuan, S.; Wang, J. Investigation of the Mechanism of Unsteady Flow Induced by Cavitation at the Tongue of a Centrifugal Pump Based on the Proper Orthogonal Decomposition Method. *Phys. Fluids* **2022**, *34*, 105113. [CrossRef]
22. Yang, G.; Zhang, D.; Shen, X.; Pan, Q.; Pang, Q.; Lu, Q. Investigation on Flow Instability in the Hump Region of the Large Vertical Centrifugal Pump under Cavitation Conditions Based on Proper Orthogonal Decomposition. *Phys. Fluids* **2024**, *36*, 115134. [CrossRef]
23. Yang, J.; Feng, X.; Liao, Z.; Pan, K.; Liu, X. Analysis on the Mechanism of Rotating Stall Inner a Pump Turbine in Pump Mode Based on the Proper Orthogonal Decomposition. *J. Fluids Eng.* **2023**, *145*, 091202. [CrossRef]
24. Xie, Q.; Chen, L.; Zhang, G.; Sun, T. Analysis of unsteady cavitation flow over hydrofoil based on dynamic mode decomposition. *Chin. J. Theor. Appl. Mech.* **2020**, *52*, 1045–1054.
25. Lu, J.; Liu, J.; Qian, L.; Liu, X.; Yuan, S.; Zhu, B.; Dai, Y. Investigation of Pressure Pulsation Induced by Quasi-Steady Cavitation in a Centrifugal Pump. *Phys. Fluids* **2023**, *35*, 025119. [CrossRef]

26. Sirovich, L. Turbulence and the dynamics of coherent structures. I. Coherent structures. *Q. Appl. Math.* **1987**, *45*, 561–571. [CrossRef]
27. Cavazzini, G.; Houdeline, J.-B.; Pavesi, G.; Teller, O.; Ardizzon, G. Unstable Behaviour of Pump-Turbines and Its Effects on Power Regulation Capacity of Pumped-Hydro Energy Storage Plants. *Renewable and Sustainable Energy Reviews* **2018**, *94*, 399–409. [CrossRef]

processes MDPI

Article

Effect of Different Voltage Frequencies of Plasma Actuators on Wind Turbine Blade Lift and Rudder Efficiency

Junjie Xu [1], Jian Zhao [2] and Jianlong Chang [1,*]

[1] College of Mechatronics Engineering, North University of China, Taiyuan 030051, China; x1239297433@163.com
[2] Jinxi Industrial Group Co., Ltd., Taiyuan 030027, China; jianqinxindan@gmail.com
* Correspondence: changjl@nuc.edu.cn

Abstract: In the field of renewable energy, wind power generation is developing rapidly. How to effectively improve the lift performance of wind turbine blades has become an urgent problem. For this reason, a plasma actuator is introduced in this paper to enhance the lift of wind turbine blades, which opens up a new path to solve this dilemma. The impact of plasma actuators with different voltage frequencies on the lift and rudder efficiency of wind turbine blades are explored in this study. Then, the flow control implications of the actuators for the airflow near the rudder are analyzed in detail. It is found that when the plasma actuator is applied to the leading edge of the rudder, the lift and rudder efficiency of the blade improve at both 0° and 15° angles of attack with increasing voltage frequency, but the effects differ significantly. At 0°, the airflow is simpler, the actuator's control effect is more pronounced, and the rudder efficiency shows a decreasing trend as the rudder reflection angle increases. In contrast, at 15°, due to complex airflow, the actuator's control capability is weaker, leading to irregular changes in rudder efficiency.

Keywords: plasma actuator; rudder efficiency; flow control; lift enhancement; vortex

1. Introduction

Energy is crucial to national development, and it is imperative to conserve energy today [1–3]. In the past, the burning of oil, coal, and natural gas was the main way to obtain electricity. However, their reserves are limited, and their burning polluted the atmosphere. With the advancement of technology, wind power has emerged as a renewable and clean way of generating electricity [4,5]. The aerodynamic performance of wind turbine blades, as the core of wind turbines, directly affects the efficiency of power generation, and it is of great significance to study the lift performance of blades [6]. However, the effect of simply changing the blade shape to improve the lift is limited, and there is an urgent need to find new methods. The rudder in an aircraft affects lift by changing the airflow state of the wing, and upward deflection increases the pressure difference between the top and bottom of the wing, which in turn increases the lift [7–9]. Inspired by this, the idea of introducing rudders into fan blades to assist in lifting force appeared. However, the traditional rudder has the problem of low deflection efficiency, which hinders its application in the lift enhancement of fan blades. Solving the problem of rudder inefficiency is the key to improving the lift performance of wind turbine blades and promoting the development of wind power generation technology, which is of great significance for realizing sustainable energy development [10,11].

To address the issue regarding how rudder efficiency can be enhanced, two schemes, namely active flow control and passive flow control, have been put forward from the perspective of flow control [12–14]. A change in wind turbine blade morphology, such as a vortex generator, is usually required for passive flow control. Control can be achieved by increasing energy. However, the driving is inefficient, and flight drag is even increased. On the other hand, active flow control is characterized by a fast response speed and high control efficiency. The flow field environment in the vicinity of the rudder can be productively enhanced through it, and the flow separation phenomenon can be reduced or even prevented [15,16]. Conventional rudders with different structural shapes have different aerodynamic characteristics. For example, high-lift rudders including wedge rudders and fishtail rudders can have their lift significantly raised throughout deflection. However, at the same time, additional drag is generated by them, and the efficiency of the rudder is diminished [17]. Furthermore, the efficiency of the rudder will also be influenced by its area. As a rule, the smaller the rudder area is, the lower its efficiency will be. Meanwhile, the airflow smoothness as well as the total rudder drag are affected by rudder construction details, for example, rudder tip shape, edge configuration, and overall streamlining design, which in turn impact rudder efficiency [18].

The decline in rudder efficiency is often strongly correlated with flow separation. When the airflow from the underside of the blade is flowing to the top of the blade through the rudder gap and is mixed with it, the normal air flow on the top surface will be influenced. Therefore, significant flow separation occurs above the rudder. As a result, the pressure differential between the upper and lower surfaces narrows, and the blade's lift decreases [19]. As blade lift decreases, the overall efficiency of the rudder decreases dramatically. Consequently, the essence of solving the problem of reduced rudder efficiency lies in how the flow separation above the rudder can be eliminated or mitigated [20,21]. If the airflow can be made to flow close to the rudder, or in cases where the degree of flow separation can be controlled to reduce the size of the flow separation area, the efficiency of the rudder can be significantly enhanced. In recent years, plasma actuators, as a new active flow control technique, have been applied extensively in the field of flow control on account of their many advantages like simple structure, flexible mounting position, and fast response [22–25].

The dielectric barrier discharge plasma actuator (DBD plasma actuator) consists of three main components: the exposed electrode, the implanted electrode, and the dielectric layer. This device is widely used because of its simple structure and the thinness of the dielectric layer that can be achieved. The aerodynamic performance of blades is significantly enhanced by DBD plasma actuators. Furthermore, the structural integrity of blades is preserved due to the thin dielectric layer employed, achieving an optimal balance between aerodynamic efficiency and safety. Two main forms of voltage, namely alternating current (AC) and pulse voltage, are predominantly included in dielectric barrier discharge systems. Low-temperature induced airflow can be generated by the barrier discharge under AC voltage, which is consequently referred to as a low-temperature discharge plasma actuator. Ablation damage to turbine blades caused by high airflow temperatures is effectively prevented by this low-temperature characteristic. In contrast, high-temperature induced airflows are generated by barrier discharges at pulsed voltages, which may lead to ablative and damaging effects being caused on the blades. However, the disadvantage of AC voltage is that its induced airflow velocity is slow, and the control effect is relatively limited. Regardless, in the application of wind turbine blades, the airflow speed can still meet the practical needs, so this paper uses the AC voltage dielectric blocking discharge plasma actuator. DBD plasma actuator's mechanism is explained as follows. When the two electrodes of the actuator are connected to the high-voltage power supply, the air near the surface of the actuator is ionized, and a plasma is formed. Under the action of electric and magnetic fields, the plasma begins to move. During the motion, the plasma collides with the surrounding gas molecules and carries out momentum transfer, thus inducing a directional motion of the gas above the actuator [26].

Currently, many scholars focus on the plasma actuator to enhance the aerodynamic performance of the blade when performing in-depth research. The core focus is on the use of plasma actuators to directly control the fluid flow state around the blade. The installation mode of the plasma actuator is modified through approaches such as series-parallel configurations and chordwise distribution adjustments [27–29]. Effective control of the blade's surrounding flow field is thereby achieved, leading to enhanced power output of the wind turbine. The dynamic stall range of the blade has been emphasized by scholars in terms of improving blade lift and optimizing aerodynamic performance [30]. By increasing the stall angle, the blade can still effectively maintain lift under the original stall angle of attack conditions. At the same time, the flow separation point of the blade is demonstrated to be significantly delayed by the plasma actuator in this study [31,32]. Consequently, the pressure difference between the upper and lower surfaces of the blade is elevated, leading to effective improvement in blade lift. Improving the aerodynamic performance of the blade is ultimately about boosting the output power of the wind turbine [33]. Through plasma actuator based flow control, significant improvements are achieved in both the self-starting capability and aerodynamic performance of wind turbines, leading to enhanced energy conversion efficiency. The active flow control methodology is demonstrated to effectively augment lift forces while simultaneously reducing drag coefficients. Existing research has primarily focused on configurations where plasma actuators are directly mounted on turbine blades. However, investigations concerning lift enhancement for blades equipped with control surfaces remain conspicuously absent in the current literature. Consequently, the development of specific solutions aimed at improving control surface efficiency is identified as an essential research direction.

To solve the problem of rudder inefficiency, the following two solutions are proposed. The first solution is that the traditional rudder structure is replaced using plasma technology. Concretely, the fluid field surrounding the blade is fulfilled to be directly controlled through plasma for the rudder. The blade's lift can be significantly raised with the application of this technology. After the roll moment generated using the maximum deflection of the aileron is compared with that on some airfoils, it has been found that the roll moment generated by the actuator even surpasses the moment caused by the aileron [34]. This demonstrated that the function of the conventional rudder can be effectively replaced by the plasma actuator. Installing the plasma actuator at the rudder's front edge in the flow separation position is the alternative, and flow control action is utilized to assist the rudder. Thus, a more significant control effect produced via the rudder deflection can be induced. By this means, the rudder deflection efficiency can be enhanced. The momentum of the blade on the rudder is directly boosted by installing a plasma actuator on it. This thus causes the blade's lift to increase at various rudder reflection angles [35,36]. In conclusion, plasma actuators are a viable option for enhancing rudder efficiency.

In this study, the second solution is adopted because it is not only applicable to a wider range of scenarios but also significantly improves the rudder efficiency while keeping the original rudder structure unchanged. In order to investigate the lift enhancement effectiveness of the plasma actuator under normal and extreme blade incoming flow conditions, 0° and 15° incoming flow angles are selected for analysis in this study. The 0° angle of attack represents the normal incoming flow condition, while the 15° angle of attack simulates the extreme situation where the turbine blades are on the verge of stalling. In this paper, with the EPPLER555 airfoil being selected as the wind turbine blade, the effects of plasma actuators with different voltage frequencies on the lift performance and rudder effectiveness of the wing under the conditions of 0° and 15° incoming flow angles are investigated. Voltage frequency is a key determinant of plasma control capability. An increase in the voltage frequency leads to a significant augmentation in plasma density due to the increased number of discharges per unit time of the plasma actuators. The momentum exchange between the plasma and air molecules is enhanced by this change, leading to further improvement in the flow field control effect of the plasma actuator. The flow separation of the boundary layer on the surface of the wind turbine blade can be effectively suppressed with the help of high-frequency flow control technology [37]. The gas flow distribution around the blade can be optimized through this technology, resulting in altered pressure distribution on both the blade ring volume and upper/lower blade surfaces. The lift and rudder efficiency of the blade are significantly improved by these changes [35]. An in-depth investigation of the regulation mechanism of the plasma actuator in the flow field near the rudder is conducted in this paper. The control effect is modified through the adjustment of voltage frequency, which is explained in detail. The resulting influence on blade lift and rudder efficiency is then systematically analyzed.

2. Computational Model

2.1. Model and Computational Domain Grid

In this study, the numerical simulations were performed using ANSYS Fluent 2020 R2 (ANSYS Inc., Canonsburg, PA, USA). The EPPLER555 airfoil is chosen as the object of investigation, with the chord length being defined as L and the maximum thickness as 0.16 L. It is noted that the plasma actuator's momentum injection and localized heating action allow for efficient control of the flow separation. The fluid apposition time on the blade surface is prolonged, and the fluid's separation point is pushed back, which in turn improves the lift of the blade [38–40]. To ensure the actuator's better control, the actuator is mounted above the rudder at the point where flow separation occurs, which is shown in red in Figure 1. The actuator length (C) used is 0.01 times the blade length and is located 16.5 C from the aft end of the rudder. To guarantee the accuracy of the simulation data, the flow field is large enough to ensure full fluid development. Explicitly, the inlet section's radius in front of the blade is 6 L, while the length of the outflow section behind the blade is 10 L. In the computational domain, the left semicircle is set as the velocity inlet, where the velocity direction is configured at 0° and 15° to the x-axis, and the velocity magnitude is 20 m/s. Meanwhile, in the meantime, the blade part is set up as a no-slip wall, and the top, lower, and right boundaries are positioned as pressure outlets.

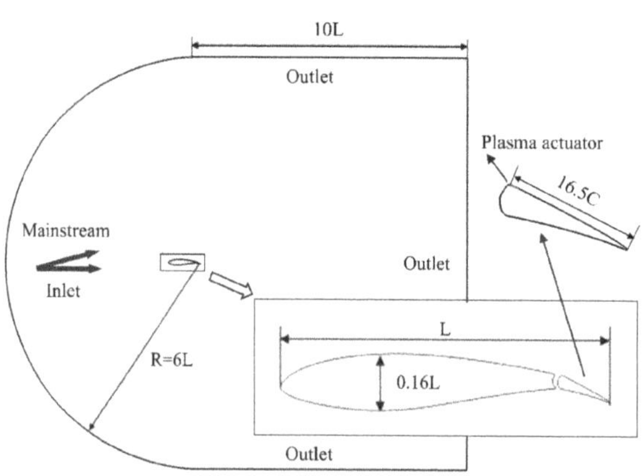

Figure 1. Flow field domain model.

The flow field mesh is an unstructured mesh, as depicted in Figure 2. In the fluid field, the grid size is progressively increased from the blade to the far-field direction. The grid size in the vicinity of the blade and in the right wake region is 1.32×10^{-5} m, while the grid size in the far-field region is expanded to 1.1×10^{-2} m. To accurately capture boundary layer flow conditions, the blade is delineated with the boundary layer mesh of 20 layers to ensure the precision of the simulation results. Meanwhile, the height of the first mesh layer on the blade surface is controlled to be 1.84×10^{-5} m, which satisfies the condition $y^+ < 1$. Meanwhile, the mesh in the wake region of the blade is encrypted to better capture the flow state of the wake of the blade.

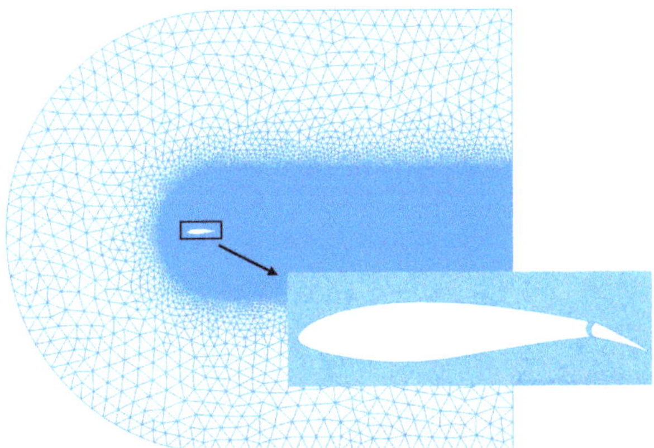

Figure 2. Computational domain mesh.

2.2. Plasma Actuator Phenomenological Model

The DBD plasma actuator discharge time is extremely short, and the specific effects of air ionization and secondary ionization on the particle motion in the flow field are not known. As a result, direct simulation of the plasma's laws of motion is not ideal because it requires an in-depth understanding of the plasma's behavioral patterns over a short period. In order to obtain a similar flow control outcome as the plasma actuator, the flow control was investigated by Shyy et al. by applying plasma excitation on a flat plate [41]. Comparing the phenomenological model with the experimental findings confirms its accuracy. Since a low-temperature plasma is produced utilizing the plasma actuator, the heat generated during plasma generation is neglected in the model. In addition, to attain control of the plasma at the macroscopic level, mean force is applied instead of particle motion. The flow control effect of the plasma is considered as an induced force, and the flow control effect of the plasma is modeled. This study's control model is derived from the phenomenological model proposed by Shyy, which is schematically revealed in Figure 3. The AOB triangular region exhibited in Figure 3 is the effective region of plasma control action, and the direction indicated by the arrow represents the direction of the electric field force, i.e., the direction of plasma movement. During directional movement, the plasma collides with neutral gas molecules, transferring momentum to them and thereby exerting a macroscopic volume force (body force) on the fluid. The body force reaches its maximum at point O and gradually decreases to its minimum along line AB. The electric field intensity at point O of the plasma actuator, denoted as E_0, represents the maximum electric field strength, while E_b signifies the cutoff electric field intensity along the sloping edge AB. In the region where the electric field strength is equal to or less than the cut-off electric field strength, it is assumed that there is no body force.

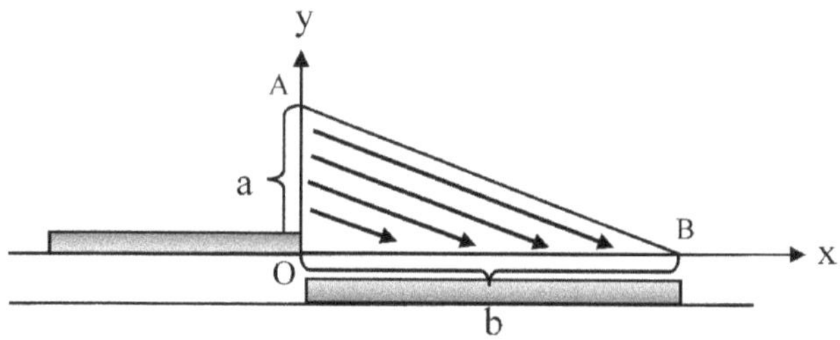

Figure 3. Schematic diagram of the working principle of the plasma actuator.

The maximum electric field strength E_0 is expressed as follows:

$$E_0 = U_0/d, \tag{1}$$

where U_0 is the maximum voltage applied between the electrodes, and d is the shortest distance between the two electrodes.

The electric field strength in the vicinity of the actuator is indicated in Equation (2).

$$|E| = E_0 - k_1x - k_2y, \tag{2}$$

Here, $k_1 = (E_0 - E_b)/b$, $k_2 = (E_0 - E_b)/a$.

The body force is expressed as follows:

$$F = \theta\alpha\rho_c e_c \Delta t E\delta, \tag{3}$$

where θ is the voltage frequency, ρ_c is the charge density, e_c is the charge of the charge, Δt is the electrode discharge time, E is the electric field strength vector, and $\alpha = 1$ is the elastic collision effective coefficient. In addition, $\delta = 1$ when the electric field force is greater than E_b, and $\delta = 0$ when the electric field force is less than E_b.

The vector of the electric field strength is known from the above equation as follows:

$$E = ((E \cdot k_2)/\sqrt{k_1^2 + k_2^2}, (E \cdot k_1)/\sqrt{k_1^2 + k_2^2}), \tag{4}$$

The phenomenological model of the DBD (dielectric barrier discharge) plasma actuator is the equivalent of the induced force to an electric field force that replaces the volumetric force source term in the N-S equations, thus providing flow control in the Fluent 2020 R2. Here, the continuous equation, momentum equation, and energy equations are as follows:

$$\frac{\partial \rho}{\partial t} + \nabla \cdot (\rho U) = 0, \tag{5}$$

$$\frac{\partial \rho U}{\partial t} + \nabla \cdot [\rho UU + PI] - \nabla \cdot \tau = F, \tag{6}$$

$$\frac{\partial \rho e}{\partial t} + \nabla \cdot [(\rho e + \rho)U - (U \cdot \tau) - Q_{ht}] = F \cdot U, \tag{7}$$

where U, ρ, P, e, and t are the flow velocity, fluid density, static pressure, internal energy per unit mass of fluid, and flow time, respectively. τ is the shear stress tensor, and Q_{ht} is the heat transfer term.

In Equation (6), F is the volumetric force source term of the momentum equation, i.e., the electric field force on the plasma. In Equation (7), $F \cdot U$ is the energy source term of the energy equation, i.e., the work done by the electric field force. Because of the low effect of the energy action of the low-temperature plasma, only the momentum source term F is added into the calculation of the flow field.

2.3. Rudder Efficiency and Turbulence Modeling

Rudder efficiency is measured as the amount of moment produced when the rudder is deflected at a specific angle. On the condition that more moments can be produced by the rudder at the same angle of deflection, it is more efficient. As the model is two-dimensional, the magnitude of the roll moment can be reflected by the lift on the blade cross-section. Therefore, the lift coefficient of the blade can be measured to evaluate the rudder efficiency [42]. The rudder efficiency is expressed in Equation (8).

$$\Delta\eta = \frac{(Cl_b - Cl_0) - (Cl_a - Cl_0)}{Cl_a - Cl_0}, \tag{8}$$

Here, Cl_b is the lift coefficient generated when the rudder is deflected and the plasma actuator is applied, Cl_0 is the lift coefficient before the rudder is deflected without the plasma actuator adopted. Cl_a is the lift coefficient produced when the rudder is deflected without the plasma actuator adopted.

The Reynolds-averaged SST K-ω turbulence model is employed in this study. The fundamental principle of the SST model lies in its hybrid formulation. The K-ω model is utilized in the near-wall region, while the K-ε model is adopted at the boundary layer edge and in free shear layers. Additionally, the transport of Reynolds shear stress is incorporated through Bradshaw's hypothesis (the ratio of shear stress to turbulent kinetic energy). This formulation effectively combines the advantages of the K-ω, K-ε, and JK models. Beyond these merits, the SST model demonstrates superior capability in handling the transport of turbulent shear stresses within adverse pressure gradients and separated boundary layers. Consequently, more accurate predictions are achieved for complex flow phenomena including adverse pressure gradients and boundary layer separation.

When compared with the standard K-ω model, several enhancements are implemented in the SST K-ω formulation. Specifically, cross-diffusion originating from the ω equation is incorporated, and the turbulent viscosity is modified by the propagation of turbulent shear stresses. These improvements contribute to the enhanced accuracy and reliability of simulation results. Therefore, the SST K-ω turbulence model is selected to mitigate the effects of shear stress and complex flow conditions. The governing equations for the SST K-ω two-equation turbulence model are presented as follows:

$$\frac{\partial \vec{U}}{\partial t} + \vec{U} \cdot \nabla \vec{U} = -\nabla P + \frac{1}{Re^{2^2}}, \tag{9}$$

$$\nabla \cdot \vec{U} = 0, \tag{10}$$

The SST K-ω turbulence model equations are expressed as follows:

$$\frac{D\rho k}{Dt} = \tau_{ij}\frac{\partial U_i}{\partial x_i} - \beta^* \rho \omega k + \frac{\partial}{\partial x_j}\left[(\mu + \sigma_k \mu_t)\frac{\partial k}{\partial x_j}\right], \tag{11}$$

$$\frac{D\rho\omega}{Dt} = \frac{\gamma}{U_t}\tau_{ij}\frac{\partial U_i}{\partial x_i} - \beta\rho\omega^2 + \frac{\partial}{\partial x_j}\left[(\mu + \sigma_\omega \mu_t)\frac{\partial \omega}{\partial x_j}\right] + z, \tag{12}$$

$$z = 2\rho(1 - F_1)\sigma_{\omega 2}\frac{1}{\omega}\frac{\partial k}{\partial x_j}\frac{\partial \omega}{\partial x_j}, \tag{13}$$

where Ω is the vortex volume.

$$F_2 = \tanh(\arg_2^2), \arg_2 = \max(2\frac{\sqrt{k}}{0.09\omega y}; \frac{500U}{y^2\omega}), \tag{14}$$

Here, y is the distance from the wall, and the parameter value is detailed below:

$$\phi = F_1\phi_1 + (1 - F_1)\phi_2, \tag{15}$$

where mix function are presented as follows:

$$F_1 = \tanh(\arg_1^4), \tag{16}$$

$$\arg_1 = \min(\max(\frac{\sqrt{k}}{0.09\omega y}; \frac{500U}{y^2\omega}); \frac{4\rho\sigma_{\omega 2}k}{CD_{k\omega}y^2}), \tag{17}$$

$$CD_{K\omega} = \max(2\rho\sigma_{\omega 2}\frac{\partial k}{\partial x_j}\frac{\partial \omega}{\partial x_j}; 10^{-20}), \tag{18}$$

2.4. Model Accuracy Validation

In experimental and simulation research, the variation curves of lift coefficients were obtained by Wei et al. [43], and it is discovered that there was a significant degree of agreement between the simulation and the experimental findings. It is suggested that the flow field around the point of stall can be calculated using the LES method. To verify the accuracy of the model, the airfoil is chosen, and the same angle of rudder reflection is set for the simulation calculation. As indicated in Figure 4, our simulation results are contrasted with those of Wei et al., and the three curves are in high agreement. Thus, it can be said that this paper's approach is acceptable.

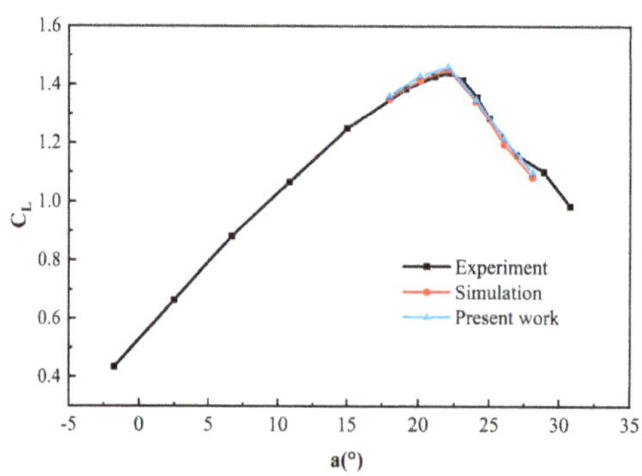

Figure 4. Comparison of lift coefficients for model accuracy validation.

To enhance computational efficiency while ensuring result accuracy, a comprehensive grid independence study is conducted. Three distinct grid configurations are systematically evaluated: Grid 1 features a minimum cell size of 3.12×10^{-5} m with 153,100 total elements; Grid 2 is refined in the blade proximity and wake regions, achieving a minimum size of 1.32×10^{-5} m in these critical zones while maintaining 582,100 elements; and Grid 3 undergoes further refinement with a 6×10^{-6} m minimum cell size around the blade and wake, coupled with enhanced far-field resolution, totaling 1,360,000 elements. The lift coefficient distributions obtained from these grid configurations are comparatively analyzed (Figure 5). It is observed that the results from Grids 2 and Grids 3 demonstrate superior agreement in lift coefficient trends when compared with Grid 1. Based on this systematic evaluation, Grid 2 is selected as the optimal configuration, providing balanced computational accuracy and resource requirements.

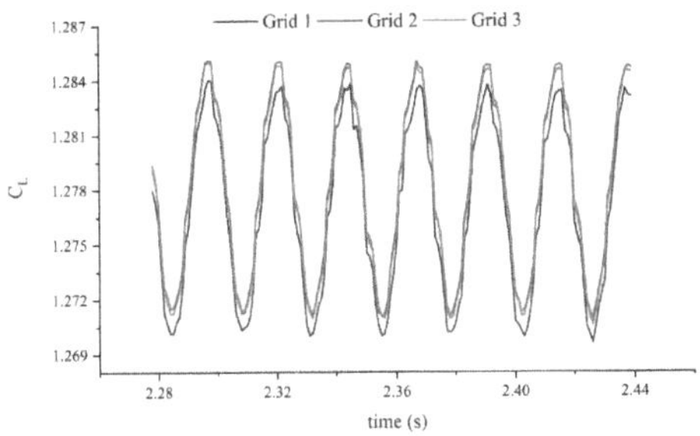

Figure 5. Comparison of lift coefficients of different mesh blades.

3. Results and Analysis

The control effectiveness of the plasma actuator is significantly influenced by the voltage frequency. On the whole, the actuator can be controlled more effectively at higher voltage frequencies. Nevertheless, in complex flow field environments, the controlling impact of the plasma may be significantly impacted. Therefore, in this study, different voltage frequency actuators are selected to investigate changing lift coefficient and rudder efficiency, due to rudder deflection under different angles of attack. Moreover, research is being performed to assess how the actuators' flow control affects the surrounding fluid field.

The Reynolds number for the study in this paper is 410,752 to determine how the lift coefficient is affected by the plasma actuator's voltage frequency, the voltage amplitude is constantly 5 kV, and the voltage frequency is set sequentially at 4 kHz, 5 kHz, 6 kHz, 7 kHz, and 8 kHz. The angle of attack, rudder deflection, and plasma actuator parameters are contained in Table 1. The angles of attack α are 0° and 15°, and the rudder deflection angles β are 3°, 4°, 5°, 6°, 7°, 8°, and 9°.

Table 1. Plasma actuator, angle of attack, and rudder deflection parameters.

Voltage Frequency	Voltage Amplitude	Angle of Attack (α)	Angle of Rudder Reflection (β)
4 kHz, 5 kHz, 6 kHz, 7 kHz, 8 kHz	5 kV	0°, 15°	3°, 4°, 5°, 6°, 7°, 8°, 9°

To facilitate the description of the parameters for the use of the plasma actuator, the following naming rule is developed. When the angle of rudder reflection is 4° and no plasma actuator is used on the rudder, the case is named 4 na (na: no actuator). As a plasma actuator with a voltage frequency of 6 kHz and a voltage amplitude of 5 kV is used on a rudder with an angle of rudder reflection of 4°, this condition is named 4a-6-5.

3.1. *Effect of Actuator Voltage Frequency on Lift Coefficient*

3.1.1. Effect of Voltage Frequency on Lift at α = 0°

The average values of lift coefficients for α = 0° are exhibited in Table 2. It is revealed from the findings that the utilization of a plasma actuator can significantly improve the blade's lift coefficient regardless of the change in the angle of rudder reflection. Especially when the angle of rudder reflection is raised from 3° to 9°, the lift coefficient can be considerably enhanced under a voltage-frequency plasma actuator of 8 kHz. Specifically, as the rudder reflection angle is raised from 8° to 9°, just a 2.9% improvement in the lift coefficient without the use of the actuator is obtained. Otherwise, when an 8 kHz plasma actuator is used, the lift coefficient at 9° rudder declination reaches 0.5526, which is a 19.6% lift coefficient improvement compared to that at 8° rudder declination. The influence of the actuator is most notable at β = 4°, particularly at 8 kHz. The actuator voltage frequency is 8 kHz, and the angle of rudder reflection is enhanced from 3° to 4°. The lift coefficient is prominently enhanced by about 82.7%. In contrast, the same change in the angle of rudder reflection without the plasma actuator increases the lift coefficient by only 6.2%.

Table 2. Average value of lift coefficient for α = 0°.

Voltage Frequency	β = 3°	β = 4°	β = 5°	β = 6°	β = 7°	β = 8°	β = 9°
No actuator	0.3603	0.3826	0.3959	0.4277	0.4425	0.4621	0.4755
4 kHz	0.4337	0.4606	0.4489	0.4767	0.4887	0.5045	0.5145
5 kHz	0.4654	0.4962	0.4654	0.4893	0.5009	0.5159	0.5240
6 kHz	0.5501	0.6104	0.4837	0.5029	0.5128	0.5270	0.5335
7 kHz	0.5888	0.6395	0.5058	0.5181	0.5237	0.5378	0.5428
8 kHz	0.6119	0.6583	0.5369	0.5360	0.5358	0.5494	0.5526

Without the actuator, the lift coefficient increases as the angle of rudder reflection rises. Yet, when the same voltage-frequency actuator is employed, the lift coefficients still rise with the enhancement of the angle of rudder reflection, except for a transient drop in the lift coefficient when rudder reflection angle goes up from 4° to 5°. It is further observed that the maximal value of the lift coefficient appears at β = 9°, as the voltage frequencies are 4 kHz and 5 kHz. Nevertheless, the lift coefficient reaches highest value at β = 4°, and the voltage frequency is from 6 kHz to 8 kHz. It is indicative of this phenomenon that the higher the voltage frequency is, the more effective the control of low rudder deflection. Based on this finding, by varying the actuator's voltage frequency at various rudder reflection angles, energy savings and the desired lift may be attained.

Lift coefficient curves of different voltage frequencies at α = 0° are shown in Figure 6a–f. Every 1° of rudder deflection, with or without the actuator, enhances blade lift. However, as the plasma actuator is applied, the lift is significantly elevated in contrast to the case without the actuator. In addition, the larger the voltage frequency is, the more obvious the enhancement effect of the blade lift is. Some of the lift coefficient curves remain stable with time, while some of the curves show small vibrations. This shows that the plasma actuator is not only effective in increasing the lift but also in avoiding the blade chattering problem caused by the drastic change of the lift. Upon detailed analysis of Figure 6, it can be seen that there is an abrupt rise in the lift coefficient when the voltage frequency reaches a specific value. To be specific, both the case of 4 kHz and 6 kHz frequencies result in substantial jumps in lift coefficients, when the rudder is deflected from 3° to 4°. A jump in the lift coefficient also results from the 4 kHz frequency when the rudder is deflected from 8° to 9°. Evidently, to attain the goals of a high lift coefficient and low energy consumption, voltage-frequency plasma actuators possessing these characteristics can be used.

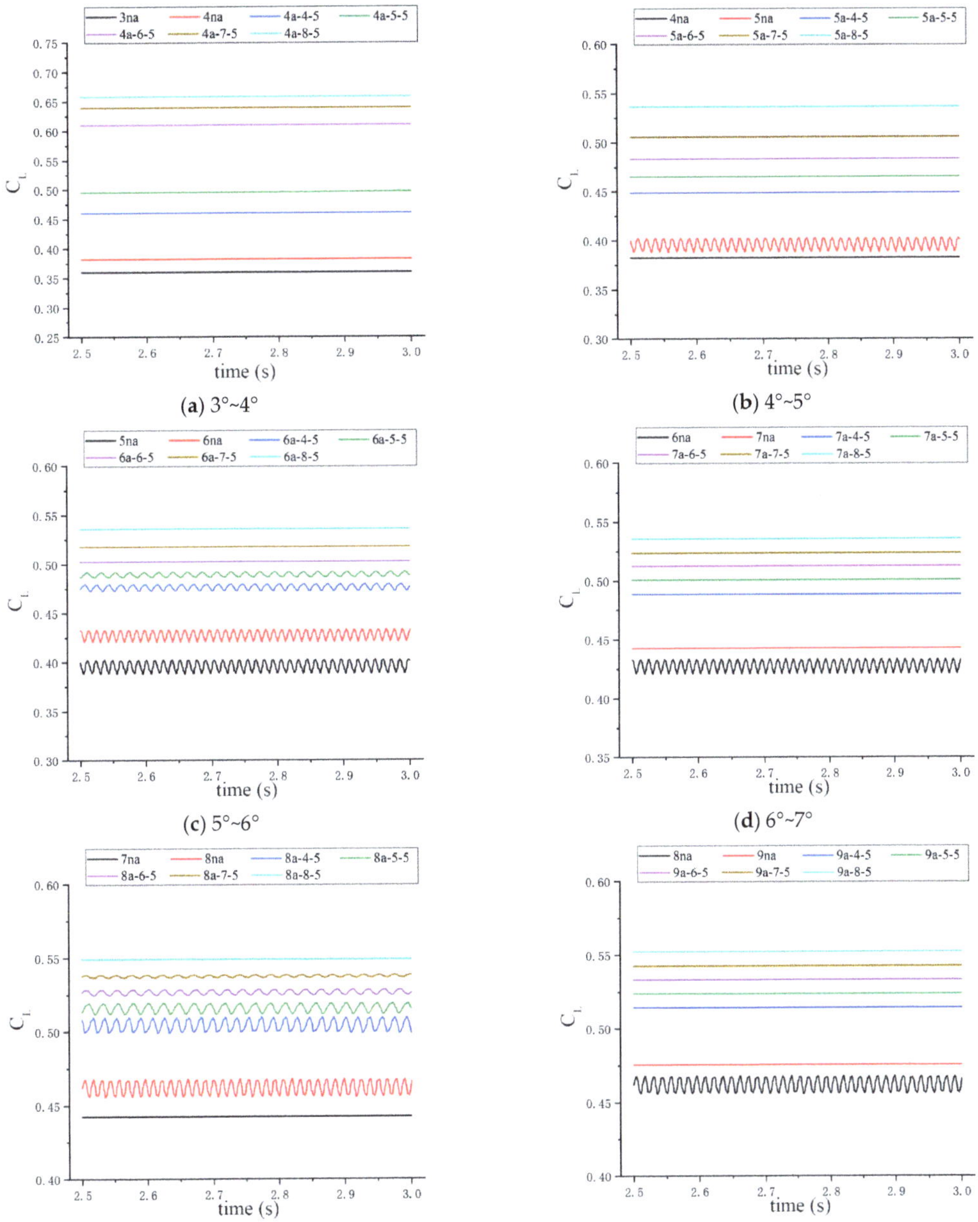

(**a**) 3°~4° (**b**) 4°~5°

(**c**) 5°~6° (**d**) 6°~7°

(**e**) 7°~8° (**f**) 8°~9°

Figure 6. Lift coefficient curves of different voltage frequencies at $\alpha = 0°$.

3.1.2. Effect of Voltage Frequency on Lift at $\alpha = 15°$

Average value of lift coefficient for $\alpha = 15°$ are demonstrated in Table 3. The lift coefficient of the blade at different angles of rudder reflections is obviously enhanced through the use of actuators. In most cases, the lift coefficient increases as the voltage frequency increases. However, when $\beta = 4°$ and the voltage frequency increases from 7 kHz to 8 kHz, the lift coefficient exhibits a decrease. In addition to this special case, all other rudder deflections adhere to this rule. Particularly at $\beta = 8^\circ$ and a voltage frequency close to 8 kHz, the plasma actuator demonstrates the optimal flow regulation efficiency. In the absence of the plasma actuator, the lift coefficient increased by merely 1.6% after the angle of rudder reflection is altered from 7° to 8°. Nevertheless, when an actuator with a frequency of 8 kHz is employed, the identical change in angle of rudder reflection leads to a 10.9% increase in the lift coefficient. To be specific, the disparity between the lift coefficients at 4 kHz and 8 kHz voltage frequencies is only 0.0124 when $\beta = 4^\circ$. In contrast, the disparity between the lift coefficients at these two frequencies raise markedly to 0.0959 at $\beta = 7^\circ$. It is revealed that the larger the angle of rudder reflection is, the more pronounced the effect of the enhancement in voltage frequency on the rudder efficiency. When the angle of rudder reflection increases from 8° to 9°, at the same voltage frequency, all lift coefficient averages decrease. There is also a diminish in the mean value of the lift coefficient after an increasing angle of rudder reflection. For instance, when the voltage frequency is 4 kHz and the angle of rudder reflection is increased from 5° to 6°, the average lift coefficient exhibits the same decreasing tendency. Nonetheless, the average lift coefficients are generally enhanced versus the situation without the actuator. It is obvious that the variation in the average lift coefficient is more complex and lacks a distinct regularity as the voltage frequency and angle of rudder reflection grows in comparison with the 0°angle of attack.

Table 3. Average lift coefficient values at $\alpha = 15^\circ$.

Voltage Frequency	$\beta = 3^\circ$	$\beta = 4^\circ$	$\beta = 5^\circ$	$\beta = 6^\circ$	$\beta = 7^\circ$	$\beta = 8^\circ$	$\beta = 9^\circ$
No actuator	1.2589	1.2635	1.2817	1.2821	1.3021	1.3235	1.3340
4 kHz	1.2708	1.3293	1.2903	1.2879	1.3072	1.4179	1.4036
5 kHz	1.3297	1.3400	1.2926	1.3600	1.3844	1.4257	1.4117
6 kHz	1.3375	1.3484	1.3610	1.3687	1.3915	1.4328	1.4181
7 kHz	1.3456	1.3559	1.3682	1.3748	1.3979	1.4379	1.4231
8 kHz	1.3538	1.3417	1.3747	1.3798	1.4031	1.4443	1.4282

The curves of the influence of various voltage frequencies upon the lift coefficient after a 1° increase in the rudder at $\alpha = 15^\circ$ are illustrated in Figure 7. It is evident that the fluctuations of all the lift coefficient curves grow in magnitude and show a cyclic variation compared to the case at $\alpha = 0^\circ$. By comparing the curves in Figure 7, three main patterns of lift coefficient curves can be concluded: delta function type, as displayed in 3 na; irregular oscillation type, as illustrated in 9 na; and oscillation delta function type, as depicted in 5a-6-7. To be specific, in Figure 7b, the curves corresponding to 5a-6-4 are of the delta function type. The formation of these three characteristic curves is intrinsically related to vortex shedding mechanisms. The delta function type curve is generated by periodic shedding of vortices with consistent intensity. When the shed vortices maintain comparable strength, the amplitude of lift coefficient fluctuations remains relatively uniform. The irregular oscillation type curve results from a non-uniform vortex strength distribution. Stronger vortices induce more pronounced lift fluctuations during shedding, while weaker vortices produce negligible effects, collectively leading to the irregular oscillatory pattern. The oscillation delta function type curve is attributed to the coexistence of high-intensity vortices with shorter shedding periods and persistent low-intensity vortex shedding. The characteristic oscillatory delta composite pattern is created by this superposition. This curve exhibits similarity to the lift coefficient curve in the case where no actuator is employed, while it notably augments the lift of the blade. The curve representing 5 na, on the other hand, belongs to the type of irregular oscillations with a smaller average fluctuation amplitude and is one of the better lift coefficient waveforms produced by the blade in the absence of plasma excitation. As for the lift coefficient curves generated by 5a-6-5, 5a-6-6, and 5a-6-7, they are of the oscillation delta function type. In this type of curve, although the amplitude of each oscillation is small, the curve of the simple delta function type has a substantially smaller variation in the highest and lowest points throughout a whole trigonometric cycle. Accordingly, the oscillatory effects of the blade can be amplified by such curves with adverse impacts. So, when selecting voltage frequencies, frequencies that may lead to the appearance of such curves should be avoided as much as possible.

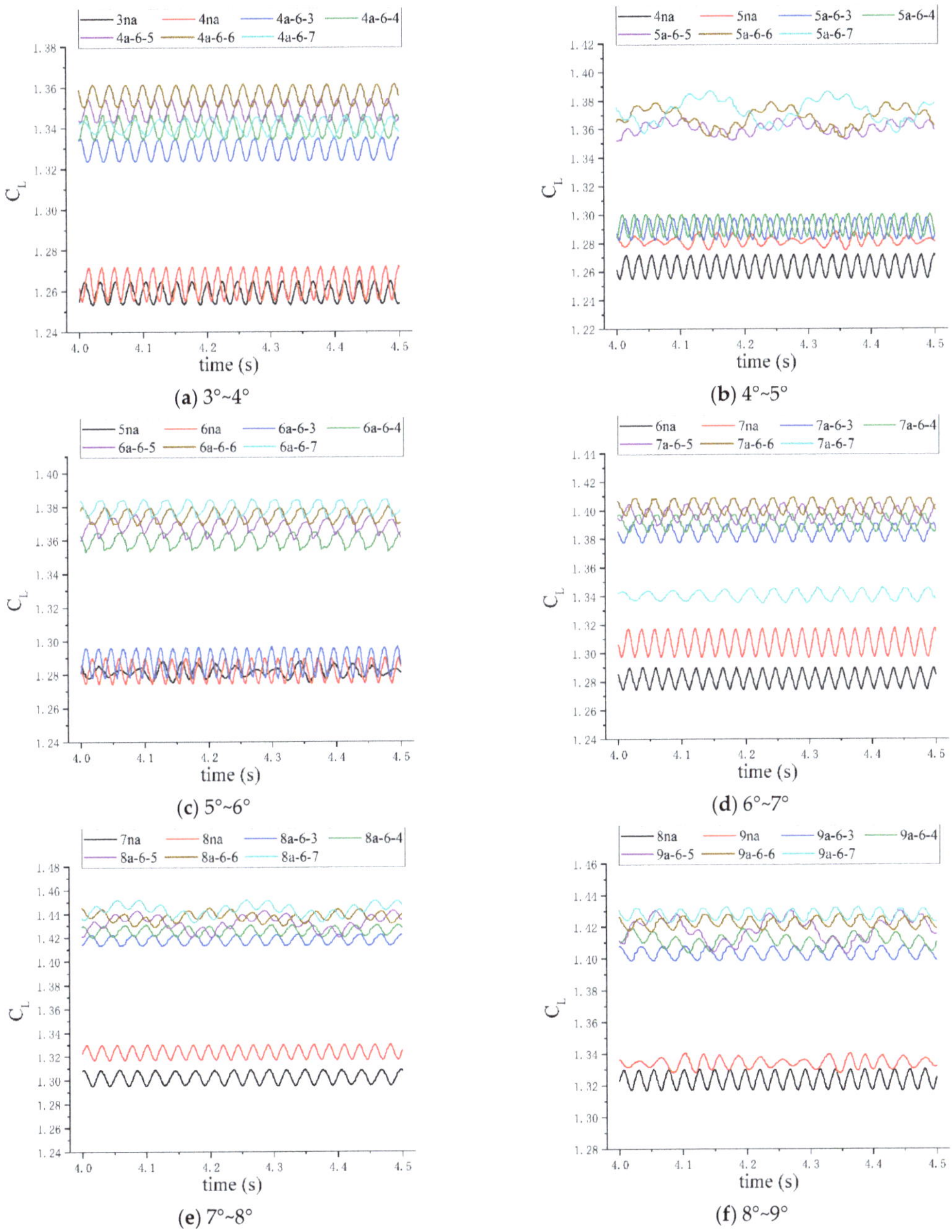

Figure 7. Lift coefficient curves of different voltage frequencies at $\alpha = 15°$.

3.2. *Effect of Actuator Voltage Frequency on Rudder Efficiency*

3.2.1. Effect of Voltage Frequency on Rudder Efficiency at $\alpha = 0°$

The impact of actuator on rudder efficiency at different voltage frequencies is revealed in Table 4. There is evidence that the use of a plasma actuator can have a very positive impact on the efficiency of the rudder. When the angle of rudder reflection is increased from 3° to 4°, the rudder efficiency is only 21.649% at a voltage frequency of 4 kHz. However, as the voltage frequency is increased to 8 kHz, the rudder efficiency surges to 76.52%, which is 54.871% higher than that at 4 kHz. It is illustrated that the improvement in rudder efficiency is considerably affected by the augmentation in voltage frequency. At the same voltage frequency, the larger the angle of rudder reflection is, the lower the rudder efficiency usually is. As an illustration, when the voltage frequency is 8 kHz and the rudder is deflected from 8° to 9°, the efficiency is reduced by 59.835% compared to that when the rudder is deflected from 3° to 4°. In addition, when the angle of rudder reflection is smaller, its efficiency improvement is faster while the actuator frequency is raised, and vice versa. Specifically, when the rudder is deflected from 3° to 4°, the rudder efficiency can be advanced by about 9.88% if the voltage frequency is enhanced from 4 kHz to 5 kHz. Nonetheless, when the rudder deflection is changed from 8° to 9°, the increase in rudder efficiency amounts to merely 2.056%, even though the voltage frequency is simultaneously elevated from 4 kHz to 5 kHz. Therefore, when the angle of rudder reflection is relatively small, a higher voltage frequency can be chosen to enhance efficiency. Conversely, when the angle of rudder reflection is relatively large, a lower voltage frequency should be adopted to conserve energy.

Table 4. Rudder efficiency for different voltage frequencies at $\alpha = 0°$.

Voltage Frequency	3°~4°	4°~5°	5°~6°	6°~7°	7°~8°	8°~9°
4 kHz	21.649%	13.853%	12.377%	10.802%	9.582%	8.440%
5 kHz	31.529%	18.165%	15.559%	13.654%	12.158%	10.496%
6 kHz	63.225%	22.948%	18.995%	16.437%	14.667%	12.551%
7 kHz	71.302%	28.725%	22.834%	18.985%	17.107%	14.564%
8 kHz	76.520%	36.853%	27.355%	21.814%	19.729%	16.685%

3.2.2. Effect of Voltage Frequency on Rudder Efficiency at $\alpha = 15°$

Upon comparing the data in Table 5, it is found that the actuator's rudder efficiency is substantially higher at $\alpha = 15°$ than at $\alpha = 0°$. Exactly, the maximum value of rudder efficiency is 76.52% at $\alpha = 0°$, whereas the maximum value of rudder efficiency attains 24,425% at $\alpha = 15°$, representing an increase of 24,348.48% compared with $\alpha = 0°$. It is indicated that the plasma at $\alpha = 15°$ is more effectively regulated with the actuator. At $\alpha = 0°$, the maximum value of rudder efficiency emerges when the voltage frequency is 8 kHz and the angle of rudder reflection is altered from 3° to 4°, and the rudder efficiency declines with the augmentation of the angle of rudder reflection. Nevertheless, a clear regularity is not exhibited in the impact of the angle of rudder reflection being increased or decreased on the rudder efficiency at $\alpha = 15°$. The complex flow within the flow field in the absence of the actuator's action results in an irregular variation in the rudder efficiency. Still, upon the actuator being incorporated, the efficiency of the rudder is remarkably enhanced in comparison with that without the actuator, albeit the variation of the rudder efficiency remains complex. It is suggested that an important role in improving rudder efficiency is played through plasma actuators. However, it is worthy of note that this effect fails to exhibit a distinct regularity with the augmentation of the angle of rudder reflection. In contrast, the rudder efficiency declines with the increase in the angle of rudder reflection at $\alpha = 0°$.

Table 5. Rudder efficiency for different voltage frequencies at $\alpha = 15°$.

Voltage Frequency	3°~4°	4°~5°	5°~6°	6°~7°	7°~8°	8°~9°
4 kHz	1430%	47%	1450%	25%	441%	663%
5 kHz	1663%	60%	19475%	412%	478%	740%
6 kHz	1846%	436%	21650%	446%	509%	801%
7 kHz	2009%	475%	23175%	479%	535%	849%
8 kHz	1700%	511%	24425%	505%	564%	897%

3.3. *Flow Field Analysis*

3.3.1. Flow Field Analysis at $\alpha = 0°$

The streamlines of different voltage frequencies at $\alpha = 0°$ and $\beta = 4°$ are presented in Figure 8. The findings show that the existence of the plasma actuator substantially raises the kinetic energy of the fluid on the rudder's upper surface. This, in turn, speeds up the fluid flow velocity and leads to the downstream movement of associated rudder vortices over the top surface. As the plasma actuator's voltage frequency grows, the area of adhering vortices is reduced, until they vanish completely at 7 kHz and 8 kHz. At this point, the gas can be made to flow closely to the upper surface of the rudder, which effectively augments the lift of the blade.

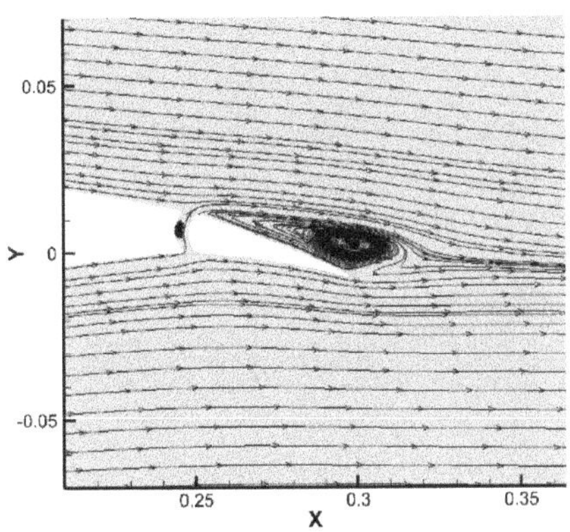

(**a**) no actuator

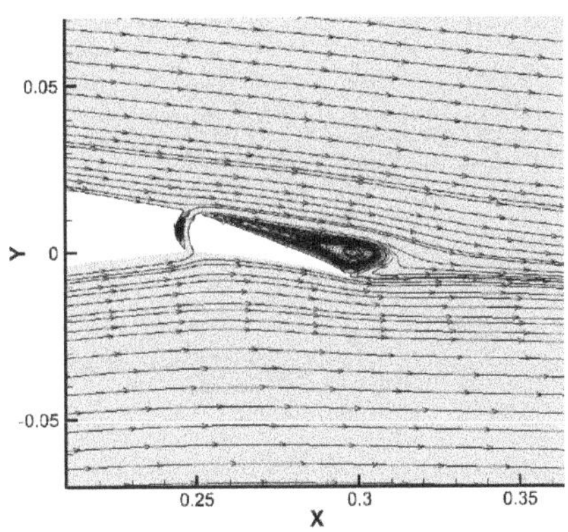

(**b**) 4 kHz

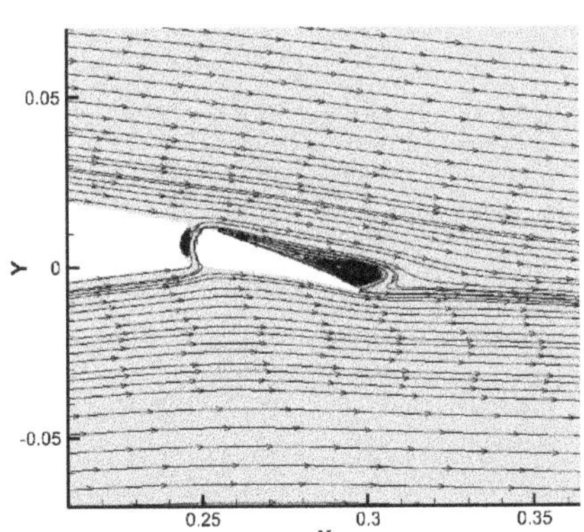

(**c**) 5 kHz

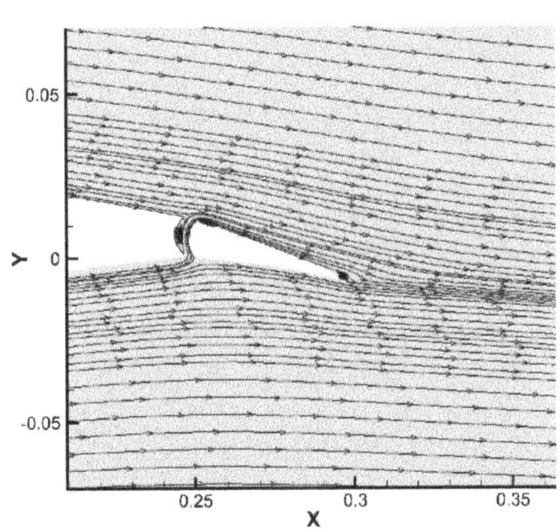

(**d**) 6 kHz

Figure 8. *Cont.*

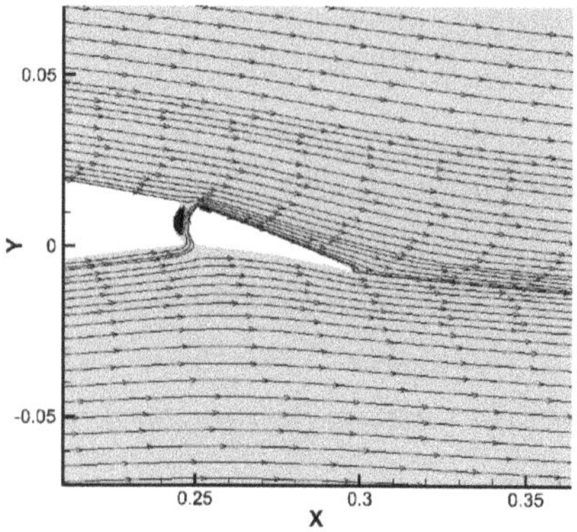

(e) 7 kHz

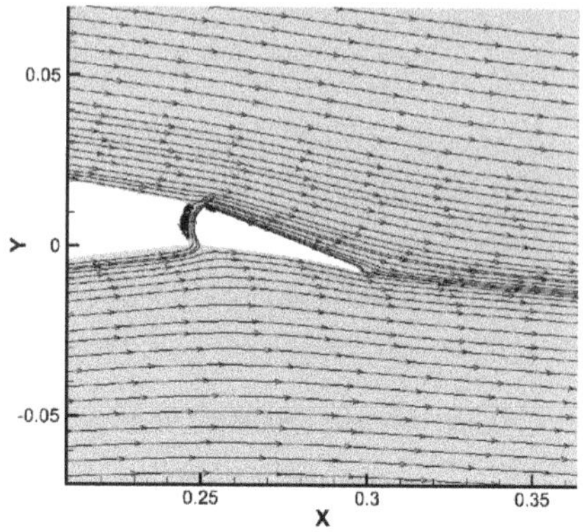

(f) 8 kHz

Figure 8. Streamlines of different voltage frequencies at $\alpha = 0°$ and $\beta = 4°$.

As shown on Figure 9a, the flow at the front edge of the rudder is quite complex. Due to the gap between the main blade and the rudder, the airflow flowing faster on the lower surface will ascend to the top surface and interact with the fluid at top surface. Fluid detachment from the blade top surface at the rudder leading edge is caused by this interaction, which in turn produces flow separation as well as the development of huge attached vortices above the rudder. To achieve optimal flow control, the actuator is situated close to the point where the flow separates [44]. However, at voltage frequencies of 4 kHz and 5 kHz, the influence of the actuator is not visible and does not cause the fluid above the rudder to flow near the blade surface. However, the attached vortex area is significantly reduced, and the tendency of the fluid to flow near the wall surface gradually emerges with the increase in the voltage frequency. When the voltage frequency is further increased to the range of 6 kHz to 8 kHz, the generation of vortices can be clearly observed at the installation location of the plasma actuator. These vortices help to improve the energy transfer among the entering flow and the boundary layer, and the fluid is induced to flow close to the wall, which markedly improves the flow field capability to resist the backpressure gradient as well as suppress flow separation.

The velocity nephograms of different voltage frequencies at $\alpha = 0°$ and $\beta = 4°$ are illustrated in Figure 9. While in the case with no actuator, a large blue area above the rudder with a velocity lower than 8 m/s is presented. This blue area is gradually reduced as the voltage frequency of the plasma actuator rises. When the voltage frequency is 8 kHz, only a very small blue low-velocity region is visible on the rudder tail. During the operation of the plasma actuator, a high-velocity fluid is generated at the actuator installation position, and the incoming flow rate is exceeded by its flow rate, which is approximately 26 m/s. In addition, this high-velocity fluid is not only present at the front edge of the rudder, but also covers the seam area between the main blade and the rudder. When the voltage frequency of the plasma actuator increases, the area of the high-velocity fluid region will be correspondingly enlarged, and the flow velocity will also be gradually augmented, so as to achieve a better flow control effect.

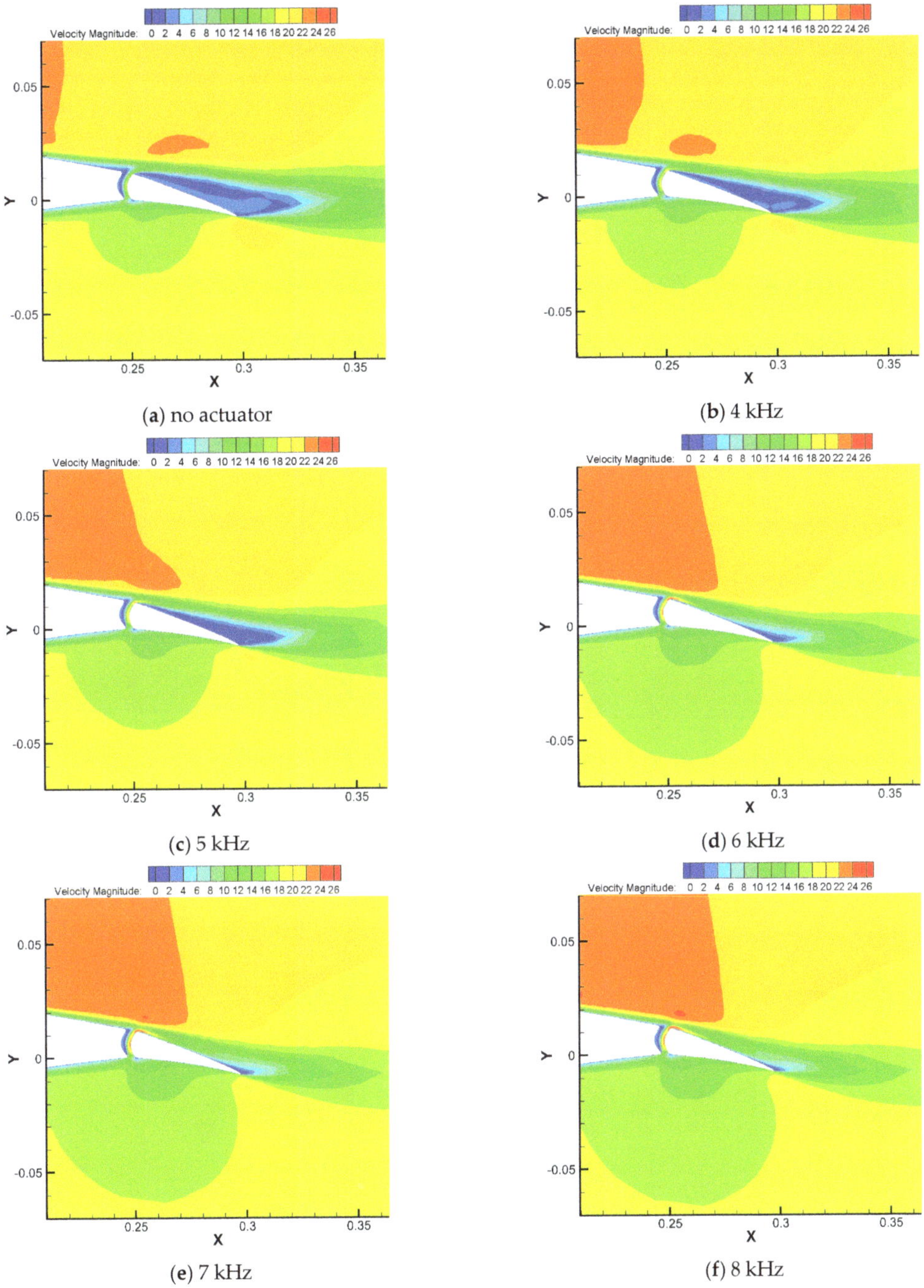

Figure 9. Velocity nephograms of different voltage frequencies at $\alpha = 0^\circ$ and $\beta = 4^\circ$.

3.3.2. Flow Field Analysis at $\alpha = 15^\circ$

The streamlines of different voltage frequencies at $\alpha = 15^\circ$ and $\beta = 4^\circ$ are depicted in Figure 10. Flow separation at the main blade is noted, and this separation persists across the entire blade. In Figure 10a–e, four vortices are generated above the blade, including the flow separation vortex above the main blade, the vortex generated by the velocity difference between blade and rudder, the vortex above rudder, and the vortex behind rudder. Under the combined effects of these vortices, the flow speed above the blade surface is reduced, and the blade's lift is significantly decreased. In Figure 10f, three vortices are generated over the blade. Among them, the flow separation vortex above the main blade shifts above the rudder, developing a vortex with a larger area. Simultaneously, the vortices above and behind the rudder progressively approach the rudder under the compression of the larger vortex, and the area of the vortex is significantly diminished. The flow trajectory of the lower edge fluid flowing from the gap to the upper edge after coupling with the upper edge fluid is shown by the red line. The angle of the red line gradually grows as the voltage frequency rises, as shown by the red line. It is indicated that an increase in voltage frequency helps the fluid flow closer to the control surface, thereby demonstrating that effective control is provided by the plasma actuator at the leading edge of the rudder. Especially in Figure 10f, a significant change is observed in the red line, indicating that the fluid is able to flow closely along the upper wall of the rudder. This indicates that the induction effect is optimized when the 8 kHz plasma actuator is employed.

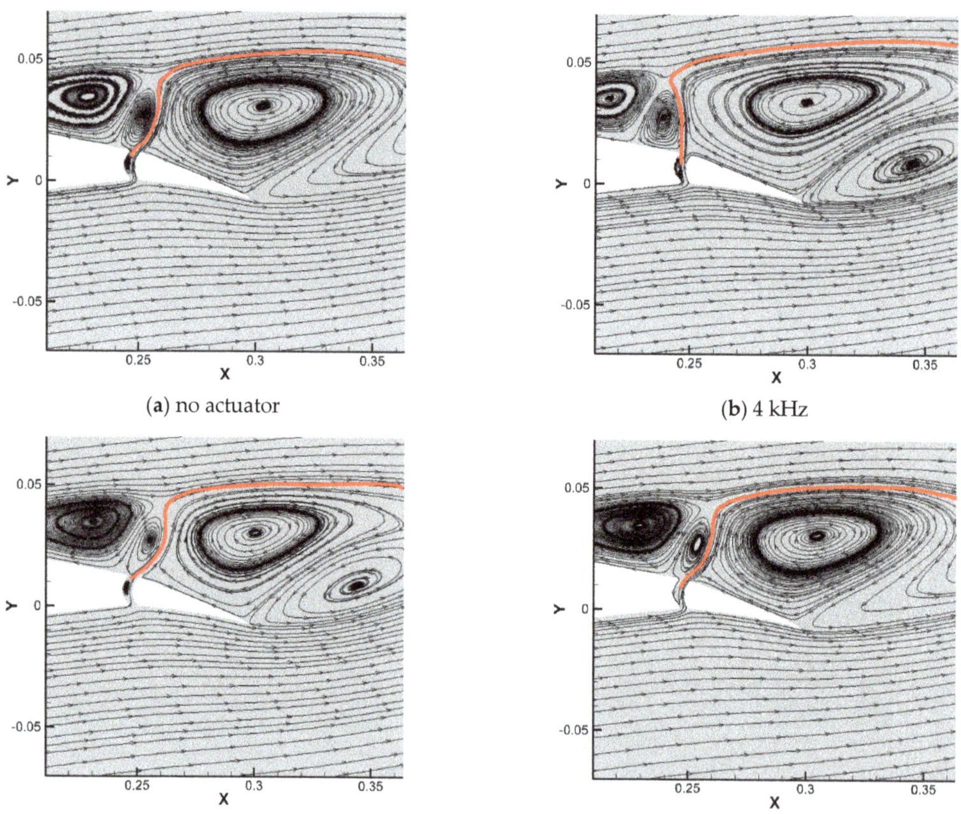

(a) no actuator (b) 4 kHz

(c) 5 kHz (d) 6 kHz

Figure 10. *Cont.*

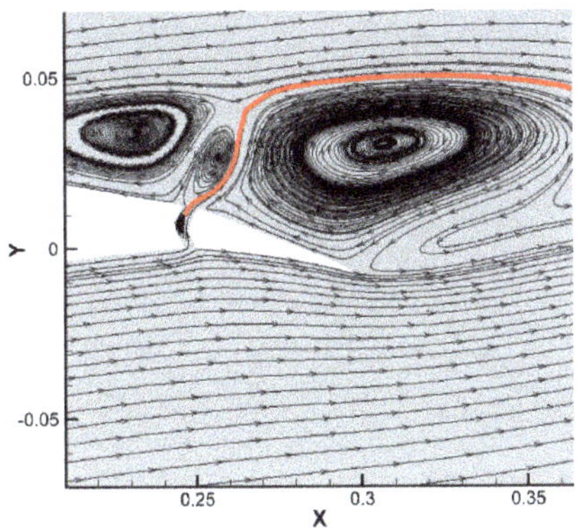

(e) 7 kHz

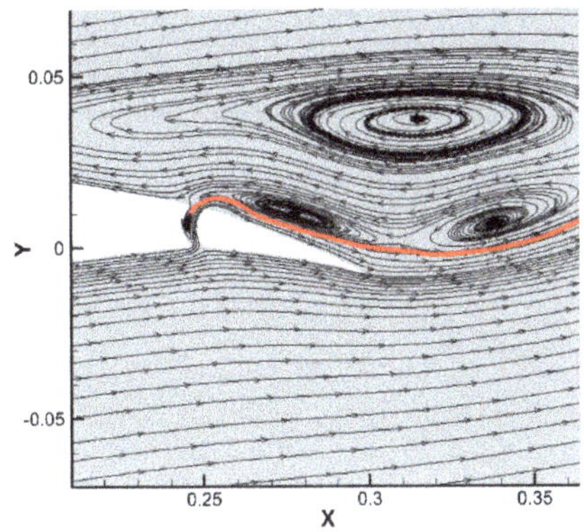

(f) 8 kHz

Figure 10. Streamlines of different voltage frequencies at $\alpha = 15°$ and $\beta = 4°$.

The velocity nephograms of different voltage frequencies at $\alpha = 15°$ and $\beta = 4°$ are presented in Figure 11. Compared with the 0° angle, the area of the blue low-speed region is significantly increased. Large low-speed areas are formed over the rudder and main blade. The low velocity region coincides with the region where vortices are present in Figure 10, and the correlation between the decrease in fluid flow velocity and the presence of vortices is further confirmed. Simultaneously, the red area at the front edge of the rudder is markedly different from the condition at $\alpha = 0°$. At $\alpha = 0°$, the upper edge of the rudder is where the red portion of the actuator is stretched. However, when $\alpha = 15°$, the red area is connected to the upper edge of the rudder only at 8 kHz. In other cases, the red high-velocity fluid will flow along the gap to the upper right of the blade. This indicates that the flow from the gap to the upper edge is not effectively controlled by the plasma actuator at the low voltage frequency at $\alpha = 15°$. At the same time, it also further indicates that under the complex flow field conditions at $\alpha = 15°$, the regulating effect of the actuator is relatively low, and the voltage frequency needs to be enhanced to resist the original flow state, which is significantly reduced compared with that at $\alpha = 0°$.

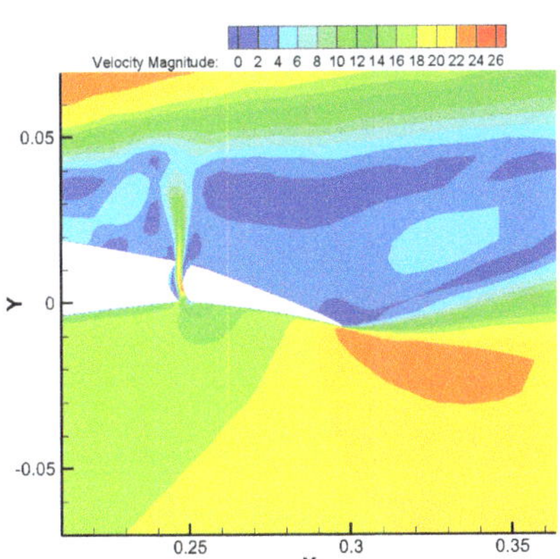

(a) no actuator

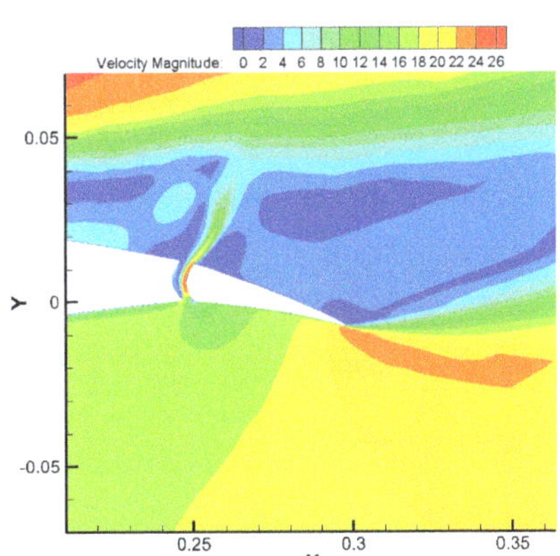

(b) 4 kHz

Figure 11. *Cont.*

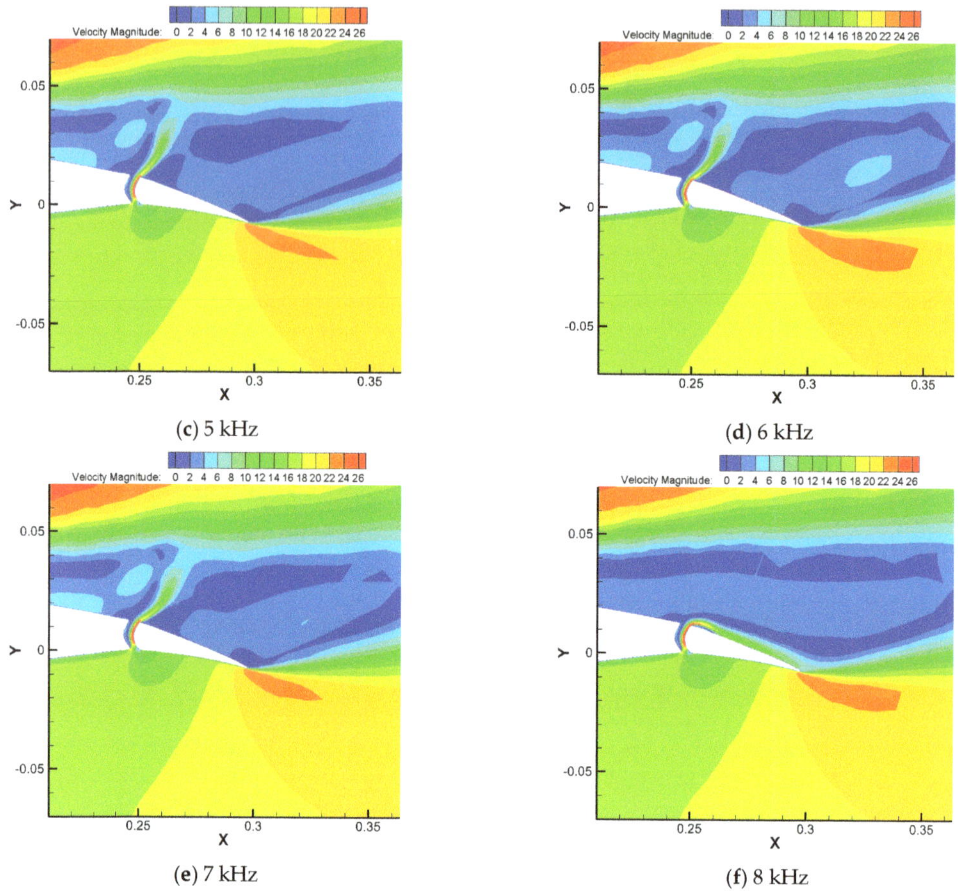

Figure 11. Velocity nephograms of different voltage frequencies at $\alpha = 15°$ and $\beta = 4°$.

4. Conclusions

In this study, the investigations are conducted on the lift coefficients, rudder efficiency, and flow field conditions of the wind turbine blades of the EPPLER555 airfoil at different rudder deflections after being controlled with different plasma actuators at two angles of attack. The following conclusions are obtained:

(a) The lift and rudder efficiency of the blade at both 0° and 15° angles of attack can be effectively improved by the actuator, and then the efficiency of wind turbine power generation is effectively improved. The actuator's influence upon the rudder efficiency is increased as the rudder deflection angle decreases at $\alpha = 0°$. However, at $\alpha = 15°$, the complex flow of airflow around the blade leads to the fact that the enhancement effect of the actuator on the rudder efficiency does not show a regular change with the rudder deflection angle. The enhancement effect of the actuator on the rudder efficiency is amplified with the rise of the voltage frequency for both angles of attack.

(b) When $\alpha = 0°$, flow separation is only caused at the leading edge of the rudder, and only attached vortices are generated above the rudder. When $\alpha = 15°$, flow separation is caused when the airflow passes through the main blade, and multiple vortices are formed above the blade, making the flow situation more complicated. Moreover, at

$\alpha = 15°$, three types of lift coefficient curves are generated by applying the plasma actuator: delta function type, irregular oscillation type, and oscillation delta function type. To reduce the issue of blade flutter created by the drastic change in the lift coefficient, the oscillating delta function type of curve should be avoided when selecting the voltage frequency. At the same time, to accomplish high lift coefficient and low energy consumption, voltage-frequency plasma actuators with jumping characteristics can be used.

(c) As the excitation frequency of the plasma actuator grows at $\alpha = 0°$, the attached vortex area above the rudder gradually reduces. As a result, the fluid flow velocity is gradually accelerated, causing the flow to move closer to the rudder. As the excitation frequency of the plasma actuator increases at $\alpha = 15°$, the angle of the flow trajectory of the fluid flowing from the gap at the lower edge to the upper edge is gradually increased. It is indicated that the increase in the voltage frequency contributes to the fluid flow being moved closer to the rudder. Especially when the voltage frequency reaches 8 kHz, the original flow state can be effectively changed, causing the fluid to be closer to the rudder.

Author Contributions: Conceptualization, J.X. and J.C.; Data curation, J.Z.; Formal analysis, J.C.; Funding acquisition, J.C.; Investigation, J.X., J.Z. and J.C.; Methodology, J.X. and J.Z.; Project administration, J.Z.; Resources, J.X.; Software, J.X.; Supervision, J.C.; Validation, J.X., J.Z. and J.C.; Visualization, J.X.; Writing—original draft, J.X.; Writing—review & editing, J.Z. and J.C. All authors have read and agreed to the published version of the manuscript.

Funding: This research was funded by the National Natural Science Foundation of China (grant number 12202408) and the Aviation Science Foundation (grant number 2019ZA0U0001).

Data Availability Statement: The data used to support the findings of this study are available from the corresponding author (changjianlong1989@126.com) upon request.

Conflicts of Interest: Author Jian Zhao was employed by Jinxi Industrial Group Co., Ltd. The remaining authors declare that the research was conducted in the absence of any commercial or financial relationships that could be construed as a potential conflict of interest.

Abbreviations

The following abbreviations are used in this manuscript:

DBD	Dielectric barrier discharge
na	No actuator

References

1. Jureczko, M.; Mrowka, M. Multiobjective Optimization of Composite Wind Turbine Blade. *Materials* **2022**, *15*, 4649. [CrossRef] [PubMed]
2. Teng, H.; Li, S.; Cao, Z.; Li, S.; Li, C.; Ko, T.J. Carbon Fiber Composites for Large-Scale Wind Turbine Blades: Applicability Study and Comprehensive Evaluation in China. *J. Mar. Sci. Eng.* **2023**, *11*, 624. [CrossRef]
3. Zhang, Y.; Zhang, C.; You, H.; Liang, K.; Wei, C.; Sun, A. Analysis of performance improvement methods for offshore wind turbine blades affected by leading edge erosion. *Ocean Eng.* **2024**, *310*, 118773. [CrossRef]
4. Chu, Y.-J.; Lam, H.-F.; Peng, H.-Y. Numerical investigation of the power and self-start performance of a folding-blade horizontal axis wind turbine with a downwind configuration. *Int. J. Green Energy* **2022**, *19*, 28–51. [CrossRef]
5. Zhang, X.; Zheng, M. Numerical Simulation of Fluid-Structure Coupling for a Multi-Blade Vertical-Axis Wind Turbine. *Appl. Sci. Basel* **2023**, *13*, 8612. [CrossRef]
6. Yan, P.; Li, Y.; Gao, Q.; Lian, S.; Wu, Q. Design and Analysis of an Adaptive Dual-Drive Lift-Drag Composite Vertical-Axis Wind Turbine Generator. *Energies* **2023**, *16*, 7529. [CrossRef]
7. Sato, M. Passive Fault-Tolerant Flight Control Design Example for Elevator Efficiency Reduction Using Structured H∞ Control. In Proceedings of the 29th Mediterranean Conference on Control and Automation (MED), Puglia, Italy, 22–25 June 2021; pp. 627–632.

8. Kim, S.; Horspool, K.R. Nonlinear Controller Design for Non-minimum Phase Flight System Enhanced by Adaptive Elevator Algorithm. In Proceedings of the AIAA Science and Technology Forum and Exposition (AIAA SciTech), Orlando, FL, USA, 6–10 January 2020.
9. Singh, Y.; Bagri, K.; Jayakumar, A.; Rizzoni, G. Fault Diagnostics for Oscillatory Failure Case in Aircraft Elevator Servos. In Proceedings of the 22nd World Congress of the International Federation of Automatic Control (IFAC), Yokohama, Japan, 9–14 July 2023; pp. 408–415.
10. Singh, U.; Rizwan, M.; Malik, H.; Marquez, F.P.G. Wind Energy Scenario, Success and Initiatives towards Renewable Energy in India-A Review. *Energies* **2022**, *15*, 2291. [CrossRef]
11. Urabe, C.T.; Ikegami, T.; Ogimoto, K. Mitigation of Short-Term Fluctuations in Wind Power Output in a Balancing Area on the Road Toward 100% Renewable Energy. *EEE Access* **2022**, *10*, 111210–111220. [CrossRef]
12. Esmaeili, A.; Sousa, J.M.M. Combined passive and active flow control for fixed-wing micro air vehicles. *Int. J. Micro Air Veh.* **2023**, *15*, 17568293231197127. [CrossRef]
13. Gupta, S.; Kumar, S.; Kumar, R. Control of leading-edge vortices over delta wing using flow control methods: A review. In Proceedings of the 2nd International Conference on Functional Material, Manufacturing and Performances (ICFMMP), Phagwara, India, 17–18 September 2022; pp. 2189–2193.
14. Rashid, S.; Nawaz, F.; Maqsood, A.; Salamat, S.; Riaz, R. Review of wave drag reduction techniques: Advances in active, passive, and hybrid flow control. *Proc. Inst. Mech. Eng. Part G J. Aerosp. Eng.* **2022**, *236*, 2851–2884. [CrossRef]
15. Spens, A.; Pisano, A.P.; Bons, J.P. Leading-Edge Active Flow Control Enabled by Curved Fluidic Oscillators. *AIAA J.* **2023**, *61*, 1675–1686. [CrossRef]
16. Zhao, G.; Zhao, Q.; Li, P.; Wang, B. Numerical Investigations for Passive and Active Flow Control on Tiltrotor-Wing Aerodynamic Interactions. *J. Aerosp. Eng.* **2021**, *34*, 04021094. [CrossRef]
17. Hai-long, S.; Obwogi, E.O.; Yu-min, S. Scale effects for rudder bulb and rudder thrust fin on propulsive efficiency based on computational fluid dynamics. *Ocean Eng.* **2016**, *117*, 199–209. [CrossRef]
18. Liu, J.; Hekkenberg, R. Sixty years of research on ship rudders: Effects of design choices on rudder performance. *Ships Offshore Struct.* **2017**, *12*, 495–512. [CrossRef]
19. Benedetti, D.M.; Veras, C.A.G. Wind-Tunnel Measurement of Differential Pressure on the Surface of a Dynamically Inflatable Wing Cell. *Aerospace* **2021**, *8*, 34. [CrossRef]
20. Carstensen, B.; Krüger, S. Coupling of a Boundary Element Method With a Boundary Layer Method for Accurate Rudder Force Calculation Within the Early Design Stage. In Proceedings of the 40th ASME International Conference on Ocean, Offshore and Arctic Engineering (OMAE), Electr Network, Virtual, 21–30 June 2021.
21. Lin, M.; Yang, F.; Wang, C. Numerical investigation on the generation mechanism of aero-heating of rudder shaft from three-dimensional flow separation and vortices. *AIP Adv.* **2022**, *12*, 045228. [CrossRef]
22. Abdullaev, A.; Kotvitskii, A.; Moralev, I.; Ustinov, M. On the Possibility of Cross-Flow Vortex Cancellation by Plasma Actuators. *Aerospace* **2023**, *10*, 469. [CrossRef]
23. Chen, S.; Shi, Z.; Zhao, Z.; Geng, X.; Chen, Z. Investigation of vertical tail buffeting alleviation controlled by nanosecond plasma actuators. *Phys. Fluids* **2021**, *33*, 087109. [CrossRef]
24. Manuilovich, S.V. Stability and Receptivity of the Swept-Wing Boundary-Layer Flow Controlled by a Periodic Sequence of Plasma Actuators. *Fluid Dyn.* **2021**, *56*, 630–644. [CrossRef]
25. Manuilovich, S.V. Resonance Impact of a Periodic Sequence of Plasma Actuators in Flow Control in a Swept-Wing Boundary Layer. *Fluid Dyn.* **2023**, *58*, 542–554. [CrossRef]
26. Yu, J.; Liu, H.; Xu, D.; Chen, F. Investigation of the DBD Plasma Effect on Flat Plate Flow. *Plasma Sci. Technol.* **2014**, *16*, 197–202. [CrossRef]
27. Benmoussa, A.; Páscoa, J.C. Enhancement of a cycloidal self-pitch vertical axis wind turbine performance through DBD plasma actuators at low tip speed ratio. *Int. J. Thermofluids* **2023**, *17*, 100258.
28. Omidi, J. DBD plasma actuator effect on mid-blade aerodynamics enhancement of a NREL 5-MW wind turbine. In Proceedings of the 2022 International Conference on Futuristic Technologies in Control Systems & Renewable Energy (ICFCR), Malappuram, India, 21–22 July 2022; pp. 1–6.
29. Omidi, J.; Mazaheri, K. Aerodynamic Enhancement and Improving the Performance of a Six-Megawatt DOWEC Wind Turbine by Micro-Plasma Actuator. *Int. J. Mech. Sci.* **2021**, *195*, 106228. [CrossRef]
30. Li, G.; Zhang, W.; Jiang, Y.; Yang, P. Experimental investigation of dynamic stall flow control for wind turbine airfoils using a plasma actuator. *Energy* **2019**, *185*, 90–101. [CrossRef]
31. Govindan, S.S.; Santiago, A.E.X. Optimized dielectric barrier discharge-plasma actuator for active flow control in wind turbine. *Struct. Control. Health Monit.* **2019**, *26*, e2454. [CrossRef]

32. Papadopoulos, C.; Kaparos, P.; Vlahostergios, Z.; Misirlis, D.; Yakinthos, K. 2D optimization of a Small Horizontal Axis Wind Turbine blade using flow control techniques. In *IOP Conference Series: Materials Science and Engineering, Proceedings of the 10th EASN International Conference on Innovation in Aviation & Space to the Satisfaction of the European Citizens (10th EASN 2020), Virtual, 2–4 of September 2020*; IOP Science: Bristol, UK, 2021; p. 012041.
33. Omidi, J. Advances and opportunities in wind energy harvesting using plasma actuators: A review. *Clean Energy* **2024**, *8*, 197–225.
34. Wang, X.; Yan, J.; Dhupia, J.S.; Zhu, X. Active Flow Control Based on Plasma Synthetic Jet for Flapless Aircraft. *IEEE Access* **2021**, *9*, 24305–24313. [CrossRef]
35. Almuina Pica, D.E.; Keener, K.; Lubitz, W.D. Control of Crosswind Force on Aircraft Vertical Tail Models using Plasma Actuators. In Proceedings of the AIAA Scitech 2022 Forum, San Diego, CA, USA, 3–7 January 2022; p. 0832.
36. Cai, J.; Li, C.; Li, H.; Meng, X. Design and Test of Plasma Control Surface on Unmanned Aerial Vehicle. In *The Proceedings of the 2018 Asia-Pacific International Symposium on Aerospace Technology (APISAT 2018) 9th*; Zhang, X., Ed.; Springer: Berlin/Heidelberg, Germany, 2019; pp. 1298–1311.
37. Greenblatt, D.; Schulman, M.; Ben-Harav, A. Vertical axis wind turbine performance enhancement using plasma actuators. *Renew. Energy* **2012**, *37*, 345–354. [CrossRef]
38. Chern, M.-J.; Wang, C.-C.; Wei, Z.-H.; Lu, P.-C. Numerical Investigation of a Pitching NACA 0012 Wing with Plasma-Based Flow Control Using Prediction-Correction Direct-Forcing Immersed Boundary Method. *J. Aerosp. Eng.* **2023**, *36*, 04023066. [CrossRef]
39. Chernyshev, S.; Gadzhimagomedov, G.; Kuryachiy, A.; Sboev, D.; Tolkachev, S. 3D Turbulent Boundary Layer Separation Control by Multi-Discharge Plasma Actuator. *Aerospace* **2023**, *10*, 73–85. [CrossRef]
40. Zafarghandi, F.M.; Soltani, M.R. Experimental Investigation of the Active Flow Control over a Cranked-Delta Wing. *J. Aircr.* **2024**, *61*, 73–85. [CrossRef]
41. Shyy, W.; Jayaraman, B.; Andersson, A. Modeling of glow discharge-induced fluid dynamics. *J. Appl. Phys.* **2002**, *92*, 6434–6443.
42. Zhao, Y.-J.; Zhou, Y.; Ji, K.-X.; Zhao, X.-Y.; Shao, X.; Chang, J.-L. Influence of Rudder Deflection Angle on Airfoil Lift Characteristics Under the Effect of Synthetic Jet Actuator. *J. Nanoelectron. Optoelectron.* **2021**, *16*, 1475–1484. [CrossRef]
43. Wei, B.; Wu, Y.; Liang, H.; Su, Z.; Li, Y. Flow control on a high-lift wing with microsecond pulsed surface dielectric barrier discharge actuator. *Aerosp. Sci. Technol.* **2020**, *96*, 105584. [CrossRef]
44. Abdollahzadeh, M.; Pascoa, J.C.; Oliveira, P.J. Comparison of DBD plasma actuators flow control authority in different modes of actuation. *Aerosp. Sci. Technol.* **2018**, *78*, 183–196. [CrossRef]

processes

Article

A CFD Study of Thermodynamics and Efficiency Metrics in a Hydrogen-Fueled Micro Planar Combustor Housing Dual Heat-Recirculating Cylindrical Combustors for MTPV Applications

Faisal Almutairi

Department of Mechanical Engineering, College of Engineering, King Faisal University, Al-Ahsa 31982, Saudi Arabia; falmutairi@kfu.edu.sa

Abstract: The micro combustor is the energy source of micro-thermophotovoltaic systems; thus, optimizing its design is one of the key parameters that lead to an increase in output energy. Therefore, to enhance the system's overall efficiency, this numerical work introduces a new design configuration for parallel-flow (PF) and counter-flow (CF) hydrogen-fueled micro cylindrical combustors integrated into a micro planar combustor. To overcome the short residence time in micro combustor applications causing high heat dissipation, the micro cylindrical combustors house heat-recirculating channels to allow more heat to transfer to the external walls. In pursuit of this target, simulations are carried out to analyze the thermodynamic and system efficiency parameters. In addition, different initial operating conditions are varied to optimize the system, including inlet velocity and equivalence ratio. The results reveal that the PF and CF structures result in significantly higher wall temperatures and more uniform wall temperature variations than the conventional design (CD). Despite the high entropy generations, the exhaust gas temperatures of the PF and CF are 591 K and 580 K lower than the CD, respectively, and both the PF and CF result in 14% increases in radiation efficiency. Increasing the inlet velocity improves the key thermal parameters in the new designs; however, the system efficiency experiences a drastic reduction. The power output density highlights the unity equivalence ratio as optimal. The PF and CF designs yield roughly identical findings, but the CF exhibits more uniform wall temperatures in most cases due to the equal thermal energy from opposite sides.

Keywords: hydrogen combustion; renewable energy; micro-thermophotovoltaic system; micro combustion; thermal performance; energy conversion

1. Introduction

Recently, micro-electro-mechanical systems have been advanced tremendously due to the technological revolution empowering high-performance applications in various domains, such as biomedicine [1], wireless communications [2], micro aircraft [3], and hydrogen production [4]. The importance of developing micro- or even nano-technologies lies in their long service lives, small sizes, easy integration, and high energy outputs [5,6]. The combustion-based micro power generation systems, particularly micro-thermophotovoltaic (MTPV), have become an active research field as they feature low harmful emissions, static design, and easy maintenance [7,8]. In addition, the components of MTPV technology are amongst its focal advantages, as it consists of a micro combustor, the filter, and the photovoltaic cell [9]. In MTPV systems, the thermal energy released by the combustion of a gaseous fuel in the micro combustor is radiated to photovoltaic cells to generate electricity [10]. The reliability, stability, and efficiency of MTPV technology rely on the micro combustor's output power density, as the latter is the energy source for the former. At a fixed energy input, the key variables that determine the micro combustor's energy output are the wall temperature and the uniformity thereof. The former determines the amount of energy radiated from the micro combustor to the PV cells, whereas the latter ensures the uniformity of radiation energy with respect to the area of the external walls. Several parameters restrict the heat transfer capacity and system efficiency of micro combustors, including the high ratio of surface area-to-volume, extensive energy dissipation, and short dwell time [11,12].

To address these challenges, scholars have proposed different methods of boosting the energy conversion efficiency of micro combustors, such as employing catalysts [13,14], bluff bodies [15,16] and porous media [17,18]. Moreover, the addition of hydrogen has demonstrated great potential for improving the heat transfer mechanisms of combustion-based applications [19,20] owing to its high flame speed, high diffusion coefficient, and high energy content [20,21]. For instance, Zhao et al. [20] investigated the potential impacts of hydrogen additions on the thermal and efficiency parameters in an ammonia-fueled micro planar combustor featuring a heat recirculating structure. It was found that enriching the ammonia–air mixture with hydrogen causes the mean wall temperature and exergy efficiency to significantly increase, while the heat dissipation resulting from the total entropy generation is decreased from 46.58 W to 36.24 W for the mass blended hydrogen ratio of 0.4 and 0.7, respectively. Similar findings were reported in [22], highlighting that the high hydrogen substitution intensifies the combustion process, leading to further improvements in the exergy and radiation efficiencies. Nieto-Londoño et al. [23] studied the effects of a hydrogen-enriched methane–air mixture in millimeter-scale reactors on flame dynamics and stabilization. The results reveal that using more hydrogen leads to more intense combustion and a reduction in CO_2 emissions, due to the high reactivity coefficient of hydrogen and the fact that it contains zero carbon atoms in its chemical formula, respectively. As hydrogen has proven to be more than 300 times as efficient as batteries [22], the utilization of a green alternative fuel like hydrogen advances the energy and technology sectors in the transition toward dependency on renewable energy sources to achieve sustainability and decarbonization.

Maximizing the benefits of hydrogen in microscale systems requires certain factors to be taken into account to accommodate the high flame propagation of hydrogen and the compact dimensions of the micro devices. Thus, many new design configurations for micro combustors have been proposed in the literature. For example, Dai et al. [24] conducted a numerical investigation on the effects of a multi-baffle structure inserted into a hydrogen-fueled micro planar combustor on energy conversion efficiency. The authors reported that incorporating multi-baffles improves outer wall temperature and wall temperature uniformity, leading to enhanced total energy outputs, particularly when the multi-baffles are placed in the center of the geometry. Li et al. [25] examined the performance of a micro cylindrical combustor featuring center-cleared twisted tapes powered by hydrogen. The outcomes reveal that the new design promotes turbulence, lowers the resistance of flow, and increases the surface area, resulting in improved heat transfer coefficients and, accordingly, key thermal parameters. Cai et al. [22] investigated the system efficiency of a hydrogen–ammonia-fueled micro cylindrical combustor featuring periodic wall structures. They highlighted that wall periodicity leads the external walls to absorb more heat and, consequently, increases the radiation efficiency and mean wall temperature. Zhang et al. [26] examined the non-premixed combustion behavior and exergy of a hydrogen-fueled micro combustor with a vase-shaped configuration and tangential inlet. It was found that the tangential inlet improves the mixing process, whereas the vase shape leads to highly increased exergy efficiency and extends the blow-out limit. Sheykhbaglou et al. [27] studied the key thermal parameters of an ammonia–hydrogen-fueled micro combustor employing a novel bluff-body and swirl-stabilized structure. As highlighted by the authors, the bluff-body with a hemisphere-shaped configuration results in a 44% radiation efficiency. Furthermore, the vane angle, ranging between 15 and 60 degrees, decreases the NO_x emissions and increases the mean wall temperature, the level of wall temperature uniformity, and radiation efficiencies.

In addition to the outlined studies, researchers have reported that separating the high-temperature flame in a micro planar combustor is an effective means of improving thermal flow dynamics and system efficiency. Tang et al. [28] performed a numerical investigation of the effects of inserting parallel separating plates into a micro planar combustor on the heat transfer process. They outlined that, compared to the conventional design, the separation of the hot-gas flow is conducive to promoting the coefficient of heat transfer by ~141 $W/(m^2 \cdot K)$, which consequently improves the mean temperature over the external walls by more than 100 K. Su et al. [29] studied the key thermal parameters of a multiple-channel micro planar combustor. The findings demonstrate that employing several flow paths results in improvements in the mean wall temperature, radiation energy, and radiation efficiency by 38.9 K, 6.17 W, and 2.17%, respectively, in comparison to the single-channel configuration.

These investigations brought to attention above, along with others in the literature, have highlighted the importance of optimizing the design configuration of the micro combustor and separating the flow in enhancing the heat transfer from the combustion to the external walls. Additionally, with the same volume of the combustion chamber and cross-sectional area, it has been noted that the micro planar combustor radiates more thermal energy than the cylindrical shape [30], making it the favorite geometry for MTPV systems. Consequently, there remains significant room for improvement in such applications in terms of developing highly efficient systems in the energy sector. Thus, as improving the micro combustor structure, partitioning the inlet stream channel, and utilizing a planar shape for micro combustors promote radiation efficiency and output power density, this paper introduces a micro planar combustor with parallel-flow (PF) and counter-flow (CF) heat-recirculating micro cylindrical combustors built in, without altering their external lengths and widths. In a previous work [31], a novel structure for a hydrogen-fueled micro cylindrical combustor with a single-path inlet and double-path outlet featuring a heat-recirculating configuration was proposed. The results reveal that, compared to the traditional design, the new design lowers the exhaust gas temperature by 171.6 K, and improves the mean wall temperature, exergy efficiency, and radiation efficiency by 63.4 K, 4.31%, and 9.92%, respectively. For this, the current numerical work aims to demonstrate the effects of a micro planar combustor housing dual micro cylindrical combustor geometries, as proposed in [31], on the thermal and efficiency variables. Section 2 describes the dimensions of the proposed design configuration, simulation modeling approaches, initial and boundary conditions, the mesh independence test, and the validation process. Section 3 compares the new geometric structure to the conventional design with respect to the thermodynamic parameters and system efficiency, and discusses the effects of varying the inlet velocity and equivalence ratio. Section 4 highlights an overview of the core outcomes of the numerical investigation.

2. Materials and Methods

2.1. Geometric Model

Figure 1 illustrates the newly proposed configuration of a micro planar combustor with two micro cylindrical combustors embedded inside, which was constructed using ANSYS Design Modeler R2 2023. The dimensions of geometry are listed in Table 1.

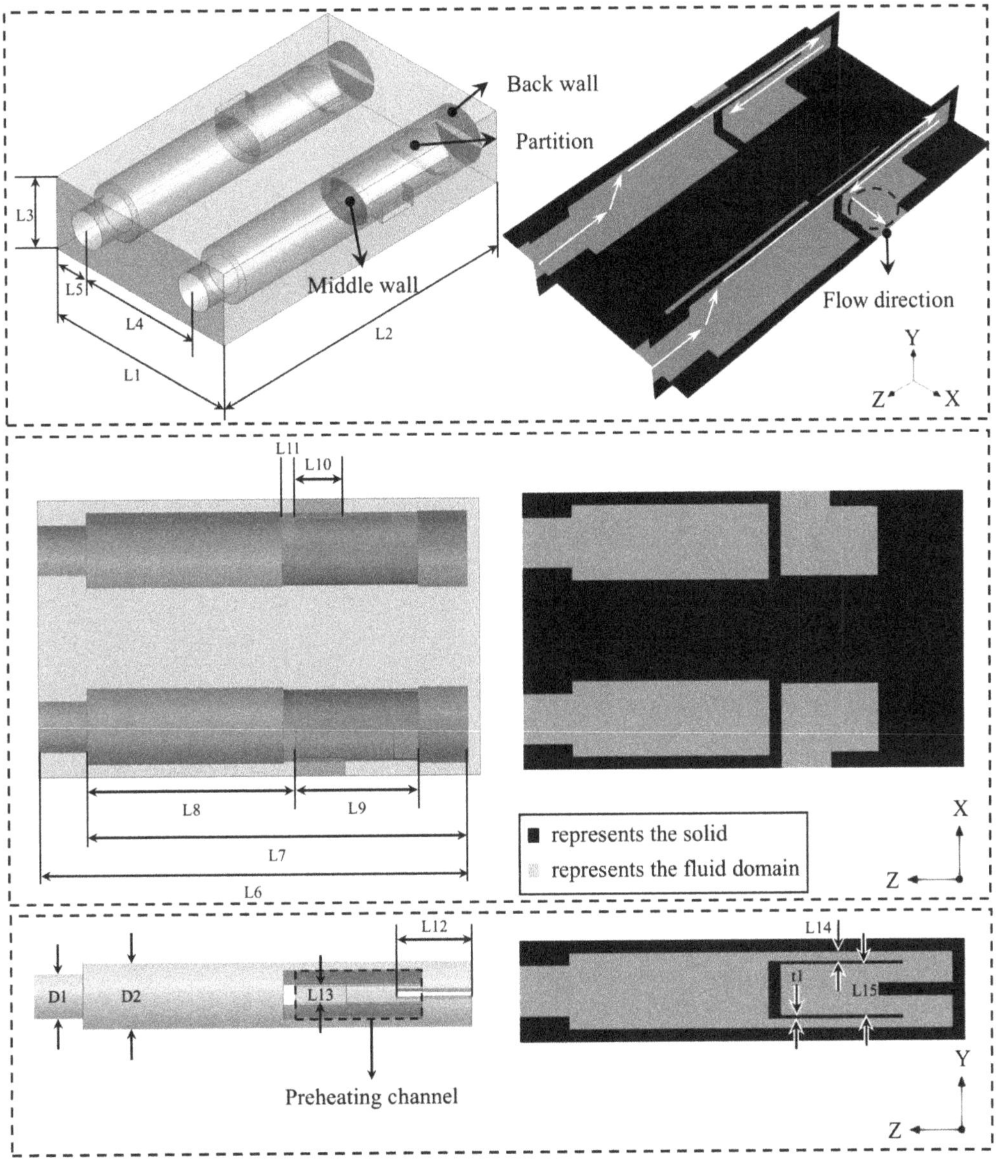

Figure 1. Schematic diagrams of two micro cylindrical combustors, featuring a heat-recirculating structure, embedded within a micro planar combustor.

Table 1. Dimensions of the new design configuration.

Variables		**Value (mm)**		
		CD	**PF**	**CF**
Length	L1	11	11	11
	L2	18	18	18
	L3	4	4	4
	L4	NA	7	7
	L5	NA	2	2
	L6	NA	17.5	17.5
	L7	NA	15.5	15.5
	L8	NA	8.5	8.5
	L9	NA	5	5
	L10	NA	2	2
	L11	NA	0.5	0.5
	L12	NA	3	3
	L13	NA	0.94	0.94
	L14	NA	0.4	0.4
	L15	NA	1.95	1.95
Diameter	D1	NA	2	2
	D2	NA	3	3
Thickness	t1	NA	0.125	0.125

In contrast to the previous investigation [31], a slight modification is here made to the micro cylindrical combustor—one outlet is eliminated due to the different widths between the micro planar and cylindrical combustors. In each micro cylindrical combustor, Figure 1 illustrates that the new design consists of one inlet and one outlet channel, a middle wall, a back wall, a heat-recirculating room, and a partition. Inserting a middle wall is important for holding the high-temperature flame, allowing more heat to transfer to the external walls. The back wall also reverses the hot gases toward the preheating structure, which increases the residence time of the flow and preheats the inlet mixture flowing from the inlets. To avoid the backflow phenomenon, a partition is employed to force the heat-containing streams to flow into the preheating channels. It is noteworthy that this investigation aims to assess how the PF and CF directions affect the thermal fluid dynamics and performance parameters. The layout of the outlined configurations is shown in Figure 2 along with the conventional design.

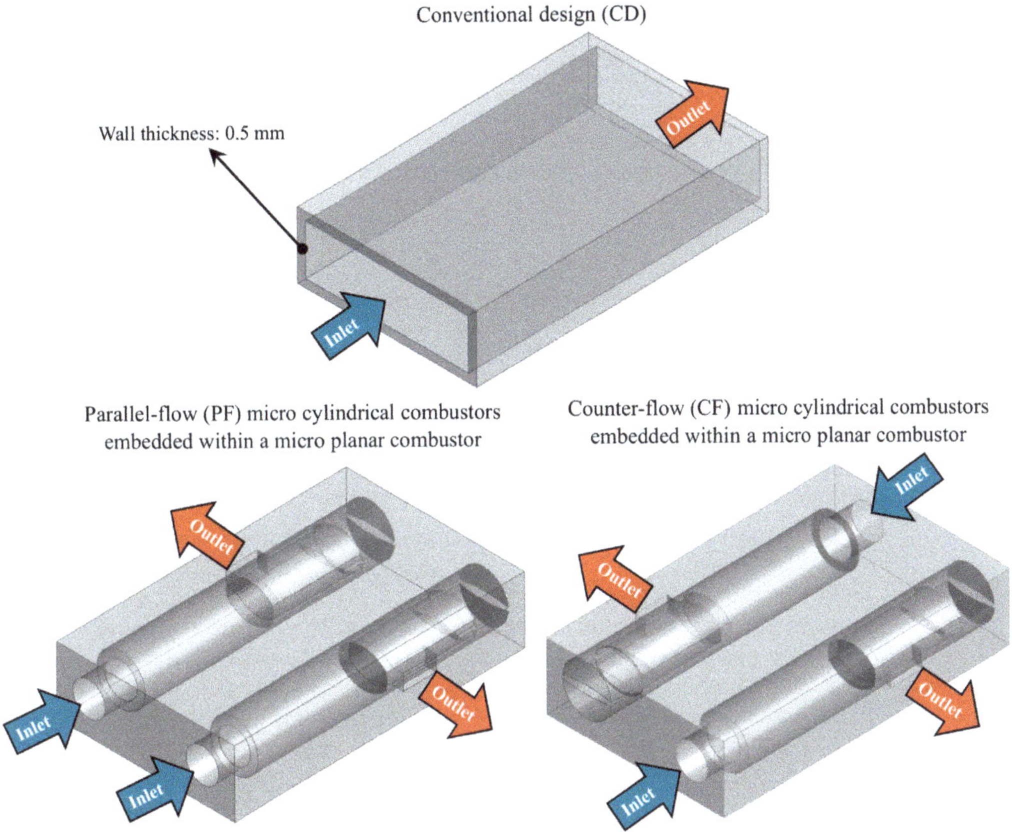

Figure 2. Layouts of the conventional design (CD) of the micro planar combustor, and parallel-flow (PF) and counter-flow (CF) micro planar combustors with two cylindrical combustors embedded inside.

2.2. Govering Equations

AMSYS Fluent R2 2023 [32] software was used to solve the conservation equations for the numerical investigations. These equations are mass, momentum, energy and 1-N species, where N donates the total number of species in the chemically reacting mixture, for which the transport equations read as follows.

Mass:

$$\nabla \cdot \left(\rho \vec{v} \right) = 0 \tag{1}$$

Momentum:

$$\rho \vec{v} \cdot \nabla \vec{v} = -\nabla P + \nabla \cdot \left(\vec{\tau} - \vec{\tau}' \right) \tag{2}$$

where

$$\vec{\tau} = \mu \left[\nabla \vec{v} + \left(\nabla \vec{v} \right)^T - \frac{2}{3} \nabla \vec{v} I \right] \tag{3}$$

Energy:

$$\nabla \cdot \left(\vec{v} (\rho E + p) \right) = \nabla \cdot \left(k_{eff} \nabla \mathrm{T} - \sum_j h_j \vec{J}_j + \left(\vec{\tau} \cdot \vec{v} \right) \right) + S_h \tag{4}$$

Species:

$$\nabla \cdot \left(\rho \vec{v} Y_i + \vec{J}_i \right) = R_i \tag{5}$$

where ρ denotes the density, μ the molecular viscosity, P the static pressure, S_h the enthalpy source term of the fluid, $\vec{v}$ the velocity vector, R_i the reaction net rate of production of species i, $\vec{\tau}'$ the Reynolds stress, $\vec{J}_i$ the diffusion flux of species i, $\vec{\tau}$ the viscous stress, Y_i the local mass fraction of species i, I the unit tensor, $\vec{J}_j$ the diffusion flux of species j, T the temperature, k_{eff} the effective conductivity, h_j the enthalpy of species j, and E the total energy of the fluid.

The gravity, surface reaction, and Dufour effect were not incorporated in the current steady-state numerical work due to their negligible effects [20,33,34]. Moreover, the Mach number is low, implying that the flow is incompressible. However, the Reynolds number of non-reacting flow calculations exceeds 500 [35]; thus, the turbulent modeling approach, namely, the Realizable k-ϵ model [36], is utilized to compute the effects of turbulence on combustion.

2.3. Numerical Setup and Boundary Conditions

The turbulence–chemistry interaction is computed using the eddy dissipation concept (EDC) [37–39] to incorporate the chemical mechanism of hydrogen combustion, which consists of 9 species and 19 chemical reactions, as listed in Table 2 [40]. The coupling between pressure and velocity is performed by means of the SIMPLE algorithm approach. The second-order upwind scheme is implemented to spatially discretize the transport equations. All residuals undergo a convergence criterion of 10^{-6}. A temperature of 300 K is set for the initial inlet temperature of the premixed mixture. In addition, this work applies velocity inlet and pressure outlet boundary conditions with a turbulent intensity of 5% and a hydraulic diameter of 4 mm at the specified boundary conditions. Zero and no-slip boundary conditions are used for diffusive flux species and interior surfaces, respectively. It is important to mention that the area-weighted average method is applied to evaluate the parameters in this study.

The calculation of the total heat loss (Q) through the outer walls reads as follows [41,42]:

$$Q = Q_{con} + Q_{rad} \rightarrow h\, A_c (T_w - T_\infty) + \varepsilon \sigma A_c \left(T_w^4 - T_\infty^4 \right) \tag{6}$$

The evaluation of the pressure drop $\left(P_{drop} \right)$ resulting from the friction and heating effects reads as [43]

$$P_{loss} = P_{in} - P_{out} \tag{7}$$

The Peclet number (Pe), which measures the relative importance of advection to diffusion effects [44,45], is computed as

$$Pe = \frac{\rho V d_h C_p}{\lambda} \tag{8}$$

The diffusion $\left(S_{gen}^{diffussion} \right)$, conduction $\left(S_{gen}^{conduction} \right)$, chemical $\left(S_{gen}^{chemical} \right)$ and total $\left(S_{gen}^{total} \right)$ entropy generations [46,47] are evaluated as follows:

$$S_{gen}^{diffussion} = \sum_i^n (\rho R D_i) \cdot \frac{\nabla \omega_i \nabla \chi_i}{\chi_i} \tag{9}$$

$$S_{gen}^{chemical} = -\sum_i^n \frac{1}{T} \mu_i \gamma_i \tag{10}$$

where

$$\mu_i = \overline{h}_i^o(T) - T\overline{s}_i^o(T) + RTln\left(\frac{\chi_i P}{P_{atm}}\right) \quad (11)$$

$$S_{gen}^{conduction} = \frac{1}{T^2} k_{mix} (\nabla \mathrm{T})^2 \quad (12)$$

$$S_{gen}^{total} = \int_V S_{gen}^{diffussion} dV + \int_V S_{gen}^{chemical} dV + \int_V S_{gen}^{conduction} dV \quad (13)$$

Table 2. Reversible chemical reactions of H_2–air combustion [40]. A_k is the pre-exponential factor of reaction rate, β_k is the activation energy of the reaction, E_k is the temperature exponent, and M is a third body efficiency.

Reactions	A_k (m kmol s)	β_k	E_k (J/mol)
1. O_2 + H = OH + O	5.10×10^{13}	−0.82	6.91×10^7
2. H_2 + O = OH + H	1.80×10^7	1.00	3.70×10^7
3. H_2 + OH = H_2O + H	1.20×10^6	1.30	1.52×10^7
4. OH + OH = H_2O + O	6.00×10^6	1.30	0.00
5. H_2 + O_2 = OH + OH	1.70×10^{10}	0.00	2.0×10^8
6. H + OH + M = H_2O + M [a]	7.50×10^{17}	−2.60	0.00
7. O_2 + M = O + O + M	1.90×10^8	0.50	4.001×10^8
8. H_2 + M = H + H + M [b]	2.20×10^9	0.50	3.877×10^8
9. H + O_2 + M = HO_2 + M [c]	2.10×10^{12}	−1.00	0.00
10. H + O_2 + O_2 = HO_2 + O_2	6.70×10^{13}	−1.42	0.00
11. H + O_2 + N_2 = HO_2 + N_2	6.70×10^{13}	−1.42	0.00
12. HO_2 + H = H_2 + O_2	2.50×10^{10}	0.00	2.90×10^6
13. HO_2 + H = OH + OH	2.50×10^{11}	0.00	7.90×10^6
14. HO_2 + O = OH + O_2	4.80×10^{10}	0.00	4.20×10^6
15. HO_2 + OH = H_2O + O_2	5.00×10^{10}	0.00	4.20×10^6
16. HO_2 + HO_2 = H_2O_2 + O_2	2.00×10^9	0.00	0.00
17. H_2O_2 + M = OH + OH + M	1.30×10^{14}	0.00	1.905×10^8
18. H_2O_2 + H = H_2 + HO_2	1.70×10^9	0.00	1.57×10^7
19. H_2O_2 + OH = H_2O + HO_2	1.0×10^{10}	0.00	7.50×10^6

[a] Enhancement factors: H_2O = 20.0. [b] Enhancement factors: H_2O = 6.0, H = 2.0, and H_2 = 3.0. [c] Enhancement factors: H_2O = 21.0, H2 = 3.3, O_2 = 0.0, and N_2 = 0.0.

The second law of thermodynamics concept [48–50] is applied to account for exergy efficiency. This investigation calculates inlet exergy $\left(E_x^{in}\right)$ and total exergy losses $\left(E_x^{eg}\right)$ as follows:

$$E_x^{in} = \dot{m}_{fuel} \times Q_{LHV} \quad (14)$$

and

$$E_x^{eg} = E_x^{loss} + \dot{m}_{inlet} \times T_\infty \times c_{p,outlet} \times \ln \frac{T_\infty}{T_{eg}} \quad (15)$$

The uncounted exergy destruction $\left(E_x^{des}\right)$ is computed as follows:

$$E_x^{des} = E_x^{in} - E_x^{eg} \quad (16)$$

The exergy efficiency (η_{exergy}) is calculated as

$$\eta_{exergy} = \left(1 - \frac{E_x^{des}}{E_x^{in}}\right) \times 100\% \quad (17)$$

The radiation efficiency ($\eta_{radiation}$) is evaluated as

$$\eta_{radiation} = \left(\frac{Q_{rad}}{\dot{m}_{fuel} \times Q_{LHV}}\right) \times 100\% \quad (18)$$

With respect to the area-weighted mean technique (T_{mw}), the wall temperature is evaluated as [51]

$$T_{mw} = \frac{\sum_{i=1}^{n} T_i A_i}{\sum_{i=1}^{n} A_i} \quad (19)$$

The uniformity of wall temperature (R_T) is computed as

$$R_T = \left(\frac{\sum_{i=1}^{n}[|T_i - T_{mw}|A_i]}{T_{mw}\sum_{i=1}^{n} A_i}\right) \times 100\% \quad (20)$$

2.4. Grid Independence and Model Validation

For all numerical studies, the mesh independence test is paramount to balance accuracy and computing power. Therefore, this subsection examines coarse, medium, and fine mesh resolutions, which encompass 301,911 cells (Mesh-I), 748,163 cells (Mesh-II), and 1,240,631 (Mesh-III), respectively. The PF configuration is selected for the mesh sensitivity analysis, and the inlet velocity and equivalence ratio are set at 15 m/s and 1, respectively. The Grid Convergence Method (GCI) [52] was employed in the grid sensitivity analysis to measure the discretization error. This technique is used to assess the accuracy of Mesh-I and Mesh-II results with respect to the results of Mesh-III, as the latter is the fine mesh resolution.

Figure 3 depicts a comparison between the three mesh densities highlighted above in terms of the temperature variation over the external walls with respect to the dimensionless length (x/L), T_{mw} and R_T. As seen in Figure 3a, the coarse mesh (Mesh-I) demonstrates notable under-predictions of the wall temperature in the inlet and back-wall regions, whereas the wall temperature of the medium-resolution mesh (Mesh-II) is roughly identical to that of the high-resolution case (Mesh-III). This is confirmed by Figure 3b, as the maximum discretization errors of Mesh-I and Mesh-II are 9.30% and 2.05%, respectively, at the geometry inlet. In general, the error bars shown in Figure 3b indicate that using Mesh-II ensures accurate results while maintaining computing power within an acceptable range. Similar findings are shown in Figure 3c, as the discretization errors of T_{mw} and R_T for Mesh-I (Mesh-II) are 2.61% and 43.7% (0.1% and 2.76%), respectively. Therefore, Mesh-II is chosen for all simulations presented in this research, as it provides precise outcomes within a reasonable computational time.

To ensure the robustness of the numerical settings and the accuracy of the simulations, a validation process is performed against experimental and numerical studies of both micro planar and cylindrical combustors. Figure 4a shows a comparison of the current work's outcomes with prior experimental [28] and numerical [53] results in terms of the wall temperature variation at 600 and 900 mL/min volume flow rates of hydrogen in a micro planar combustor design. The validation demonstrates that the present work accurately predicts the wall temperature in the inlet region at a 600 mL/min hydrogen discharge, as confirmed by the experiment; however, it underestimates the wall temperature in the outlet region by an error of roughly 3.92% compared to the experimental results. The maximum error for the same volume flow rate is found to be around 7.45% in comparison with the numerical work. For a 900 mL/min hydrogen flow rate, the maximum discrepancies of the wall temperature between the current work and the experimental and numerical data occur at the inlet zone, with 2.25% and 5.45% error rates.

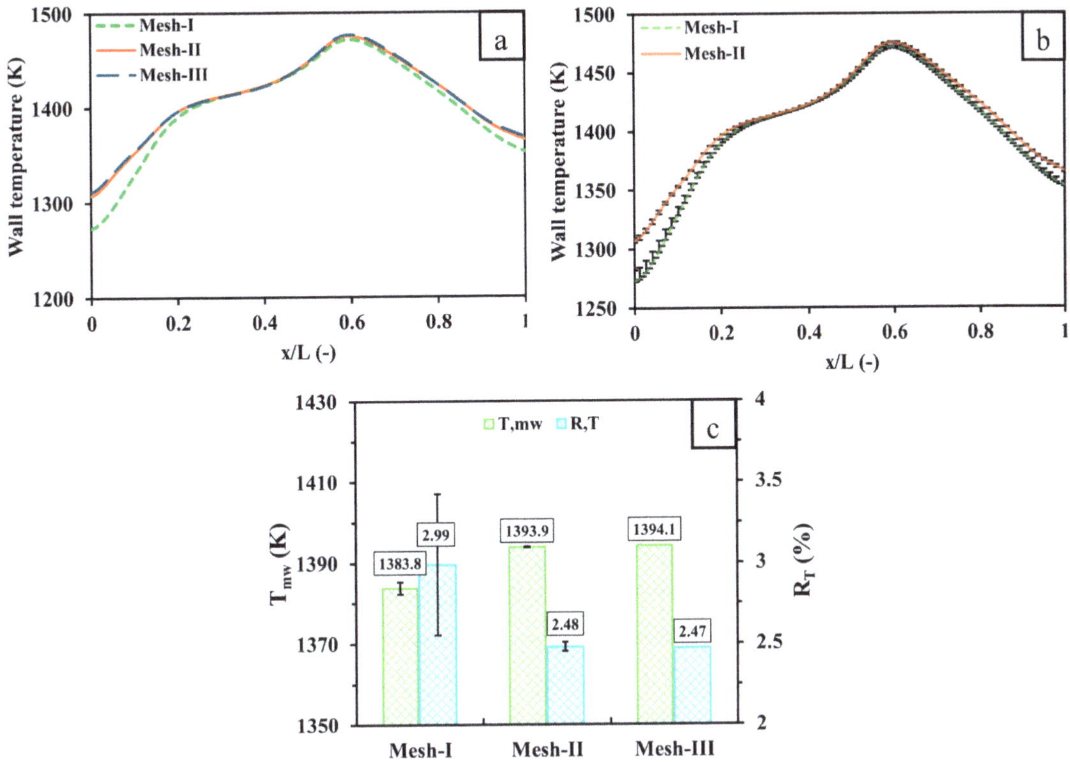

Figure 3. A comparison between three different mesh densities, which are 301,911 cells (Mesh-I), 748,163 cells (Mesh-II), and 1,240,631 (Mesh-III), with respect to (**a**) the temperature over the external wall, (**b**) the wall temperatures of Mesh-I and -II with their error bars, and (**c**) T_{mw} and R_T with error bars for the results of Mesh-I and -II. The error bars for the results of Mesh-I and -II are generated using the data of Mesh-III. x and L are the distance between inlets and the distance from the inlet to the outlet.

Figure 4b compares the validation of the mean wall temperature of the present work using the micro cylindrical combustor with the experimental [54] and simulated [55] data at a velocity of 12 m/s. As observed, the current work exhibits over-predictions of the mean wall temperature in all test cases, where the maximum errors for the experiment and simulation are approximately 6.19% and 5.96% at equivalence ratios of 1 and 0.6, respectively. Such discrepancies, as shown in Figure 4a,b, can result from measurement inaccuracies of the experiment and the employment of different mesh densities for the simulation. Overall, the percentages of error are acceptable and rational, indicating the applicability and viability of the numerical settings and modeling approaches.

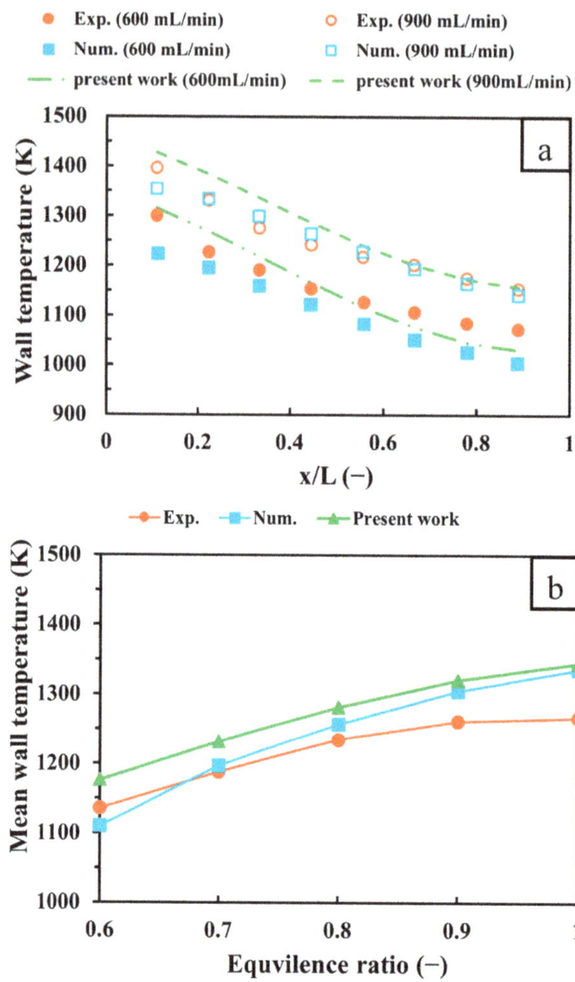

Figure 4. Comparisons of (**a**) the wall temperature distribution with respect to (x/L) of the present results against the experimental [28] and numerical [53] findings for a micro planar combustor configuration and (**b**) the mean wall temperature at different equivalence ratios of the present results against the experimental [54] and numerical [55] findings for a micro cylindrical combustor configuration.

3. Results

3.1. Effects of Two Micro Cylindrical Combustors Embedded Within a Micro Planar Combustor

This sub-section discusses the impacts of the new design on the thermal and thermodynamic parameters, and demonstrates its potential in comparison with the traditional design of the micro planar combustor. In addition, the effects of parallel and counter flows on system efficiency are examined here. As the inlet areas between the CD configuration and PF and CF configurations are different, fixing the inlet velocity leads to a scientifically invalid comparison due to the resulting different power inputs. For this, the comparisons of CD, PF, and CF structures are presented here at an input power of 275 W and a unity equivalence ratio.

Figure 5 illustrates the variations of T_{mw} and R_T of CD, PF, and CF structures at an input power and equivalence ratio of 275 W and 1, respectively. As observed, PF and

CF exhibit an effective means of enhancing T_{mw} compared to CD by roughly more than 118.3 K. Furthermore, the newly proposed designs demonstrate more evenly distributed temperature over the outer walls of the micro combustor as R_T is reduced in the PF and CF configurations, in comparison to the CD configuration, by 3.7% and 3.65%, respectively. The parameters presented in Figure 5 are considered as two of the most important thermal performance parameters, as higher values of T_{mw} indicate an improvement in the heat transfer mechanisms from the combustion to the micro combustor's walls, while lower levels of R_T refer to a very uniform promotion of heat transfer throughout the geometry. On the other hand, T_{mw} and R_T are roughly identical between PF and CF, implying that the direction of inlet flows has no notable effects.

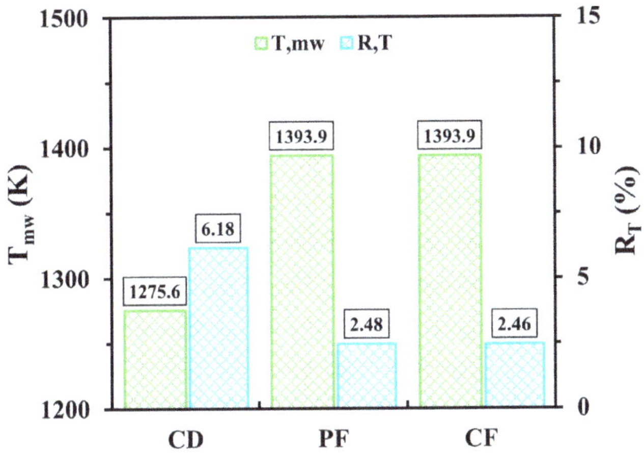

Figure 5. A comparison of the T_{mw} and R_T values between CD, PF, and CF configurations at an input power of 275 W and an equivalence ratio of unity.

Figure 6 displays the effects of CD, PF, and CF cases on the distribution of temperature throughout the domain at a 275 W input power and a unity equivalence ratio. As can be seen in Figure 6, the high-temperature flame occupies a wider range in CD as compared to PF and CF. However, the dwell time of flow is significantly shorter in the former case, whereas establishing the middle and back walls in the two latter cases increases the residence period of the great thermal energy stream, increasing the heat transfer capacity to the external surfaces. This is brought about as the middle wall holds a high thermal energy, while the back wall redirects the hot flows toward the preheating channels to act as a thermal energy source for the inlet flow. The highlighted effects confirm the considerable improvements in T_{mw} and R_T, as shown in Figure 5.

Figure 7 shows the impacts of CD, PF, and CF on how the Pe number varies at a 275 W input power and a unity equivalence ratio. For all test cases, high degrees of Pe number distributions are confined in the inlet region owing to the high density of the unburnt gases. In contrast to the CD case, the contour representations of the Pe number reveal that the PF and CF cases demonstrate intermediate variations of Pe number in the outlets of the micro combustor and in the narrow channel between the preheating room and external walls. This illustrates that the outlined regions experience a dominance of advection effects over those of diffusion, yielding a greater transfer of heat in the new designs.

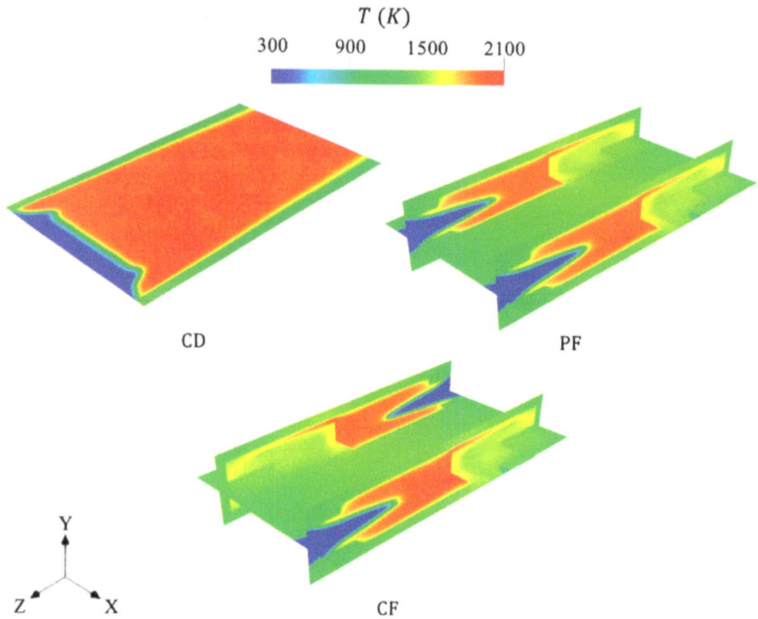

Figure 6. Variations of temperature across CD, PF, and CF configurations at a 275 W input power and a unity equivalence ratio.

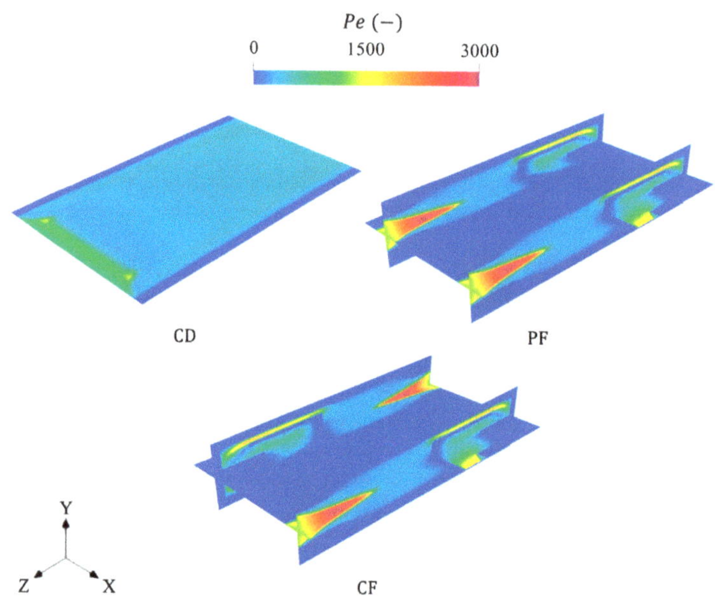

Figure 7. Distributions of Pe number throughout the CD, PF, and CF structures at a 275 W input power and a unity equivalence ratio.

Figure 8 depicts representations of how the total entropy generation varies in the CD, PF, and CF test cases. As expected, the areas occupied by high entropy generation are confined within the region of the onset of combustion in all cases. The causes of this lie in that the aforementioned region experiences high chemical reaction rates, high temperature gradients, and high concentration gradients of species, which govern the entropy productions in terms of chemistry, heat conduction, and diffusion, respectively. The direction of flow plays no critical role in the entropy generation, as the spatial variations of total entropy in the PF and CF cases are quite similar to each other.

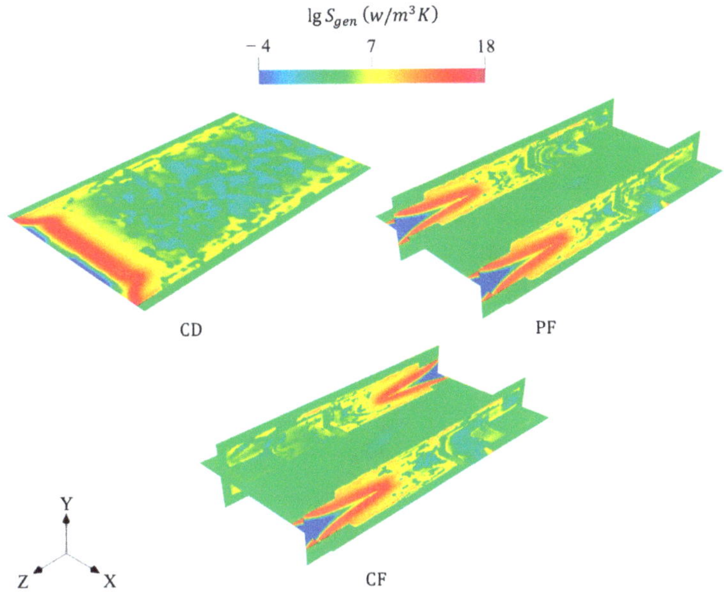

Figure 8. Spatial variations of total entropy generations under CD, PF, and CF configurations.

Figure 9 provides volume integral forms of total entropy production with the illustration of the percentage participation of each mechanism under different design configurations. The PF and CF structures are conducive to dissipating a higher amount of energy, unlike the CD structure, as the percentages of S_{gen} increase between the two former cases and the latter case are 827.1% and 824.1%, respectively. These high increases can be attributed to the high temperature gradients between the high-temperature flame and the walls of the preheating channel, which notably increase the heat conduction mechanism of entropy. In addition to this, the high intensities of combustion in the new designs are caused by higher chemical reaction rates and, therefore, lead to higher rates of production and destruction of species, which consequently promote chemical and diffusion mechanisms of entropy, respectively.

Figure 10 compares between CD, PF, and CF in terms of the pressure loss, exhaust gas temperature (T_{eg}), and exergy and radiation efficiencies. As seen in Figure 10a, PF and CF, in contrast to CD, show significantly increased pressure loss due to the notably higher rate of collision between the stream flowing from the inlets and the middle and back barriers, which raises the necessity of additional pumping power. However, such walls elongate the dwell period of flow inside the micro combustor, and thus improve the heat transfer to the outer walls. This leads to benefits from extremely high thermal energy, as the T_{eg} values of PF and CF are much lower than that of CD by 591 K and 580 K, respectively. This is despite the fact that the higher entropy generation in PF and CF, as shown Figure 9, capitalizing on greater thermal energy (as confirmed by the trends of T_{eg}), increases the exergy (radiation) efficiency of PF and CF compared to CD, by 7.2% (14%) and 7.3% (14%), respectively, as illustrated in Figure 10b. On the other hand, no remarkable discrepancies are observed in relation to the operation conditions of a 275 W input power and a unity equivalence ratio.

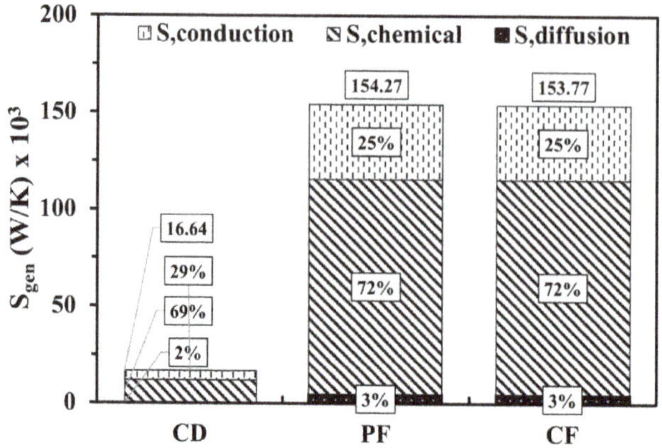

Figure 9. Volume integrations of different entropy mechanisms of CD, PF, and CF configurations.

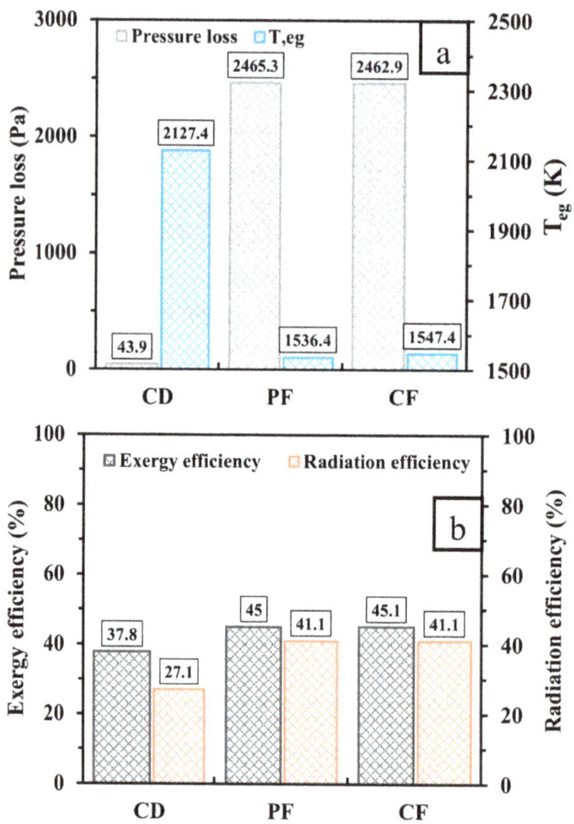

Figure 10. Comparisons of (**a**) pressure loss and exhaust gas temperature (T_{eg}) and (**b**) exergy and radiation efficiencies for CD, PF, and CF test cases.

3.2. Effects of Inlet Velocity

The initial operation conditions are critical to optimizing the system performance. Thus, the variation of inlet velocity, which controls the input power, and its effects on the thermal parameters and output power density are discussed here. As the new designs

(PF and CF configurations) notably improve T_{mw}, R_T, energy and radiation efficiencies compared to the traditional structure of the micro planar combustor (CD configuration), as previously presented, only PF and CF are investigated in this sub-section at an equivalence ratio of 1 and inlet velocities of 6, 9, 12, and 15 m/s, which are equivalent to input powers of 109.7 W, 164.5 W, 219.3 W, and 275 W, respectively.

Figure 11 shows the effects of inlet velocities on T_{mw} and R_T for PF and CF designs. For both cases, altering V_{in} from low to high is conducive to increasing T_{mw} due to feeding the systems with more energy, and therefore absorbing more thermal energy into the walls. In contrast, reducing V_{in} results in a fluctuating trend of R_T, as it decreases and then increases. The optimal uniform wall temperatures of PF and CF are recorded at 12 m/s and 9 m/s, respectively. It is worth noting that the CF case demonstrates more equal variations of temperature over the micro combustor's walls for all V_{in} values when compared to the PF case, owing to the production of thermal energy from opposite sides of the domain.

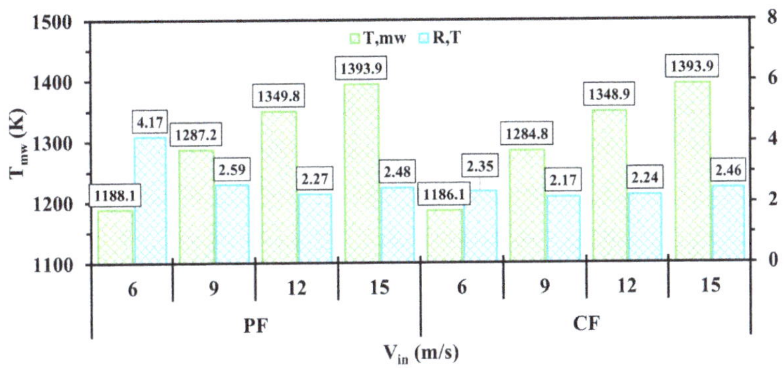

Figure 11. Variations of T_{mw} and R_T at different inlet velocities for PF and CF configurations.

Figure 12 illustrates the variations in temperature for PF and CF configurations at different levels of V_{in}. As can be seen in Figure 12, increasing V_{in} leads to the enrichment of the combustor with more fuel, and consequently increases the in-combustor temperature. This results in the release of more thermal energy, thus transferring more heat to the external walls, which confirms the increase in T_{mw} at higher V_{in}. In addition, escalating V_{in} causes the onset of combustion to shift to a greater distance from the inlet due to the high diffusion of hydrogen. This effect controls the flame position and how long the flame propagates throughout the micro combustor, which accordingly determines the uniformity of wall temperature. As shown in Figure 12, the amount of energy entrained into the micro combustor at an inlet velocity of 9 m/s for the CF case results in the ideal flame shape for achieving the best R_T (See Figure 11), particularly when the flame heats the walls equally from opposing sides.

Figure 13 provides a contour representation of Pe number distributions in PF and CF with respect to various V_{in} values. Disregarding the high Pe number near the inlets, caused by the high density of unburnt flowing mixture, the increase in V_{in} in all test cases shows high Pe numbers in the outlets, and in the narrow channel separating the outer walls and the preheating path. This indicates that the advection effects dominate in the highlighted regions, implying an improvement in the heat transfer mechanisms.

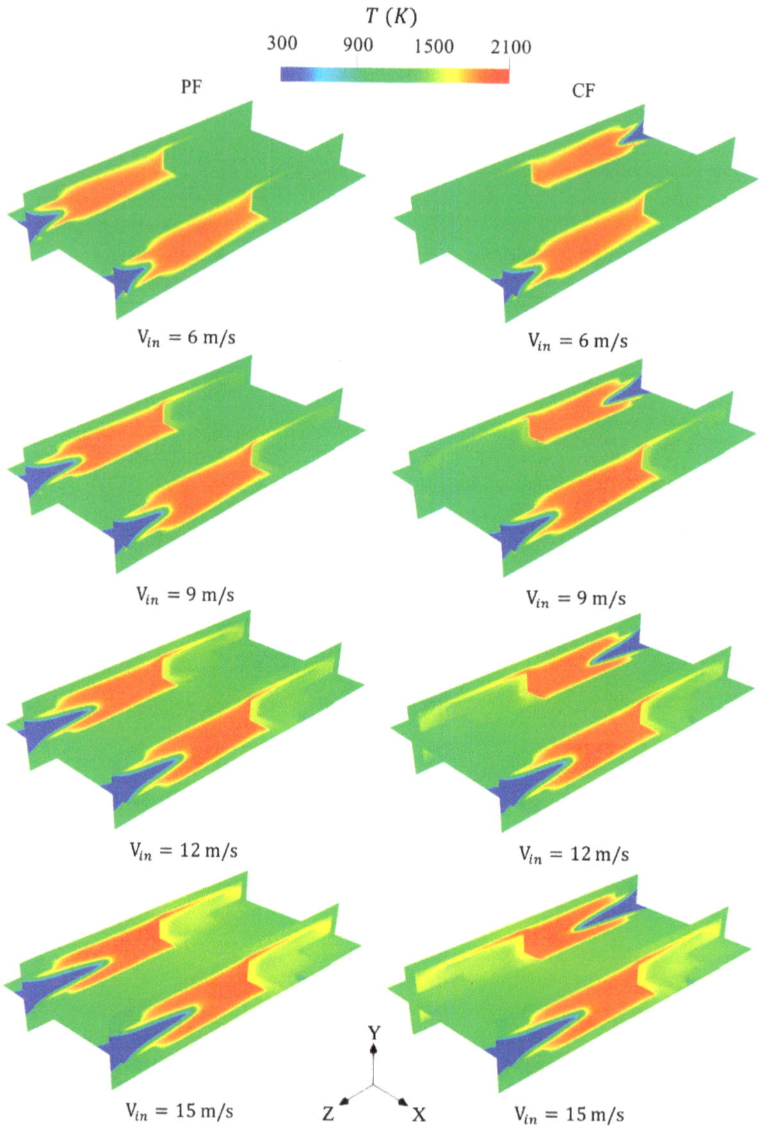

Figure 12. Spatial distributions of temperature for PF and CF configurations at various inlet velocities.

Figure 14 depicts how the total entropy generation spatially varies over the micro combustor domain in the PF and CF configurations under various values of V_{in}. As observed in all cases, the area occupied by the highest values in both PF and CF closely resembles the shape of the flame in the region of combustion initiation. This is mainly caused by the great chemical activities occurring in the outlined area, which result in tremendously high rates of generation and consumption of species. For this, supplying more fuel to the system at higher V_{in} enhances chemistry and, hence, elongates the regions of high entropy generation. Reversing the flow results in no notable effects in the entropy generation, as the PF and CF configurations are identical in terms of the promotion of entropy.

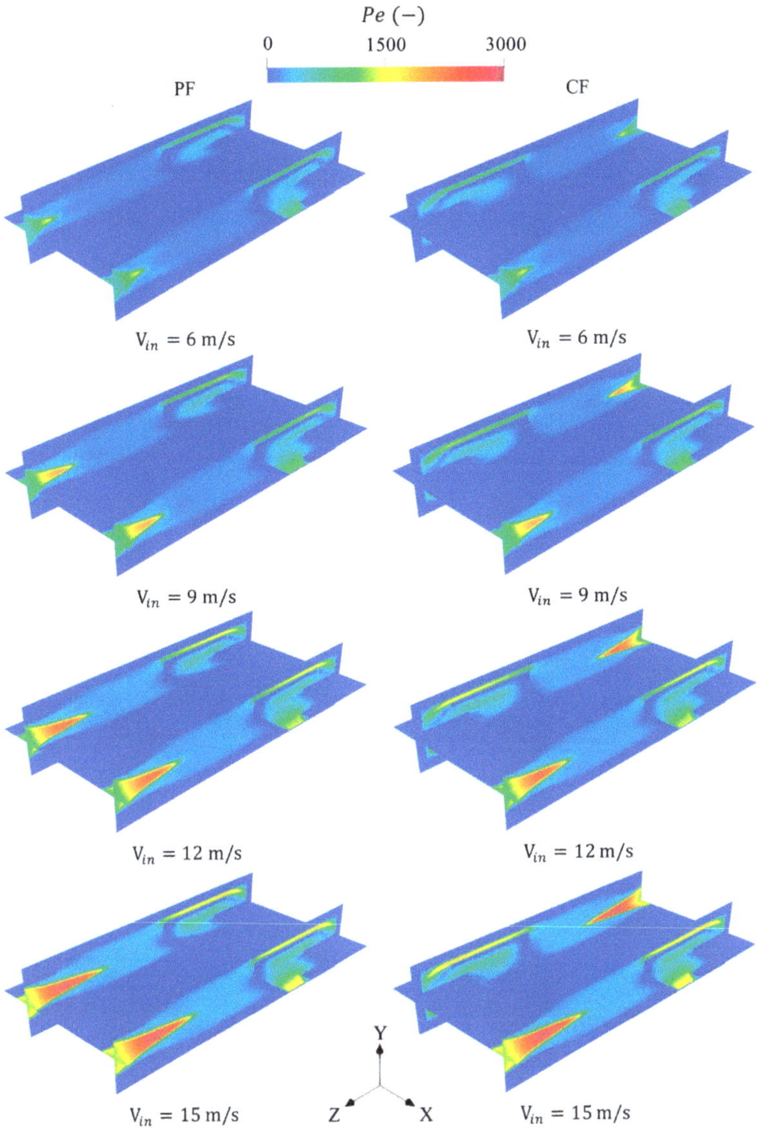

Figure 13. Spatial variations of Pe number in the PF and CF cases for different levels of V_{in}.

Figure 15 demonstrates the magnitude of S_{gen} and the participation of conduction, chemical, and diffusion entropy generation for PF and CF at different values of V_{in}. As highlighted before, increasing V_{in} leads to increases in the input power of the system, resulting in intensive combustion. Thereby, approaching greater degrees of V_{in} causes the magnitude of entropy generation to significantly increase. All mechanisms of entropy production are found to be proportionally escalated when increasing V_{in}; however, the chemical mechanism seems to prevail owing to the vigorous chemical rates at such conditions. The values of total entropy generation between PF and CF are similar, proving that reversing one flow direction does not impact entropy generation.

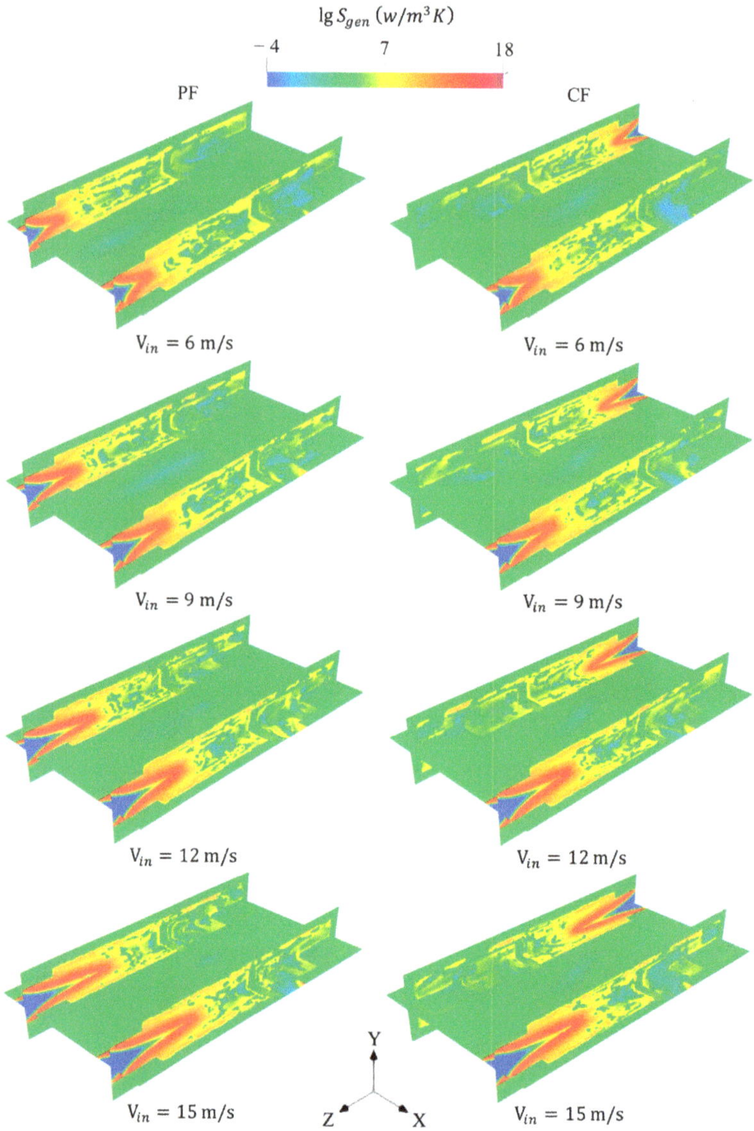

Figure 14. Variations of total entropy generations in PF and CF at different values of V_{in}.

Figure 16 shows the variations in trends of pressure loss, T_{eg}, and exergy and radiation efficiencies for PF and CF configurations in a range of values of V_{in}. Increasing the flow velocity of the inlet mixture remarkably promotes the collision rate of the hot gases with the middle and back walls, leading to significant escalations in the pressure loss in PF and CF cases, as shown in Figure 16a. Moreover, the transition of V_{in} from low to high in all test cases feeds the system with more energy and, therefore, increases T_{eg}, indicating a greater waste of energy from the system. This suggests that the capacity of the investigated application is limited in terms of benefiting from high energy content. This conclusion is supported by the variations of exergy and radiation efficiencies demonstrated in Figure 16b, which reveal decreasing trends when approaching higher V_{in}. Trivial discrepancies are recorded in the system's efficiency when comparing PF and CF, referring to the negligible effects of the flow direction.

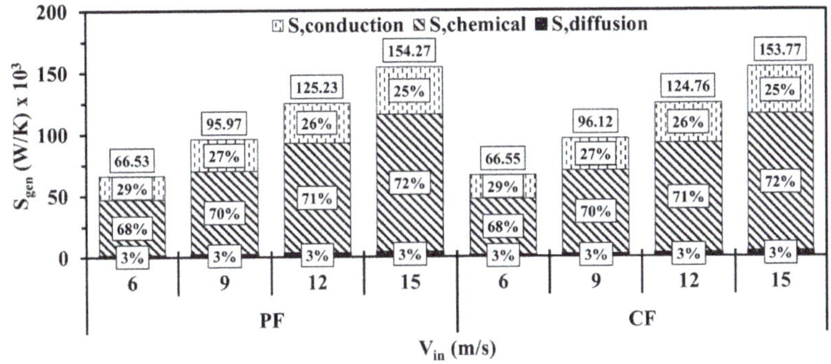

Figure 15. Volume integrations of entropy generation mechanisms for PF and CF at different V_{in}.

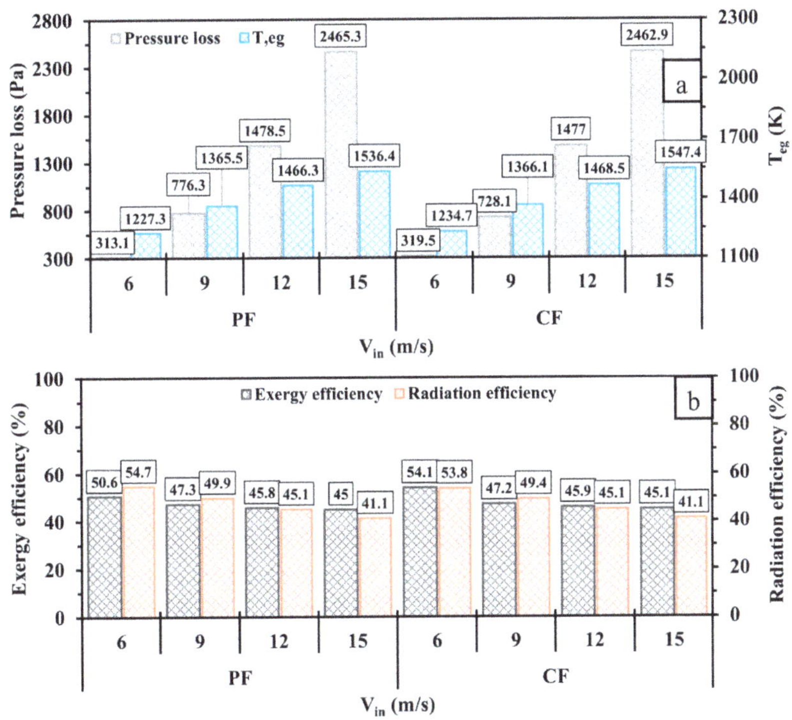

Figure 16. Comparisons of (**a**) pressure loss and T_{eg} and (**b**) exergy and radiation efficiencies for PF, and CF test cases under different conditions of V_{in}.

3.3. Effects of Inlet Equivalence Ratio

Complete combustion depends on a wide variety of variables. One of these parameters is the inlet equivalence ratio (Φ), which controls the ratio of fuel and air and therefore plays a critical role in the fuel economy. This part discusses the effects of varying Φ from 0.6 to 1.2 for PF and CF cases at a 275 W input power.

Figure 17 outlines the effects of fuel-lean and fuel-rich conditions on T_{mw} and R_T for PF and CF. As can be seen in Figure 17, T_{mw} drastically increases from the low-fuel condition

($\Phi < 1$) to the stoichiometric condition ($\Phi = 1$), and then decreases when transitioning to the high-fuel condition ($\Phi > 1$), which trends are caused by insufficient amounts of fuel and oxygen, respectively. However, inconsistent patterns of R_T are found between PF and CF configurations, as the former and latter achieve the optimal R_T values of 2.15% and 1.97% at Φ of 0.8 and 0.6, respectively.

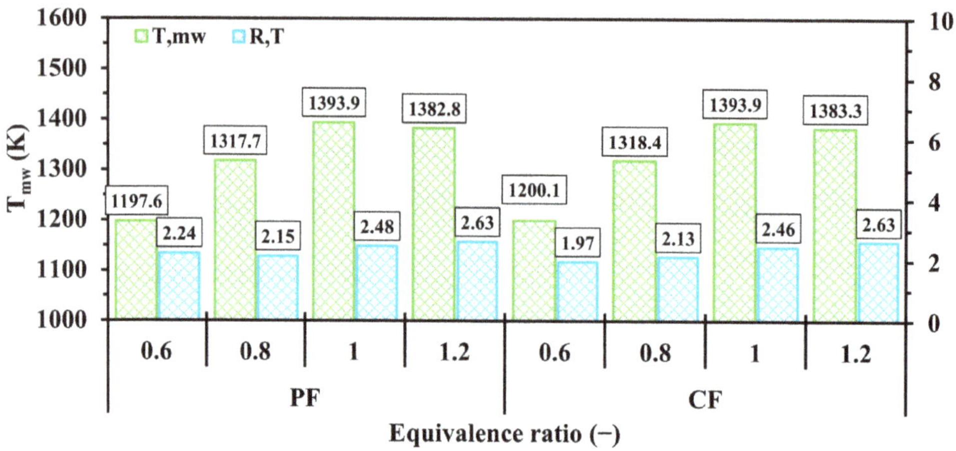

Figure 17. Comparisons of T_{mw} and R_T at different Φ between 0.6 and 1.2 for PF and CF configurations.

Figure 18 displays the variation in temperature across the micro combustor at different levels of Φ for PF and CF test cases. As observed in Figure 18, the fuel-lean conditions ($\Phi < 1$) foster low temperatures and a further initiation of combustion from the inlets due to the low quantity of hydrogen and difficulties in triggering the combustion, respectively. Hydrogen's high diffusivity is expected to shift the onset of combustion further downstream; however, combustion in the fuel-rich condition ($\Phi > 1$) occurs relatively closer to the inlet compared to the stoichiometric condition ($\Phi = 1$), suggesting that the strength of the mixture to ignition at $\Phi > 1$ dominates over the high diffusion mobility of hydrogen.

Figure 19 presents the contours of Pe number distributions at different Φ values, between 0.6 and 1.2, for PF and CF structures. The Pe number distributions show a pronounced decrease after the onset of combustion with an increasing Φ. This can be attributed to the high diffusivity of hydrogen, as a higher Φ implies an increase in hydrogen content in the premixed mixture, thereby increasing the diffusion effect over that of advocation at such conditions.

Figure 20 depicts how the total entropy generations are spatially varied with respect to different levels of Φ for PF and CF structures. It is shown in Figure 20 that high entropy production covers a wider area in the micro combustor domain at lower values of Φ. The causes of this lie in the fact that the fuel-lean mixture is difficult to ignite, and therefore a longer distance is required to initiate the combustion. The line separating the unburnt and burnt gases is known for its intense chemical processes, which notably generate entropy at higher degrees.

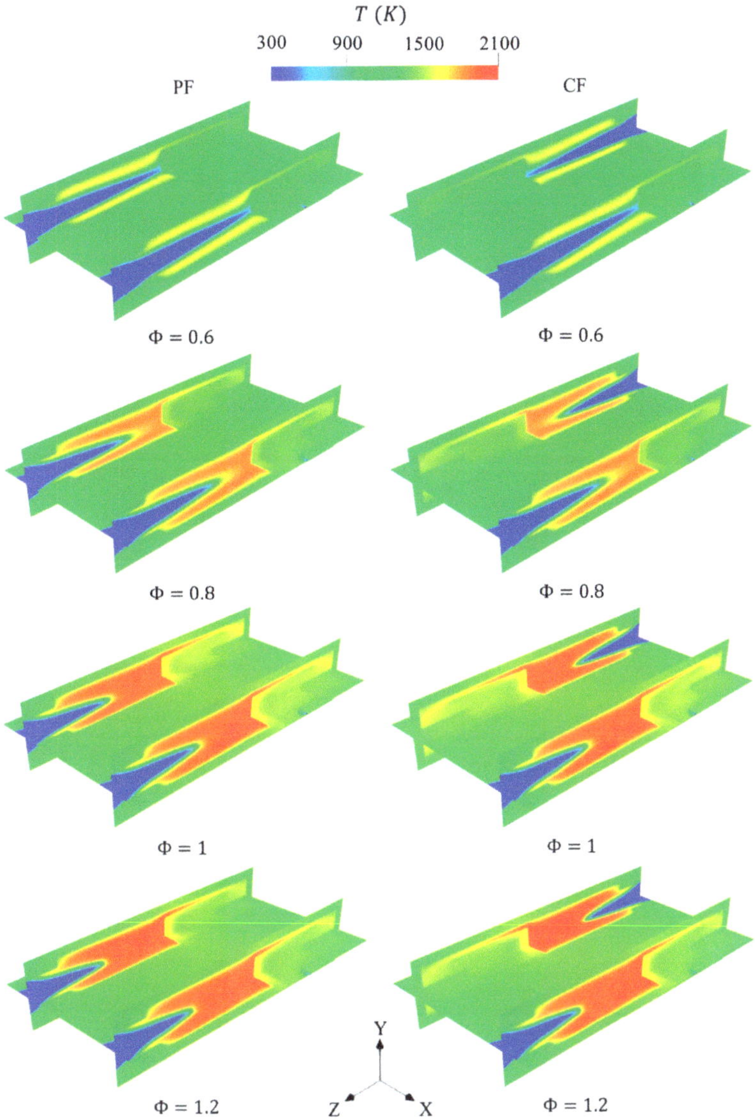

Figure 18. Spatial distributions of in-combustor temperature at various values of Φ for PF and CF configurations.

Figure 21 demonstrates the magnitude of entropy generation along with the contributions of each entropy process at a range of Φ from 0.6 to 1.2 for PF and CF configurations. As seen in Figure 21, entropy generation exhibits increasing and decreasing trends from lean to stoichiometric conditions and from stoichiometric to rich conditions, respectively, suggesting that the $\Phi = 1$ case results in a higher combustion intensity. In addition, the outcomes reveal that a higher Φ increases the participation rate of the chemical mechanism, whereas the percentage of heat conduction mechanism proportionally decreases, indicating the dominance of the chemistry rate over the temperature gradients under such conditions.

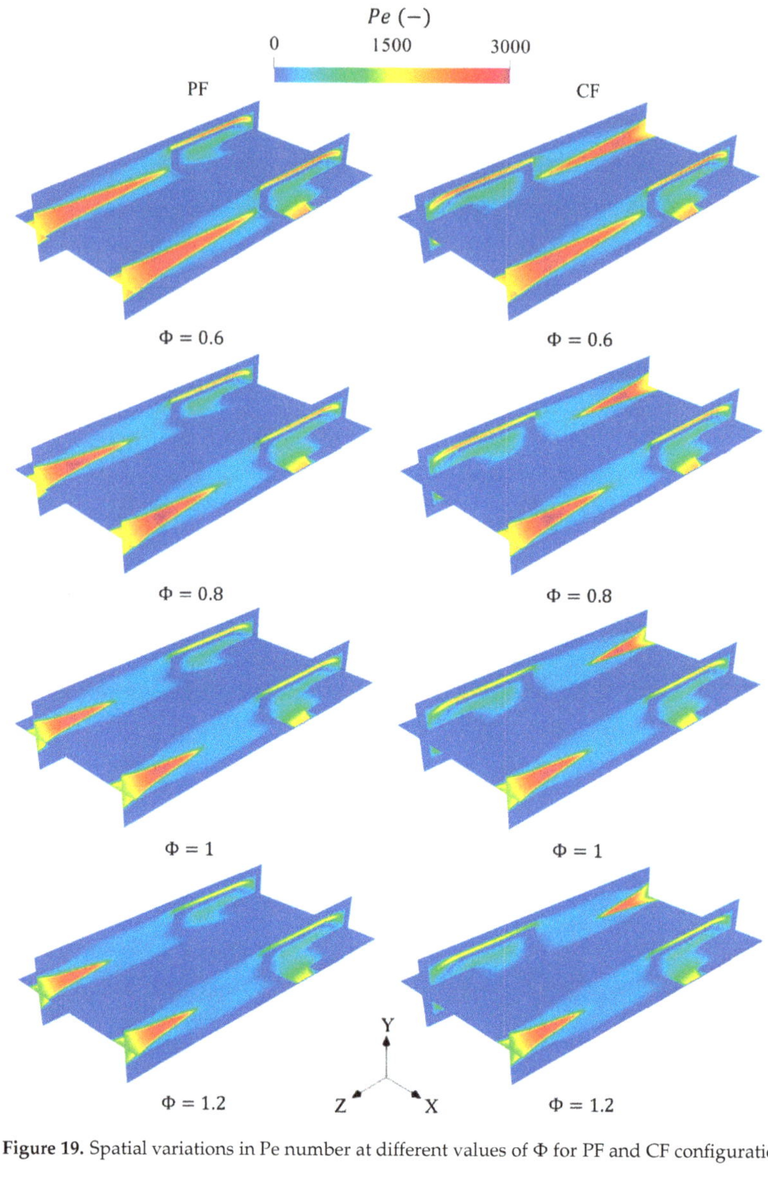

Figure 19. Spatial variations in Pe number at different values of Φ for PF and CF configurations.

Figure 22 outlines the impacts of varying Φ values on pressure loss and T_{eg}, as well as exergy and radiation efficiencies. The pressure loss findings, as shown in Figure 22a, demonstrate gradual reductions for both configurations as Φ increases, indicating that the fuel-rich conditions require less pumping power. However, T_{eg} encounters an increase in Φ from 0.6 to 1 and then a decrease from 1 to 1.2, which trends are caused by the low consumption rates of oxygen and hydrogen, respectively. Regarding system efficiency, both exergy and radiation efficiencies achieve their peaks at a Φ of 1. Despite the T_{eg} trend, the unity equivalence ratio case provides the optimal balance between fuel and air, as it exhibits the greatest power output density and T_{mw}.

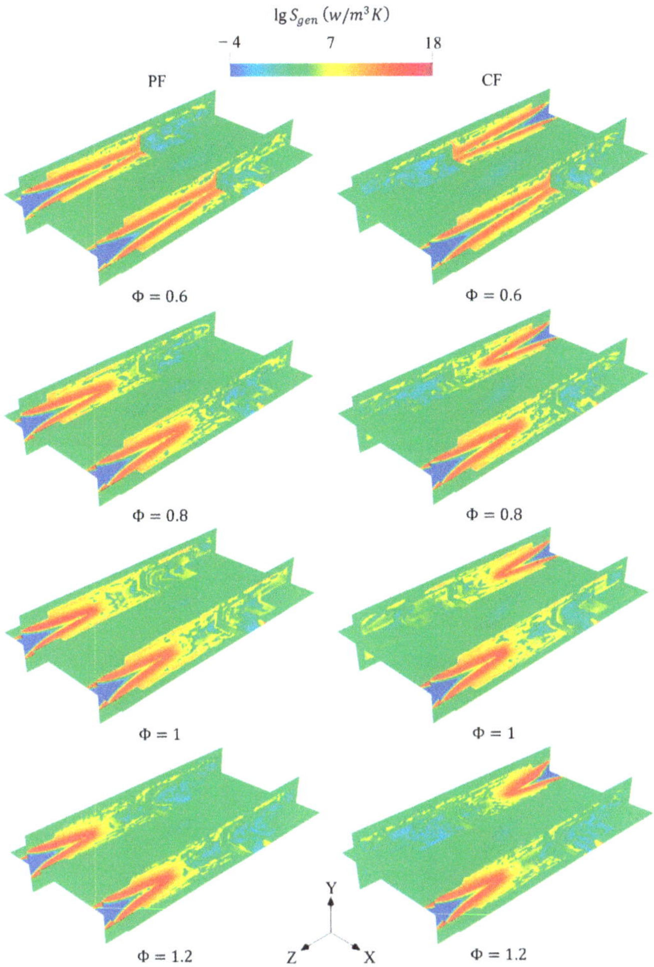

Figure 20. Spatial variations in total entropy generation at a variety of Φ values for PF and CF configurations.

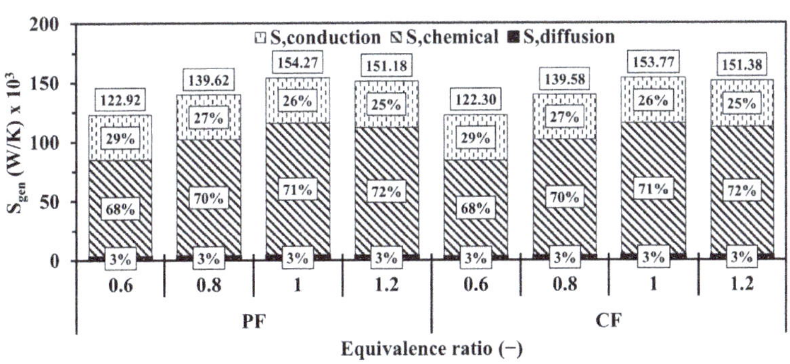

Figure 21. Volume integrations of entropy generation processes at various Φ for PF and CF.

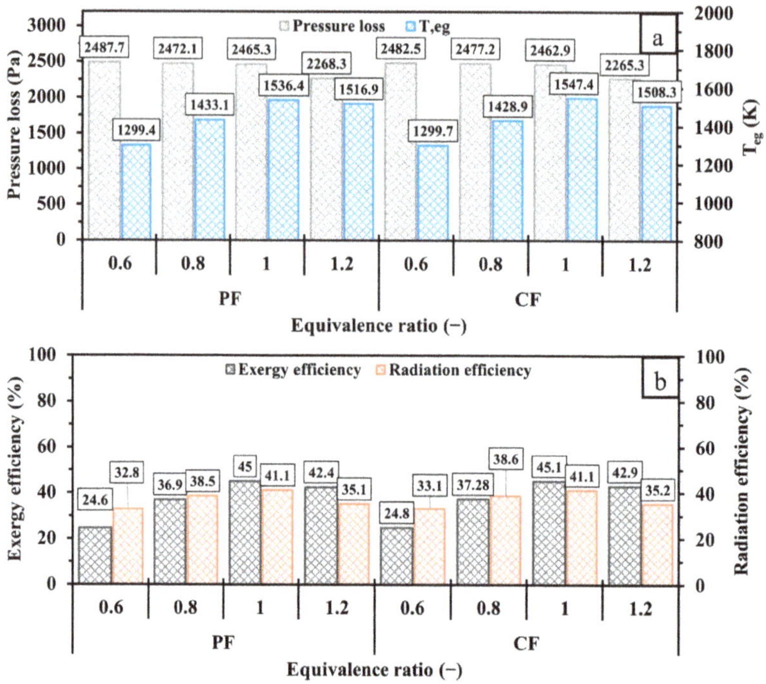

Figure 22. Variations of (**a**) pressure loss and T_{eg} and (**b**) exergy and radiation efficiencies at different levels of Φ for PF and CF test cases.

4. Conclusions

This numerical work investigates a novel micro combustor consisting of a PF and CF micro planar combustor with two micro cylindrical combustors embedded inside. The micro cylindrical combustors feature heat-recirculating channels to increase the dwell time of flow, and also to act as a preheating source for inlet premixed mixtures. The thermodynamic and performance parameters of the proposed configuration compared to the conventional one are analyzed to demonstrate the potential of the former. In addition, the inlet velocity and equivalence ratio are varied to optimize the system's output power density. The main conclusions drawn from the simulations are as follows:

- The PF and CF structures significantly improve T_{mw} and R_T compared to CD structure, indicating enhancements in the rate of absorption of heat produced by combustion by the external walls. This is confirmed by the greater variations in Pe number for the two former cases, implying that the advocation effects dominate over those of diffusion. It is noteworthy that both PF and CF result in pronounced increases in the magnitude of S_{gen} compared to the CD configuration, indicating high dissipations of energy. However, the PF and CF configurations save more thermal energy, as the T_{eg} values are reduced by 591 K and 580 K compared to the CD configuration, respectively, resulting in higher exergy and radiation efficiencies;
- Increasing V_{in} leads to notably advanced T_{mw} values for PF and CF; however, reversing the direction of flow from one inlet in CF provides the optimal R_T at a V_{in} of 9 m/s. Furthermore, the increase in V_{in} shifts the onset of combustion further downstream, and significantly boosts the S_{gen} magnitude. It is important to highlight that the chemical entropy generation mechanism is escalated when approaching higher V_{in}, suggesting intense combustion under such conditions. Nevertheless, altering the

V_{in} value from low to high increases T_{eg} and decreases the exergy and radiation efficiencies due to the low capacity of a combustor at the micro scale to utilize greater input powers;

- For both PF and CF test cases, the T_{mw} and the exergy and radiation efficiencies are improved when increasing Φ from lean to stoichiometric, and are reduced under rich conditions, due to the low consumption rates of oxygen and hydrogen, respectively. This implies that the unity equivalence ratio optimizes the fuel and air ratio. Further, the vigorous chemistry in the highlighted case enhances S_{gen} and T_{eg}. In comparison with PF, CF shows relatively better R_T in most states of Φ owing to the equal amounts of thermal energy released from opposite sides of the micro combustor.

Funding: This work was supported by the Deanship of Scientific Research, King Faisal University, Saudi Arabia [Grant No.: KFU251361].

Data Availability Statement: The data presented in the study are included in the article; further inquiries can be directed to the corresponding author.

Acknowledgments: The author gratefully thanks the Deanship of Scientific Research at King Faisal University (Saudi Arabia) for supporting this research as a part of the Research Grants Program [Raed Grant No.: KFU251361]. The author also would like to gratefully acknowledgment King Abdullah University of Science and Technology (KAUST) for their support in permitting the use of the high-end computing facilities to conduct the numerical investigations.

Conflicts of Interest: The author declares that he has no known competing financial interests or personal relationships that could have appeared to influence the work reported in this paper.

Abbreviations

The following abbreviations are used in this manuscript:

E	Total total energy of the fluid $\left(\text{J}\cdot\text{kg}^{-1}\right)$
Pe	Peclet number $(-)$
R_i	Reaction net rate of production of species i
h	Natural convection heat transfer coefficient $\left(\text{W}\cdot\text{m}^{-2}\cdot\text{K}^{-1}\right)$
$\vec{v}$	Vector velocity $(\text{m}\cdot\text{s}^{-1})$
Q_{rad}	Heat losses by radiation (W)
Q_{con}	Heat losses by convection (W)
k_{eff}	Effective thermal conductivity $\left(\text{W}\cdot\text{m}^{-2}\cdot\text{K}^{-1}\right)$
Y_i	Local mass fraction of species i $(-)$
S_h	Source term of enthalpy $(\text{W}\cdot\text{m}^{-3})$
k_{mix}	Mixture thermal conductivity $\left(\text{W}\cdot\text{m}^{-1}\ \text{K}^{-1}\right)$
A_c	External wall surface area (m^2)
A_i	External wall area of cell i (m^2)
$\vec{J}_j$	Diffusion flux of species j $(\text{mol}\cdot\text{m}^{-2}\cdot\text{s}^{-1})$
$S_{gen}^{diffussion}$	Mass diffusion mechanism of entropy generation $\left(\text{W}\cdot\text{K}^{-1}\right)$
$S_{gen}^{chemical}$	Chemical reaction of entropy generation $\left(\text{W}\cdot\text{K}^{-1}\right)$
$S_{gen}^{conduction}$	Heat conduction of entropy generation $\left(\text{W}\cdot\text{K}^{-1}\right)$
S_{gen}^{total}	Total entropy generation $\left(\text{W}\cdot\text{K}^{-1}\right)$
$\bar{s}_n^o$	Entropy of species i at reference conditions $\left(\text{J}\cdot\text{kg}^{-1}\cdot\text{K}^{-1}\right)$
$\bar{h}_n^o$	Enthalpy of species i at reference conditions $\left(\text{J}\cdot\text{kg}^{-1}\right)$
E_x^{in}	Inlet exergy (W)
E_x^{des}	Uncounted exergy destruction (W)
E_x^{eg}	Total exergy losses (W)

E_x^{loss}	Energy loss from the combustion exhaust gas (W)
R_T	Wall temperature uniformity (%)
P	Pressure (Pa)
P_{atm}	Atmospheric pressure (101,325 Pa)
T_i	Outer wall temperature of cell i (K)
T_{eg}	Exhaust gas temperature (K)
$T_{w,m}$	Area-weighted mean wall temperature (K)
T_∞	Ambient temperature (K)
T_w	Temperature of external wall (K)
R	Gas constant $\left(\text{J}\cdot\text{kg}^{-1}\ \text{K}^{-1}\right)$
D_i	Mass diffusivity of species i $(\text{m}^2\cdot\text{s}^{-1})$
C_p	Specific heat capacity $\left(J\cdot\text{kg}^{-1}\ \text{K}^{-1}\right)$
V	Velocity $(\text{m}\cdot\text{s}^{-1})$
Q_{LHV}	Lower heating value $\left(\text{MJ}\cdot\text{kg}^{-1}\right)$
$\dot{m}_{fuel}$	Fuel mass flow rate $(\text{kg}\cdot\text{s}^{-1})$
$\dot{m}_{inlet}$	Inlet stream mass flow rate $(\text{kg}\cdot\text{s}^{-1})$
h_j	Specific nthalpy of species j $\left(\text{J}\cdot\text{kg}^{-1}\right)$
Greek letters	
ρ	Mixture gas density $(\text{kg}\cdot\text{m}^{-3})$
$\vec{\tau}$	Viscous stress $(\text{N}\cdot\text{m}^{-2})$
$\vec{\tau}'$	Reynolds stress $(\text{N}\cdot\text{m}^{-2})$
μ_i	Chemical potential of species i $\left(\text{J}\cdot\text{kg}^{-1}\right)$
ω_i	Mass fraction of species, i
η_{exergy}	Exergy efficiency (%)
$\eta_{radiation}$	Radiation efficiency (%)
σ	Stephan–Boltzmann constant $\left(5.67 \times 10^{-8}\ \text{W}\cdot\text{m}^{-2}\text{K}^{-4}\right)$
ε	Emissivity of the solid surface
λ	Thermal conductivity $\left(\text{W}\cdot\text{m}^{-1}\ \text{K}^{-1}\right)$
γ_i	Production rate of species i $\left(\text{mol}\cdot\text{L}^{-1}\cdot\text{s}^{-1}\right)$
Φ	Equivalence ratio $(-)$
χ_i	Mole fraction of species, i

References

1. Chircov, C.; Grumezescu, A.M. Microelectromechanical Systems (MEMS) for Biomedical Applications. *Micromachines* **2022**, *13*, 164. [CrossRef]
2. Percy, J.J.; Kanthamani, S. Revolutionizing Wireless Communication: A Review Perspective on Design and Optimization of RF MEMS Switches. *Microelectron. J.* **2023**, *139*, 105891. [CrossRef]
3. Bai, S.; Ding, R.; Chirarattananon, P. A Micro Aircraft with Passive Variable-Sweep Wings. *IEEE Robot. Autom. Lett.* **2022**, *7*, 4016–4023. [CrossRef]
4. Ding, J.; E, J.; Cai, L.; Luo, B.; Li, J. Development of a Reaction Mechanism of Hydrogen Production through Rich Methane-Acetylene Blending in a Porous Medium Micro-Combustor. *Energy Convers. Manag.* **2025**, *325*, 119360. [CrossRef]
5. Song, L.; Zhang, S.; Wang, Q.; Chen, W.; Liu, B.; Zhao, Y.-D. Light-Controlled Microbots Gathering as a Sterilization Platform for Highly Efficient Capturing, Concentrating and Killing Targeted Bacteria. *Chem. Eng. J.* **2022**, *435*, 135067. [CrossRef]
6. Tang, A.; Cai, T.; Li, C.; Zhou, C.; Gao, L. Flame Visualization and Spectral Analysis of Combustion Instability in a Premixed Methane/Air-Fueled Micro-Combustor. *Energy* **2024**, *294*, 130793. [CrossRef]
7. Zhao, H.; Zhao, D.; Sun, D.; Semlitsch, B. Electrical Power, Energy Efficiency, NO and CO Emissions Investigations of an Ammonia/Methane-Fueled Micro-Thermal Photovoltaic System with a Reduced Chemical Reaction Mechanism. *Energy* **2024**, *305*, 132248. [CrossRef]
8. Cai, S.; Yang, W.; Ding, Y.; Zeng, Q.; Wan, J. Hydrogen-Air Premixed Combustion in a Novel Micro Disc-Burner with an Annular Step. *Fuel* **2022**, *313*, 123015. [CrossRef]

9. Cai, L.; E, J.; Li, J.; Ding, J.; Luo, B. A Comprehensive Review on Combustion Stabilization Technologies of Micro/Meso-Scale Combustors for Micro Thermophotovoltaic Systems: Thermal, Emission, and Energy Conversion. *Fuel* **2023**, *335*, 126660. [CrossRef]
10. E, J.; Mei, Y.; Feng, C.; Ding, J.; Cai, L.; Luo, B. A Review of Enhancing Micro Combustion to Improve Energy Conversion Performance in Micro Power System. *Int. J. Hydrogen Energy* **2022**, *47*, 22574–22601. [CrossRef]
11. Amani, E.; Daneshgar, A.; Hemmatzade, A. A Novel Combined Baffle-Cavity Micro-Combustor Configuration for Micro-Thermo-Photo-Voltaic Applications. *Chin. J. Chem. Eng.* **2020**, *28*, 403–413. [CrossRef]
12. He, Z.; Yan, Y.; Fang, R.; Ou, Z.; Zhang, Z.; Yang, Z.; Zhang, Z. Numerical Investigation of a Novel Micro Combustor with a Central and Bilateral Slotted Blunt Body. *Int. J. Hydrogen Energy* **2021**, *46*, 23564–23579. [CrossRef]
13. Kaisare, N.S.; Di Sarli, V. The Effect of Catalyst Placement on the Stability of a U-Bend Catalytic Heat-Recirculating Micro-Combustor: A Numerical Investigation. *Catalysts* **2021**, *11*, 1560. [CrossRef]
14. Yan, Y.; Wu, G.; Huang, W.; Zhang, L.; Li, L.; Yang, Z. Numerical Comparison Study of Methane Catalytic Combustion Characteristic between Newly Proposed Opposed Counter-Flow Micro-Combustor and the Conventional Ones. *Energy* **2019**, *170*, 403–410. [CrossRef]
15. Yilmaz, H. Investigation of Combustion and Emission Performance of a Micro Combustor: Effects of Bluff Body Insertion and Oxygen Enriched Combustion Conditions. *Int. J. Hydrogen Energy* **2019**, *44*, 25985–25999. [CrossRef]
16. Zhang, Y.; Lu, Q.; Fan, B.; Long, L.; Quaye, E.K.; Pan, J. Effect of Multiple Bluff Bodies on Hydrogen/Air Combustion Characteristics and Thermal Properties in Micro Combustor. *Int. J. Hydrogen Energy* **2023**, *48*, 4064–4072. [CrossRef]
17. Yan, Y.; Zhang, C.; Wu, G.; Feng, S.; Yang, Z. Numerical Study on Methane/Air Combustion Characteristics in a Heat-Recirculating Micro Combustor Embedded with Porous Media. *Int. J. Hydrogen Energy* **2022**, *47*, 20999–21012. [CrossRef]
18. Li, J.; Xiao, H.; Li, Q.; Shi, J. Heat Recirculation and Heat Losses in Porous Micro-Combustors: Effects of Wall and Porous Media Properties and Combustor Dimensions. *Energy* **2021**, *220*, 119772. [CrossRef]
19. Nauman, M.; Pan, J.; Lu, Q.; Zhang, Y.; Liu, C.; Li, F.; Quaye, E.K. Analyzing the Combustion Characteristics of Premixed Methane-Oxygen with Different Hydrogen Addition Ratios in a Catalytic Micro-Combustor. *J. Energy Inst.* **2024**, *114*, 101655. [CrossRef]
20. Zhao, H.; Zhao, D.; Becker, S. Thermal Performances Investigation on an Ammonia-Fuelled Heat-Recirculating Micro-Combustor with Reduced Chemical Mechanism. *Appl. Therm. Eng.* **2024**, *236*, 121685. [CrossRef]
21. Ni, S.; Zhao, D.; Wu, W.; Guan, Y. NOx Emission Reduction Reaction of Ammonia-Hydrogen with Self-Sustained Pulsating Oscillations. *Therm. Sci. Eng. Prog.* **2020**, *19*, 100615. [CrossRef]
22. Cai, L.; E, J.; Zhao, D. Numerical Investigation on NOx Emissions and Thermal Performance of Hydrogen/Ammonia Fueled Micro-Combustors with Periodic Wall Structures. *Renew. Energy* **2025**, *241*, 122347. [CrossRef]
23. Nieto-Londoño, C.; Silva-López, W.; Gómez-Velásquez, N. Numerical Assessment of the Combustion of Methane–Hydrogen–Air Mixtures in Micro-Scale Conditions. *Processes* **2025**, *13*, 794. [CrossRef]
24. Dai, C.; Zuo, W.; Li, Q.; Zhou, K.; Huang, Y.; Zhang, G.; E, J. Energy Conversion Efficiency Improvement Studies on the Hydrogen-Fueled Micro Planar Combustor with Multi-Baffles for Thermophotovoltaic Applications. *Energy* **2024**, *313*, 134099. [CrossRef]
25. Li, J.; E, J.; Ding, J.; Cai, L.; Luo, B. Investigation on Thermophotovoltaic Performance Enhancement of the Hydrogen Fueled Micro Combustors with the Center-Cleared Twisted Tapes for Micro-Thermophotovoltaic System. *Appl. Therm. Eng.* **2025**, *264*, 125483. [CrossRef]
26. Zhang, Y.; Lu, Q.; Fan, B.; Chen, W.; Jiang, C.; Pan, J.; Jia, R. Study on the Non-Premixed Combustion Characteristics and Exergy Analysis of H2/Air in Vase-Shaped Micro Combustor. *Appl. Therm. Eng.* **2023**, *233*, 121104. [CrossRef]
27. Sheykhbaglou, S.; Dimitriou, P. Numerical Analysis of Combustion and Thermal Performance of a Bluff-Body and Swirl-Stabilized Micro-Combustor with Premixed NH3/H2/Air Flames. *Energies* **2025**, *18*, 780. [CrossRef]
28. Tang, A.; Pan, J.; Yang, W.; Xu, Y.; Hou, Z. Numerical Study of Premixed Hydrogen/Air Combustion in a Micro Planar Combustor with Parallel Separating Plates. *Int. J. Hydrogen Energy* **2015**, *40*, 2396–2403. [CrossRef]
29. Su, Y.; Cheng, Q.; Song, J.; Si, M. Numerical Study on a Multiple-Channel Micro Combustor for a Micro-Thermophotovoltaic System. *Energy Convers. Manag.* **2016**, *120*, 197–205. [CrossRef]
30. Kang, X. Micro/Mesoscale Combustion Based Portable Power Generating System. Ph.D. Thesis, The University of Queensland, Brisbane, QLD, Australia, 2017.
31. Almutairi, F. A Numerical Study on Key Thermal Parameters and NOx Emissions of a Hydrogen-Fueled Double-Channel Outlet Micro Cylindrical Combustor Employing a Heat-Recirculating Configuration for Thermophotovoltaic Applications. *Processes* **2024**, *12*, 1848. [CrossRef]
32. *ANSYS Fluent User's Guide*; Release 23.2; ANSYS, Inc.: Canonsburg, PA, USA, 2023.

33. Ni, S.; Zhao, D.; Sellier, M.; Li, J.; Chen, X.; Li, X.; Cao, F.; Li, W. Thermal Performances and Emitter Efficiency Improvement Studies on Premixed Micro-Combustors with Different Geometric Shapes for Thermophotovoltaics Applications. *Energy* **2021**, *226*, 120298. [CrossRef]
34. Yang, X.; Yu, B.; Peng, X.; Zhou, H. Investigation of Thermal Performance and Energy Conversion in a Novel Planar Micro-Combustor with Four-Corner Entrances for Thermo-Photovoltaic Power Generators. *J. Power Sources* **2021**, *515*, 230625. [CrossRef]
35. Kuo, C.H.; Ronney, P.D. Numerical Modeling of Non-Adiabatic Heat-Recirculating Combustors. *Proc. Combust. Inst.* **2007**, *31*, 3277–3284. [CrossRef]
36. Shih, T.-H.; Liou, W.W.; Shabbir, A.; Yang, Z.; Zhu, J. A New K-ϵ Eddy Viscosity Model for High Reynolds Number Turbulent Flows. *Comput. Fluids* **1995**, *24*, 227–238. [CrossRef]
37. Zhao, H.; Zhao, D.; Becker, S.; Zhang, Y. NO Emission and Enhanced Thermal Performances Studies on Counter-Flow Double-Channel Hydrogen/Ammonia-Fuelled Microcombustors with Oval-Shaped Internal Threads. *Fuel* **2023**, *341*, 127665. [CrossRef]
38. Mousavi, S.M.; Sotoudeh, F.; Jun, D.; Lee, B.J.; Esfahani, J.A.; Karimi, N. On the Effects of NH_3 Addition to a Reacting Mixture of H_2/CH_4 under MILD Combustion Regime: Numerical Modeling with a Modified EDC Combustion Model. *Fuel* **2022**, *326*, 125096. [CrossRef]
39. Cai, T.; Zhao, D.; Karimi, N. Optimizing Thermal Performance and Exergy Efficiency in Hydrogen-Fueled Meso-Combustors by Applying a Bluff-Body. *J. Clean. Prod.* **2021**, *311*, 127573. [CrossRef]
40. Yang, W.M.; Chua, K.J.; Pan, J.F.; Jiang, D.Y.; An, H. Development of Micro-Thermophotovoltaic Power Generator with Heat Recuperation. *Energy Convers. Manag.* **2014**, *78*, 81–87. [CrossRef]
41. Cai, T.; Zhao, D. Effects of Fuel Composition and Wall Thermal Conductivity on Thermal and NOx Emission Performances of an Ammonia/Hydrogen-Oxygen Micro-Power System. *Fuel Process. Technol.* **2020**, *209*, 106527. [CrossRef]
42. Li, J.; Huang, J.; Chen, X.; Zhao, D.; Shi, B.; Wei, Z.; Wang, N. Effects of Heat Recirculation on Combustion Characteristics of N-Heptane in Micro Combustors. *Appl. Therm. Eng.* **2016**, *109*, 697–708. [CrossRef]
43. He, Z.; Yan, Y.; Zhao, T.; Feng, S.; Li, X.; Zhang, L.; Zhang, Z. Heat Transfer Enhancement and Exergy Efficiency Improvement of a Micro Combustor with Internal Spiral Fins for Thermophotovoltaic Systems. *Appl. Therm. Eng.* **2021**, *189*, 116723. [CrossRef]
44. Fan, Y.; Schlick, C.P.; Umbanhowar, P.B.; Ottino, J.M.; Lueptow, R.M. Modelling Size Segregation of Granular Materials: The Roles of Segregation, Advection and Diffusion. *J. Fluid Mech.* **2014**, *741*, 252–279. [CrossRef]
45. Figueroa, A.; Meunier, P.; Cuevas, S.; Villermaux, E.; Ramos, E. Chaotic Advection at Large Péclet Number: Electromagnetically Driven Experiments, Numerical Simulations, and Theoretical Predictions. *Phys. Fluids* **2014**, *26*, 013601. [CrossRef]
46. Ni, S.; Zhao, D.; Zhu, X. Heat Transfer and Entropy Production Evaluation on Premixed Hydrogen/Air-Fuelled Micro-Combustors with Internal Threads. *Fuel* **2021**, *303*, 121325. [CrossRef]
47. Sun, Y.; Zhao, D.; Ni, S.; David, T.; Zhang, Y. Entropy and Flame Transfer Function Analysis of a Hydrogen-Fueled Diffusion Flame in a Longitudinal Combustor. *Energy* **2020**, *194*, 116870. [CrossRef]
48. Sahoo, B.B.; Saha, U.K.; Sahoo, N. Diagnosing the Effects of Pilot Fuel Quality on Exergy Terms in a Biogas Run Dual Fuel Diesel Engine. *Int. J. Exergy* **2012**, *10*, 77–93. [CrossRef]
49. Stepanov, V.S. Chemical Energies and Exergies of Fuels. *Energy* **1995**, *20*, 235–242. [CrossRef]
50. Tong, J.; Cai, T. Enhancing Thermal Performance, Exergy and Thermodynamics Efficiency of Premixed Methane/Air Micro-Planar Combustor in Micro-Thermophotovoltaic Systems. *Energies* **2023**, *16*, 118. [CrossRef]
51. Peng, Q.; Yang, W.; E, J.; Li, Z.; Xu, H.; Fu, G.; Li, S. Investigation on H2/Air Combustion with C3H8 Addition in the Combustor with Part/Full Porous Medium. *Energy Convers. Manag.* **2021**, *228*, 113652. [CrossRef]
52. Ismail, B.C.; Urmila, G.H.I.A.; Patrick, J.R.; Christopher, J.F.; Hugh, C.; Raad, P.E. Procedure for Estimation and Reporting of Uncertainty Due to Discretization in CFD Applications. *J. Fluids Eng.* **2008**, *130*, 078001.
53. Rong, H.; Zhao, D.; Cai, T.; Becker, S. Enhancing Thermal and Exergy Performances in a CO_2-Free Micro-Combustor with Reverse Flow Double-Channel Outlet Structure. *Appl. Therm. Eng.* **2023**, *233*, 121180. [CrossRef]
54. Yang, W.; Chou, S.; Shu, C.; Xue, H.; Li, Z. Effect of Wall Thickness of Micro-Combustor on the Performance of Micro-Thermophotovoltaic Power Generators. *Sens. Actuators A Phys.* **2005**, *119*, 441–445.
55. Zuo, W.; Zhang, Y.; Li, Q.; Li, J.; He, Z. Numerical Investigations on Hydrogen-Fueled Micro-Cylindrical Combustors with Cavity for Micro-Thermophotovoltaic Applications. *Energy* **2021**, *223*, 120098. [CrossRef]

processes

MDPI

Article

Experimental Study and Numerical Simulation of Sediment's Promoting Effect on Cavitation Based on Flow Field Analysis

Zeyuan Yang, Gang Yu and Qin Zhao *

Key Laboratory of Fluid and Power Machinery (LFPM), Ministry of Education, Xihua University, Chengdu 610039, China; 15723036297@163.com (Z.Y.); 18090237765@163.com (G.Y.)
* Correspondence: zhaoqin@mail.xhu.edu.cn

Abstract: Sediment-laden water significantly exacerbates the cavitation damage in hydraulic machinery compared to clear water, underscoring the importance of investigating the effects of sediment on cavitation. This study examines cavitation in sediment-laden water using a Venturi flow channel and a high-speed camera system. Natural river sand samples with a median diameter of 0.05, 0.07, and 0.09 mm are selected, and sediment-laden water with a concentration of 10, 30, and 50 g/L is prepared. The results indicate that increasing the sediment concentration or reducing the sediment size intensifies cavitation, and the influence of the sediment concentration is significantly greater than that of the sediment size. Meanwhile, the numerical simulation is conducted based on a gas–liquid–solid phase mixing model. The findings indicate that a higher sediment concentration corresponds to a greater shearing effect near the wall, which raises the drag on the attached sheet-like cavitation clouds and enhances the re-entrant jet which can intensify the shedding of cavitation clouds. Furthermore, sediment particles contribute to more vortices. Therefore, for hydraulic machinery operating in sediment-laden water of high concentration, the relative velocity should be reduced to mitigate the shearing effect, thereby weakening the synergy of cavitation and sediment erosion at the turbine runner.

Keywords: sediment-laden water; cavitation; high-speed photography; gas–liquid–solid phase mixing model; shearing effect

1. Introduction

Hydropower is one of the most important types of renewable energy, and its power generation accounts for 70% of the total energy production. The excessive river sedimentation in regions such as the Himalayas and Alps poses a severe challenge to sustainable hydropower development [1]. China has the highest hydropower development potential in the world, accounting for 43% of the total number of dams in the world. China's installed hydropower capacity accounts for 27% of the world's hydropower installed capacity [2]. At the same time, China is facing the problem of sedimentation. Moreover, hydraulic machinery (i.e., pumps, turbines, and reversible pump turbines) often operate under cavitation conditions. The synergistic effect of cavitation and sediment erosion can cause more serious damage as shown in Figure 1 [3,4].

Among the hydropower stations that have been built in China, 40% of the turbines have been severely affected by the synergistic effect of cavitation and sediment erosion, which will reduce the turbine efficiency, shorten the service life, and increase the operation and maintenance costs. However, the synergistic wear mechanisms are barely understood, and the conclusions reached by studies differ from each other. The synergistic bubble particle effect is manifested in the dual inhibition or promotion mechanism [5,6]. There are many factors affecting the promotion or inhibition of the synergistic effect between abrasion and cavitation, including the solid–liquid mixture [7,8] and particle properties [9,10].

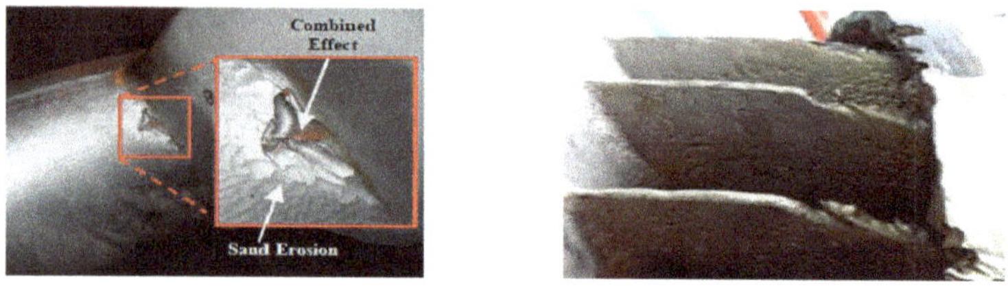

(a) (b)

Figure 1. Erosion in the hydro turbines. (**a**) Turbine runner with synergy of cavitation and sediment erosion, (**b**) synergistic damage caused by particles and cavitation at the leading edge of the runner blades.

Raichenko, Byakova [11], and Morch [12] suggested that inhomogeneities within the liquid can function as nucleation centers for bubbles and developed models to analyze the bubble formation process. Shen et al. [13] found that the effect on single cavitation bubble dynamics could be negligible for small viscosities, but it should be considered for relatively high viscosities. Liu et al. [14] found that fine particles increased the fluid's viscosity, and the strong liquid viscosity produced higher resistance to the growth and collapse of bubbles, lowering the particle velocity and suppressing cavitation.

The properties of particles including size, concentration, shape, and type are key influencing factors for the damage from cavitation and abrasion.

Lian et al. [10] reported that the critical particle diameter was approximately 0.035–0.048 mm. When the sediment had a particle size smaller than the critical value, the damage from the combined wear was slightly less severe than that from cavitation in clear water. Using a specialized vibrating device, Wu and Gou [15] discussed the combined wear under three sediment concentration cases (25, 50, and 85 kg/m^3) and five particle diameter scenarios (0.531, 0.253, 0.063, 0.042, 0.026 mm). They revealed that when the particle diameter was smaller than the critical value (0.048 mm), the wear effect weakened as the concentration increased. Chen et al. [16] found that the generated damage was less severe than that in clear water when the average sediment size was smaller than 0.120 mm, and Zhong and Minemura [17] suggested that for sediment with a size less than 1 mm, the flow would only cause cavitation erosion. In general, when the particle size is smaller than the threshold, cavitation is suppressed; otherwise, it is intensified [18–20]. For particles smaller than the threshold size, the medium sand and clay in the particle-laden flow have non-Newtonian fluid features.

In the presence of sand, the cavitation pressure threshold rises proportionately to the sediment concentration. Zhu et al. [21] indicated that the sediment concentration was linearly related to the critical cavitation value, with higher particle concentrations corresponding to more possible occurrences of cavitation. This was because many gas nuclei with different sizes existed within the particle surface cracks; increasing the sediment concentration inevitably generated more gas nuclei in the flow [22,23]. However, the conclusion that an increased particle concentration monotonically promotes or inhibits the combined erosion was contradicted by some other reports, which suggested a particle concentration threshold. As the concentration increased under a fixed particle diameter, the combined wear initially weakened and then strengthened [24,25].

In addition, the shape and type of particle will affect the cavitation [26–30]. However, the size and concentration of particles are crucial factors affecting cavitation, and their different combinations can inhibit or promote cavitation. Based on scholars' previous research [10,15], natural river sand with a median diameter of 0.05, 0.07, and 0.09 mm is selected, and sediment-laden water with a concentration of 10, 30, and 50 g/L is prepared for the study of how different sediment conditions promote cavitation in a Venturi structure flow channel in this paper.

The mechanism of the inhibition or promotion of cavitation has been studied. Li and Zhang et al. [31,32] noticed that the particle shooting effect (acceleration of particles when bubbles expanded and after bubbles and particles detached) was more pronounced as the particle-to-bubble size ratio was reduced. Li et al. [33] claimed that bubble–particle interactions were heavily dependent on the particle-to-bubble size ratio and particle-to-liquid density ratio. Poulain et al. [34] reported a quantitative relation between sand movement due to cavitation bubbles and their distance. Arora et al. [35,36] discovered that cavitation bubbles' collapse would accelerate the nearby sand particles, possibly aggravating abrasion and damage.

The current research on the synergistic mechanisms focuses on the theoretical analyses of individual cavitation bubbles or the combined effects of sediment and cavitation through material erosion assessments. Meanwhile, significant achievements have been made in the impact of cavitation on particles. However, there is a dearth of studies that directly analyze sediment-laden flow fields to understand the sediment–bubbles relationship. This study explores the effect of sediment-laden flow on cavitation evolution. The viscosity of sediment-laden water varies with the sediment size and concentration, and has influences on the flow field and will affect the formation, development, and collapse of cavitation. We obtained the viscosity of sediment-laden water through a viscometer and apply it to subsequent numerical calculations and analysis of the flow field.

There are relatively few visual experimental studies on the cavitation characteristics of sediment-laden water in the existing research. Due to its simple structure, the cavitation mechanism of the Venturi is consistent with hydraulic machinery such as pumps and turbines, making it important experimental equipment for studying the cavitation mechanisms and characteristics. This study employed a high-speed camera system to capture the morphology of cavitation and the complete developmental process within a Venturi structure flow channel containing sediment-laden water. Meanwhile, some scholars have used the three-phase flow (solid–liquid–vapor) model to analyze the erosion of the wall by sediment particles in the presence of cavitation bubbles. However, there is a lack of research on the cavitation flow field affected by sediment. In this paper, based on sediment–water two-phase flow, we consider the occurrence and development of cavitation, and compare the impact of sediment particles on the cavitation cycle.

2. Visual Cavitation Experiment

2.1. Experimental Apparatus

The experimental apparatus is a small desktop-integrated self-circulating system, equipped with a 40 L water storage tank, a pump with a head of 20 m, a T4-type light source, a display screen, and a Venturi flow channel made of Seiko transparent acrylic material as shown in Figure 2a. The dimensions of the equipment are 550 mm in length, 300 mm in width, and 1100 mm in height. The working voltage is 220 volts, and the power is 300 watts. The throat pressure can be read through a vacuum gauge connected to the throat. The experimental apparatus has been modified to read the inlet and outlet pressures: a hole is drilled in the display screen at a position 1 cm downstream of the inlet, and a pressure gauge with a measuring range of 0 to 0.16 MPa is attached. A hole is drilled in the display screen at a position 3 cm away from the outlet, and a vacuum gauge with a measuring range of −0.1 to 0 MPa is attached.

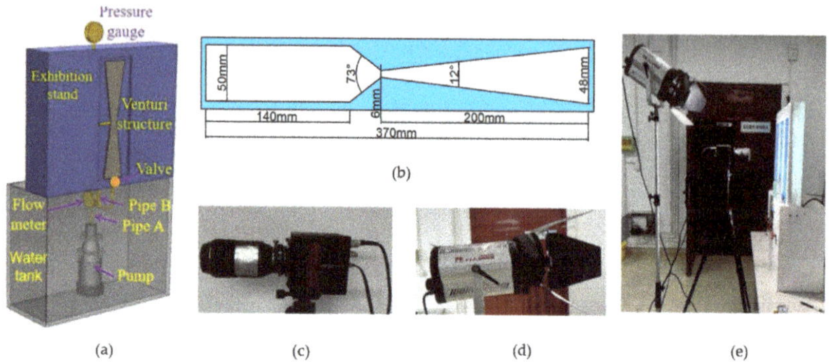

Figure 2. Visual cavitation experimental system. (**a**) Schematic of the cavitation mechanism experiment apparatus, (**b**) schematic of Venturi structure dimensions, (**c**) high-speed camera, (**d**) light source, (**e**) experimental field diagram.

The Venturi flow channel comprises four distinct sections: an inlet straight section, a contraction section, a throat, and a diffusion section. As the fluid traverses the throat of the Venturi channel, the reduction in cross-sectional area results in an increase in fluid velocity, leading to a decrease in static pressure. This pressure drop induces the vacuum condition, creating the requisite environment for cavitation formation. The dimensions of the Venturi channel are illustrated in Figure 2b. A Fastcam AX-100 high-speed camera (Photron, Tokyo, Japan) equipped with a Nikon lens (Nikon, Tokyo, Japan) with a focal length of 105 mm is used to record the cavitation phenomena as shown in Figure 2c. To enhance the quality of the cavitation cloud images, a photographic light source is utilized for additional illumination and is shown in Figure 2d. The experimental site is shown in Figure 2e. The Photron FASTCAM Viewer software (version 3.6.9.1) is utilized to precisely control the frame rate, recording duration, and resolution of the high-speed photography system, which can achieve a 20,000 fps frame rate at the maximum. During the experiments, the frame rate was set to 4000 fps to ensure that the captured images maintain sufficient detail, with an image resolution of 1024 × 1024 pixels.

2.2. Measurement of Viscosity

The viscosity of a liquid significantly influences its flow behavior and cavitation characteristics. In sediment-laden water, viscosity differences arise from the variations in particle concentration and size compared to clear water. These viscosity variations impact the cavitation behavior, which is the primary concern of the present study. To measure the sediment-laden water's viscosity, an NDJ-8S digital viscometer (Shanghai Zhongchen Digital Technology Equipment Co., Ltd., Shanghai, China) was employed, as illustrated in Figure 3a. This instrument drives the rotor to rotate at a constant speed via an electric motor. When the rotor spins in the liquid, the liquid exerts a viscous torque on the rotor. A sensor detects and collects data on this viscous torque, which are then processed by a computer to determine the viscosity of the measured liquid. Assuming laminar flow and negligible slip between the liquid and the measuring spindle's surface, the viscosity of the liquid can be calculated using the appropriate formula.

$$\mu = \frac{M}{4\pi h\omega}\left(\frac{1}{r^2} - \frac{1}{R^2}\right) \quad (1)$$

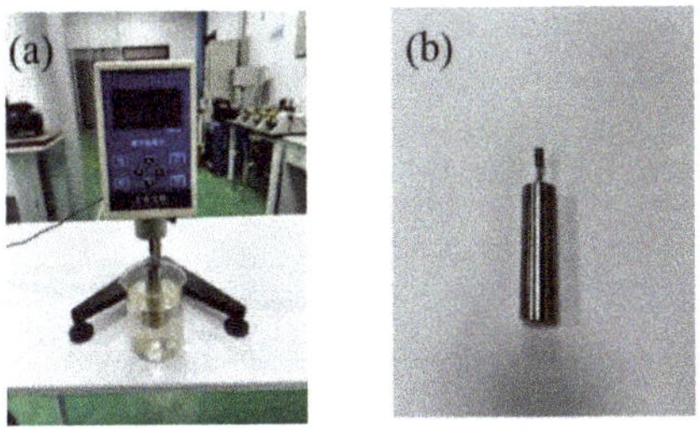

Figure 3. Viscosity measuring instrument. (**a**) NDJ-8S digital display viscometer, (**b**) zero rotor.

In this context, M represents the torque exerted by the internal friction of the liquid on the rotating spindle, r denotes the inner cylinder radius, and R is the outer cylinder radius. h indicates the immersed height of the liquid surrounding the inner cylinder, and ω signifies the angular velocity of the rotating spindle. This instrument is equipped with an over range alarm function: when the test value exceeds 100% of the measuring range, the display will show "over". To ensure the measurement accuracy, the percentage of the measuring range is controlled between 10% and 90%. Rotor 0 has the smallest measuring range, with a full scale of 20 mPa·s at a rotation speed of 60 rpm. Therefore, Rotor 0 was selected in this study.

The viscosity of clear water measured with this instrument is 1.13 mPa·s when the water temperature is 15 degrees Celsius, and the viscosity measurements for sediment-laden water are summarized in Table 1. Notably, at a low sediment concentration (S = 10 g/L), the viscosity matches that of clear water, regardless of the particle size. However, as the sediment concentration increases, the viscosity rises, with the particle size playing a more significant role. Smaller particles contribute to a greater increase in viscosity. For example, with a particle diameter of d = 0.09 mm, the viscosity rises from 1.132 mPa·s to 1.219 mPa·s, reflecting a 7.68% increase. In contrast, with a particle diameter of d = 0.05 mm, the viscosity escalates from 1.139 mPa·s to 1.756 mPa·s, a notable increase of 54.17%. Thus, an increase in the sediment concentration with finer particles leads to a substantial rise in the viscosity compared to clear water, underscoring the importance of the cavitation behavior under such conditions.

Table 1. Viscosity of sand water μ (mPa·s).

d (mm) \ S (g/L)	10	30	50
0.09	1.132	1.141	1.219
0.07	1.134	1.153	1.275
0.05	1.139	1.165	1.756

2.3. *Experiment Results*

2.3.1. Comparison of the Cavitation Intensity

Since cavitation initiates and then develops at the wall of the throat, the cavitation number at the throat is adopted to reflect the intensity of the cavitation. In the experimental Venturi structure flow channel, the transition from the converging flow section to the diverging flow section occurs abruptly at a certain cross-section, and the throat is exactly this abrupt cross-section. The pressure measuring points are arranged on the inner wall of the throat, and the pressure at the measuring points is used to reflect the pressure at the throat. Based on the experimentally measured average pressure at the throat, the cavitation numbers for various sediment-laden water samples are presented in Table 2. The findings demonstrate that the cavitation number decreases at higher sediment concentrations, indicating the enhanced cavitation intensity. As the water flow decelerates from the throat to the diffusion section, the drag effect of the sediment particles slows the decrease in flow velocity and leads to a more gradual increase in pressure. Consequently, this results in a lower pressure at the throat, effectively promoting cavitation. A higher concentration corresponds to a more pronounced particle drag effect.

Table 2. Cavitation number of sand water.

d (mm)	S (g/L)	P_{throat} (Pa)	σ
0.09	10	22,325	0.27
	30	14,325	0.17
	50	8325	0.08
0.07	10	19,325	0.24
	30	11,825	0.14
	50	7325	0.07
0.05	10	17,325	0.21
	30	10,325	0.12
	50	6325	0.06

The effect of particle dimension on the cavitation number is notably less significant than that of the sediment concentration. For example, at d = 0.09 mm, increasing the sediment concentration from 10 g/L to 50 g/L results in a reduction in the cavitation number from 0.27 to 0.08, a decrease of 70.37%. Conversely, in sediment-laden water with a concentration of 50 g/L, reducing the particle size from d = 0.09 mm to d = 0.05 mm only decreases the cavitation number from 0.08 to 0.05, reflecting a reduction of only 12.5%.

The relationship between the cavitation number and viscosity is illustrated in Figure 4, which is supported by the data in Table 1. The graph demonstrates that, in sediment-laden water, an increased viscosity correlates with a smaller cavitation number and an increased cavitation intensity. When the sediment size exceeds the critical value, an increase in concentration or a decrease in sediment size will promote cavitation, and the promoting effect of concentration is more significant than that of the sediment size. Therefore, the cavitation of sediment-laden water with a high concentration (S = 50 g/L) should be taken seriously.

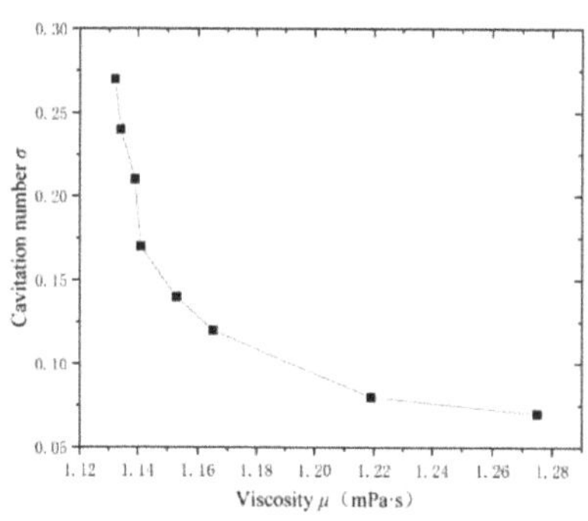

Figure 4. Change in cavitation number with the viscosity of sand water.

2.3.2. Evolution Process of the Cavitation Cloud

During the cavitation experiments, high flow rates revealed a distinct phenomenon of periodic cavitation bubble shedding at the cloud tail. For instance, at a flow rate of Q = 294 mL/s and a sediment concentration of S = 50 g/L with different sediment sizes, the cavitation cloud evolution over a selected cycle, denoted as T, is analyzed and depicted. The captured raw images underwent MATLAB (version 9.13 R2022b) processing including background subtraction, binarization, and denoising, as demonstrated in Figure 5.

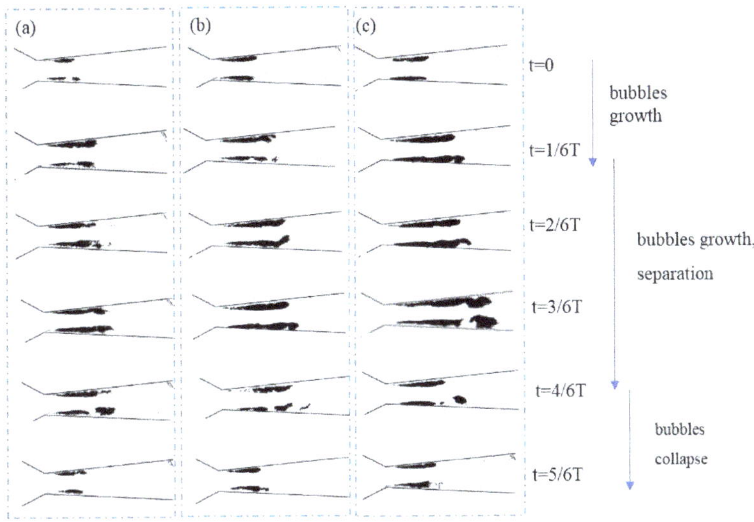

Figure 5. Cavitation evolution after binarization of sand water. (**a**) d = 0.05 mm, (**b**) d = 0.07 mm, (**c**) d = 0.09 mm.

The cavitation bubbles at the tail of the cloud undergo three distinct stages: growth, separation, and collapse. During cavitation growth, the cloud extends downstream on both the upper and lower sides. During the separation stage, noticeable breaking and detachment of bubbles occur at both sides of the tail. Finally, in the collapse stage, the separated bubbles rapidly collapse. The cloud length before detaching from the wall is defined as the cavitation cloud length (L), and its periodic variation over time is illustrated in Figure 6.

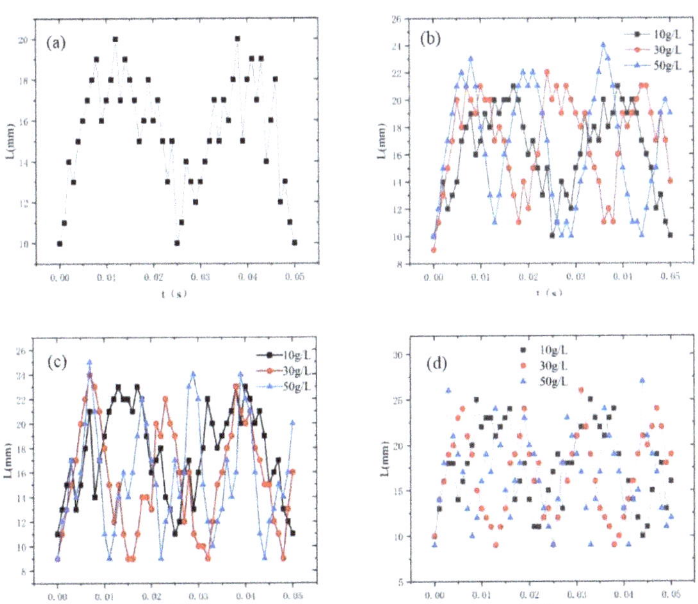

Figure 6. Periodic variation in the cavitation cloud length (Q = 294 mL/s). (**a**) Clear water, (**b**) d = 0.09 mm, (**c**) d = 0.07 mm, (**d**) d = 0.05 mm.

To examine the development cycle of cavitation clouds in both clear water and sediment-laden water, ten development cycles were selected for analysis. An arithmetic average was computed to determine the average period. The cavitation period for clear water is T = 0.0255 s. The impacts of the sediment size and concentration on the cavitation cloud evolution are presented in Figure 7.

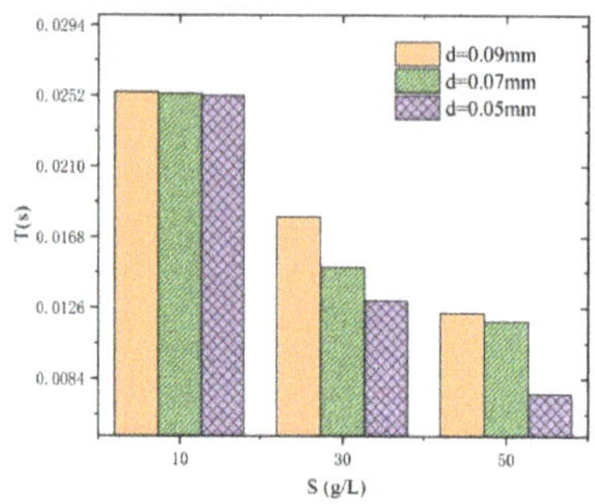

Figure 7. Effects of sediment size and concentration on cavitation period.

Notably, when the sediments have a very low concentration (S = 10 g/L), the cavitation cloud's evolution cycle closely resembles that of clear water and is minimally affected by the particle size. However, as the sediment concentration increases, shorter cycles become more pronounced with smaller particle sizes. For instance, when the sediment concentration S = 50 g/L and the particle diameter d = 0.09 mm, the cycle is T = 0.0123 s; with d = 0.05 mm, the cycle is T = 0.0075 s, showing a 39% reduction. The impact of the sediment concentration on the cavitation cloud's evolution cycle is even more significant. With a particle diameter of d = 0.09 mm, increasing the sediment concentration from S = 10 g/L to 50 g/L reduces the cycle from T = 0.0254 s to 0.0123 s, a 51.57% decrease. When the particle diameter is d = 0.05 mm, the cycle decreases from T = 0.0252 s to 0.0075 s, a reduction of 70.24% when the sediment concentration grows from 10 g/L to 50 g/L. It is found that a decreasing sediment size or increasing sediment concentration shortens the development cycle of cavitation, and the effect of concentration is more significant than that of size.

This phenomenon can be attributed to the cavitation cloud shedding induced by the near-wall re-entrant jet. Sediment particles, characterized as non-wettable solids, exhibit a greater tensile stress than water fractures. Consequently, the re-entrant jet transporting the sediment particles possesses enhanced tearing and destructive capabilities compared to that in clear water, thereby accelerating the shedding of the cavitation cloud. The destructive potential of the re-entrant jet becomes increasingly pronounced at higher sediment concentrations.

2.3.3. Maximum Length of the Wall-Bounded Cavitation Cloud

The average maximum cavitation cloud length L_{max} before shedding was calculated over 10 cycles at a flow of 294 mL/s, yielding L_{max} = 19 mm in clear water. The influences of the sediment size and sediment concentration on this length are illustrated in Figure 8. It is evident that both increasing sediment concentration and decreasing sediment size contribute to an expansion of the cavitation cloud, as indicated by the increased length. Notably, the magnitude of these increases is comparable, suggesting that the sediment size and concentration exert similar influences on the cavitation cloud length.

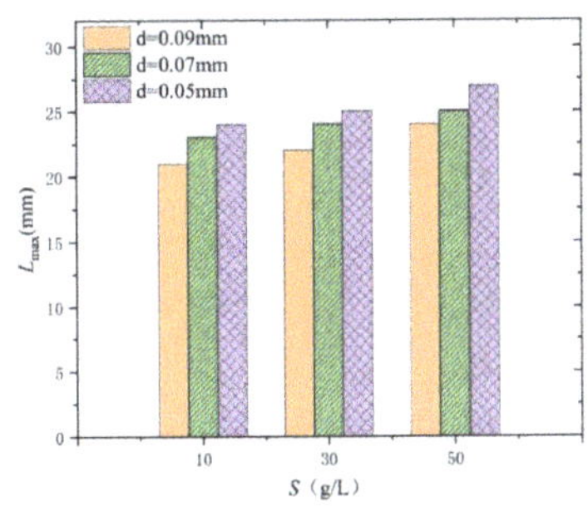

Figure 8. Variations in the maximum cavitation cloud length with particle size and sediment content.

As shown in Table 2 and Figure 8, L increases with the sediment-laden water's viscosity. The sediment facilitates the formation of additional cavitation nuclei compared to a clear water medium, which subsequently leads to the generation of more cavitation bubbles and promotes cavitation effectively. Smaller sediment generates a greater number of cavitation nuclei, further enhancing the development of cavitation.

3. Numerical Calculation

3.1. Mathematical Model

3.1.1. Solid–Liquid Mixture Model

The mixture model is employed to manage solid–liquid flow by accounting for the mixed states of multiple substances (phases), which is used to analyze the movement of sediment-laden water flow in this paper. In the model, the average velocity is defined as follows:

$$\vec{u}_m = \frac{\alpha_w \rho_w \vec{u}_w + \alpha_s \rho_s \vec{u}_s}{\rho_m} \tag{2}$$

where $\rho_m = \alpha_w \rho_w + \alpha_s \rho_s$ is the density of sediment-laden water, $\vec{u}_m$ is the velocity of sediment-laden water, α_w and α_s denote the water's and sediment's volume fraction, respectively, ρ_w and ρ_s represent the water's and sediment's density, respectively, and $\vec{u}_w$ and $\vec{u}_s$ signify the water's and sediment's velocity, respectively.

The continuity equation is:

$$\frac{\partial \rho_m}{\partial t} + \nabla \cdot \left(\rho_m \vec{u}_m\right) = 0 \tag{3}$$

The momentum equation is:

$$\begin{aligned} &\frac{\partial}{\partial t}\left(\rho_m \vec{u}_m\right) + \nabla \cdot \left(\rho_m \vec{u}_m \vec{u}_m\right) \\ &= -\nabla P + \nabla \cdot \left[\mu_m \left(\nabla \vec{u}_m + \nabla \vec{u}_m^{\,T}\right)\right] + \rho_m \vec{g} + \nabla \cdot \left(\sum_{i=1}^{n} \alpha_p \rho_p \vec{u}_{drt} \vec{u}_{drt}\right) \end{aligned} \tag{4}$$

where n represents the number of phases, μ_m is the viscosity of sediment-laden which can be obtained from Table 1, α_p, ρ_p, and $\vec{u}_p$ represent the volume fraction, density and velocity of phase p, respectively, and $\vec{u}_{drt} = \vec{u}_p - \vec{u}_m$ represents the drift velocity of phase p.

The volume fraction equation is:

$$\frac{\partial(\alpha_p \rho_p)}{\partial t} + \nabla \cdot \left(\alpha_p \rho_p \vec{u}_p\right) = 0 \tag{5}$$

3.1.2. Solid–Liquid–Vapor Mixture Model

After completing the numerical calculation of the water–sand flow, cavitation is taken into account, and a three-phase calculation of solid–liquid–vapor is carried out. Here, the sediment-laden water is regarded as a whole, and it is combined with vapor to form a mixture model. Equation (2) is rewritten as:

$$\vec{u}_{mv} = \frac{\alpha_{ws}\rho_{ws}\vec{u}_{ws} + \alpha_v\rho_v\vec{u}_v}{\rho_{mv}} \tag{6}$$

where $\rho_{mv} = \alpha_{ws}\rho_m + \alpha_v\rho_v$ is the mixed density of sediment-laden water and vapor, $\vec{u}_{mv}$ is the velocity of sediment-laden water and vapor, α_{ws} and α_v denote the sediment-laden water's and vapor's volume fraction, respectively, ρ_{ws} and ρ_v represent the sediment-laden water's and vapor's density, respectively, and $\vec{u}_{ws}$ and $\vec{u}_v$ signify the sediment-laden water's and vapor's velocity, respectively.

The continuity equation is rewritten as:

$$\frac{\partial\rho_{mv}}{\partial t} + \nabla\cdot\left(\rho_{mv}\vec{u}_{mv}\right) = 0 \tag{7}$$

The momentum equation is:

$$\begin{aligned}&\frac{\partial}{\partial t}\left(\rho_{mv}\vec{u}_{mv}\right) + \nabla\cdot\left(\rho_{mv}\vec{u}_{mv}\vec{u}_{mv}\right)\\&= -\nabla P + \nabla\cdot\left[\mu_{mv}\left(\nabla\vec{u}_{mv} + \nabla\vec{u}_{mv}^{\,T}\right)\right] + \rho_{mv}\vec{g} + \nabla\cdot\left(\sum_{p=1}^{n}\alpha_p\rho_p\vec{u}_{drt}\vec{u}_{drt}\right)\end{aligned} \tag{8}$$

where n represents the number of phases, $\mu_{mv} = \alpha_{ws}\mu_m + \alpha_v\mu_v$ is the mixed viscosity of sediment-laden and vapor, the phase p denotes the sediment-laden and vapor, and $\vec{u}_{drt} = \vec{u}_p - \vec{u}_{mv}$ represents the drift velocity of the phase p.

The Zwart-Gerber-Belamri cavitation model is used to determine the volume fraction of cavitation bubbles. The transport equation for the mass transfer between vapor and sediment-laden water is:

$$\frac{\partial(\alpha_v\rho_v)}{\partial t} + \nabla\cdot\left(\alpha_v\rho_v\vec{u}_v\right) = \dot{R} \tag{9}$$

where the subscript v is the vapor phase, α_v is the volume fraction of the vapor, ρ_v is the density of the vapor, $\vec{u}_v$ represents the velocity of the vapor, and $\dot{R}$ is the net mass transfer rate between the vapor and the sediment-laden water.

$$\dot{R}_e = F_{vap}\frac{\rho_v\rho_{ws}}{\rho_{mv}}\alpha_v(1-\alpha_v)\frac{3}{R_B}\sqrt{\frac{2(p_v-p)}{\rho_{ws}}}(p_v > p) \tag{10}$$

$$\dot{R}_c = F_{cond}\frac{\rho_v\rho_{ws}}{\rho_{mv}}\alpha_v(1-\alpha_v)\frac{3}{R_B}\sqrt{\frac{2(p_v-p)}{\rho_{ws}}}(p_v < p) \tag{11}$$

where R_B is the radius of the cavitation bubble, ρ_{ws} is the density of sediment-laden water, F_{vap} is the evaporation coefficient, and $F_{vap} = 50$, F_{cond} is the condensation coefficient, $F_{cond} = 0.2$.

In the early stage of our research, the SST k-ω and RNG k-ε models were used to carry out the numerical calculation of cavitation in clear water. It was found that the maximum length value of the cavitation cloud calculated by the RNG k-ε model was closer to the experimental result. Therefore, when calculating the cavitation of sand-laden water, the RNG k-ε model is still adopted.

3.2. Grid Generation

Considering the symmetry of the flow channel, a half geometric model was established as shown in Figure 9a. The geometric model presented herein features a regular structure and employs a structured mesh. Given that cavitation occurs near the wall, and reverse flow is present in that region, it is essential to refine the mesh near the wall during the meshing process to enhance the simulation accuracy. The first layer of mesh nodes closest to the wall was set to $y^+ = 1$, comprising a total of 27 boundary layer mesh layers. Furthermore, since cavitation typically occurs in the throat of the Venturi structure, the mesh in this area was also refined, as depicted in Figure 9b.

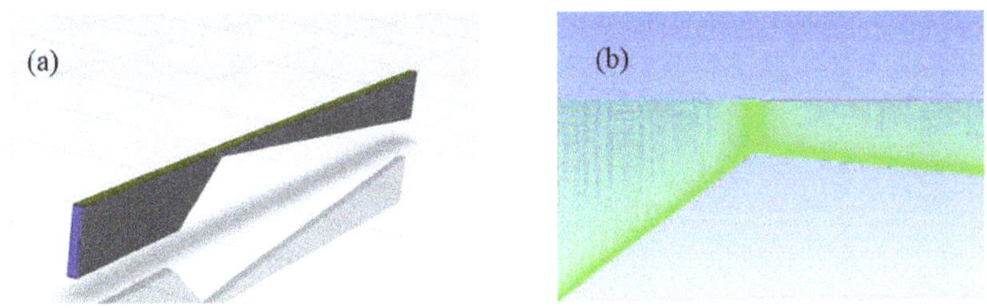

Figure 9. Numerical geometric model. (**a**) Three-dimensional modeling of the channel, (**b**) mesh throat encryption diagram.

Grid independence verification was performed to reduce the dependency of the calculation results on the grid. Numerical calculations were carried out using five grid schemes for a clear water flow rate of 294 mL/s, with the grid configurations listed in Table 3. The experimental measurements obtained inlet and outlet pressures of 144,325 Pa and 92,325 Pa, respectively. Seven monitoring points were selected starting from the throat, 0.2 mm away from the wall, and at 3 mm intervals along the flow direction, as shown in Figure 10a. The pressures at these seven points at time t = T/2 were plotted to show the pressure distribution along the flow path, as depicted in Figure 10b. According to the results in Figure 10b, the pressure values of Grid 4 and Grid 5 were very close, with their pressure curves basically overlapping. Further increasing the number of grids led to almost no noticeable change in the pressure at the monitoring points. Therefore, the grid with 7,670,400 cells was selected.

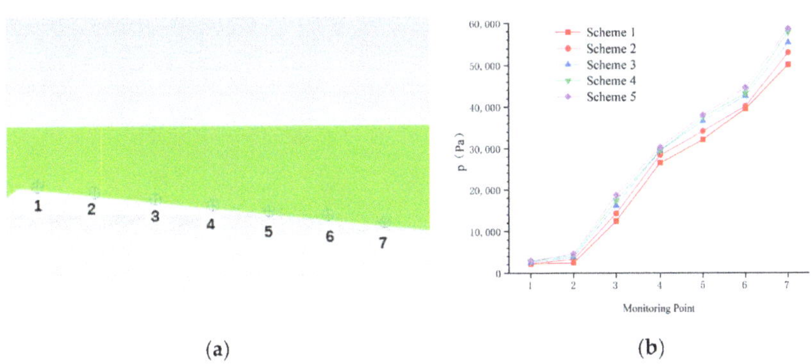

Figure 10. Grid independence verification. (**a**) Monitoring points, (**b**) pressure distribution.

Table 3. Grid scheme.

Scheme	**1**	**2**	**3**	**4**	**5**
Grid number	2,351,584	3,569,521	5,665,841	7,670,400	8,641,851

3.3. Boundary Conditions and Parameter Settings

In the cavitation experiments of sediment-laden water, the outlet pressure remains constant while the inlet pressure varies with the particle size and sediment concentration. When d = 0.05 mm and Q = 294 mL/s, for the sediment concentrations of 30 g/L and 50 g/L, the measured corresponding inlet pressures were 146,360 Pa and 152,340 Pa, respectively. The inlet boundary condition adopts a velocity inlet, with vin = 1.44 m/s, and the inlet pressure is taken as the experimental value. The outlet boundary condition is set as a pressure outlet, with pout = 92,325 Pa. The turbulent intensities at the inlet and outlet were set according to the hydraulic diameter. The turbulent intensity at the inlet was 5%, and the turbulent viscosity ratio was 10. The backflow turbulent intensity was 5%, and the backflow turbulent viscosity ratio was 10. The time step was set to be 1×10^{-3} s. The pressure–velocity coupling equations were solved based on the SIMPLEC algorithm, and the residual was set to be 1×10^{-3}. The discretization format of pressure is PRESTO!, the discretization format of the gradient is Least Squares Cell-Based, and the discretization format of the remaining parameters was selected as the first-order upwind scheme.

The vapor density $\rho_v = 0.017 \text{kg/m}^3$, vapor viscosity $\mu_v = 1.00 \times 10^{-2} \text{mPa} \cdot \text{s}$, saturated vapor pressure p_v= 2236Pa, bubble radius $R_B = 1 \times 10^{-5}$ m, temperature T = 293.15 K, the sand density $\rho_s = 2600 \text{ kg/m}^3$, and the viscosity of the sediment-laden water were determined based on the experimental values.

3.4. Analysis of the Calculation Results

3.4.1. Comparison of the Cavitation Cloud Morphology and Periodicity

The experimental photos and numerical simulation images of cavitation alongside the vapor volume fraction distribution from numerical simulations of sediment-laden water, with sediment sizes of d = 0.05 mm and sediment concentrations of S = 50 g/L and S = 30 g/L over one cycle are shown in Figure 11. Cavitation initiates at the throat, and as the cycle progresses, the wall-attached cavitation cloud's length and thickness increase, thereby expanding the cavitation region. During the interval from 2/6T to 5/6T, shedding cavitation clouds form at the tail, moving with the flow before gradually collapsing. Alongside the periodic development of cavitation, the vapor–liquid re-entrant jet at the tail during cavitation collapse exhibits similar periodicity. The static pressure near the collapsing cavitation bubbles at the tail increases rapidly, resulting in the formation of a concave jet structure. A portion of the liquid at the tail, driven by the concave jet, moves in the opposite direction to the main liquid flow, generating a vapor–liquid coexisting re-entrant jet. The numerical simulations capture the periodicity of the cavitation process, which undergoes cyclical phases of contraction, expansion, and retraction. Notably, the shape of the wall-attached cavitation region in the simulations aligns with the non-shed cavitation cloud observed in the experiments, and the extent of the cavitation region closely resembles the experimental findings.

The length of the cavitation cloud L at the corresponding moment can be found from Figure 11. The ratio of this length L to the maximum length of the cavitation cloud Lmax that appears within a cycle is defined as Lc (i.e., Lc = L/Lmax). The Lc values from the experiment and numerical simulation are compared, as shown in Table 4. The error range is between 2.6% and 13.43%, and it is considered that the numerical simulation is consistent with the variation in the actual cavitation cloud length.

The values of L_{max} in sediment-laden water before shedding, were compared experimentally and numerically, as shown in Figure 12. It is found that the numerically calculated cavitation cloud length closely aligns with the experimental results; however, the evolution cycle of the cavitation cloud in the simulation is slightly longer than that observed in the experiments. For a sediment concentration of S = 50 g/L, the experimental cycle is T = 0.0075 s, while the calculated cycle is T = 0.0089 s, indicating an 18.6% difference. Conversely, at a sediment concentration of S = 30 g/L, the experimental cycle is T = 0.0130 s, and the calculated cycle is T = 0.0151 s, resulting in a 16.1% difference. As the sediment concentration increases, the cavitation cloud cycle shortens while the cavitation region expands. The numerical results agree with the experiment outcomes.

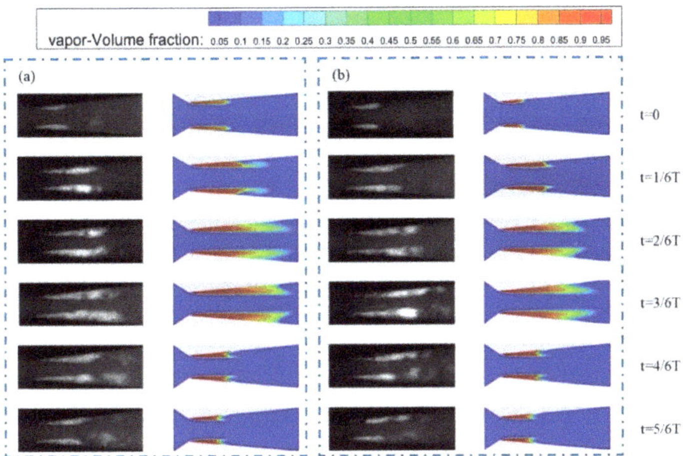

Figure 11. Comparison of the cavitation morphology between the experimental and numerical calculation. (**a**) S = 50 g/L, (**b**) S = 30 g/L.

Table 4. Comparison of *Lc* between experiment and numerical simulation.

S (g/L)	t	Experiment Lc	Numerical Simulation Lc	Error (%)
50	0	0.36	0.39	8.33
50	1/6T	0.59	0.56	−5.08
50	2/6T	0.87	0.98	12.64
50	4/6T	0.65	0.43	−12.31
50	5/6T	0.41	0.36	−12.20
30	0	0.35	0.28	−8.57
30	1/6T	0.67	0.58	−13.43
30	2/6T	0.83	0.80	−3.61
30	4/6T	0.77	0.75	−2.60
30	5/6T	0.34	0.32	−5.88

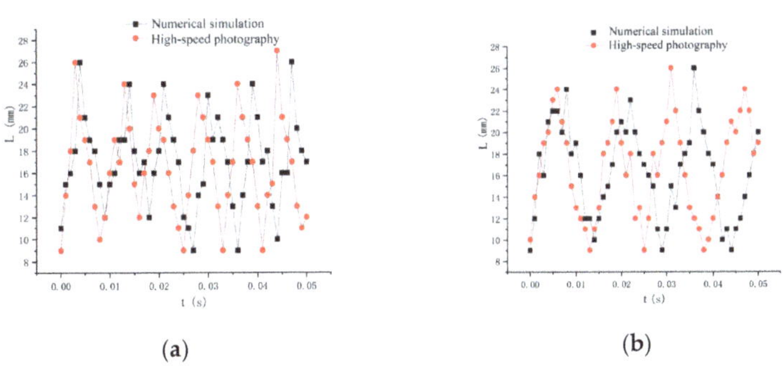

Figure 12. Comparison of the cavitation cloud evolution period between the experiment and numerical calculation. (**a**) S = 50 g/L, (**b**) S = 30 g/L.

3.4.2. Re-Entrant Jet and Reflow

The reverse pressure gradient of the diffusion section generates a re-entrant jet which causes the detachment of bubbles at the tail end of the cavitation cloud. The streamline diagrams of sediment-laden water cavitation at a sediment concentration of S = 50 g/L are shown in Figure 13. At t = 1/6 T, a re-entrant jet emerges at the wall near the cavitation cloud tail, resulting in an upward lift of the tail. At t = 2/6 T, as the cavitation cloud expands and the tail continues to rise, the reflow region is obvious. By t = 3/6T, the cavitation cloud attains its maximum length, with the reflow region moving further away and expanding to its greatest extent. At t = 4/6T, due to the collapse of the detached cavitation cloud, the reflow region is pushed back and the cavitation cloud near the wall begins to contract. By t = 5/6T, the contraction of the cavitation cloud persists, and the reflow region continues to shift upstream toward the tail, further reducing in size, and it finally becomes a re-entrant jet.

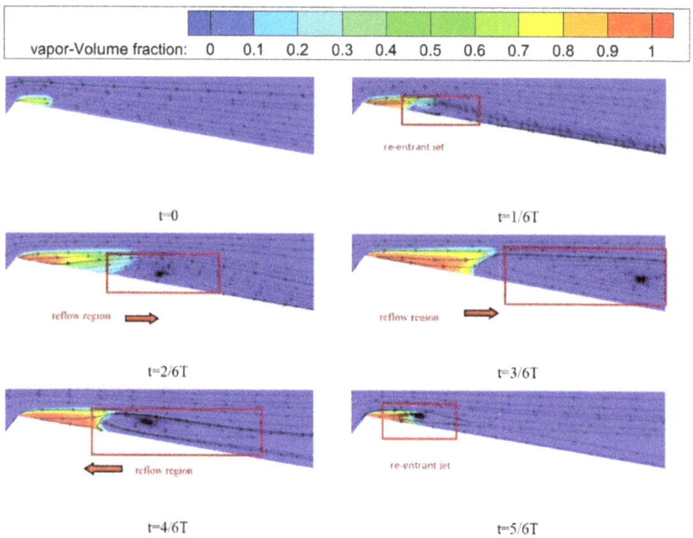

Figure 13. Streamline of the cavitation of sand water.

A comparative analysis of the streamlines of cavitation flows, maintaining the same flow rate for sediment-laden water at a sediment concentration of S = 30 g/L and S = 50 g/L (Figure 14) reveals that an increased sediment concentration correlates with a larger reflow region at the tail of the cavitation cloud.

From the throat to the downstream diffusion section, the flow velocity decreases along the way. The density and inertia of the sediment-laden water increase when the concentration is raised. Under the same flow rate conditions, the trend of decreasing flow velocity slows down. Therefore, the velocity gradient from the edge wall to the central flow area rises accordingly.

Meanwhile starting from the throat, a small amount of sediment settles and accumulates on the side walls, while most of the sediment particles are suspended in the water and carried downstream by the water flow, forming the main flow area. There are strip-shaped areas with a low sediment content between the subsidence zone and the mainstream. As the overall sediment concentration decreases, this strip-shaped area approaches clear water (Figure 15).

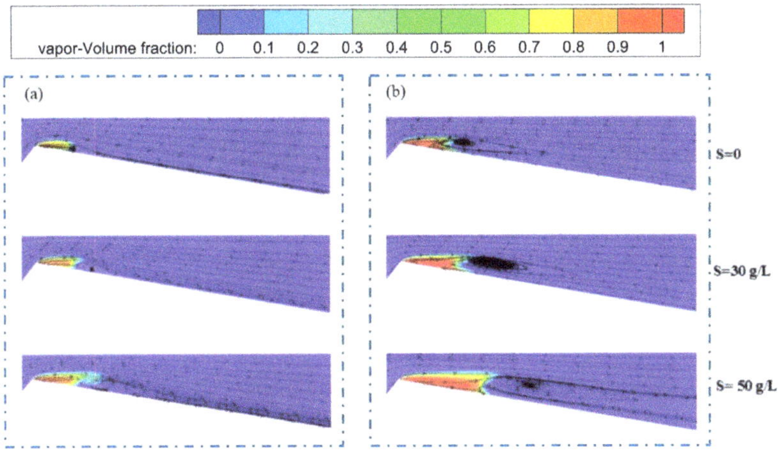

Figure 14. Evolution of streamline with different sediment contents. (**a**) t = 1/6T, (**b**) t = 4/6T.

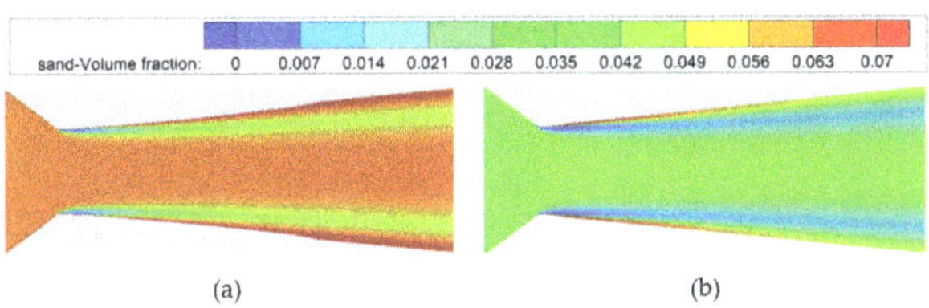

Figure 15. Sediment distribution near the throat. (**a**) S = 50 g/L, (**b**) S = 30 g/L.

Based on the analysis of the velocity gradient and sediment distribution, the shearing stress $\tau = \mu \frac{\partial v_x}{\partial y}$ near the wall enlarges, and the drag on bubbles enhances with the concentration increasing, which results in an increase in the length of the sheet-like cavitation cloud. Therefore, the reverse pressure gradient at the cavitation cloud tail increases, leading to an expanded re-entrant jet as shown in Figure 14a. The collapse of detached bubbles pushes the reflow region upstream, causing the cavitation cloud to contract. The higher the sediment content, the more obvious the driving effect, and the reflow region increases significantly as shown in Figure 14b.

3.4.3. Vortex Distribution and Evolution

The cavitation behavior significantly affects the vortex structure. Because of the growth and shedding process of bubbles, the vortex structure changes over time, and the absolute value of vorticity near the wall is relatively large. This originates from the flow's strong shearing effect near the wall. The smooth vortex structure grows with the development of bubbles, and there is a high vorticity value at the position where the reflow advances, which disturbs the bubbles, enhances turbulent fluctuations, and causes the vortex structure to lift up. This indicates that the reflow impacts the surface of the bubbles, ultimately leading to the detachment of cloud bubbles.

The vorticity diagrams for cavitation in sediment-laden water with S = 50 g/L—calculated using the Q-criterion—are shown in Figure 16. The sheet-like vortices at the front of the cavitation cloud are relatively smooth and stable, whereas the vortices at the tail exhibit considerable periodic fluctuations due to the influence of the reflow, undergoing a cycle of shedding, downstream movement, return, and merging. From t = 1/6T to t = 3/6T, the reflow impacts the surfaces of the sheet-like cavitation bubbles, ultimately leading to cavitation cloud shedding. As the cloud progresses downstream, large-scale vortices are generated, complicating the vorticity characteristics. After t = 4/6T, the reflow propels the vortices to return, with larger vortex structures being supplanted by smaller-scale vortices.

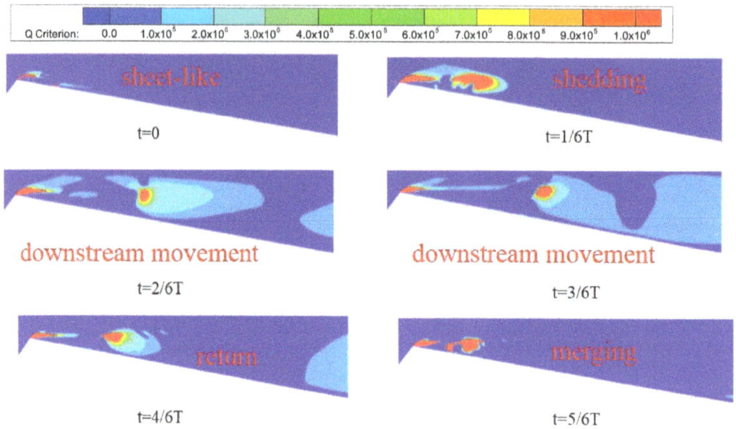

Figure 16. Vorticity structure evolution of sandy water cavitation.

Figure 17 compares the cavitation vorticity of clear water and sediment-laden water with a sediment concentration of 30 g/L and 50 g/L, all under identical flow rates. It is observed that during the vortices shedding phase, higher sediment concentrations enhance the shear effect near the wall, leading to an increase in vorticity. The re-entrant jet causes the shedding of vortices, and higher sediment concentrations lead to a stronger impact of the jet on the sheet-like vortices, causing more vortices to detach and the boundaries of the vortices to become more turbulent as shown in Figure 17a. During the vortices return phase, larger-scale vortices transform into small-scale vortices which exhibit greater energy consumption due to energy dissipation. In sediment-laden water with a higher sediment concentration, viscous damping is more pronounced, leading to more reduction in vorticity and more small-scale vortices as shown in Figure 17b.

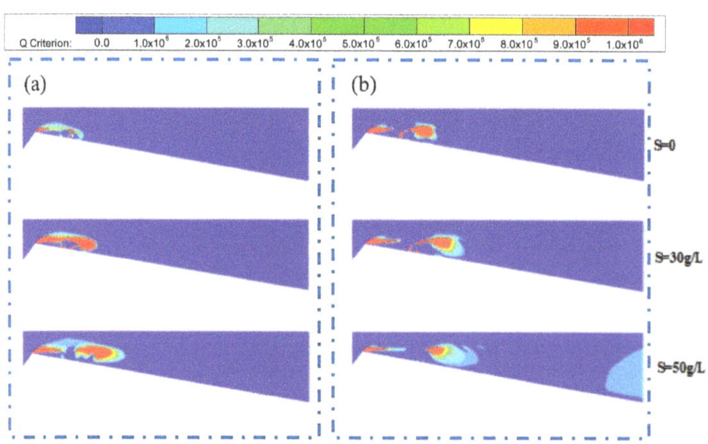

Figure 17. Vorticity evolution with different sediment contents. (**a**) t = 1/6T, (**b**) t = 4/6T.

4. Conclusions

This study investigates the cavitation phenomenon in sediment-laden water using a Venturi-based hydraulic cavitation device, integrating both numerical simulations and experimental methods. By comparing the numerical results with the experimental data, the study investigates cavitation patterns, analyzes the flow field characteristics associated with cavitation in sediment-laden water, and explores the influences of the sediment size and concentration on cavitation, drawing the following conclusions:

1. The impact of the sediment concentration on the viscosity is more pronounced than that of the particle size. As the sediment concentration increases, the viscosity rises significantly, with this trend becoming more accentuated as the particle size decreases.
2. The results of the experiment and numerical simulation show that an increase in the sediment concentration or a decrease in the sediment size leads to a shorter evolution cycle of the cavitation cloud and a higher frequency of cavitation cloud shedding, with the sediment concentration exerting a more substantial influence. The cavitation number decreases as well, with the sediment concentration having a more significant effect. When the concentration reaches 50 g/L, the cavitation number is less than 0.1. Therefore, attention should be paid to the cavitation generated when the sediment concentration is high during the operation of hydraulic machinery.
3. The numerical analyses of the flow field of the cavitation in sediment-laden water reveal that the sediment expands the reflow region and enhances the re-entrant jet, thereby promoting cavitation cloud shedding due to the more pronounced shear effect. Additionally, the sediment contributes to a more complex vortex structure: on one hand, it enhances the interaction with smooth, sheet-like vortices attached to the wall, leading to increased vortex shedding; on the other hand, it consumes more energy during the vortex return, thus forming more small-scale vortices.
4. When the sediment concentration increases, it can increase the viscosity of the sediment-laden water. On the one hand, it increases the pressure gradient, leading to the enhancement of the re-entrant jet and accelerating the shedding of the cavitation cloud. At the same time, the viscous resistance is strengthened, which expands the range of the cavitation zone. In engineering, the geometric shape of the runner blades of a turbine can be optimized to reduce the relative flow velocity of the fluid, thereby decreasing the pressure gradient and weakening the shear effect.

The channel depth in this article is only 4 mm, which is much smaller than the channel length and width, so a sidewall effect will occur. Sidewall effects can boost the flow resistance and pressure differentials, facilitating the occurrence of cavitation. In future research, we need to analyze the impact of depth and refine the results accordingly.

Since the fluid is regarded as incompressible in the calculation process, in fact, the shedding and collapse of cavitation bubbles should be a compressible flow process. In the follow-up, the compression factor should be taken into account to refine the calculation model. At the same time, due to the narrow channel, the influence of the roughness of the side wall on cavitation should also be considered, which also needs to be optimized and improved in the subsequent calculations.

Author Contributions: Conceptualization, Q.Z.; methodology, Q.Z.; software, Z.Y.; validation, Z.Y. and G.Y.; investigation, Q.Z.; resources, Q.Z.; data curation, G.Y.; writing—original draft preparation, Z.Y.; writing—review and editing, Q.Z.; visualization, Z.Y. and G.Y. All authors have read and agreed to the published version of the manuscript.

Funding: This research was funded by the National Natural Science Foundation of China (Grant No. 52379092).

Data Availability Statement: The datasets used and/or analyzed during the current study are available from the corresponding author on reasonable request.

Conflicts of Interest: The authors declare no conflicts of interest.

Abbreviations

Symbol	Physical meaning
μ	dynamic viscosity
M	torque
S	sediment concentration
d	particle diameter
σ	cavitation number
Q	flow rate
L	cavitation cloud length
$L_{\max}$	maximum cavitation cloud length
L_c	L/Lmax
T	cavitation cycle
ρ_m	density of sediment-laden water
ρ_w	water's density
ρ_s	sediment's density
ρ_p	density of phase p
ρ_v	vapor's density
ρ_{mv}	mixed density of sediment-laden water and vapor
$\vec{u}_m$	velocity of sediment-laden water
$\vec{u}_w$	water's velocity
$\vec{u}_s$	sediment's velocity
$\vec{u}_p$	velocity of phase p
$\vec{u}_v$	vapor's velocity
$\vec{u}_{mv}$	mixed velocity of sediment-laden water and vapor
$\vec{u}_{drt}$	drift velocity
α_w	water's volume fraction
α_s	sediment's volume fraction
α_{ws}	volume fraction of sediment-laden water
α_v	vapor's volume fraction
α_p	volume fraction of phase p
μ_m	dynamic viscosity of sediment-laden water
μ_v	dynamic viscosity of vapor
μ_{mv}	mixed dynamic viscosity of sediment-laden water and vapor
$\dot{R}$	net mass transfer rate
R_B	bubble radius
F_{vap}	evaporation coefficient
F_{cond}	condensation coefficient
p_v	saturated vapor pressure

References

1. Noon, A.A.; Kim, M.-H. Sediment and Cavitation Erosion in Francis Turbines—Review of Latest Experimental and Numerical Techniques. *Energies* **2021**, *14*, 1516. [CrossRef]
2. Jia, J.A. Technical Review of Hydro-Project Development in China. *Engineering* **2016**, *2*, 302–312. [CrossRef]
3. Thapa, B.; Chaudhary, P.; Dahlhaug, O.G.; Upadhyay, P. Study of combined effect of sand erosion and cavitation in hydraulic turbines. In Proceedings of the International Conference on Small Hydropower (Hydro Sri Lanka: Fluid Dynamics), Kandy, Sri Lanka, 22–24 October 2007; pp. 22–24.
4. Teran, L.A.; Rodriguez, S.A.; Laín, S.; Jung, S. Interaction of particles with a cavitation bubble near a solid wall. *Phys. Fluids* **2018**, *30*, 123304. [CrossRef]
5. Kumar, P.; Saini, R. Study of cavitation in hydro turbines—A review. *Renew. Sustain. Energy Rev.* **2010**, *14*, 374–383. [CrossRef]
6. Felix, D.; Albayrak, I.; Abgottspon, A.; Boes, R.M. Hydro-abrasive erosion of hydraulic turbines caused by sediment—A century of research and development. *IOP Conf. Ser. Earth Environ. Sci.* **2016**, *49*, 122001. [CrossRef]

7. Sun, J.; Ge, X.F.; Zhou, Y.; Liu, D.M.; Liu, J.; Li, G.Y.; Zheng, Y. Research on synergistic erosion by cavitation and sediment: A review. *Ultrason. Sonochemistry* **2023**, *95*, 106399. [CrossRef]
8. Karimi, A.; Martin, J.L. Cavitation erosion of materials. *Int. Met. Rev.* **1986**, *31*, 1–26. [CrossRef]
9. Arora, M.; Ohl, C.D.; Lohse, D. Effect of nuclei concentration on cavitation cluster dynamics. *J. Acoust. Soc. Am* **2007**, *121*, 3432–3436. [CrossRef]
10. Lian, J.; Gou, W.; Li, H.; Zhang, H. Effect of sediment size on damage caused by cavitation erosion and abrasive wear in sediment-water mixture. *Wear* **2018**, *398–399*, 201–208. [CrossRef]
11. Raichenko, I.A.; Byakova, V.A. Theory of initial growth of a microcavity in a liquid metal around a gas-releasing particle. Part 1. Physical and mathematical models. *Powder Metall. Met. Ceram.* **2021**, *40*, 606–611. [CrossRef]
12. Morch, K.A. Cavitation nuclei and bubble formation—A dynamic liquid-solid interface Problem. *J. Fluids Eng.* **2000**, *122*, 494–498. [CrossRef]
13. Shen, Y.; Yasui, K.; Zhu, T.; Ashokkumar, M. A model for the effect of bulk liquid viscosity on cavitation bubble dynamics. *Phys. Chem. Chem. Phys.* **2017**, *19*, 20635–20640. [CrossRef] [PubMed]
14. Liu, X.M.; He, J.; Lu, J.; Ni, X.W. Effect of liquid viscosity on a liquid jet produced by the collapse of a laser-induced bubble near a rigid boundary. *Jpn. J. Appl. Phys.* **2009**, *48*, 016504. [CrossRef]
15. Wu, J.-H.; Gou, W.-J. Critical size effect of sand particles on cavitation damage. *J. Hydrodyn. Ser. B* **2013**, *25*, 165–166. [CrossRef]
16. Chen, S.Y.; Xu, W.L.; Luo, J.; Li, J.B.; Zhai, Y.W. Experimental study on the mesoscale causes of the effect of sediment size and concentration on material cavitation erosion in sandy water. *Wear* **2021**, *488*, 204114. [CrossRef]
17. Zhong, Y.; Minemura, K. Measurement of erosion due to particle impingement and numerical prediction of wear in pump casing. *Wear* **1996**, *199*, 36–44. [CrossRef]
18. Wang, Y.; Wu, J.; Ma, F. Cavitation–silt erosion in sand suspensions. *J. Mech. Sci. Technol.* **2018**, *32*, 5697–5702. [CrossRef]
19. Huang, S.; Ihara, A.; Watanabe, H.; Hashimoto, H. Effects of Solid Particle Properties on Cavitation Erosion in Solid-Water Mixtures. *J. Fluids Eng.* **1996**, *118*, 749–755. [CrossRef]
20. Wang, J.; Chen, H.; Qin, L.; Li, Y.; Chen, D. Key roles of micro-particles in water on occurrence of cavitation-erosion of hydro-machinery. *Chin. Sci. Bull.* **2008**, *53*, 1603–1607. [CrossRef]
21. Wang, L.; Zhu, R.; Chang, J. Effect of sand concentration in water on cavitation pressure in Qingtongxia and Bapanxia hydropower stations. *J. Hydroelectr. Eng.* **2008**, *27*, 44–47.
22. Zhang, Y.; Qian, Z.; Ji, B.; Wu, Y. A review of microscopic interactions between cavitation bubbles and particles in silt-laden flow. *Renew. Sustain. Energy Rev.* **2016**, *56*, 303–818. [CrossRef]
23. Chang, J. Cavitation and cavitation erosion of pump and turbine with silt-laden water as working medium. *J. Drain. Irrig. Mach. Eng.* **2010**, *28*, 93–99.
24. Hu, H.X.; Zheng, Y.G. The effect of sand particle concentrations on the vibratory cavitation erosion. *Wear* **2017**, *384–385*, 95–105. [CrossRef]
25. Xu, W.; Zhang, Y.; Luo, J.; Arong; Zhang, Q.; Zhai, Y. The impact of particles on the collapse characteristics of cavitation bubbles. *Ocean. Eng.* **2017**, *131*, 15–24. [CrossRef]
26. Romero, R.; Teran, L.; Coronado, J.; Ladino, J.; Rodríguez, S. Synergy between cavitation and solid particle erosion in an ultrasonic tribometer. *Wear* **2019**, *428–429*, 395–403. [CrossRef]
27. Yan, D.; Wang, J.; Liu, F. Inhibition of the ultrasonic microjet-pits on the carbon steel in the particles-water mixtures. *AIP Adv.* **2015**, *5*, 077159. [CrossRef]
28. Bahadur, S.; Badruddin, R. Erodent particle characterization and the effect of particle size and shape on erosion. *Wear* **1990**, *138*, 189–208. [CrossRef]
29. Levy, A.V.; Chik, P. The effects of erodent composition and shape on the erosion of steel. *Wear* **1983**, *89*, 151–162. [CrossRef]
30. Chen, H.; Wang, J.; Chen, D. Cavitation damages on solid surfaces in suspensions containing spherical and irregular microparticles. *Wear* **2009**, *266*, 345–348. [CrossRef]
31. Li, S.; Khoo, B.C.; Zhang, A.M.; Wang, S. Bubble-sphere interaction beneath a free surface. *Ocean. Eng.* **2018**, *169*, 469–483. [CrossRef]
32. Li, S.; Zhang, A.M.; Han, R.; Liu, Y.Q. Experimental and numerical study on bubble sphere interaction near a rigid wall. *Phys. Fluids* **2017**, *29*, 092102. [CrossRef]
33. Li, S.; Zhang, A.M.; Wang, S.; Han, R. Transient interaction between a particle and an attached bubble with an application to cavitation in silt-laden flow. *Phys. Fluids* **2018**, *30*, 082111. [CrossRef]
34. Poulain, S.; Guenoun, G.; Gart, S.; Crowe, W.; Jung, S. Particle motion induced by bubble cavitation. *Phys. Rev. Lett.* **2015**, *114*, 214501. [CrossRef] [PubMed]

35. Arora, M.; Ohl, C.D.; Morch, K.A. Cavitation inception on microparticles: A self-propelled particle accelerator. *Phys. Rev. Lett.* **2004**, *92*, 174501. [CrossRef]
36. Borkent, B.M.; Arora, M.; Ohl, C.D.; De Jong, N. The acceleration of solid particles subjected to cavitation nucleation. *J. Fluid Mech.* **2008**, *610*, 157–182. [CrossRef]

processes

MDPI

Article

Theoretical Recommendations and Validation for the Generation of Realistic Irregular Waves Through the WaveMIMO Methodology

Maycon da Silveira Paiva [1], Ana Paula Giussani Mocellin [1], Phelype Haron Oleinik [1] Elizaldo Domingues dos Santos [1], Luiz Alberto Oliveira Rocha [1], Liércio André Isoldi [1,*] and Bianca Neves Machado [2]

[1] School of Engineering, Federal University of Rio Grande (FURG), Rio Grande 96203-900, RS, Brazil; mayconpaiva@furg.br (M.d.S.P.); ana.mocellin@furg.br (A.P.G.M.); phelype.oleinik@gmail.com (P.H.O.); elizaldosantos@furg.br (E.D.d.S.); luizrocha@furg.br (L.A.O.R.)

[2] Interdisciplinary Department, Federal University of Rio Grande do Sul (UFRGS), Tramandaí 95590-000, RS, Brazil; bianca.machado@ufrgs.br

* Correspondence: liercioisoldi@furg.br; Tel.: +55-53-99109-7356

Abstract: Irregular wave generation in numerical simulations is critical for accurately modeling realistic sea conditions, which is essential in coastal and offshore engineering applications, such as for wave energy conversion. Therefore, this study presents theoretical recommendations for generating realistic irregular waves using the WaveMIMO methodology and validates its accuracy against experimental data. For the parameters investigation, spectral data are processed to obtain orbital velocity profiles of wave propagation, which are imposed as boundary conditions (BCs) in a numerical wave channel. The simulations were conducted using the ANSYS-Fluent 2024 R2 software, which employs the multiphase volume of fluid (VOF) model to treat the interface between phases. Seeking theoretical recommendations for the application of this methodology, the present study investigated the discretization of the region where the prescribed velocity BC is imposed, the mesh sensitivity in the free surface region, the time step used, and the location of the velocity vector in each segment of the prescribed velocity BC imposition region. The results obtained were compared with realistic sea state data obtained from the TOMAWAC spectral model, referring to the municipalities of Rio Grande and Tramandaí, in the state of Rio Grande do Sul, Brazil. The results indicated that, compared to recommendations from the previous literature, the recommended configuration improved wave generation accuracy by 7–8% for Rio Grande and 2–3% for Tramandaí. Finally, the WaveMIMO methodology and its theorical recommendations were validated against experimental data found in the literature, reaching an excellent agreement.

Keywords: realistic irregular waves; WaveMIMO methodology; ANSYS-Fluent; TOMAWAC; computational fluid dynamics

1. Introduction

The global energy sector is undergoing a critical transition as societies confront the dual challenges of climate change and the depletion of fossil fuel reserves. Growing environmental concerns about greenhouse gas emissions and ecological degradation, coupled with the finite nature of fossil resources, have driven nations to explore and integrate renewable alternatives into their energy matrices [1]. Among these, oceans represent an underexploited renewable resource, with wave energy standing out as particularly promising. Numerous global studies have assessed and quantified this potential across different countries, including Mexico [2], Algeria [3], France [4], Indonesia [5], India [6], Argentina, Brazil, Chile, Colombia, Uruguay [7], Australia [8], China [9], among others.

The energy contained in ocean waves can be extracted through converter devices that transform kinetic energy into electricity, such as oscillating water column (OWC), oscillating bodies, overtopping devices, and submerged horizontal plates (SHP) [10,11], along with other proposed technologies. These devices have been investigated through both experimental and numerical approaches. According to the International Energy Agency [12], beyond laboratory testing, multiple prototypes are currently under development or testing worldwide, including in China, South Korea, Singapore, Australia, New Zealand, the United States, Ireland, France, and Portugal.

Numerical modeling offers a cost-effective alternative for studying wave energy converters, requiring fewer financial and human resources. Consequently, numerous numerical studies have investigated the performance of wave energy conversion devices. Examples include OWC [13–21], overtopping [22–30], SHP [11,31–33], and even hybrid devices combining multiple conversion principles [34–36]. However, to enable reliable numerical investigations, it is crucial to accurately understand and reproduce wave behavior, particularly irregular waves, which represent real ocean conditions. Thus, developing robust numerical modeling methodologies for this phenomenon remains essential.

Numerical wave generation has evolved through diverse approaches, each addressing specific challenges in wave propagation modeling. For regular waves, conventional methodologies have been rigorously tested, such as Lisboa et al. [37], who performed numerical simulations using ANSYS-Fluent software to quantify the effects of linear and quadratic damping coefficients in numerical beaches, demonstrating significantly reduced wave reflection. Zabihi et al. [38] conducted a benchmark study between ANSYS-Fluent and Flow-3D, showing both models are powerful tools for wave generation. Windt et al. [39] presented a validation of a numerical wave tank model demonstrating strong agreement between simulated and experimentally monitored free surface elevations. Jiang et al. [40] studied regular wave propagation on a sloping bed under steady wind action, analyzing nearshore wave-wind interaction mechanisms; the results indicated that wind accelerates wave breaking, increases the breaker index, and enhances turbulence.

Regarding irregular wave generation, Oh et al. [41] developed a nonlinear wave simulation code using a higher-order spectral (HOS) method, employing fourth-order Runge–Kutta time integration and anti-aliasing zero-padding for stability; the results confirmed its efficiency in predicting nonlinear wave interactions. Choi et al. [42] presented a methodology for simulating nonlinear irregular waves using the HOS method, validated against extreme wave events; the results showed strong agreement between OpenFOAM, HOS, and experimental data in predicting wave breaking and elevations. Kim et al. [43] introduced a unified nonlinear wave generation and validation procedure for numerical and experimental wave tanks; a comparative analysis of mild and extreme breaking waves demonstrated strong agreement across stochastic and deterministic metrics between HOS simulations, OpenFOAM, and experimental results.

Moreover, Kim et al. [44] presented a real-time wave forecasting system for directional seas using Lagrangian-based models, enabling efficient processing of sparse optical measurement data; experimental validation confirmed that directional wave components must be considered to achieve accuracy comparable to unidirectional forecasts. Canard et al. [45] developed a wave generation method controlling both spectrum and crest statistics, enabling precise irregular wave field reproduction beyond spectral methods, particularly for unidirectional non-breaking conditions. Kim and Ducrozet [46] advanced the complementary improved choppy wave model (CICWM) to address directional wave field limitations by incorporating essential nonlinear terms. Experimental validation demonstrated CICWM reduces linear model prediction errors in directional seas while maintaining real-time forecasting capability through simplified assimilation.

Another example is the WaveMIMO methodology, proposed and verified by Machado et al. [47], which is the focus of the present study. This methodology uses realistic sea state data to generate irregular waves, enabling reproduction of natural wave phenomena. To do so, spectral data are processed to obtain discretized wave propagation velocity profiles, imposed as prescribed velocity boundary conditions. Notably, the methodology's capability for regular wave generation was validated by Maciel et al. [48] through numerical–experimental comparisons. Several studies have employed WaveMIMO to analyze wave energy converters fluid–dynamic behavior under realistic irregular waves, including geometric evaluations of the OWC [16,17], overtopping [25,30], and SHP [33] devices. Furthermore, specific aspects of the WaveMIMO methodology have been investigated in dedicated studies: Mocellin et al. [17] examined the influence of numerical wave channel bathymetry on realistic irregular wave generation, while Paiva et al. [49] and Maciel et al. [50] presented case studies evaluating key application parameters.

In this sense, the present study aims to establish general theoretical recommendations regarding the application of the WaveMIMO methodology [47]. The importance of refining the WaveMIMO methodology lies in its potential to generate realistic irregular waves, thereby achieving a more accurate representation of natural wave phenomena. Therefore, studies were carried out to simulate realistic irregular waves based on the sea states found on the coast of the Rio Grande do Sul (RS), southern Brazil. Locations with different characteristics of significant height, mean period, and water depth were analyzed, enabling an evaluation of the methodology's performance under different wave climates. As a final contribution, this study validates the refined methodology by comparing numerical results with experimental data from the literature, a novel aspect of this paper.

2. Mathematical and Numerical Modeling

To carry out this study, numerical simulations of the generation and propagation of realistic irregular waves in a wave channel were conducted using the ANSYS-Fluent 2024 R2 software [51], based on the finite volume method (FVM). To treat the interface between the phases (water and air), the multiphase volume of fluid (VOF) [52] model was used. The representation of the phases in the computational model requires the concept of volumetric fraction (α), where the sum of the fractions in each control volume must be equal to 1. Thus, each computational cell can assume three distinct states, that is, containing only water, where [53]

$$\alpha_{\text{water}} = 1, \tag{1}$$

containing only air, where [53]

$$\alpha_{\text{air}} = 1, \tag{2}$$

or, containing the interface between both phases, which are immiscible fluids, where [53]

$$\alpha_{\text{water}} = 1 - \alpha_{\text{air}}. \tag{3}$$

Furthermore, it is important to note that, when employing the VOF model, a single set of equations is solved, comprising the conservation equations for mass, volumetric fraction, and momentum. These equations are provided, respectively, by [54]

$$\frac{\partial(\rho)}{\partial t} + \nabla\left(\rho\vec{V}\right) = 0, \tag{4}$$

$$\frac{\partial(\alpha)}{\partial t} + \nabla\left(\alpha\vec{V}\right) = 0, \quad (5)$$

$$\frac{\partial}{\partial t}\left(\rho\vec{V}\right) + \rho\left(\nabla\vec{V}\right)\vec{V} = -\nabla p + \nabla\bar{\bar{\tau}} - \rho\vec{g} + S. \quad (6)$$

where ρ is the fluid density (kg/m^3), calculated for the mixture of phases as [53]

$$\rho = \alpha_{water}\rho_{water} + (1 - \alpha_{water})\rho_{air}, \quad (7)$$

while t is the time (s); $\vec{V}$ is the velocity vector (m/s); p is the static pressure (Pa); $\vec{g}$ is the gravity acceleration vector (m/s^2); and $\bar{\bar{\tau}}$ is the strain rate tensor (N/m^2). Additionally, it is important to note that the sink term (S) refers to the numerical beach tool, which applies damping to minimize reflections, and is given by [55,56]

$$S = -\left[C_1\rho V + \frac{1}{2}C_2\rho|V|V\right]\left(1 - \frac{z - z_{fs}}{z_b - z_{fs}}\right)\left(\frac{x - x_s}{x_e - x_s}\right)^2, \quad (8)$$

where C_1 and C_2 are the linear (s^{-1}) and quadratic (m^{-1}) damping coefficients, respectively; V is the velocity along the z direction (m/s); z_{fs} and z_b are the vertical positions of the free surface (FS) and the channel bottom (m); and x_s and x_e are the starting and ending positions of the numerical beach (m). Lastly, it is important to highlight that the damping coefficients, C_1 and C_2, are set to 20 s^{-1} and 0 m^{-1}, respectively, as recommended by [37].

In addition to the governing equations, the following boundary conditions (detailed later in Section 4.1) were applied to model the numerical wave channel:

- Prescribed velocity: where discretized orbital velocity profiles are imposed (WaveMIMO methodology);
- Pressure outlet: atmospheric pressure was considered, i.e., $p = 101.30$ kPa;
- No-slip and impermeability: where null velocities were considered, $\vec{V} = 0$ m/s.

Numerical Methods Applied

To conduct numerical simulations in the ANSYS-Fluent software, specific set of configurations were defined to solve the governing equations of the problem. The pressure-implicit with splitting of operators (PISO) scheme employed for the pressure–velocity coupling, while the pressure staggering option (PRESTO) scheme was used for pressure spatial discretization. These approaches are summarized in Table 1, along with other numerical methods used in this study.

Table 1. Methods and parameters used in the present numerical simulations.

Parameters		Numerical Inputs
Solver		Pressure-Based
Pressure–Velocity Coupling		PISO
Spatial Discretization	Gradient Evaluation Pressure Momentum Volume Fraction	Green-Gauss-Cell-Based PRESTO First Order Upwind Geo-Reconstruct
Temporal Differencing Scheme		First Order Implicit
Under-Relaxation Factors	Pressure Momentum	0.3 0.7
Residual	Continuity x-velocity z-velocity	10^{-3}
Regime Flow		Laminar

It is highlighted that the flow was considered under a laminar regime, which, according to Gomes et al. [57], is a simplification that does not cause significant accuracy loss while reducing computational processing time. The simulation was initialized with inertia-dominated flow conditions. Additionally, the numerical procedures adopted were adapted from methodologies established in previous studies, such as those by [17,47–49].

3. WaveMIMO Methodology

In general, the WaveMIMO methodology involves numerically generating waves by imposing discretized wave propagation velocity profiles in the horizontal (u) and vertical (w) directions as prescribed velocity boundary conditions (BC). To do so, the Spec2Wave 1.2.1 software [58] is used to process realistic sea state data to obtain these velocity profiles. A flowchart of the WaveMIMO methodology is presented in Figure 1. It should be noted that the purple boxes indicate the user inputs (location/time selection); the blue boxes are the software processing steps (TOMAWAC, Spec2Wave); and the orange boxes are the WaveMIMO methodology's outputs (free surface elevation time series, velocity profiles).

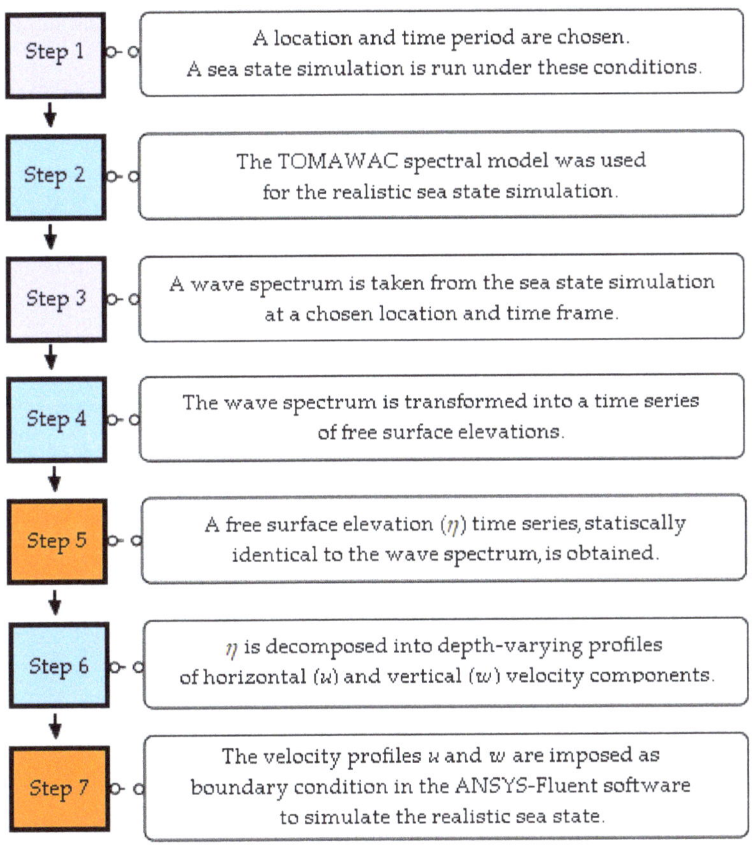

Figure 1. WaveMIMO methodology flowchart.

3.1. Realistic Sea State Data

As illustrated in Figure 1, Step 1 for applying the WaveMIMO methodology involves defining a location and a time interval. For the studies conducted in this paper, a realistic sea state database for the coast of RS, Brazil, for the year 2018 was used. This database was generated using the TOMAWAC spectral model (Step 2). This model obtains data by solving the equation that describes the general behavior of wave propagation in an unstable and inhomogeneous medium, as given by [59]

$$\frac{\partial N}{\partial t} + \frac{\partial(\dot{x}N)}{\partial x} + \frac{\partial(\dot{z}N)}{\partial z} + \frac{\partial\left(\dot{k}_x N\right)}{\partial k_x} + \frac{\partial\left(\dot{k}_z N\right)}{\partial k_z} = Q(k_x, k_z, x, z, t), \quad (9)$$

where N represents the directional spectrum of wave action density ($m^2/Hz/rad$); k_x the component x of the wave number vector (m^{-1}); k_z the component z of the wave number vector (m^{-1}). Moreover, Q is the source term (m^2/rad), which aggregates standard spectral wave processes, such as wind input, dissipation mechanisms, and nonlinear interactions. Details regarding its complete mathematical formulations can be found in the model's technical documentation [59].

The realistic sea state generated is then characterized by statistical parameters, such as the mean period (T_m) of the irregular waves. This parameter is essentially an arithmetic average of the wave periods within the spectrum, weighted by their contribution to the spectral energy [59]. The other is the significant wave height (H_s), which is widely used to describe sea states, as it represents the waves responsible for the most significant surface agitation [60].

Thus, each point in the domain contains statistical data on the observed H_s and T_m, which are analyzed to determine the most frequent sea state at each study site. In this investigation, it was considered realistic sea states from the municipalities of Rio Grande and Tramandaí, located in the southern and northern regions of the RS coast, respectively. Figure 2 presents the bivariate histogram that relates the occurrences of H_s and T_m and identifies the most frequent combination for the municipalities of (a) Rio Grande and (b) Tramandaí. It allows for the wave spectrum taken from the simulation (Step 3) to be determined.

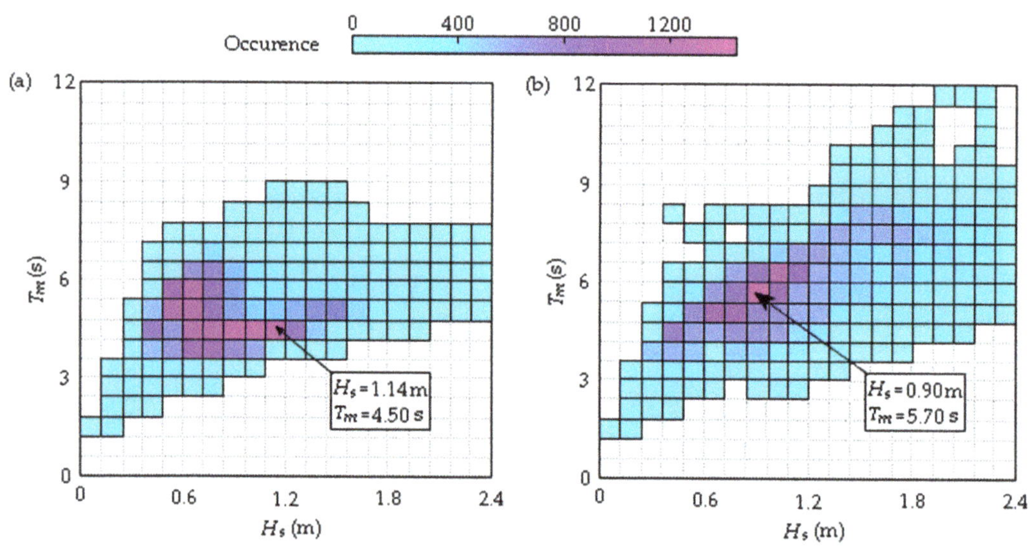

Figure 2. Bivariate histogram for the municipality of (**a**) Rio Grande and (**b**) Tramandaí.

Through these statistical parameters, along with the depth (h) found in each location, it is possible to establish the representative regular waves for the sea states addressed. The final characteristic to be determined is the wavelength, which is calculated using the dispersion equation [61]:

$$\omega^2 = g\, k tanh(kh), \tag{10}$$

where ω is the angular frequency (Hz); while k is the wave number (m^{-1}), given, respectively, by the following [61]:

$$\omega = \frac{2\pi}{T}, \tag{11}$$

$$k = \frac{2\pi}{\lambda}, \tag{12}$$

with λ being the wavelength (m); and T the wave period (s).

Thus, the characteristics of the representative regular waves are presented in Table 2, where H and T stand, respectively, for H_s and T_m (see Figure 2). One can note that these wave characteristics are similar to those found in China [9] and Spain [62], for example. In this study, these characteristics are used as reference parameters for both the spatial and temporal discretization of the computational domain as in [17,30,47,49].

Table 2. Characteristics of representative regular waves of each sea state addressed.

Municipality	Height—H (m)	Length—λ (m)	Period—T (s)	Depth—h (m)
Rio Grande	1.14	31.50	4.50	13.29
Tramandaí	0.90	45.91	5.70	10.98

Next, the database is searched for FS elevation time series corresponding to the most frequent combinations of H_s and T_m at each location. These time series are then reproduced in the ANSYS-Fluent software using the WaveMIMO methodology. Table 3 provides detailed information on the selected locations, as well as the time period to which the data correspond. It should be noted that, in the case of Rio Grande, the distance shown in Table 3 refers to the distance to the local breakwater.

Table 3. Characteristics of the selected locations and time periods for the studies.

Municipality	Geographic Coordinate	Shore Distance (m)	Date	Time
Rio Grande	32°11′24″ S, 52°04′45″ W	171.06	9 November 2018	07:15
Tramandaí	29°59′51″ S, 50°06′18″ W	2094.33	28 May 2018	10:15

3.2. Obtaining Discretized Orbital Velocity Profiles

Following the flowchart presented in Figure 1, the next step involves transforming the wave spectrum into a time series of the corresponding FS elevation (Step 4). This is achieved using the Spec2Wave software [63], based on the methodology described in Oleinik et al. [63], which applies the Inverse Fourier Transform to the selected spectrum. Once the time series of the FS elevation for realistic irregular waves is obtained, it can be approximated by a finite sum of monochromatic waves (Step 5). These waves are individually described, according to Airy's Linear Wave Theory, by [61]

$$\eta = A cos(kx - \omega t), \tag{13}$$

while the velocity potential is given by [60]

$$\phi = -A\frac{g}{\omega}\frac{cosh(kh + kz)}{cosh(kh)}sin(kx - \omega t) \tag{14}$$

Thus, the orbital velocity profiles of wave propagation can be obtained (Step 6). These profiles come from the partial derivatives in the horizontal (u) and vertical (w) directions, and are given, respectively, by [61]

$$u = \omega A \frac{cosh(kz + kh)}{sinh(kh)} cos(kx - \omega t), \tag{15}$$

$$w = \omega A \frac{sinh(kz + kh)}{sinh(kh)} sin(kx - \omega t), \tag{16}$$

where A is the wave amplitude (m), given by [61]

$$A = \frac{H}{2}. \tag{17}$$

Therefore, it is obtained the discrete data that constitute the orbital velocity profiles, which are later used as BC in the ANSYS-Fluent software (Step 7). It allows to reproduce the FS elevation data of realistic irregular waves of a given region and period of time. Moreover, it is worth noting that when a series of FS elevation data are available, the WaveMIMO methodology can be applied to extract the u and w velocity profiles directly from it, skipping Steps 1–5 (see Figure 1).

4. Description of Case Studies

4.1. Parameter Evaluation Studies

A series of four sequential studies on the generation of realistic irregular waves was conducted, with each subsequent study building upon the findings of its predecessor. Thus, investigations were carried out regarding the discretization of the prescribed velocity BC imposition region, mesh sensitivity in the FS region, time step independence, and velocity vector location. It is worth noting that these studies were also performed considering the realistic sea state found in the Mostardas (MS) municipality, referring to the location of geographic coordinates 31°09′51″ S, 50°47′27″ W [49].

To this end, it was necessary to develop two distinct computational domains for the wave channels, one for each municipality considered. This approach was necessary because the channel dimensions depend on the characteristics of the regular waves that represent the realistic sea states addressed. Figure 3 illustrates the computational domain that should be adapted for Rio Grande and Tramandaí, highlighting the principal quotas, the BC implemented, the region where the numerical beach is imposed, and the water level at rest (WLR).

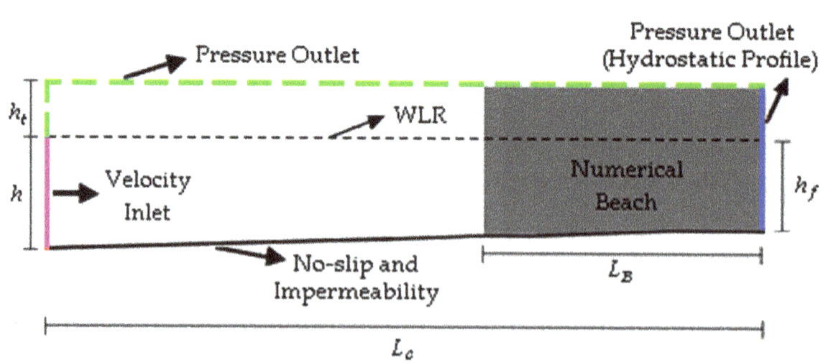

Figure 3. Illustration of the computational domain used.

As observed in Figure 3, at the bottom of the channel (black solid line), the non-slip and impermeability BC is applied where the velocities u and w are considered null. On the lower portion of the left wall (pink solid line), the prescribed velocity BC is imposed. On the upper portion of the left wall and at the top of the channel (green dashed line), atmospheric pressure is applied. Notably, the atmospheric pressure BC is also used on the right lateral wall (blue solid line) to characterize an open flow channel. Additionally, a hydrostatic profile is imposed to ensure that the mean water level remains at a constant depth equal to h_f, preventing the wave channel from emptying.

Moreover, as also illustrated in Figure 3, the channel depth varies to replicate the bathymetry observed at the study site, as indicated by Mocellin et al. [17]. To achieve this, bathymetric data from nautical charts provided by the Brazilian Navy's Hydrography and Navigation Directorate [64] were utilized. According to the recommendations of Lisboa et al. [37], the region where the numerical beach is imposed extends over a length of 2λ. It is also worth mentioning that, in both cases, the top of the channel is positioned 5.00 m above the WLR, corresponding to $h_t = 5.00$ m. The remaining dimensions for each wave channel are detailed in Table 4.

Table 4. Characteristics of the numerical wave channels employed.

Municipality	h (m)	h_f (m)	L_C (m)	L_B (m)
Rio Grande	13.29	10.54	171.06	63.00
Tramandaí	10.98	10.59	229.56	91.82

For the spatial discretization of the computational domain, the stretched mesh [56] was adopted, featuring increased refinement in the region of interest, specifically the FS region. Thus, the computational domain was divided vertically into three regions. The first region, R1, contains only air and demands less refinement; it was discretized into 20 computational cells. The second one is the FS region, R2, containing the water–air interface, which requires greater refinement to ensure numerical convergence; it was discretized into 40 cells. The last vertical region is R3, containing only water. Since phenomena in this region influence the problem more substantially than those in R1, it was discretized into 60 computational cells. Moreover, horizontally, only one region is considered, R4, where 50 computational cells were allocated per λ. Figure 4 illustrates the stretched mesh.

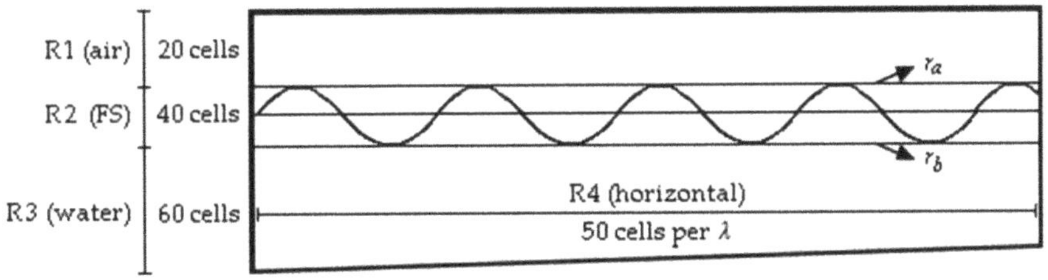

Figure 4. Illustration of the stretched mesh regions.

Figure 4 also illustrates the lines r_a and r_b, which represent the upper and lower limits of the FS region in the stretched mesh, respectively. Additionally, for temporal discretization, an initial time step of $\Delta t = 0.05$ s was used, following the recommendation of Machado et al. [47], for a total simulation time of 900 s of generation and propagation of realistic irregular waves, λ.

Aiming to analyze the results obtained, a probe was used to monitor the elevation of the FS in the wave generation zone, specifically at $x = 0$ m. These results were compared with the FS elevation derived from the TOMAWAC spectral model. To quantitatively evaluate the results, the metrics mean absolute error (MAE) and root mean square error (RMSE) were considered [65]. However, it is important to note that differently scaled data produce differently scaled metric values. Thus, the metrics were normalized following the approach adopted in [39,48,66,67]. Thus, the normalized metrics, NMAE and NRMSE, are given, respectively, as follows:

$$\text{NMAE} = \left(\frac{1}{M}\sum_{i=1}^{M} O_i - P_i\right)\frac{100}{P_{max} - P_{min}}, \tag{18}$$

$$\text{NRMSE} = \left(\sqrt{\frac{1}{M}\sum_{i=1}^{M}(O_i - P_i)^2}\right)\frac{100}{P_{max} - P_{min}}, \tag{19}$$

where O_i represents the value obtained from ANSYS-Fluent (m); P_i represents the data derived from TOMAWAC (m), with P_{max} and P_{min} being the maximum and minimum values found, respectively; and M denotes the total amount of data.

4.1.1. Prescribed Velocity Boundary Condition Imposition Region—Description

The objective here is to investigate the discretization of the numerical wave channel region where the prescribed velocity BC is applied. For the implementation of the WaveMIMO methodology, it is necessary to subdivide this region into segments, where the transient discrete data of the orbital wave propagation velocities are inserted. In this context, it is worth mentioning that the subdivision recommended by Machado et al. [47] is $h/14$, meaning the depth of the location is divided into 14 equal segments, as illustrated in Figure 5, where the segments are represented by the colors blue and pink.

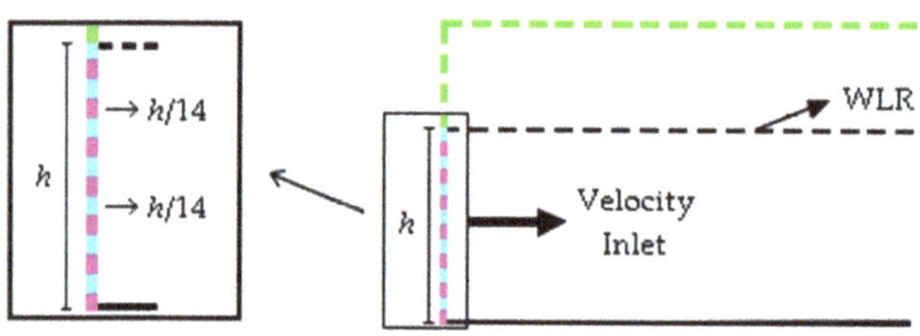

Figure 5. Illustration of the recommended subdivision.

In addition to the aforementioned approach, two other analyses were proposed, which involve relating h to the parameters of maximum wave amplitude (A_{max}) and significant wave height (H_s) of the realistic irregular waves. Thus, instead of fixing the number or size of the segments to be used, this approach varies according to the elevation series under consideration, with segments approximately equal in size to A_{max} or H_s. It should be noted that A_{max} is determined by the magnitude of the highest crest or the deepest trough found in the FS elevation series. In this context, Tables 5 and 6 present the characteristics of each case evaluated, respectively, for Rio Grande and Tramandaí.

Table 5. Characteristics of the cases evaluated for the study of the discretization in the region of the prescribed velocity BC considering the realistic sea state found in Rio Grande.

Case	Parameter	Number of Segments	Segment Dimension (m)
1	$h/14$	14	0.95
2	H_s	11	1.21
3	A_{max}	8	1.66

Table 6. Characteristics of the cases evaluated for the study of the discretization in the region of the prescribed velocity BC considering the realistic sea state found in Tramandaí.

Case	Parameter	Number of Segments	Segment Dimension (m)
1	$h/14$	14	0.78
2	H_s	12	0.92
3	A_{max}	9	1.22

One can note that the FS region size differs for each case analyzed, and each computational domain. This variation arises because this region is determined by the size of the segments used, incorporating one segment above and another below the WLR. Furthermore, the discretization of 40 computational cells vertically within the FS region was maintained for all cases analyzed.

4.1.2. Mesh Sensitivity—Description

After determining the best approach for the discretization of the region where the prescribed velocity BC is applied, a mesh sensitivity analysis was performed regarding the FS region. This study was inspired by the recommendation of Romanowski et al. [68], which suggests that the FS region should be subdivided into distinct areas, i.e., the most refined regions are those that contain the majority of the elevations, which requires a statistical analysis.

In addition to the approach adopted in the previous study (Section 4.1), which follows the recommendation of Gomes et al. [57], this study proposes evaluating three additional discretization approaches that do not require statistical analysis of the elevations of the realistic irregular waves to be numerically reproduced. The proposed new approaches involve dividing each segment of the FS region into two equal-sized segments and varying the number of computational cells applied to each one, as illustrated in Figure 6. The lines r_a and r_b shown in Figure 6 represent, respectively, the upper and lower bounds of the R2 domain region (see Figure 4).

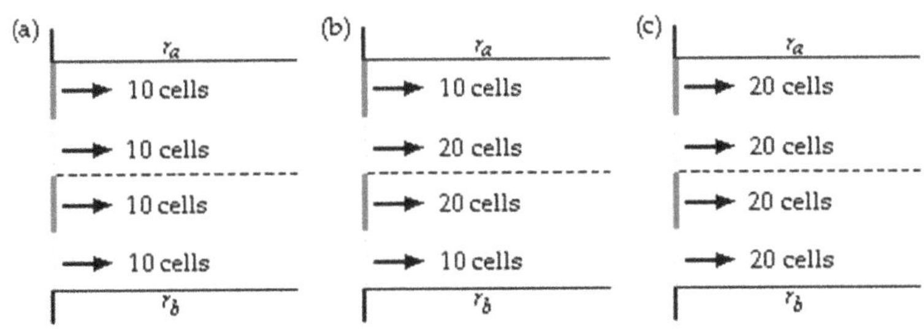

Figure 6. FS discretization: (**a**) Case 4, (**b**) Case 5, (**c**) Case 6.

Thus, the analyzed cases were as follows:

- Case 4: each segment was discretized into 10 cells, totaling 40 computational cells vertically, maintaining the refinement suggested by Gomes et al. [57];
- Case 5: the central segments (around the WLR) were discretized into 20 cells, while the remaining segments were discretized into 10, totaling 60 computational cells vertically. This approach resembles the recommendation by Romanowski et al. [68];
- Case 6: Each segment was discretized into 20 cells, doubling the refinement suggested by Gomes et al. [57], resulting in 80 computational cells vertically.

4.1.3. Time Step Independence—Description

After defining the spatial discretization aspects to be applied, a temporal discretization investigation was conducted. In this regard, a time step independence study was performed to establish a relationship between the time step (Δt) and T_m. The aim was to identify the relationship that provides the best match between the FS elevations generated in ANSYS-Fluent and those obtained through the TOMAWAC spectral model. The evaluated cases for both wave climates considered are presented in Table 7.

Table 7. Cases tested for the time step independence study.

Case	Ratio	Δt (s)—Rio Grande	Δt (s)—Tramandaí
7	$T_m/60$	0.0750	0.0950
8	$T_m/90$	0.0500	0.0630
9	$T_m/120$	0.0375	0.0475
10	$T_m/150$	0.0300	0.0380

4.1.4. Location of the Velocity Vector—Description

The final investigation proposed concerns the location of the velocity vector that contains transient discrete data forming the orbital velocity profiles of wave propagation used as a prescribed velocity BC. It is important to note that the calculation of u and w profiles is performed within the Spec2Wave software, where their positions (vertical coordinates) are indicated, ranging from 0 to h.

Following the recommendation of Machado et al. [47], the velocity vector is positioned at the center of each segment in the BC imposition region. In addition to this approach, two other methods were evaluated, as illustrated in Figure 7. These were proposed by Maciel et al. [50] in a study conducted considering the sea state for the municipality of Rio Grande in 2014. The present study aimed to identify which approach provides the best results, as follows:

- Case 11: positioning the vector at the upper part of the segments;
- Case 12: positioning the vector at the center of the segments;
- Case 13: positioning the vector at the lower part of the segments.

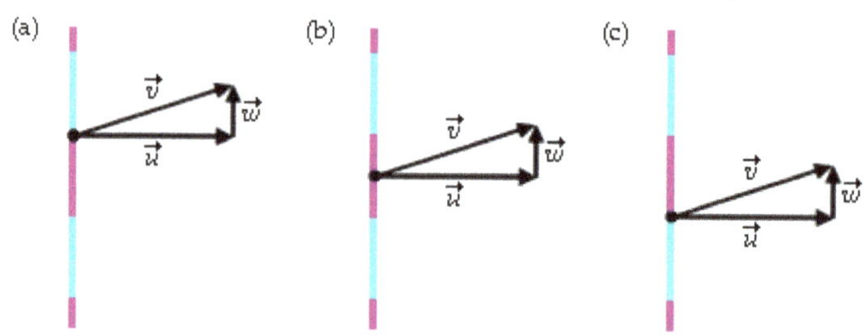

Figure 7. Illustration of the velocity vector in a segment positioned as follows: (**a**) upper part, (**b**) central part, (**c**) lower part.

4.2. *WaveMIMO Methodology Validation*

The final study aims to validate the numerical model and the theoretical recommendations for the WaveMIMO methodology. For this purpose, a numerical simulation was conducted using a two-dimensional domain at laboratory scale, replicating the experiment by Koutrouveli et al. [34]. It is important to highlight that, in previous studies (Section 4.1), spectral data were processed to obtain a time series of FS elevations. In the present study, however, the FS elevation data monitored by probe S_1 in [34] were used. Thus, Steps 1–5 for applying the WaveMIMO methodology were skipped (see Figure 1).

As previously mentioned, these data were processed using the Spec2Wave software to obtain the discretized orbital velocity profiles, which were imposed as prescribed velocity BCs. Consequently, in the present study, the wave generation zone ($x = 0$ m) was aligned with the position of probe S_1 in the experiment by Koutrouveli et al. [34]. Figure 8a presents the illustration of the physical setup used in the laboratory experiment of Koutrouveli et al. [34], while Figure 8b presents the computational domain used for model validation, highlighting the main dimensions and the BC applied.

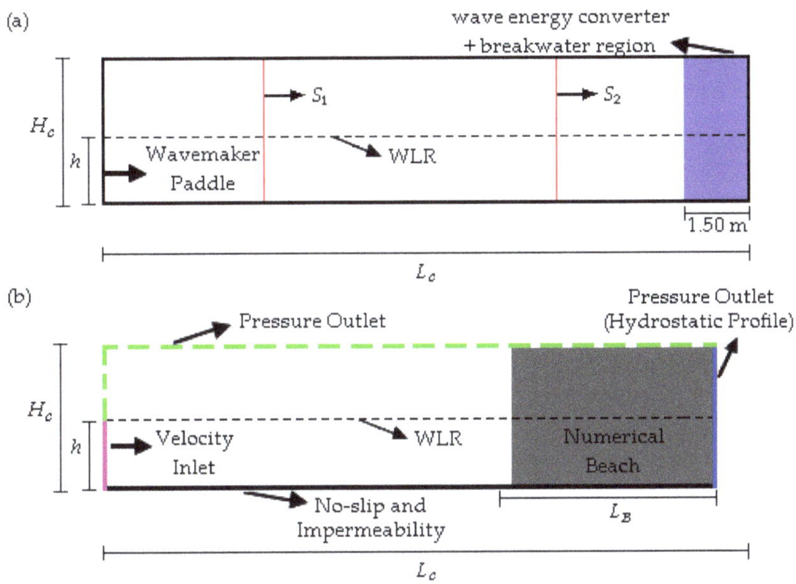

Figure 8. Illustration of (**a**) the physical setup experiment by Koutrouveli et al. [34] and (**b**) computational domains employed for the validation study.

In the laboratory experiment of Koutrouveli et al. [34] (Figure 8a), the wave channel had height of $H_C = 1.20$ m and a total length of $L_C = 14.00$ m, with the first 12.50 m dedicated to free wave propagation, while the last 1.50 m included a breakwater with a wave energy converter device attached. Two probes were used to monitor the FS elevation: the first, (S_1), positioned at $x = 3.80$ m and the second, (S_2), at $x = 9.50$ m, i.e., 5.70 m away from the first probe. By replicating this setup, the WaveMIMO methodology can be validated for both irregular wave generation and propagation. Thus, irregular waves were generated with the following characteristics: $H_s = 0.034$ m, peak period, $T_p = 1.27$ s, $\lambda = 2.20$ m, and channel depth, $h = 0.488$ m.

Due to the lack of detailed information regarding all physical characteristics of the laboratory experiment [34], certain adaptations were required. To maintain consistency with the computational model applied in previous studies, the numerical wave channel employed in the present study (Figure 8b) was defined with a height $H_C = 1.20$ m and a length $L_C = 13.10$ m. This length corresponds to the 12.50 m free wave propagation region in [34], excluding the initial 3.80 m (alignment with S_1) and including an additional 4.40 m numerical beach, following the 2λ recommendation Lisboa et al. [37]. Notably, the numerical beach was incorporated in the physical-to-numerical model adaptation because the experimental wave channel featured a dynamic reflection absorption system integrated in the wavemaker. Additionally, the physical setup had breakwaters on both sides of the hybrid wave energy converter device, enhancing wave energy dissipation toward the channel's end. According to Koutrouveli et al. [34], coupling the device with the breakwater even improved its hydraulic performance.

As in previous studies performed here, the spatial and temporal discretization of the computational domain was based on the characteristics of the generated waves. Additionally, the numerical configurations employed were maintained, as well as the theoretical recommendations obtained for the application of the WaveMIMO methodology. Thus, two probes were used to monitor the FS elevation: the first, positioned in the generation zone, at $x = 0$ m and the second, in the propagation zone, at $x = 5.70$ m, corresponding to S_2 in [34]. The results were compared with the experimental and numerical data presented by Koutrouveli et al. [34]. As before, the quantitative evaluation was carried out using the NMAE and NRMSE metrics.

5. Results

5.1. Parameter Evaluation Studies

This section presents the results obtained throughout the investigations into the WaveMIMO methodology parameters and their respective discussions. It is worth mentioning that the studies were conducted sequentially, meaning the findings from each preceding investigation provided support the execution of the subsequent one.

For the analysis of these results, the findings from Paiva et al. [49] were incorporated to the discussions, as the authors evaluated the same WaveMIMO methodology parameters using data from the municipality of Mostardas, in the central region of the RS state coast, Brazil. The integration of these findings strengthens the generalizability of the parameter evaluations, since it provides a geographical complementarity to the municipalities addressed in the present study (Rio Grande in the south and Tramandaí in north of the RS state). Additionally, it is worth noting that the aforementioned article was authored by the same researchers involved in the current study, ensuring that the same numerical model was implemented.

5.1.1. Prescribed Velocity Boundary Condition Imposition Region—Results

To enable a better analysis during the evaluation of the results, the qualitative comparison of the FS elevation (η) was divided into three time-intervals, each spanning 300 s. Furthermore, the plotted graphs feature lines parallel to the x-axis, outlining the FS region (see Figure 6). Each pair of lines corresponds to an evaluated case, as follows:

- r_{1a} and r_{1b}: solid lines, corresponding to Case 1;
- r_{2a} and r_{2b}: dashed lines, corresponding to Case 2;
- r_{3a} and r_{3b}: dotted lines, corresponding to Case 3.

It should be noted that the position of these lines on the z-axis relates to the size of the segments considered in each case (see Tables 5 and 6). Thus, Figure 9 presents the monitored results for the realistic sea state observed in Rio Grande. As shown, in all cases, the realistic irregular waves generated through the WaveMIMO Methodology accurately reproduce the realistic sea state derived from TOMAWAC. Additionally, one can note the absence of initial inertia conditions effects, which is due to the fact that the results were monitored at $x = 0$ m.

In cases where the segments used for discretizing the prescribed velocity BC imposition region are smaller than the A_{max} parameter (Case 3), wave crests and troughs extend outside the FS region, as observed in [49]. It is possible to see time instants where the crests and troughs cross the r_{1a} and r_{1b} lines in Figure 9a, near $t = 0$ s, in Figure 9b, around $t = 450$ s and $t = 600$ s, and Figure 9c, near $t = 650$ s. On the other hand, the only occurrence where the r_{2a} and r_{2b} lines were also crossed happen in Figure 9a, near $t = 0$ s. Notably, the r_{3a} and r_{3b} lines are never reached, as the trough used to define A_{max}, which occurs close to

$t = 0$ s, was not reached by the realistic irregular waves generated though the WaveMIMO methodology.

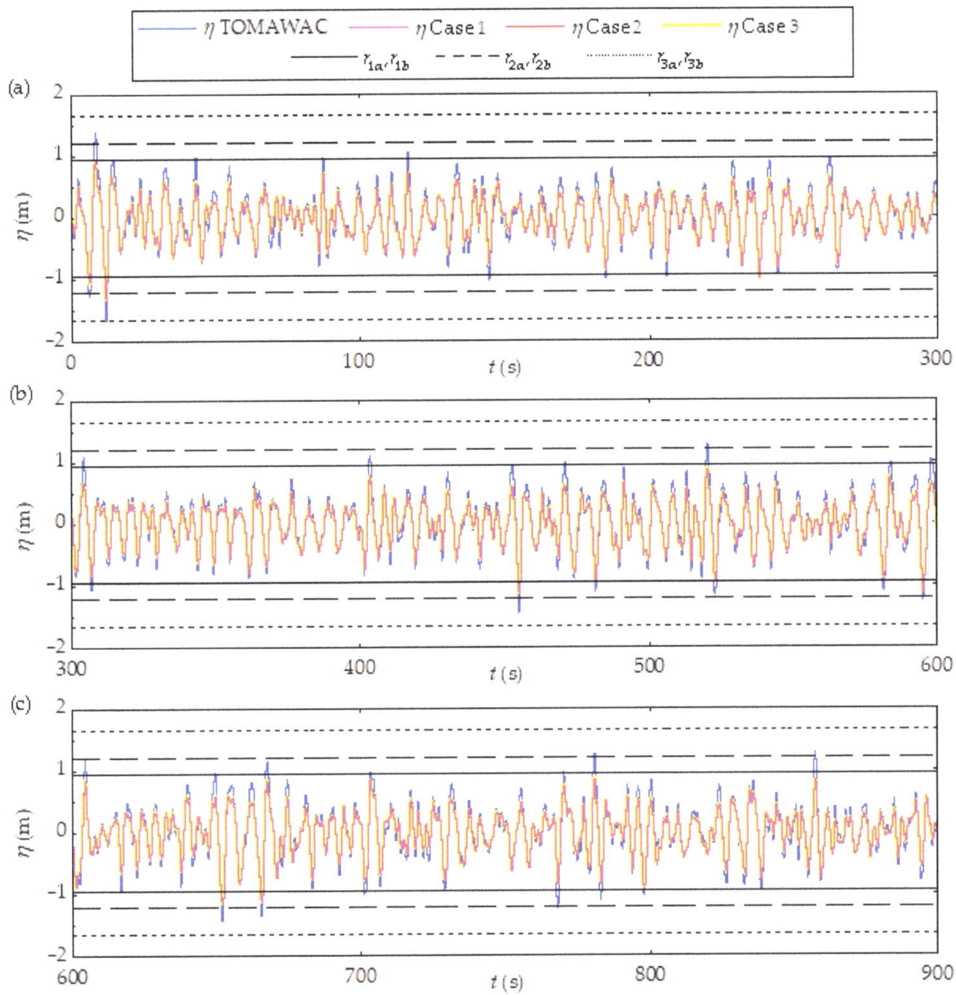

Figure 9. Qualitative results for the discretization study of the prescribed velocity BC imposition region, evaluated under Rio Grande's realistic sea state: (**a**) 0 s $\leq t \leq$ 300 s; (**b**) 300 s $\leq t \leq$ 600 s; (**c**) 600 s $\leq t \leq$ 900 s.

Therefore, most generated elevations remain within the region bounded by lines r_{1a} and r_{1b}, which contain computational cells of practically identical size. For proper evaluation of these results, quantitative analysis is required. Table 8 summarizes the NMAE and NRMSE metrics, along with the vertical dimension (Δz) of each mesh element within the FS region and the respective processing times for all cases.

Table 8. Quantitative results for the discretization study of the prescribed velocity BC imposition region, evaluated under Rio Grande's realistic sea state.

Case	Δz (m)	NMAE (%)	NRMSE (%)	Processing Time (h)
1	0.0237	3.7303	4.8421	5.92
2	0.0302	3.7730	4.9046	5.72
3	0.0415	3.8388	4.9934	5.67

As shown in Table 8, NMAE and NRMSE values reveal a consistent but gradual degradation in accuracy as the size of the segments increased. Thus, Case 1 emerges as the best-performing scenario, since it presented the lowest NMAE and NRMSE values. While the absolute differences in metrics across cases appear small, the relative degradation in accuracy is notable. Compared to Case 1, Case 2 shows increases of 1.10% (NMAE) and 1.30% (NMRSE), while Case 3, the worst-performing scenario, exhibits larger increases of 2.90% (NMAE) and 3.10% (NRMSE). In comparison to Case 1, it was noted that Case 2 has elements 27.43% larger and a processing time 3.38% shorter, whereas Case 3 has elements 75.11% larger and a processing time 4.22% shorter.

Following the same methodology applied to Rio Grande, a qualitative assessment was conducted for the results obtained considering the realistic sea state found in the municipality of Tramandaí. Thus, Figure 10 compares the FS elevation monitored against the TOMAWAC data, demonstrating that in all cases, the realistic irregular waves generated using the WaveMIMO methodology accurately replicate the sea state addressed. Moreover, no inertia conditions effects are noted since results were monitored at $x = 0$ m.

As before, there are time instants where wave crests and troughs cross the r_{1a} and r_{1b} lines, such as in Figure 10a, near $t = 70$ s and $t = 250$ s, Figure 10b, around $t = 550$ s, Figure 10c, close to $t = 600$ s. Additionally, troughs are observed to cross the r_{2a} and r_{2b} lines in Figure 10c, near $t = 600$ s. As previously noted for the realistic sea state of Rio Grande, the r_{3a} and r_{3b} lines are never reached, since the crest used to define A_{max} was not attained by the realistic irregular waves generated through the WaveMIMO methodology.

Consequently, most elevations remain within the region bounded by the r_{1a} and r_{1b} lines. Table 9 presents the NMAE and NRMSE metrics for the evaluated cases, along with the vertical dimensions of each mesh element in the FS region and the corresponding computational processing time required for each case.

Table 9. Quantitative results for the discretization study of the prescribed velocity BC imposition region, evaluated under Tramandaí's realistic sea state.

Case	Δz (m)	NMAE (%)	NRMSE (%)	Processing Time (h)
1	0.0392	4.4364	5.5955	6.03
2	0.0450	4.4455	5.6045	5.77
3	0.0582	4.4591	5.6227	5.70

The results shown in Table 9 confirm Case 1 as the best configuration, as well as the gradual degradation in accuracy previously observed. The comparative analysis reveals that, compared to Case 1, Case 2 shows increases of 1.15% (NMAE) and 1.25% (NRMSE), while Case 3, the worst case evaluated, exhibits a more substantial increment of 2.85% (NMAE) and 3.05% (NRMSE). Therefore, for both realistic sea states addressed, the metrics follow similar trends although NRMSE demonstrates slightly greater sensitivity, with consistently 0.10–0.20% higher relative differences than NMAE. Regarding Δz and the processing time, the analysis shows that compared to Case 1, Case 2 employs mesh elements 14.80% larger while requiring 4.31% less processing time, whereas Case 3 uses mesh elements 48.47% larger with 5.47% shorter computational time.

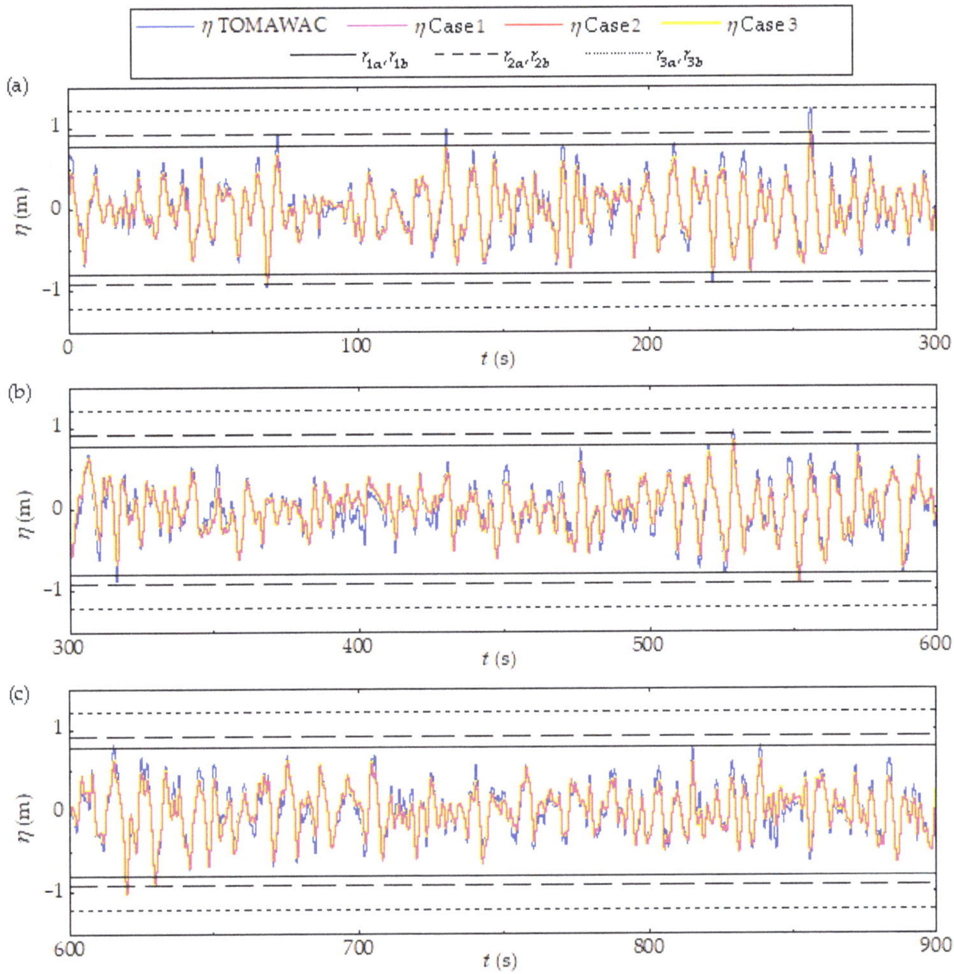

Figure 10. Qualitative results for the discretization study of the prescribed velocity BC imposition region, evaluated under Tramandaí's realistic sea state: (**a**) 0 s $\leq t \leq$ 300 s, (**b**) 300 s $\leq t \leq$ 600 s, (**c**) 600 s $\leq t \leq$ 900 s.

It is important to emphasize that processing time was not considered a determining factor in this study due to its minimal temporal variation. Given that all cases maintain a spatial discretization of 40 computational cells in the FS region, the observed Δz variation demonstrates negligible impact on processing time requirements. Consequently, Case 1 stands out as the most effective approach in the present analysis, consistent with the findings of Paiva et al. [49]. Therefore, it is recommended to maintain the indication of Machado et al. [47] to subdivide this region into segments of size $h/14$.

Additionally, it is noted that Case 1 features the smallest Δz and, as most elevation data fall within the region bounded by the r_{1a} and r_{1b}, lines, these data are computed with higher precision. Thus, the accuracy reduction from Case 1 to 3 correlates with the use of larger mesh elements in the FS region, highlighting the need for further investigation of mesh sensitivity in this region.

5.1.2. Mesh Sensitivity—Results

Following the results of Section 5.1.1, it was established that the FS region comprises two segments of size $h/14$ (Case 1). For the mesh sensitivity study, it was proposed to subdivide these FS segments into four smaller segments of size $h/28$ each. Similarly to the previous study, qualitative analysis shows the evaluated cases are virtually indistinguishable. Thus, the monitored results are presented in the Supplementary Materials Section, where Figures S1 and S2 show the FS elevation comparison for the present study under the Rio Grande and Tramandaí realistic sea states, respectively. These results demonstrate that all cases successfully reproduce the target realistic sea states through the realistic irregular waves generated via the WaveMIMO methodology.

Since visual distinction between approaches proves to be inefficient, quantitative evaluation becomes essential. The results monitored under Rio Grande's realistic sea state are presented in Table 10, featuring calculated NMAE and NRMSE metrics alongside processing times for each case. Moreover, Case 1 served as the reference for this analysis.

Table 10. Quantitative results for the study of mesh sensitivity in the FS region, evaluated under Rio Grande's realistic sea state.

Case	NMAE (%)	NRMSE (%)	Processing Time (h)
1	3.7303	4.8421	5.92
4	3.7303	4.8421	6.08
5	3.6612	4.7467	6.22
6	3.6776	4.7664	6.65

Both Cases 1 and 4 contain 40 computational cells in the FS region, which explains their identical NMAE and NRMSE values shown in Table 10. This shows that simply dividing the segments into smaller segments maintains result accuracy while slightly increasing processing time by 2.70%. Among the new approaches proposed for FS region discretization, the best-performing scenario was Case 5. In a comparative analysis, Case 4, the worst case evaluated, shows metrics that are 1.89% (NMAE) and 2.01% (NRMSE) higher, while processing time is 2.25% shorter. Considering the refinement increase is on the order of 50% (from 40 to 60 computational cells), the difference in processing time is reasonably low. When comparing Case 5 to Case 6, the metrics show increases of 0.45% (NMAE) and 0.41% (NMRSE), while processing time is 6.91% higher for a 33.33% refinement increase (from 60 to 80 computational cells).

Table 11 presents the results under Tramandaí's realistic sea state, including NMAE and NRMSE metrics with corresponding processing times. Consistent with previous analyses, Case 1 serves as the reference for comparative evaluation.

Table 11. Quantitative results for the study of mesh sensitivity in the FS region, evaluated under Tramandaí's realistic sea state.

Case	NMAE (%)	NRMSE (%)	Processing Time (h)
1	4.4364	5.5955	6.03
4	4.4364	5.5955	6.17
5	4.4318	5.5864	6.23
6	4.4636	5.6318	6.47

Table 11 confirms that Cases 1 and 4 present identical metrics, with Case 4 requiring 2.32% longer processing time. As before, Case 5 performs best; however, in this case, Case 6 is the worst-performing one, with metrics that are 0.72% (NMAE) and 0.81% (NRMSE) higher, along with a 3.86% increase in processing time due to its 33.33% refinement mesh increase. When comparing Case 4 to Case 5, the metrics are 0.10% (NMAE) and 0.16% (NRMSE) higher, while processing time is only 0.97% shorter for Case 4, despite Case 5 employing a 50% refinement increase in the FS region.

This study demonstrates that applying greater mesh refinement only in the segments closest to the WLR was effective, since Case 5 yielded better results than Case 6 for both realistic sea states addressed. Therefore, considering these evaluations along with the results for Mostardas' realistic sea state [49], a recommendation for stretched mesh application in realistic irregular wave generation was established, which corresponds to Case 5. However, it is important to note that this modifies the result defined in Section 5.1, as the prescribed velocity BC imposition region now comprises 15 segments: 2 segments of size $h/28$ in the FS region and 13 segments of size $h/14$ in the R3 domain region (see Figure 4), extending to the channel bottom.

5.1.3. Time Step Independence—Results

It is important to emphasize that this investigation adopted the mesh configuration established in prior studies (Sections 5.1.1 and 5.1.2). As before, the evaluated cases are virtually indistinguishable in qualitative analysis; these results are therefore presented in the Supplementary Materials Section (see Figures S3 and S4 for the Rio Grande and Tramandaí realistic sea states, respectively). Notably, in all cases, the WaveMIMO methodology successfully reproduced the realistic sea state addressed.

For precise quantitative analysis, the monitored results are presented in Table 12 (Rio Grande) and Table 13 (Tramandaí), including the NMAE and NRMSE metrics alongside processing times, which scale with temporal step refinement. Accordingly, these tables also include the relative differences (d_r), indicating the processing time increase. For this purpose, cases are always compared to their immediate predecessor (the row above in Tables 12 and 13). Case 5 serves as the reference for result analysis.

Table 12. Quantitative results for the study of time step independence, evaluated under Rio Grande's realistic sea state.

Case	NMAE (%)	NRMSE (%)	Processing Time (h)	d_r (%)
5	3.6612	4.7467	6.22	-
7	3.7368	4.8454	4.25	-
8	3.6612	4.7467	6.22	46.35
9	3.6316	4.7072	9.08	45.98
10	3.6217	4.6908	12.23	34.69

Table 13. Quantitative results for the study of time step independence, evaluated under Tramandaí's realistic sea state.

Case	NMAE (%)	NRMSE (%)	Processing Time (h)	d_r (%)
5	4.4318	5.5864	6.23	-
7	4.5636	5.7636	3.13	-
8	4.4727	5.6455	4.97	58.79
9	4.4364	5.6000	5.88	18.31
10	4.4182	5.5773	8.63	46.77

For both realistic sea states (Rio Grande and Tramandaí), the metrics show continuous improvement as the time step is refined from $T_m/60$ to $T_m/150$, consistent with findings for Mostardas municipality [49]. Moreover, Case 5 ($\Delta t = 0.05$ s), corresponding to $T_m/90$

(Rio Grande) and $T_m/114$ (Tramandaí), demonstrates results aligned with the other cases, for both error metrics and processing time, as seen in Tables 11 and 12.

Regarding Rio Grande's realistic sea state, metric reductions are 2.02% (NMAE) and 2.04% (NRMSE), comparing Cases 7 and 8, 0.81% (NMAE) and 0.83% (NRMSE), from Case 8 to 9, and 0.27% (NMAE) and 0.35% (NRMSE), from Case 9 to 10. Similarly, considering Tramandaí's realistic sea state, reductions are 1.99% (NMAE) and 2.05% (NRMSE), comparing Case 7 to 8, 0.81% (NMAE and NRMSE), from Case 8 to 9, and 0.41% (NMAE and NRMSE), from Case 9 to 10.

Additionally, Tables 11 and 12 confirm that these incremental accuracy gains come with more significant processing time increases. Therefore, the recommended temporal discretization uses Case 9 ($\Delta t = T_m/120$), as the reduction in NMAE and NRMSE metrics, compared to Case 10, is less than 0.50% for both realistic sea states, while the processing time increase exceeds 30.00%. These results are in agreement with those obtained for the municipality of Mostardas [49].

5.1.4. Location of the Velocity Vector—Results

Once again, it should be mentioned that the results obtained in previous studies (Sections 5.1.1–5.1.3) are taken into account in this final evaluation, which concerns the location where the velocity vector is calculated within each segment of the prescribed velocity BC imposition region. Notably, qualitative comparisons between the FS elevations monitored in ANSYS-Fluent software and those derived from the TOMAWAC spectral model are presented in the Supplementary Materials Section (see Figures S5 and S6 for the Rio Grande and Tramandaí realistic sea states, respectively). As before, for all evaluated cases, the realistic irregular waves generated through the WaveMIMO methodology accurately reproduce the target realistic sea states. Consequently, similar to previous studies, visual distinction between cases remains impossible.

Regarding quantitative evaluation, the results are presented in Table 14 (Rio Grande) and Table 15 (Tramandaí). However, processing times are omitted as they remain constant and correspond to using $\Delta t = T_m/120$ (see Tables 11 and 12). It is highlighted that Case 9 is not taken as reference for the analysis in the present study, as it matches Case 12's configuration.

Table 14. Quantitative results for the study of velocity vector' location, evaluated under Rio Grande's realistic sea state.

Case	NMAE (%)	NRMSE (%)
11	3.4474	4.4375
12	3.6316	4.7072
13	3.8783	5.0428

Table 15. Quantitative results for the study of velocity vector's location, evaluated under Tramandaí's realistic sea state.

Case	NMAE (%)	NRMSE (%)
11	4.3318	5.4591
12	4.4364	5.6000
13	4.5136	5.7045

As observed in Tables 14 and 15, a consistent pattern emerges where Case 11 (velocity vector at the upper segment position) presented the best results, followed by Case 12 (central position, as recommended by Machado et al. [47], while Case 13 (lower position) showing the poorest performance. This behavior aligns with findings for the Mostardas municipality [49], but contrasts with Maciel et al. [50], who reported superior performance with lower-position vectors.

In a comparative analysis for Rio Grande's sea state, Case 12 shows increases of 5.34% (NMAE) and 6.08% (NRMSE) relative to Case 11, while Case 13 exhibits more significant increases of 12.49% (NMAE) and 13.64% (NRMSE). Similarly, for Tramandaí's sea state, Case 12 presents increases of 2.41% (NMAE) and 2.58% (NRMSE), while Case 13 shows increases of 4.20% (NMAE) and 4.50% (NRMSE) compared to Case 11.

In addition, this study reveals the most substantial metric variations, particularly for Rio Grande's realistic sea state. Therefore, the recommendation is to consider the velocity vectors at the upper position of the segments in the prescribed velocity BC imposition region. Finally, it should be highlighted that all cases evaluated throughout the WaveMIMO methodology parameter assessments have demonstrated results that compare favorably with existing literature such as [39,66], confirming the robustness of this approach for realistic irregular wave generation.

5.2. WaveMIMO Methodology Validation

This section presents the validation results of the WaveMIMO methodology [47] for irregular wave generation and propagation in a numerical wave channel. As previously mentioned, the laboratory experiments conducted by Koutrouveli et al. [34] served as the reference for this validation. The validation begins with qualitative analysis. Thus, Figure 11 compares the FS elevation monitored using the WaveMIMO methodology against both experimental and numerical results from Koutrouveli et al. [34]. Specifically, Figure 11a displays results monitored at the wave generation zone, $x = 0$ m, while Figure 11b shows results from the wave propagation zone, $x = 5.70$ m.

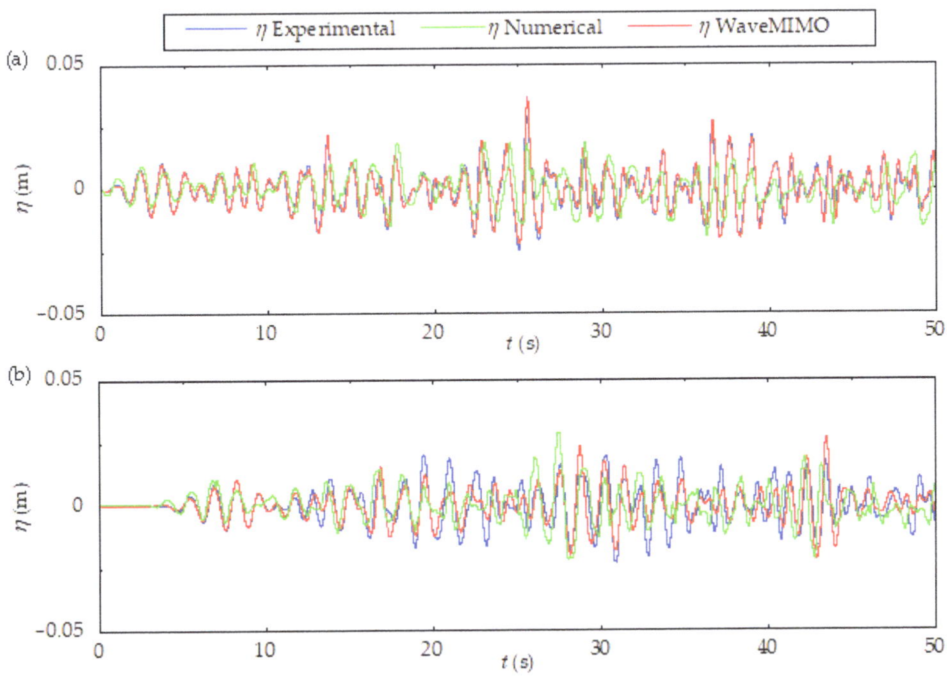

Figure 11. Qualitative comparison of the results obtained in the present study with the experimental and numerical data found in [34]: (**a**) FS elevation at $x = 0$ m and (**b**) FS elevation at $x = 5.70$ m.

As observed in Figure 11a, the WaveMIMO methodology demonstrates near-perfect agreement with the experimental data at the generation zone, with precise matching of amplitudes and phase synchronization throughout the 50 s of simulation. In contrast, the numerical results from Koutrouveli et al. [34] show phase discrepancies, including inverted crests/troughs, occurring near $t = 27$, 32, and 48 s.

In the propagation zone (Figure 11b), one can note the inertia condition effects on the flow, as there are no elevations within the first seconds of the monitored results in both numerical approaches and the physical setup. Moreover, WaveMIMO maintains superior consistency with experimental trends compared to numerical results of Koutrouveli et al. [34]. While both methods approximate the wave envelope, WaveMIMO preserves correct phase sequencing despite amplitude variations, whereas the reference model [34] shows systematic errors: crests replacing troughs, close to $t = 5$, 10, 20, 25, 35, 45, and 50 s and troughs replacing crests, near $t = 10$, 12, 35, 40, and 49 s. Notably, these discrepancies emerge within the first 10 s of simulation in [34], whereas the WaveMIMO methodology maintains near-perfect agreement. Both numerical approaches (present study and [34]) begin to diverge more significantly from experimental results after $t = 20$ s suggesting the onset of reflection effects in the laboratory setup, which neither fully replicates.

This robust agreement with the experimental data verifies the WaveMIMO capability to accurately reproduce the target irregular waves, similar to what was observed for regular waves in Maciel et al. [48]. To further validate these observations, quantitative metrics, NMAE and NRMSE, are analyzed in Table 16.

Table 16. Quantitative comparison between numerical and experimental data.

Case	$x = 0$ m		$x = 5.70$ m	
	NMAE (%)	NRMSE (%)	NMAE (%)	NRMSE (%)
Present study	2.6239	3.4785	7.6844	10.6914
Koutrouveli et al. [34]	10.6485	13.1727	13.5124	17.2615

The results presented in Table 16 corroborate the qualitative analysis performed, validating the WaveMIMO methodology for the generation and propagation of realistic irregular waves. The results monitored at both probes demonstrate consistency with previous numerical–experimental studies on wave energy [39,48,66,67,69]. Considering the same criterion applied in the validation of numerical models across various fields [34,48,66,69–75], the results obtained in the present study represent excellent agreement with experimental data, with metrics below 10%. Only the NRMSE value at $x = 5.70$ s slightly exceeds this threshold, remaining below 11%, which still indicates a good agreement.

Notably, the increase in metrics in the wave propagation zone was expected, as the numerical model employed differs from the physical model, incorporating adaptations such as the numerical beach. However, the metrics recorded by Koutrouveli et al. [34] at this probe exceed those of the WaveMIMO methodology by 75.87% (NMAE) and 61.44% (NRMSE). This can be partially explained by the use of a numerical beach in the present model, which replicates, within appropriate proportions, the breakwater used in the experimental setup of Koutrouveli et al. [34], a feature that is absent in their numerical model. These differences were even more pronounced in the wave generation zone, reaching 305.82% (NMAE) and 278.68% (NRMSE), which can be attributed to the fact that these results were also monitored in a propagation zone in [34].

Thus, it was possible to validate not only the use of the WaveMIMO methodology but also the entire numerical model employed, including the applied numerical configurations (see Table 1) and the theoretical recommendations obtained in the present study. Moreover, the modifications made to adapt the experimental model to the numerical one, such as laminar flow conditions and the use of a numerical beach, also proved effective.

Finally, the visualization of the physical phenomenon is presented through the volumetric fraction field, shown in Figure 12, where the water phase is represented in blue and the air phase in red. Figure 12 illustrates the hydrodynamic behavior in the channel at the end of the simulation, at $t = 50$ s; it highlights the beginning of the channel, where surface elevations are observed and the end of the channel, where the efficiency of the numerical beach in dissipating the energy of the waves reaching this region can be observed, as no elevations are present.

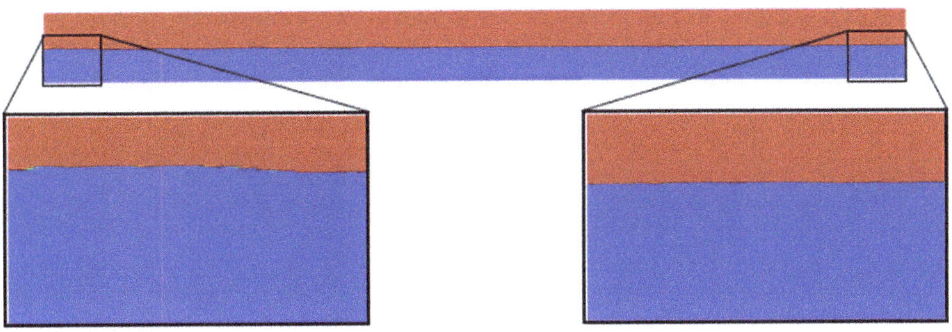

Figure 12. Physical phenomenon representation at $t = 50$ s.

6. Conclusions

The present study systematically evaluated the application parameters of the WaveMIMO methodology [47] through four sequential investigations of realistic irregular wave generation. Numerical simulations were performed using wave climate data from coastal areas of the Brazilian state of Rio Grande do Sul near the cities of Rio Grande and Tramandaí, with comparative analysis incorporating previous results from the coastal region of city of Mostardas [49]. These assessments enabled the development of theoretical recommendations for implementing the WaveMIMO methodology in the realistic irregular wave generation. Final validation was achieved through qualitative and quantitative comparison with laboratory experimental data from Koutrouveli et al. [34], confirming the methodology's effectiveness for both wave generation and propagation of these waves.

Among these studies, there are those that focus on the spatial discretization of the computational domain employed, such as the investigation into the discretization of the prescribed velocity BC imposition region. Based on the evaluations and analyses conducted, it was determined that the best results were achieved in Case 1, where it was subdivided into 14 segments. In another study, the mesh sensitivity in the FS region was analyzed. It was possible to establish Case 5 as a theoretical recommendation, where the central segments of the FS region were discretized with greater refinement, totaling 60 computational cells. Additionally, the temporal discretization was investigated, where the time step was related to the mean period. In this evaluation the processing time required was considered to determine the most suitable relation. Thus, the recommendation of using $\Delta t = T_m/120$, referred to as Case 9, was established. The final study addressed the evaluation of the velocity vector's location. As a result, Case 11 was determined as the theoretical recommendation, which consists of placing the vector at the upper part of each segment of the prescribed velocity BC imposition region.

In summary, when applying the WaveMIMO methodology for realistic irregular wave generation, the prescribed velocity BC region should comprise 15 segments: the first two, located just below the WLR, with a size of $h/28$, while the remaining 13 segments must have a size of $h/14$. Thus, the FS region on the stretched mesh consists of the first two segments below the WLR and two equally sized segments above it. For spatial discretization, segments near the WLR should contain 20 computational cells, whereas the remaining segments should use 10 cells. The temporal discretization requires a time step of $T_m/120$ and the velocity vector should be considered at the top of each segment in the prescribed velocity BC region.

At last, it is worth highlighting that, when comparing the optimized configuration (Case 12) to recommendations from the previous literature (Case 1) [47], the metrics were reduced by 7.58% (NMAE) and 8.36% (NRMSE) for Rio Grande, while Tramandaí's sea state showed more modest improvements of 2.36% (NMAE) and 2.44% (NRMSE). Thus, after performing the investigations, it was possible to validate the WaveMIMO methodology against experimental data from Koutrouveli et al. [34]. One can highlight that, in a general way, an excellent agreement between the irregular waves generated with WaveMIMO methodology and the experimental data was achieved, since most employed indicators were below 10%. For future studies, it is suggested to implement a prescribed velocity BC positioned in the upper section of the left wall of the channel. This approach could emulate the effects of wind on the FS of the channel, thereby improving the adjustment of missed crests and troughs, enhancing wave generation accuracy.

Supplementary Materials: The following supporting information can be downloaded at: https://www.mdpi.com/article/10.3390/pr13051395/s1, Figure S1: Qualitative results for the study of mesh sensitivity in the free surface region, evaluated under Rio Grande's realistic sea state: (a) 0 s $\leq t \leq$ 300 s; (b) 300 s $\leq t \leq$ 600 s; (c) 600 s $\leq t \leq$ 900 s; Figure S2: Qualitative results for the study of mesh sensitivity in the free surface region, evaluated under Tramandaí's realistic sea state: (a) 0 s $\leq t \leq$ 300 s; (b) 300 s $\leq t \leq$ 600 s; (c) 600 s $\leq t \leq$ 900 s; Figure S3: Qualitative results for the time step independence study, evaluated under Rio Grande's realistic sea state: (a) 0 s $\leq t \leq$ 300 s; (b) 300 s $\leq t \leq$ 600 s; (c) 600 s $\leq t \leq$ 900 s; Figure S4: Qualitative results for the time step independence study, evaluated under Tramandaí's realistic sea state: (a) 0 s $\leq t \leq$ 300 s; (b) 300 s $\leq t \leq$ 600 s; (c) 600 s $\leq t \leq$ 900 s, Figure S5: Qualitative results for the study of the velocity vector location, evaluated under Rio Grande's realistic sea state: (a) 0 s $\leq t \leq$ 300 s; (b) 300 s $\leq t \leq$ 600 s; (c) 600 s $\leq t \leq$ 900 s; Figure S6: Qualitative results for the study of the velocity vector location, evaluated under Tramandaí's realistic sea state: (a) 0 s $\leq t \leq$ 300 s; (b) 300 s $\leq t \leq$ 600 s; (c) 600 s $\leq t \leq$ 900 s.

Author Contributions: Conceptualization, M.d.S.P., A.P.G.M., P.H.O., L.A.I. and B.N.M.; methodology, M.d.S.P., A.P.G.M., P.H.O., E.D.d.S., L.A.O.R., L.A.I. and B.N.M.; software, P.H.O., E.D.d.S., L.A.O.R., L.A.I. and B.N.M.; validation, M.d.S.P., A.P.G.M., P.H.O., L.A.I. and B.N.M.; formal analysis, M.d.S.P., A.P.G.M., L.A.I. and B.N.M.; investigation, M.d.S.P., A.P.G.M., L.A.I. and B.N.M.; resources, E.D.d.S., L.A.O.R., L.A.I. and B.N.M.; data curation, M.d.S.P., A.P.G.M. and P.H.O.; writing—original draft preparation, M.d.S.P. and A.P.G.M.; writing—review and editing, M.d.S.P., L.A.I. and B.N.M.; visualization, L.A.O.R. and E.D.d.S.; supervision, L.A.O.R., E.D.d.S., L.A.I. and B.N.M.; project administration, E.D.d.S., L.A.O.R., L.A.I. and B.N.M.; funding acquisition, E.D.d.S., L.A.O.R., L.A.I. and B.N.M. All authors have read and agreed to the published version of the manuscript.

Funding: This research was funded by the Brazilian Coordination for the Improvement of Higher Education Personnel—CAPES (Finance Code 001), the Research Support Foundation of the State of Rio Grande do Sul—FAPERGS (Public Call FAPERGS 07/2021—Programa Pesquisador Gaúcho—PqG—21/2551-0002231-0), the Brazilian National Council for Scientific and Technological Development—CNPq (Processes: 307791/2019-0, 308396/2021-9, 309648/2021-1, and 403408/2023-7), and by the institutional research assistance program for newly hired professors at the Federal University of Rio Grande do Sul—UFRGS.

Data Availability Statement: The data presented in this study are available on request from the corresponding author. The data are not publicly available due to privacy reasons.

Acknowledgments: M.d.S.P. and A.P.G.M. thank CAPES and CNPq for their master's and doctorate scholarships. L.A.O.R., E.D.d.S. and L.A.I. thank CNPq for research grants. All authors thank FAPERGS and CNPq.

Conflicts of Interest: The authors declare no conflicts of interest. The funders had no role in the design of this study, in the collection, analyses, or interpretation of data, or in the writing of the manuscript, as well as in the decision to publish the results.

References

1. Adesanya, A.; Misra, S.; Maskeliūnas, R.; Damaševičius, R. Prospects of Ocean-Based Renewable Energy for West Africa's Sustainable Energy Future. *Smart Sustain. Built Environ.* **2020**, *10*, 37–50. [CrossRef]
2. Hernández-Fontes, J.V.; Felix, A.; Mendoza, E.; Cueto, Y.R.; Silva, R. On the Marine Energy Resources of Mexico. *J. Mar. Sci. Eng.* **2019**, *7*, 191. [CrossRef]
3. Amarouche, K.; Akpimar, A.; Bachari, N.E.I.; Houma, F. Wave energy resource assessment along the Algerian coast based on 39-year wave hindcast. *Renew. Energy* **2020**, *152*, 840–860. [CrossRef]
4. Sergent, P.; Baudry, V.; Bonviller, A.D.; Michard, B.; Dugor, J. Numerical Assessment of Onshore Wave Energy in France: Wave Energy, Conversion and Cost. *J. Mar. Sci. Eng.* **2020**, *8*, 947. [CrossRef]
5. Ribal, A.; Babanin, A.V.; Zieger, S.; Liu, Q. A high-resolution wave energy resource assessment of Indonesia. *Renew. Energy* **2020**, *160*, 1349–1363. [CrossRef]
6. Patel, R.P.; Nagababu, G.; Kachhwaha, S.S.; Arun Kumar, S.V.V.; Seemanth, M. Combined Wind and Wave Resource Assessment and Energy Extraction along the Indian Coast. *Renew. Energy* **2022**, *195*, 931–945. [CrossRef]
7. Shadman, M.; Roldan-Carvajal, M.; Pierart, F.G.; Haim, P.A.; Alonso, R.; Silva, C.; Osorio, A.F.; Almonacid, N.; Carreras, G.; Maali Amiri, M.; et al. A Review of Offshore Renewable Energy in South America: Current Status and Future Perspectives. *Sustainability* **2023**, *15*, 1740. [CrossRef]
8. Liu, J.; Meucci, A.; Liu, Q.; Babanin, A.V.; Ierodiaconou, D.; Xu, X.; Young, I.R. A High-Resolution Wave Energy Assessment of South-East Australia Based on a 40-Year Hindcast. *Renew. Energy* **2023**, *215*, 118943. [CrossRef]
9. Ye, Z.; Ma, X.; Yang, N.; Cui, L. Assessment of Wave Energy Resources in the Pearl River Estuary of China. *Desalin. Water Treat.* **2023**, *298*, 222–232. [CrossRef]
10. Pecher, A.; Kofoed, J.P. *Handbook of Ocean Wave Energy*, 7th ed.; Springer Nature: Berlin/Heidelberg, Germany, 2017.
11. Seibt, F.M.; dos Santos, E.D.; Isoldi, L.A.; Rocha, L.A.O. Constructal Design on Full-Scale Numerical Model of a Submerged Horizontal Plate-Type Wave Energy Converter. *Mar. Syst. Ocean. Technol.* **2023**, *18*, 1–13. [CrossRef]
12. IEA-OES. *Annual Report: An Overview of Ocean Energy Activities in 2021*; Executive Committee of IEA Ocean Energy Systems: Paris, France, 2022.
13. Han, M.M.; Wang, C.M. Coupled analytical-numerical approach for determining hydrodynamic responses of breakwater with multiple OWCs. *Mar. Struct.* **2021**, *80*, 103097. [CrossRef]
14. Mia, M.R.; Zhao, M.; Wu, H.; Munir, A. Numerical Investigation of Offshore Oscillating Water Column Devices. Renew. *Energy* **2022**, *191*, 380–393. [CrossRef]
15. Cannata, G.; Simone, M.; Gallerano, F. Numerical Investigation into the Performance of an OWC Device under Regular and Irregular Waves. *J. Mar. Sci. Eng.* **2023**, *11*, 735. [CrossRef]
16. Maciel, R.P.; Oleinik, P.H.; Dos Santos, E.D.; Rocha, L.A.O.; Machado, B.N.; Gomes, M.d.N.; Isoldi, L.A. Constructal Design Applied to an Oscillating Water Column Wave Energy Converter Device under Realistic Sea State Conditions. *J. Mar. Sci. Eng.* **2023**, *11*, 2174. [CrossRef]
17. Mocellin, A.P.G.; Maciel, R.P.; Oleinik, P.H.; dos Santos, E.D.; Rocha, L.A.O.; Ziebell, J.S.; Isoldi, L.A.; Machado, B.N. Geometrical Analysis of an Oscillating Water Column Converter Device Considering Realistic Irregular Wave Generation with Bathymetry. *J. Exp. Theor. Anal.* **2023**, *1*, 24–43. [CrossRef]
18. Chen, W.; Xie, W.; Zhang, Y.; Wang, C.; Wang, L.; Huang, L. Improving Wave Energy Conversion Performance of a Floating BBDB-OWC System by Using Dual Chambers and a Novel Enhancement Plate. *Energy Convers. Manag.* **2024**, *307*, 118332. [CrossRef]
19. Qian, K.; Chen, L.; Zhou, Y.; Ning, D. Hydrodynamics of an Offshore Multi-Chamber OWC Wave Energy Converter. *Energy* **2024**, *304*, 132239. [CrossRef]
20. Manimaran, R. Comparative Evaluation of Floating OWC and BBDB Wave Energy Converters by Regular and Irregular Waves Using Numerical Approach. *Ships Offshore Struct.* **2024**, 1–15. [CrossRef]

21. Fu, L.; Ning, D.; Wang, R.; Mayon, R. Numerical and Experimental Study on Hydrodynamic Performance of a Land-Based Dual-Chamber OWC Device Under Irregular Waves. Renew. *Sustain. Energy Rev.* **2025**, *207*, 114895. [CrossRef]
22. Contestabile, P.; Crispino, G.; Russo, S.; Gisonni, C.; Cascetta, F.; Vicinanza, D. Crown Wall Modifications as Response to Wave Overtopping under a Future Sea Level Scenario: An Experimental Parametric Study for an Innovative Composite Seawall. *Appl. Sci.* **2020**, *10*, 2227. [CrossRef]
23. Musa, M.A.; Roslan, M.F.; Ahmad, M.F.; Muzathik, A.M.; Mustapa, M.A.; Fitriadhy, A.; Mohd, M.H.; Rahman, M.A.A. The Influence of Ramp Shape Parameters on Performance of Overtopping Breakwater for Energy Conversion. *J. Mar. Sci. Eng.* **2020**, *8*, 875. [CrossRef]
24. Jungrungruengtaworn, S.; Thaweewat, N.; Hyun, B.S. Three-dimensional effects on the performance of multi-level overtopping wave energy converter. *IOP Conf. Ser. Mater. Sci. Eng.* **2021**, *113*, 012016. [CrossRef]
25. Hubner, R.G.; Fragassa, C.; Paiva, M.S.; Oleinik, P.H.; Gomes, M.N.; Rocha, L.A.O.; Santos, E.D.; Machado, B.N.; Isoldi, L.A. Numerical Analysis of an Overtopping Wave Energy Converter Subjected to the Incidence of Irregular and Regular Waves from Realistic Sea States. *J. Mar. Sci. Eng.* **2022**, *10*, 1084. [CrossRef]
26. Izzul, M.; Fadhli, M.; Azlan, M.; Noor, M. Parametric study on wave overtopping due to wedge angle and freeboard of wavecat wave energy converter. *UMT J. Undergrad. Res.* **2022**, *4*, 37–50. [CrossRef]
27. An, S.-H.; Kim, G.-G.; Lee, J.-H. Optimal Design of the Overtopping Wave Energy Converter Based on Fluid–Structure Interaction Simulation. *J. Coast. Res.* **2024**, *116*, 578–582. [CrossRef]
28. Liu, Z.; Zhang, G. Overtopping Performance of a Multi-Level CROWN Wave Energy Convertor: A Numerical Study. *Energy* **2024**, *294*, 130795. [CrossRef]
29. Cao, D.; Zeng, H.; He, J.; Liang, H.; Yang, Z.; Ancha, V.R.; Chen, H. A Numerical Investigation of Irregular Wave Overtopping on a 2D Fixed Overtopping Wave Energy Converter. *Eng. Appl. Comput. Fluid. Mech.* **2025**, *19*, 2446658. [CrossRef]
30. Paiva, M.d.S.; Mocellin, A.P.G.; Oleinik, P.H.; dos Santos, E.D.; Rocha, L.A.O.; Isoldi, L.A.; Machado, B.N. Geometrical Evaluation of an Overtopping Wave Energy Converter Device Subject to Realistic Irregular Waves and Representative Regular Waves of the Sea State That Occurred in Rio Grande—RS. *Processes* **2025**, *13*, 335. [CrossRef]
31. Fang, Q.; Yang, C.; Guo, A. Hydrodynamic Performance of Submerged Plates During Focused Waves. *J. Mar. Sci. Eng.* **2019**, *7*, 389. [CrossRef]
32. Wang, C.; Zhang, Y. Hydrodynamic Performance of an Offshore Oscillating Water Column Device Mounted over an Immersed Horizontal Plate: A Numerical Study. *Energy* **2021**, *222*, 119964. [CrossRef]
33. Thum, G.Ü.; Maciel, R.P.; Oleinik, P.H.; Rocha, L.A.O.; dos Santos, E.D.; Seibt, F.M.; Machado, B.N.; Isoldi, L.A. Numerical Analysis of the Submerged Horizontal Plate Device Subjected to Representative Regular and Realistic Irregular Waves of a Sea State. *Fluids* **2024**, *9*, 188. [CrossRef]
34. Koutrouveli, T.I.; Di Lauro, E.; das Neves, L.; Calheiros-Cabral, T.; Rosa-Santos, P.; Taveira-Pinto, F. Proof of concept of a breakwater-integrated hybrid wave energy converter using a composite modelling approach. *J. Mar. Sci. Eng.* **2021**, *9*, 226. [CrossRef]
35. Simonetti, I.; Esposito, A.; Cappietti, L. Experimental Proof-of-Concept of a Hybrid Wave Energy Converter Based on Oscillating Water Column and Overtopping Mechanisms. *Energies* **2022**, *15*, 8065. [CrossRef]
36. Wan, C.; Yang, C.; Bai, X.; Bi, C.; Chen, H.; Li, M.; Jin, Y.; Zhao, L. Numerical Investigation on the Hydrodynamics of a Hybrid OWC Wave Energy Converter Combining a Floating Buoy. *Ocean. Eng.* **2023**, *281*, 114818. [CrossRef]
37. Lisboa, R.C.; Teixeira, P.R.; Didier, E. Regular and irregular wave propagation analysis in a flume with numerical beach using a Navier-stokes based model. *Defect. Diffus. Forum* **2017**, *372*, 81–90. [CrossRef]
38. Zabihi, M.; Mazaheri, S.; Rezaee Mazyak, A. Wave Generation in a Numerical Wave Tank. *Int. J. Coast. Offshore Environ. Eng.* **2017**, *2*, 33–43. [CrossRef]
39. Windt, C.; Davidson, J.; Ransley, E.J.; Greaves, D.; Jakobsen, M.; Kramer, M.; Ringwood, J.V. Validation of a CFD-Based Numerical Wave Tank Model for the Power Production Assessment of the Wavestar Ocean Wave Energy Converter. *Renew. Energy* **2020**, *146*, 2499–2516. [CrossRef]
40. Jiang, C.; Yang, Y.; Deng, B. Study on the Nearshore Evolution of Regular Waves under Steady Wind. *Water* **2020**, *12*, 686. [CrossRef]
41. Oh, S.; Jung, J.H.; Cho, S.K. Higher-Order Spectral Method for Regular and Irregular Wave Simulations. *J. Ocean. Eng. Technol.* **2020**, *34*, 406–418. [CrossRef]
42. Choi, Y.M.; Bouscasse, B.; Ducrozet, G.; Seng, S.; Ferrant, P.; Kim, E.S.; Kim, Y.J. An Efficient Methodology for the Simulation of Nonlinear Irregular Waves in Computational Fluid Dynamics Solvers Based on the High Order Spectral Method with an Application with OpenFOAM. *Int. J. Nav. Archit. Ocean. Eng.* **2023**, *15*, 100510. [CrossRef]
43. Kim, Y.J.; Canard, M.; Bouscasse, B.; Ducrozet, G.; Le Touzé, D.; Choi, Y.-M. High-Order Spectral Irregular Wave Generation Procedure in Experimental and Computational Fluid Dynamics Numerical Wave Tanks, with Application in a Physical Wave Tank and in Open-Source Field Operation and Manipulation. *J. Mar. Sci. Eng.* **2024**, *12*, 227. [CrossRef]

44. Kim, I.-C.; Ducrozet, G.; Bonnefoy, F.; Leroy, V.; Perignon, Y. Real-Time Phase-Resolved Ocean Wave Prediction in Directional Wave Fields: Enhanced Algorithm and Experimental Validation. *Ocean. Eng.* **2023**, *276*, 114212. [CrossRef]
45. Canard, M.; Ducrozet, G.; Bouscasse, B. Generation of Controlled Irregular Wave Crest Statistics in Experimental and Numerical Wave Tanks. *Ocean. Eng.* **2024**, *310*, 118676. [CrossRef]
46. Kim, I.-C.; Ducrozet, G. Real-Time Phase-Resolved Ocean Wave Prediction in Directional Wave Fields: Second-Order Lagrangian Wave Models. *Ocean. Eng.* **2024**, *313*, 119316. [CrossRef]
47. Machado, B.N.; Oleinik, P.H.; Kirinus, E.P.; Dos Santos, E.D.; Rocha, L.A.O.; Gomes, M.; Das, N.; Conde, J.M.P.; Isoldi, L.A. WaveMIMO Methodology: Numerical Wave Generation of a Realistic Sea State. *J. Appl. Comput. Mech.* **2021**, *1*, 1–20. [CrossRef]
48. Maciel, R.P.; Fragassa, C.; Machado, B.N.; Rocha, L.A.O.; Dos Santos, E.D.; Gomes, M.; Das, N.; Isoldi, L.A. Verification and Validation of a Methodology to Numerically Generate Waves Using Transient Discrete Data as Prescribed Velocity Boundary Condition. *J. Mar. Sci. Eng.* **2021**, *9*, 896. [CrossRef]
49. Paiva, M.d.S.; Mocellin, A.P.G.; Oleinik, P.H.; Santos, E.D.d.; Rocha, L.A.O.; Isoldi, L.A.; Machado, B.N. Geração de Ondas Irregulares Realísticas Através da Metodologia WaveMIMO: Um Estudo de Caso para Avaliação de Parâmetros. *Rev. Thema* **2024**, *23*, 777–802. (In Portuguese) [CrossRef]
50. Maciel, R.P.; Machado, B.N.; Oleinik, P.H.; dos Santos, E.D.; Gomes, M. das N.; Isoldi, L.A. Investigation of numerical irregular wave generation using transient discrete data as boundary conditions of prescribed velocity. *Defect. Diffus. Forum* **2023**, *427*, 227–235. [CrossRef]
51. ANSYS. *ANSYS Fluent Theory Guide, Release 20.2*; ANSYS Inc.: Canonsburg, PA, USA, 2017.
52. Hirt, C.W.; Nichols, B.D. Volume of fluid (VoF) method for the dynamics of free boundaries. *J. Comput. Phys.* **1981**, *39*, 201–225. [CrossRef]
53. Srinivasan, V.; Salazar, A.J.; Saito, K. Modeling the disintegration of modulated liquid jets using volume-of-fluid (VoF) methodology. *Appl. Math. Model.* **2011**, *35*, 3710–3730. [CrossRef]
54. Versteeg, H.K.; Malalasekera, W. *An Introduction to Computational Fluid Dynamics—The Finite Volume Method*; Pearson Education Limited: London, UK, 2007.
55. Park, J.C.; Kim, M.H.; Miyata, H.; Chun, H.H. Fully nonlinear numerical wave tank (NWT) simulations and wave run-up prediction around 3-D structures. *Ocean. Eng.* **2003**, *30*, 1969–1996. [CrossRef]
56. Foyhirun, C.; Kositgittiwong, D.; Ekkawatpanit, C. Wave Energy Potential and Simulation on the Andaman Sea Coast of Thailand. *Sustainability* **2020**, *12*, 3657. [CrossRef]
57. Gomes, M.N.; Lorenzini, G.; Rocha, L.A.O.; Santos, E.D.; Isoldi, L.A. Constructal Design Applied to the Geometric Evaluation of an Oscillating Water Column Wave Energy Converter Considering Different Real Scale Wave Periods. *J. Eng. Thermophys.* **2018**, *27*, 173–190. [CrossRef]
58. Oleinik, P.H. *O programa Spec2Wave: Manual do usuário*; Oleinik, P. H.: Rio Grande, Brazil, 2022. (In Portuguese)
59. Benoit, M.; Marcos, F.; Becq, F. *TOMAWAC: Software for Sea State Modelling on Unstructured Grids over Oceans and Coastal Seas (Release 6.1)*; EDF R&D: Paris, France, 2011.
60. Holthuijsen, L.H. *Waves in Oceanic and Coastal Waters*; Cambridge University Press: Cambridge, UK, 2007.
61. McCormick, M.E. *Ocean Engineering Mechanics: With Applications*; Cambridge University Press: New York, NY, USA, 2010.
62. Arguilé-Pérez, B.; Ribeiro, A.S.; Costoya, X.; deCastro, M.; Gómez-Gesteira, M. Suitability of wave energy converters in northwestern Spain under the near future winter wave climate. *Energy* **2023**, *278*, 127957. [CrossRef]
63. Oleinik, P.H.; Maciel, R.P.; dos Santos, E.D.; Rocha, L.A.O.; Machado, B.N.; Isoldi, L.A. Numerical method for the characterization of sea states using realistic irregular waves on computational fluid dynamics simulations for application on wave energy converters. *Sustain. Energy Technol. Assess.* **2025**, *73*, 104093. [CrossRef]
64. Cardoso, S.D.; Marques, W.C.; Kirinus, E.d.P.; Stringari, C.E. Levantamento batimétrico usando cartas náuticas. In *13ª Mostra da Produção Universitária, Rio Grande*; Federal University of Rio Grande: Rio Grande, Brazil, 2014. (In Portuguese)
65. Chai, T.; Draxler, R.R. Root mean square error (RMSE) or mean absolute error (MAE)? Arguments against avoiding RMSE in the literature. *Geosci. Model. Dev.* **2014**, *7*, 1247–1250. [CrossRef]
66. López, I.; Pereiras, B.; Castro, F.; Iglesias, G. Optimisation of Turbine-Induced Damping for an OWC Wave Energy Converter Using a RANS–VOF Numerical Model. *Appl. Energy* **2014**, *127*, 105–114. [CrossRef]
67. Katsidoniotaki, E.; Shahroozi, Z.; Eskilsson, C.; Palm, J.; Engström, J.; Göteman, M. Validation of a CFD Model for Wave Energy System Dynamics in Extreme Waves. *Ocean. Eng.* **2023**, *268*, 113320. [CrossRef]
68. Romanowski, A.; Tezdogan, T.; Turan, O. Development of a CFD methodology for the numerical simulation of irregular sea-states. *Ocean. Eng.* **2019**, *192*, 106530. [CrossRef]
69. Windt, C.; Faedo, N.; García-Violini, D.; Peña-Sanchez, Y.; Davidson, J.; Ferri, F.; Ringwood, J.V. Validation of a CFD-Based Numerical Wave Tank Model of the 1/20th Scale Wavestar Wave Energy Converter. *Fluids* **2020**, *5*, 112. [CrossRef]
70. Jamieson, P.D.; Porter, J.R.; Wilson, D.R. A Test of the Computer Simulation Model ARCWHEAT1 on Wheat Crops Grown in New Zealand. *Field Crops Res.* **1991**, *27*, 337–350. [CrossRef]

71. Heinemann, A.B.; van Oort, P.A.J.; Fernandes, D.S.; Maia, A.H.N. Sensitivity of APSIM/ORYZA Model Due to Estimation Errors in Solar Radiation. *Bragantia* **2012**, *71*, 572–582. [CrossRef]
72. Li, M.-F.; Tang, X.-P.; Wu, W.; Liu, H.-B. General Models for Estimating Daily Global Solar Radiation for Different Solar Radiation Zones in Mainland China. *Energy Convers. Manag.* **2013**, *70*, 139–148. [CrossRef]
73. Nyang'au, W.O.; Mati, B.M.; Kalamwa, K.; Wanjogu, R.K.; Kiplagat, L.K. Estimating Rice Yield under Changing Weather Conditions in Kenya Using CERES Rice Model. *Int. J. Agron.* **2014**, *2014*, 849496. [CrossRef]
74. Bounoua, Z.; Chahidi, L.O.; Mechaqrane, A. Estimation of Daily Global Solar Radiation Using Empirical and Machine-Learning Methods: A Case Study of Five Moroccan Locations. *Sustain. Mater. Technol.* **2021**, *28*, e00261. [CrossRef]
75. El Alaoui, M.; Chahidi, L.O.; Rougui, M.; Lamrani, A.; Mechaqrane, A. Prediction of Energy Consumption of an Administrative Building Using Machine Learning and Statistical Methods. *Civ. Eng. J.* **2023**, *9*, 1007–1022. [CrossRef]

MDPI AG
Grosspeteranlage 5
4052 Basel
Switzerland
Tel.: +41 61 683 77 34

Processes Editorial Office
E-mail: processes@mdpi.com
www.mdpi.com/journal/processes

www.ingramcontent.com/pod-product-compliance
Lightning Source LLC
Chambersburg PA
CBHW061557220926
51850CB00005B/362
9783725857197